Theory of Neutron Star Magnetospheres

Theoretical Astrophysics

David N. Schramm, series editor

Theory of Neutron Star Magnetospheres

F. Curtis Michel

THE UNIVERSITY OF CHICAGO PRESS

Chicago and London

F. CURTIS MICHEL is the Andrew Hays Buchanan Professor of Astrophysics at Rice University. He has contributed more than 150 articles to the scientific literature and was a scientific astronaut with NASA for several years.

The University of Chicago Press, Chicago 60637
The University of Chicago Press, Ltd., London
© 1991 by The University of Chicago
All rights reserved. Published 1991
Printed in the United States of America
99 98 97 96 95 94 93 92 91   5 4 3 2 1

Michel, F. Curtis.
    Theory of neutron star magnetospheres / F. Curtis Michel.
        p.   cm. — (Theoretical astrophysics)
    Includes index.
    Includes bibliographical references.
    ISBN 0-226-52330-6. — ISBN 0-226-52331-4 (pbk.)
    1. Neutron stars—Atmospheres. 2. Magnetosphere. 3. Astrophysics.
    I. Title. II. Series.
    QB843.N4M53   1991   90-35810
    523.8′874—dc20   CIP

⊚ The paper used in this publication meets the minimum requirements of the
American National Standard for Information Sciences—Permanence of Paper
for Printed Library Materials, ANSI Z39.48-1984

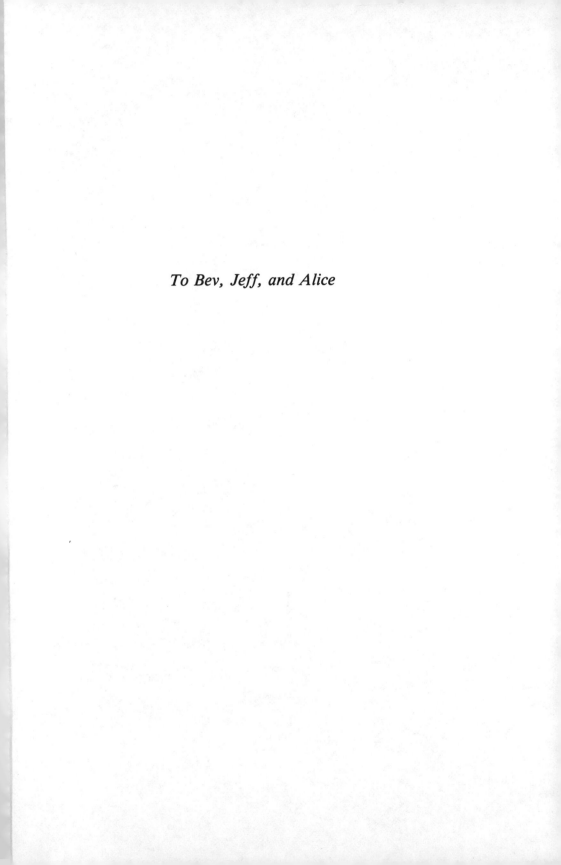

To Bev, Jeff, and Alice

Contents

Figures

Tables

Preface

First came the discovery of the radio pulsars. These objects are now almost universally believed to be rotating magnetized neutron stars. Next came the discovery of pulsating X-ray sources, even more energetic objects radiating in an entirely different part of the electromagnetic spectrum but nevertheless believed also to be rotating magnetized neutron stars. Today we have the gamma-ray burst sources and considerable circumstantial evidence that these bursts originate near neutron stars. Moreover, there is a smattering of peculiar objects such as SS 433, Sco X-1, and Cyg X-3 which may involve neutron stars.

The literature on the above phenomena is extensive and widely scattered. Manchester and Taylor (1977) and Smith (1977) have summarized the observational data for pulsars. I undertook to summarize the theoretical situation for pulsars some years ago, when it appeared that a standard model was well in hand. Ironically, by the time I finally finished writing the review for *Reviews of Modern Physics* (1982), the entire theoretical foundation of that standard model had collapsed. Although this monograph is not intended as a review, it still retains a flavor of that original effort. As before, I attempt to emphasize the problems and uncertainties as well as the considerable progress that have been made to date. Nothing is gained in science by hiding the skeletons.

I hope this book is useful to a wide readership, but particularly to students interested in an introduction to the subject and also to my colleagues who may value as much as I a compact reference source.

The word "magnetosphere" in the title is a bit too restrictive, but convenient. We will also be interested in relativistic winds, the interactions of these winds, and other related phenomena. It is presumably activity within the neutron star magnetosphere that drives such winds in the first place. Although in one sense neutron stars are exotic objects, in another sense they are central to issues of what happens at the end points of stellar evolution, what exactly happens in a supernova explosion, where certain types of nucleosynthesis take place, etc. Are neutron stars precursors to black holes? And of course there is a wide range of fundamental physics concerning everything from the properties of matter 10 times more dense than an atomic nucleus to quantum electrodynamics of intense magnetic fields, which become opaque to energetic photons. Strange states of matter involving "liquids" of pure electrons may play an important role. The mere existence of pulsars has provided important probes of the interstellar medium, its structure, and magnetic fields. A binary pulsar finally has, for the first time, provided observational support for the existence of gravitational radiation, and more extreme binary systems may be discovered to give even more stringent observational information on the strong-field limit of gravitational theory.

For some theorists, the data are casual acquaintances and not old friends, so the observational results quoted within should not be taken as definitive.

16 February 1990
Houston, Texas

1

Pulsars as Neutron Stars

1.1 Historical Notes

One of several important discoveries of recent years has been that of pulsars, astrophysical objects that produce sharp pulses of relatively low frequency radio emission of considerable intensity.

1. Discovery

On 28 November 1967 the Mullard Radio Astronomy Observatory array (2048 dipoles at 81.5 MHz) observed a train of pulses of varying amplitudes but quite regular spacing near 19^h19^m right ascension and 21 degrees of northern (plus) declination. It was not at first realized what exactly had been observed (given especially that an automobile ignition could produce such a regular pattern of interference), but Jocelyn Bell, a research student, eventually found the observation to be genuine and was assigned the task of identification. Ultimately a Nobel Prize went to the observatory director, Antony Hewish. Like many astrophysical objects, pulsars are identified by their rough location in the sky, and this first pulsar is now known as PSR 1919 + 21, where the right ascension and declination have been incorporated into the designation. It is now generally believed that pulsars are rapidly rotating neutron stars. In addition to being radio sources, pulsars radiate even more energy in the form of invisible "winds." The Crab Nebula is a supernova remnant that is lit up by the pulsar in its center, as best as one can tell.

2. Early Ideas

Fritz Zwicky (1938) had long ago proposed that such extreme objects might be found in supernova remnants, and the very term *neutron star* seems to have been coined by Baade and Zwicky (1934a). The modern quantitative theory of such objects began with Oppenheimer and Volkoff (1939). Just before the discovery of pulsars, it was theorized that a rotating magnetized neutron star could be the source of the nebular energy output (Pacini 1967). John Wheeler (1966) and Lev Landau (1932) had also predicted the existence of something like neutron stars before the pulsar phenomenon came to apparently require such objects. (See also Harrison et al. 1958; Cameron 1959a,b; Salpeter 1960; Hamada and Salpeter 1961; Saakyan 1963; and Morton 1964.) Piddington (1957) suggested the relatively "modern" interpretation that the Crab Nebula's magnetic field was that of a wound-up magnetic field from a central star. The Piddington (1957) estimate is interesting. First he assumes a star like the Sun, so he takes a radius of 10^{11} cm and a magnetic field of 10^3 gauss. He then imagines the magnetic field to be dragged out radially to the size of the nebula, which only leaves 10^{-11} gauss. To get instead a 10^{-3} gauss field then requires an "amplification factor" of 10^8, which, if due to rotation, would require the star to rotate once every 5 minutes (his guess for the period) for 900 years. To my knowledge, this is the earliest estimate of the Crab pulsar period! It was suggested in the discovery paper (Hewish et al. 1968) that a neutron star might be responsible for the phenomenon, and this interpretation was pursued energetically by Gold (1968). Even observation in a sense preceded observation: hard X-ray data establishing the existence of a pulsar in the Crab Nebula were in hand even before the discovery of pulsars themselves as radio sources (Fishman et al. 1969a,b). Only a few radio pulsars emit detectable levels of X-rays. The Crab Nebula itself has been an active center of attention, and we should mention Dewhirst's (1983) reproduction of Messier's drawings from 1843, one of which fairly accurately reproduces the central continuum region.

It is this rotating magnetized neutron star model that we will discuss. The rotator model is intrinsically interesting in its own right in that it combines in a nontrivial way the interaction of rotation with magnetic fields in an astrophysical context. Thus it is a useful starting point whatever pulsars might be, even if they were to prove to be something else. To begin with, we will discuss the magnetic field and charged-particle distribution thought to exist about a rotating neutron star, its *magnetosphere*.

We give in this chapter a minimal review of the pulsar phenomenon, which can easily be skipped if one is already familiar with the subject or already convinced that we are in fact dealing with objects having magnetic fields of order 10^{12} gauss and having escape velocities and sometimes rotational velocities at their surfaces that are both a significant fraction of the speed of light.

1.2 Nature of Pulsars

Pulsars are astronomical objects that populate the plane of our galaxy and therefore appear to be concentrated along the Milky Way. Hence they are at stellar distance (hundreds to thousands of light-years), and the inverse-distance-squared decline in apparent brightness favors observation of those at hundreds of light-years. Pulsars have been observed as far away as in our nearby companion galaxy, the Large Magellanic Cloud (sec. 1.5), but they have not yet been observed in neighboring distinct galaxies such as Andromeda, although such observation is, in principle, possible (Bahcall et al. 1970). Optical searches for Crab-like pulsars have so far been unsuccessful (Middleditch and Kristian 1984). Pulsars emit regular pulses of radio-frequency emission (i.e., at typical

UHF television broadcast frequencies). At higher frequencies their intensities typically drop off as the second to third power of frequency, and they rapidly become faint and ultimately unobservable. At lower frequencies they are difficult to observe for a variety of technical reasons (intrinsic turnover of the spectrum, scattering, ionospheric absorption, etc.). The pulsational periodicity (P of order of 1 s) is extremely regular with a "Q" ($Q \equiv \Delta P/P$) of typically 10^{11}, apparently a very pure tone. Indeed, some millisecond pulsars are more stable than existing atomic clocks. However, the pulses are better described in terms of a highly periodic "window" somewhere within which a pulse might or might not be observed; any given pulse might have a wide variety of shapes and amplitudes, as one can see in figure 1.1. This "window" is open a very short time compared to the time between successive windows (i.e., the "duty cycle" is small, about 3% typically). All pulsars are observed to be either slowing down with apparent time scales of around a million years (this slowing down would limit their Q values to about 10^{13} were it not for a small but detectable level of "timing noise") or changing period too slowly for detection. For these and a number of other reasons, it is thought that the basic pulse period is due to the rotation of the neutron star. Moreover, the loss of rotational energy can be estimated from the slowing-down rate and in all cases is much larger than needed to supply the radio luminosity.

The sharp pulses emitted by pulsars enable one to estimate their distances directly because the electrons in interstellar space cause the pulses to travel at different velocities depending on the frequency at which they are being observed. Thus, to reconstruct the original pulse, a frequency-dependent delay must be used, and that delay is proportional to the path-length integrated electron density, which is termed the *dispersion measure* (sec. 2.6.2.b). The electron density can be estimated using pulsars associated with objects at relatively well known distances. The galactic distribution shown in figure 1.2 (Manchester and Taylor 1981).

1. Rotation versus Oscillation versus Orbital Motion

Periodic astronomical phenomena generally result from one of three mechanisms: rotation, orbital motion, or macroscopic oscillation of the entire object. The only known candidate for a 1 s period is a rotating neutron star, namely a star that has collapsed to the point that only nuclear degeneracy pressure supports it against self-gravitation and prevents it from collapsing to a point (black hole). As for the other possibilities, orbiting objects would generally have to speed up to lose energy, contrary to the observed slowing-down behavior. (Exception: tidally dominated systems such as the Earth-Moon system where the orbital rate is actually decreasing.) Oscillating (e.g., radially pulsating) objects rarely have very high Q values because even small nonlinearities will couple energy from a pure single-frequency mode into other modes of oscillation having other frequencies. In any event, white dwarfs oscillate too slowly (tens of seconds) and neutron stars oscillate too rapidly (milliseconds) to cover anything

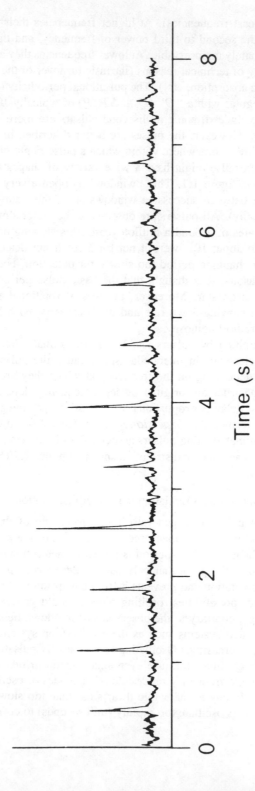

FIG. 1.1. Pulse train. Chart record of individual pulses from the 0.714 s pulsar PSR 0329 + 54 at 410 MHz. From *Pulsars*, by R. N. Manchester and J. H. Taylor. Copyright © 1977 by W. H. Freeman and Company. Reprinted by permission.

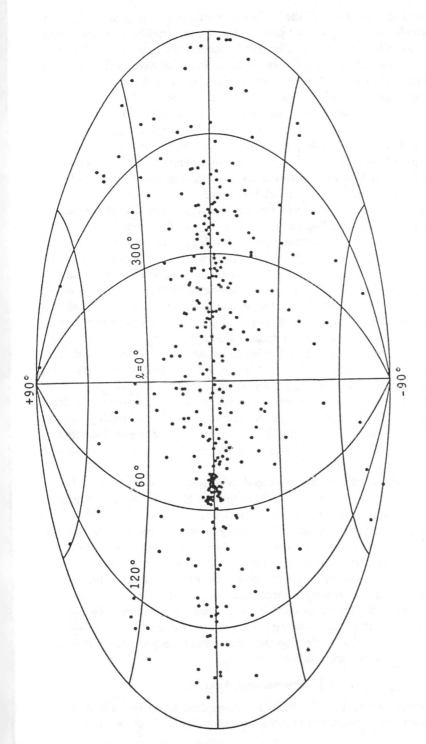

FIG. 1.2. Galactic distribution of pulsars. The location of 330 pulsars in galactic coordinates. Galactic center is in middle, with galactic latitude (unlabeled except at poles) in 30° increments. The dense cluster near 60° longitude represents the high sensitivity (but limited coverage) of the Arecibo telescope. From *Pulsars*, by R. N. Manchester and J. H. Taylor. Copyright © 1977 by W. H. Freeman and Company. Reprinted by permission.

but the most extreme limits of the known spectrum of pulsar periods. The idea of a star held up by degeneracy pressure is already familiar from the Chandrasekhar mass limit (Chandrasekhar 1957), namely the maximum stellar mass that electron degeneracy pressure can support. Stars that have exhausted all internal sources of energy and can no longer be supported by thermal pressure are destined to become first white dwarfs, then neutron stars, and finally (perhaps) black holes. It is an interesting "accident" of nature that the maximum mass of a neutron star exceeds that of a white dwarf. Consequently adding mass to a white dwarf can in principle push it over the Chandrasekhar limit and allow it to collapse to the neutron star stage. If the maximum neutron star mass were less than the maximum white dwarf mass, a collapsing star would pass by the neutron star stage and on to the black hole stage.

There was a famous controversy between Eddington and Chandrasekhar regarding the mass limit. Eddington did not buy the idea that a star could collapse to zero size and felt that "something else" must happen to ward off this eventuality. Unfortunately, Eddington seized on the idea that the equation of state for degenerate relativistic matter still had an adiabatic index of 5/3 rather than 4/3, because that would lead to stabilization (Eddington 1935a). This value would not have prevented stars from collapsing at a high enough mass in any case, but would have moved the limit to much larger values than likely for white dwarfs. Møller and Chandrasekhar (1935) refuted Eddington's argument, but he persisted (1935b), and finally Peierls (1936) came out against the argument. This led to a long silence until 1940, when Eddington presented another argument for this nonstandard equation of state for degenerate relativistic matter, at which point the entire issue was apparently politely ignored until Eddington's death (ibid., 1945, 105, 1). Oddly enough, Eddington was right. Something else *did* happen: either a neutron star being formed or a type I supernova disintegrated the white dwarf. It is not clear that a white dwarf can ever directly collapse to form a black hole.

Typical neutron star radii are estimated to be 10 km, and the canonical mass is taken to be just over the Chandrasekhar limit of about 1.4 times the mass of our own Sun ($M_{\odot}$). The thinking here is that neutron stars are formed from massive stars that have converted the hydrogen in their cores to helium and consequently have essentially become white dwarfs surrounded by an extended envelope of matter (the rest of the star). As it evolves, such a star in effect accretes matter onto a white dwarf-like core. The maximum mass of the neutron star is estimated to be somewhat larger than the 1.4 $M_{\odot}$ figure, perhaps by a factor of about 2. Somewhat by default, then, it has come to be widely assumed that pulsars are neutron stars rotating at the observed radio emission periodicity. Nevertheless, we will also mention the other models. Some pulsars that have prototypical properties are listed in table 1.1.

1.3 Dimensional Analysis

In this section we review the basic reasoning leading to the widely quoted estimates for pulsar parameters such as magnetic fields of 10^{12} gauss.

TABLE **1.1 Some Important Pulsars**

Designation	Popular Name	Notable Feature(s)
0531 + 21	Crab pulsar (NP 0532) (NP 0531)	Fastest known for many years (.033 s) pulsed emission from radio to γ-ray obvious supernova association giant pulses glitches variations in dispersion measure extremely "inefficient" (10^{-9} of spin-down energy converted to radio-emission)
0540 − 693	LMC pulsar	Very similar to Crab pulsar, but no glitches! 55 kpc away in LMC pulsed optical
0833 − 45	Vela pulsar	Supernova association glitches pulsed optical and γ-ray
1913 + 16	Hulse-Taylor binary	First binary pulsar discovered evidence of gravitational radiation consistent with general relativity
1937 + 214	Millisecond pulsar	Fastest known: 642 rotations/s
1957 + 20	Eclipsing pulsar	Second fastest millisecond pulsar eclipsed by 0.02 $M_\odot$ companion every 9.2 hours
1951 + 32	CTB 80 Pulsar	Pulsar inside a wind-blown (?) shell
1845 − 19	—	Presently slowest known: 4.3082 s period
1237 + 25	—	Exceptionally complicated pulse shape (5 components)
1641 − 45	—	Slow pulsar also having glitches
0809 + 74	—	Both nulls and drifts
0826 − 34	—	Exceptionally wide pulse profile ($\approx 145°$) numerous subpulses drifting back and forth
1919 + 21	CP 1919	First pulsar discovered (CP = Cambridge pulsar, a largely discontinued designation)
1821 − 24	Pulsar in M28	First pulsar discovered in a globular cluster also a millisecond pulsar

Let us address the properties of neutron stars. Neutron stars are held up not by thermal pressure but by the degeneracy pressure of the nucleons. Thermal support cannot be important because the very high temperatures required would lead to rapid neutrino emission, which would deplete the thermal energy. Indeed, these are presumably just the neutrinos seen from SN 1987A (see for example Bruenn 1987). Nucleons do not become degenerate until nuclear densities are reached, namely densities of the order of 10^{15} g/cc. Consequently a 1.4 $M_\odot$ sphere ($\equiv 3 \times 10^{33}$ g) would be only about 10 km in radius and would have a moment of inertia (I) of about 10^{45} g cm^2. Models of nuclear matter give values quite close to these rough estimates.

For a pulsar to be readily detectable by existing radio telescopes it must have a radio luminosity of more than about 10^{25} ergs/s, compared to the 4×10^{33} ergs/s of essentially blackbody radiation from the Sun, very little of which falls in the radio band. A parsec (pc) is the distance a star (in the plane of the ecliptic) would have to be from an observer to appear to shift back and forth compared to the "fixed" (much more distant) stars with an amplitude of 1 second of arc over 1 year; more simply it is the length of the long sides of an equilateral triangle with a narrow angle of 1 arcsecond and a base of 1 astronomical unit; most simply, it is 3.0568×10^{18} cm. Because the periods of all known radio pulsars are increasing (not true for the binary X-ray pulsars), something is removing the rotational energy of the neutron star. This energy loss is the major one, with the radio luminosity representing only a small fraction ($\approx 10^{-5}$) of the loss in rotational energy. A typical slowing-down rate is given by the dimensionless period derivative, $\dot{P} \approx 10^{-15}$, so for a pulsar period of 1 s, the rotational energy-loss rate is

$$\dot{W} = I\Omega\dot{\Omega} = I(2\pi)^2 \dot{P}/P^3 \approx 4 \times 10^{31} \text{ ergs/s} \qquad (1)$$

(here $\Omega \equiv 2\pi/P$ is the frequency of rotation in radians/s). For an object as small as 10 km in radius to experience such a large torque indicates that it must have a strong coupling to the external universe, and the best guess so far is that a strong intrinsic dipole magnetic field, when rotated this rapidly, would account for the torque. One way of estimating this magnetic field is to equate the above power output with that from classical magnetic dipole radiation from a rotating magnetized sphere (e.g., Jackson 1975; see sec. 5.2). This estimate was first given by Ostriker and Gunn (1969a). We shall see later that other models for estimating the field lead to similar results because this is essentially a dimensional scaling argument.

Calculating the radiation rate from a rotating or oscillating dipole is a standard textbook exercise, so let us estimate this loss rate using typical "astrophysical" approximations. This approach has the advantage that the overall physics is not overshadowed by mathematical detail (available in sec. 5.2). A sphere of radius a with a uniform magnetization B_0 has an external dipole field which declines in strength with the third power of distance; hence, at large distances r compared to a, the field will be

$$B = B_0(a/r)^3, \qquad (2)$$

where we will ignore the factor of 2 difference in whether the field strength is taken to be at the equator or at the pole. Although the field could well have higher multipoles, they drop off at successively larger powers of $(a/r)^2$ and become unimportant at large distances. If the sphere rotates at angular rate Ω, rigid rotation of the external dipole field equals the speed of light, c, at the distance

$$R_L = c/\Omega. \qquad (3)$$

At this distance there must be a transition from the quasi-static dipole field near the rotator which drops off with distance as $1/r^3$ to the wave field far from the rotator, which drops off as $1/r$. The two fields must here be comparable, and one can simply estimate

$$B_{\text{wave}} = B_0(a/R_L)^3, \tag{4}$$

where B_{wave} is the wave field (falling off as r^{-1}) at R_L while the right-hand side is just the stellar magnetic dipole field (dropping off as r^{-3}). The energy-loss rate is just the energy density B_{wave}^2/μ_0 times the wave velocity c times the area $4\pi R_L^2$, or

$$\dot{W} = 4\pi B_0^2 a^6 \Omega^4 / \mu_0 c^3. \tag{5}$$

The exact result is smaller by a factor of $2/3$, simply due to averaging over the $\cos^2(\theta)$ radiation pattern. The quantity $B_0 a^3$ is the magnetic moment, and for $P = 1$ s and $B_0 a^3 = 10^{20}$ Weber-meters one obtains (including now the factor of $2/3$)

$$\dot{W} = 0.96 \times 10^{31} \text{ ergs/s.} \tag{6}$$

The reader may note that we use MKS units for Maxwell's equations, but quote results in cgs units. As one can see, the two estimates are in the same ballpark, and this has led to the conviction that such huge magnetic fields actually exist (plus independent support from observations of what might be electron cyclotron lines corresponding to such fields from other neutron star candidates). This simple calculation emphasizes the very general nature of the argument. For example, the existence of plasma about the object cannot easily change the above estimate. "How much" plasma one has is broadly parameterized by the Alfvén velocity:

$$V_A = B/(\mu_0 \rho)^{1/2}, \tag{7}$$

where ρ is the mass density. In this form, the Alfvén velocity is the proper velocity ("proper" in the special relativistic sense is $\gamma\beta c$, where $\beta = v/c$ and $\gamma = 1/(1 - \beta^2)^{1/2}$) and can be as large or small as one wishes. It is evident that our estimate is mainly sensitive to where the light-cylinder distance is taken to be, which in turn would be "corrected" to $R_L = \beta c/\Omega$ to include inertial effects from the plasma. We will see later that $V_A \gg c$ in the conventional model; hence R_L is expected to be essentially independent of plasma density because β would be so close to unity.

In practical units, the surface (polar) magnetic field is generally taken to be

$$B = 3.2 \times 10^{19} (P\dot{P})^{1/2} \text{ gauss,} \tag{8}$$

or about 10^{12} gauss for nominal parameters ($P \approx 1$ s; $\dot{P} \approx 10^{-15}$). We will use equation 8 as a standard estimate because it is what is used by the observers, and it is convenient if such formal field estimates are internally consistent, regardless of what the actual field strengths might be. As noted, this total energy output is much larger than the 10^{25} ergs/s of radio energy output, because at most only about 10^{-5} of the energy seems to be emitted in the form of radio waves (there is actually a wide range of apparent efficiencies of converting spin-down power to radio from 10^{-3} to about 10^{-9}).

The total spin-down energy output is largely invisible for most pulsars. The important exception is the Crab pulsar, which seems to be illuminating the entire nebula and seems to be the source of the otherwise rather large magnetic field in the nebula. Because these nebulae are expected to dissipate in a time scale much shorter than the pulsar lifetime, most pulsars are found to be isolated objects.

It is interesting to note that while it is difficult to model the very short period pulsars as anything other than neutron stars, the converse need not be true. Rotating white dwarfs could in principle radiate significant amounts of rotational energy and could constitute a separate class of long-period (but still quite rapid) objects.

1.4 Comparison with Planetary Magnetospheres

Much research has been done on the magnetospheres of the Earth, Jupiter, and Saturn. Indeed, Jupiter functions as a pulsar itself, albeit a very weak one (see sec. 7.2 for a discussion of the possible parallelism). Nevertheless, a number of assumptions made about planetary magnetospheres are thought to be inapplicable to the pulsar case.

1. Confinement by Mirroring

The inner regions of planetary magnetospheres are typically characterized by trapped particles (e.g., the Van Allen belts). These particles are trapped by magnetic mirroring, by which they spiral back and forth along the dipolar magnetic field lines, the spiral becoming tighter and finally reversing as the field strength increases toward each pole. The particle trajectories satisfy the adiabatic invariant condition

$$B \sin^2\alpha = B_0, \tag{1}$$

where B is the local magnetic field and α is the "pitch angle" between the magnetic vector ($\mathbf{B}$) and the velocity vector ($\mathbf{v}$). The maximum value of B that the particle can penetrate to is therefore B_0. At these points the particle velocity (the speed is a constant in a pure magnetic field owing to conservation of energy) is entirely perpendicular to the local field vector, and the velocity component parallel to $\mathbf{B}$ is going through zero as the particle "bounces" off this "mirror point." In a dipole field there are two points along every field line, one over each magnetic pole, where this condition is met, and charged particles can then be trapped in between. The particles therefore are trapped in a "minimum B" geometry (a nomenclature more common in controlled thermonuclear fusion). In addition to this bounce motion between magnetic mirror points, gradient, curvature, and gravitational drifts cause the particles to drift around the magnetic dipole axis. In general the net drift is unrelated to either the direction or speed of the planetary rotation. The particles themselves may be quite energetic ($\approx$MeV). It is often convenient to speak of "parallel" and "perpendicular energy" (relative to the magnetic field direction) because the particle motion conserves the sum of the two but shifts back and forth from having pure perpendicular energy at the mirror point to having (possibly) almost pure parallel energy at the field minimum. It is possible to simply have an equatorial orbit with no bouncing (i.e., the mirror points coincide). Energy is not, of course, a spatial vector quantity, a point not entirely pedantic given that one student was encouraged, in going to the relativistic case, to try to introduce parallel and perpendicular *Lorentz factors!* (A fallacy worth pondering, but see sec. 2.5.2.)

The conventional view of a pulsar magnetosphere simply boosts the magnetic field from about 1 gauss at the surface of a planet (a more or less typical order of magnitude within the solar system) to 10^{12} times stronger at the surface of a neutron star, but this shift has a profound effect on the particle motion because radiative losses, which are relatively unimportant in the planetary magnetosphere, become a dominant consideration. The spiraling particle then almost immediately radiates away its perpendicular energy (sec. 2.5), and without this component, the particle no longer mirrors in the magnetic field (even if it did, it would quickly radiate away even its parallel energy). To first order, then, any "radiation belts" would be precipitated to the surface, leaving a vacuum. Now, however, stellar rotation comes to play an unfamiliar role as well, by inducing an intense electrostatic field that the particles are pulled from the surface to neutralize. Our neutron star magnetosphere is now filled with a relatively "cold" ($kT \ll GMm/r$, hence $T \ll 10^{12}$ K) plasma that has been electrostatically lofted and which now corotates with the star, rather than consisting of particles having their own independent drift patterns. In the simplest case, this plasma will *not* be quasi-neutral but will have large volumes of essentially one sign of charge (to neutralize the huge electrostatic fields, these field lines must terminate on nonzero charge densities). In the case of planetary magnetospheres, this rotationally induced electric field is acknowledged but is generally dismissed as being small compared to other sources of electric field such as that seen in the flowing magnetized solar wind. In pulsar theory, just the opposite assumption is made.

2. Landau Orbitals

The above nonneutral plasma will also be a nonclassical one in the sense that the cyclotron frequency is so large in the strong magnetic field that the particle energies are importantly quantized into *Landau levels* having energy $\hbar\omega \approx 12$

keV (electrons). Indeed, the energy levels are

$$E = (\hbar\omega_c/2)(2n - m + |m| + 1), \tag{2}$$

where $\hbar\omega_c = 11.5 \times B_{12}$ keV, with B_{12} the magnetic field in units of 10^{12} gauss. The quantum numbers n (radial) and m (magnetic) are integers starting at zero, so the ground state has the usual "zero-point" energy of $\hbar\omega/2$ with the states equally spaced thereafter. If the plasma radiates until only the ground state is populated (i.e., $T < 10^8$ K), one has not only a nonneutral plasma but also a completely polarized electron plasma.

3. Temperature

Let us then ask what constraints there might be on the temperature of particles in a neutron star magnetosphere. Temperature is usually an independent parameter under "ordinary" astrophysical and laboratory conditions. In a relativistic system, however, it is very difficult for the thermal forces (plasma pressure) to be important. By definition, in a relativistic system the thermal velocities must be comparable to c, which corresponds to temperatures of the order of 10^{10} K just for electrons. Such temperatures cannot be maintained if the plasma is optically thick (blackbody radiation going as T^4), is magnetized (owing to the large synchrotron radiation rates if optically thin), or is in thermal contact with the neutron star surface (which cannot be much hotter than about 10^6 K either observationally or theoretically).

The internal temperature of a neutron star is similarly constrained, but for somewhat different physical reasons. For one, in a neutron star the nucleons would have to be near-relativistic for thermal pressure to be significant, which boosts the required temperatures to around 10^{13} K. But at the much lower temperature of 10^{10} K, a significant background density of electron-positron pairs is required for thermodynamic equilibrium, which in turn (owing to the weak interactions) requires a background of neutrinos. However, the neutrino "background" leaks out of the neutron star within seconds, providing a very effective radiation mechanism to carry off internal energy above 10^{10} K. The surface of the neutron star will not be that hot, of course, because the blackbody radiation above 10^{10} K would be incredible, so the surface temperature is limited to about 10^6 K, at which luminosity photons from the hot interior can just diffuse out fast enough to maintain the surface temperature.

4. Density

The final distinction we will make between planetary and pulsar magnetospheres is that in the pulsar case the number of particles is also controlled by the electrostatic field rather than being a more or less free parameter. This follows because the rotationally induced electric field will only loft negative particles if it is directed at the surface (more precisely, if the component parallel to the local magnetic field, $\mathbf{E} \cdot \mathbf{B}$, is directed at the surface) and vice versa. If these particles accumulate in the magnetosphere, they will eventually produce electric fields of their own. Thus their concentration is limited to the point that the self-fields become comparable to the accelerating fields. Consequently the particle concentration presumably cannot be too different from that required for such neutralization (which we will estimate below). Particle concentrations in planetary magnetospheres are typically many orders of magnitude larger and must typically be measured rather than theoretically inferred.

Even more surprising, there may even be regions essentially devoid of particles in a pulsar magnetosphere (sec. 4.2.2). For a planetary magnetosphere such considerations are almost entirely missing. The trapped particle concentrations are typically many orders of magnitude larger than the above limit because the particle distribution is nearly neutral on the bulk average, and if it were for some reason not neutral (MeV particles could in principle maintain potential differences of the order of millions of volts), low-energy plasma particles from the planetary ionosphere would be accelerated up into the magnetosphere to provide the neutralization.

A pattern of thinking has evolved for the pulsar magnetosphere that is quite different from that for planetary magnetospheres, and hopefully the above comments may be useful in preventing confusion between the two views.

1.5 Observational Situation

This section is provided only as a brief sketch of the observed properties of pulsars. More detailed discussion than the thumbnail sketch here is to be found in the books by Smith (1977) and Manchester and Taylor (1977), and a more up-to-date source is provided in the proceedings of the IAU Symposium No. 95, *Pulsars*. (See also Manchester 1974 for a brief assessment of pulsar properties and problems of continuing interest. Taylor and Stinebring 1986 have reviewed recent developments to that date. Revised editions of both the Manchester and Taylor book and the Smith book are reportedly in the works.)

1. New Discoveries

The discovery of pulsars as such was followed by a number of other startling discoveries concerning pulsars. Which might be considered the most important is a matter of personal taste, but table 1.2 is a serviceable short list. The significance of some of these discoveries will be the topic of sections to follow.

2. Pulsar Distances

The immediate question asked about any newly discovered astronomical object is, "How far away is it?" If the source moves across the sky relative to the background stars at an appreciable rate (its proper motion), it must be quite close. The same follows if it moves back and forth relative to very distant stars as the Earth orbits the Sun (its parallax). Except for a handful of exceptions, pulsars have neither detectable parallax nor proper motion, which immediately places them out among the "fixed" stars, i.e., at a distance of hundreds of light-

TABLE 1.2—Some Important Pulsar Discoveries

Date Published	Discovery	Authors
1968	Radio pulsars	Hewish, Bell, Pilkington, Scott, and Collins
1968	Crab pulsar	Staelin and Reifenstein
1969	Crab optical pulses	Cocke, Disney, and Taylor
1975	Binary pulsar	Hulse and Taylor
1982	Millisecond pulsar	Backer, Kulkarni, Heiles, Davis, and Goss
1987	Pulsar in M28	Lyne, Brinklow, Middleditch, Kulkarni, Backer, and Clifton
1988	Eclipsing pulsar	Fruchter, Stinebring, and Taylor

years. Very tiny proper motions have been observed for a few pulsars, which is consistent with these large distances and also suggests unusually high velocities (Manchester et al. 1974; Cordes 1986 gives a larger compilation). Another distance indicator is whether the pulsar is in association with other objects of known distance. The best example of such an association is the pulsar in the Crab Nebula, PSR 0531+21, which is in the center of that supernova remnant. The distance to this remnant is known since it is young enough (936 years since its apparent origin in the supernova of 1054) to expand perceptibly in a few years, and the Doppler shift of radiating filaments moving toward and away from us can also be measured. These two pieces of information immediately give the distance to the nebula (within some significant uncertainty over its exact shape), which comes out to be about 2000 pc (Trimble 1968). Yet another indicator is the statistical distribution of pulsars in the sky. They are strongly concentrated about the Milky Way. In other words, they are sources in the disk of our spiral galaxy and are seen at distances comparable to the thickness of that disk, namely several hundred pc. The last piece of information does not permit one to determine the distance to any particular pulsar, however.

a. *Dispersion Measure*

However, the pulsed nature of the pulsar emission together with the slightly dispersive nature of the interstellar medium combine to provide an index (the dispersion measure, DM), which allows distance estimates to be made for individual pulsars. The index of refraction of a plasma (see also sec. 2.6) is just

$$n = \left(1 - \frac{\omega_p^2}{\omega^2} \right)^{1/2} \approx 1 - \frac{\omega_p^2}{2\omega^2}, \tag{1}$$

where ω_p is the plasma frequency:

$$\omega_p^2 \equiv e^2 n_e / \epsilon_0 m_e, \tag{2}$$

and in practical units

$$\omega_p = 5.64 \times 10^4 n_e^{1/2} \text{ rad/s},$$

or

$$f_p = 9000 \, n_e^{1/2} \text{ Hz},$$

with n_e the electron concentration in electrons/cm^3. This form for the index of refraction neglects finite temperature and magnetic fields. It is, however, an excellent approximation for the low densities and weak field strengths appropriate for the interstellar medium. Typical values for n_e are now determined to be about 0.03 cm^{-3}, giving $\omega_p \approx 10^4$ rad/s. Since the index of refraction is less than unity (see sec. 2.6), the phase velocity slightly exceeds the velocity of light and the group velocity is slightly less,

$$V_g = nc \approx \left(1 - \frac{\omega_p^2}{2\omega^2}\right) c. \tag{3}$$

Consequently a sharp pulse of radio emission is dispersed so that the high-frequency components reach the Earth before the low-frequency components. Pulsars are therefore interstellar whistlers. By measuring the pulse arrival times at different frequencies, one measures the accumulated time-difference caused by the difference in group velocity over the path length; thus one measures directly the dispersion measure,

$$\text{DM} = \int n_e dl, \tag{4}$$

which is conveniently expressed in the mixed units of pc per cc (a large distance to fold into a small cube). Thus a DM of 30, combined with the above estimate for n_e, implies a distance of 1000 pc (1 kpc). The DM for the Crab pulsar, for example, is 57. The good agreement with independently determined distances to the nebula, about 2 kpc, is only slightly tempered by the fact that this pulsar is 1 of about 30 used to estimate the average value for n_e. Workers now attempt to correct even for spatial variations in n_e. Manchester and Taylor (1977, see also Manchester and Taylor 1981), for example, identify corrections for height above the galactic plane and for distance from the galactic center. For most pulsars, the DM is almost fully attributable to interstellar electrons. The one exception is the Crab pulsar, wherein very small variations have been detected (Rankin and Roberts 1971; Rankin et al. 1988a). Evidently there is a variable contribution from the nebula (e.g. wisps moving across the line of sight: Apparao 1974) or from plasma even nearer to the pulsar. Theoretical considerations suggest that intrinsic corrections from electrons near the pulsar contribute less than a few hundredths of one unit of dispersion measure. Observed dispersion measures range roughly from about 3 to over 1000, with 100 being near the median.

Modern models of the interstellar medium are at contrast with a quasi-uniform value for the electron density (McKee and Ostriker 1977), wherein 95% is occupied with very low ionization ($n_e \approx 10^{-3}$) and the DM is provided by the relatively smaller volumes intervening. Thus rather large statistical variations would be implied.

b. *Rotation Measure*

Another index, the rotation measure, is a measure of the Faraday rotation and is used to estimate the path-averaged line-of-sight component and distribution of the galactic magnetic field (of order 10^{-6} gauss: Manchester 1972; Michel and Yahil 1973; Simard-Normandin and Kronberg 1980). The rotation measure, unlike the dispersion measure, is quoted in *measured* units, namely radians of rotation per wavelength of observation (meters) squared. To convert to practical units, one must multiply by a conversion factor that turns out to be essentially unity, namely

$$\int n_e B \cos \theta dl = 1.232 \times \text{RM}, \tag{5}$$

where B is in microgauss (n_e and l are in electrons/cc and pc, respectively, as with dispersion measure). Positive RM means the magnetic field points on the average at the observer. Not only are DM and RM quoted in discordant units, but they arise from discordant effects, DM from differences in the group velocity and RM from differences in the phase velocity of the waves.

c. *Photon Mass Limits*

We can use pulsar dispersion to put a limit on the mass of a photon. Equation 1 can be rewritten, using $n \equiv ck/\omega$, in relativistic form, $ck = n\omega$, $\omega^\alpha = \omega(1, \mathbf{n})$, $\omega_\alpha = \omega(1, -\mathbf{n})$, where $\mathbf{n}$ is the propagation-direction unit-vector, to give

$$\sum_\alpha \omega^\alpha \omega_\alpha = \{\omega^p\}^2. \tag{6}$$

If the photon had a mass $m_\gamma = \hbar \omega_\gamma$, we would have instead

$$\sum_\alpha \omega^\alpha \omega_\alpha = \omega^{\gamma 2} + \omega_p^2, \tag{7}$$

and therefore ω_γ must be less than the inferred plasma frequency; hence $m_\gamma \leq \hbar \omega_p \approx 10^{-9}$ eV. (For a fuller discussion and refined estimates, see Feinberg 1969, Warner and Nather 1969, Synge 1969, Rawls 1972, and Goldhaber and Nieto 1971. Cole 1976 shows that pulsar observations provide [negative] tests of ether theories. Sadeh et al. 1968 similarly show that no "mass effect on frequency" is produced by the Sun.)

3. Periods

One of the slowest pulsars is PSR 0525+21 at 3.745 s (the slowest to date is PSR 1845-19 at 4.308 s), while for a long time the fastest known was the Crab pulsar (PSR 0531 + 21) at 0.033 s, almost companions in the sky (also having nearly the same apparent distance). A number of authors have suggested a common origin of these two pulsars (0525 being ejected in the event: Gott et al. 1970; Morris et al. 1978; but see Trimble and Rees 1971a,b). Wright (1979) discusses the general case for pulsar pair associations. High spatial velocities for pulsars have been proposed to arise either in the collapse event (Michel 1970a) or by radiation reaction (Tademaru and Harrison 1975; Harrison and Tademaru 1975; Tademaru 1976, 1977; Morris et al. 1976). However, a surprising breakthrough discovery revealed several new pulsars, "millisecond" pulsars, with periods up to 20 times faster! The fast pulsars are rare and the slow ones are common, of course, if the fast ones slow down rapidly. It follows almost automatically that the millisecond pulsars *must*, for whatever reasons, have weak magnetic fields to allow them to survive long enough to still be observed spinning rapidly, an expectation confirmed by their very small values of $\dot{P}$. The distribution of pulsars with period is shown in figure 1.3. These data do not include the millisecond pulsars or other pulsars discovered much after the 1981 Manchester and Taylor compilation. (Some data sets are slightly larger thanks to the kind efforts of J. H. Taylor, who supplied data tapes on the 368 pulsars known at that time.) The turnover at about 0.5 s leading to a paucity of long-period pulsars is widely interpreted by the observational community as evidence for magnetic field decay. However, it must be stressed that there are important observational selection effects. For one, there is inevitably some lower limit to the apparent radio flux that can be detected. If the radio luminosity is proportional to the the total power output (theorized to be $L \approx P^{-4}$), for example, then slow pulsars accumulate, but all are too dim to be seen. A straightforward numerical modeling (sec. 2.4) produces similar distributions to those observed (figure 1.4) simply from magnetic spin-down (no magnetic field decay). A linear rise at short periods is also predicted by such spin-down, consistent with figure 1.3. Even the improving designs of radio telescopes introduce complications. The most sensitive telescope is that at Arecibo, and searches have been made down to about 1 mJy (Hulse and Taylor 1975). However, this telescope is fixed in a limestone sinkhole. As the Earth rotates, this dish sweeps across the galactic plane twice. Because the pulsars populate the plane and more objects of all types are toward the galactic center, a large number of pulsars with designations PSR 19nn are known because that is the right ascension at which Arecibo views the inner part of the Milky Way. Another selection effect is that distant or heavily obscured pulsars have radio signals difficult to recognize as being pulsed owing to the effects of a large dispersion measure, as can be seen in figure 1.5. In addition to bias against short-period and high-dispersion pulsars, there can be

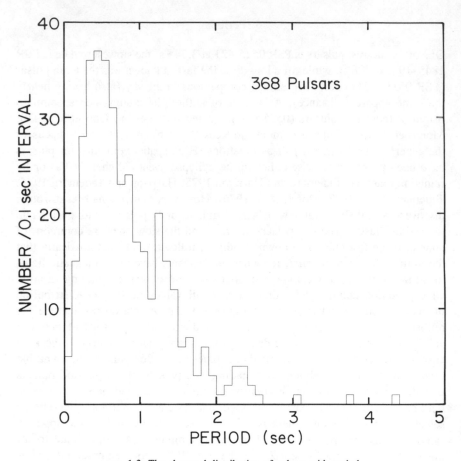

FIG. 1.3. The observed distribution of pulsars with period.

problems from variations in pulsar intensities from interstellar scintillation or intrinsic variability (a pulsar might be dim during a survey).

a. *Millisecond Pulsars*

It was a standard assumption among pulsar theorists that neutron stars with periods as fast as a millisecond (i.e., the breakup rotation speed) could be formed. It was also realized that a slowing-down law scaling as $B^2 \Omega^4$ together with a "standard" magnetic field of about 10^{12} gauss would result in a millisecond pulsar that was spun down well before it was likely to be observed if formed in a supernova; it takes years for the debris from a supernova to clear sufficiently for radio signals to be seen, and supernovae are themselves centuries apart. Thus a new nearby supernova might contain such a pulsar, and the recent supernova in the LMC may provide just such a test, but unfortunately the pulsar may be shrouded by the supernova ejecta for several years (Michel et al. 1987).

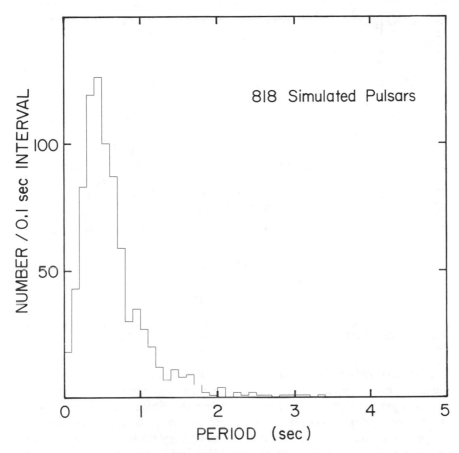

FIG. 1.4. Simulated period distribution of pulsars. Pulsars created randomly in space and allowed to spin down owing to magnetic dipole torque (details in sec. 2.4).

Alternatively, one might have hoped that some pulsars would be directly formed with weak fields, in which case the slowing down would be small and permit a millisecond pulsar to last long enough to be observed. Moreover, there is an important selection effect against detecting very rapid pulsars (Vivekanand et al. 1982).

Interestingly, when the first millisecond pulsar was discovered (Backer et al. 1982), an entirely different scenario was put forward (Alpar et al. 1982). Rather than appeal to selection effects as above, it was proposed that the pulsar was (1) originally an old binary X-ray pulsar whose magnetic field had largely decayed away, which was then (2) spun up by rapid accretion from the companion, and finally was (3) liberated from the companion when the latter underwent a supernova explosion (or some equivalent catastrophic disruption), which liberated this rapidly rotating but weakly magnetized pulsar. Although this scenario is seemingly more complicated than the selection effect picture, it is perhaps significant

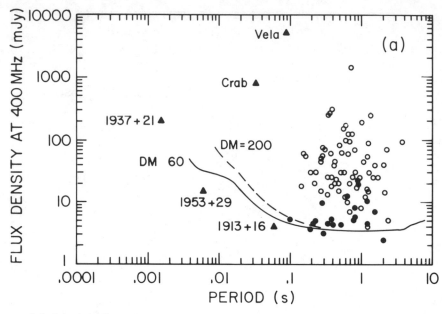

FIG. 1.5. Selection effects. Results of a pulsar survey showing period and dispersion measure restrictions (dashed and solid lines). G. Stokes, D. J. Segelstein, J. H. Taylor, and R. J. Dewey, 1986, *Ap. J.*, 311, 694.

that the next millisecond pulsar discovered (Boriakoff et al. 1983) was indeed found in a binary system. The more obvious view (formation with a weak field) was also put forward (Arons 1983b; Brecher and Chanmugam 1983; Michel and Dessler 1983). Indeed, that view would be consistent with models of type I supernovae (sec. 10) that appeal to contact-binary white-dwarf progenitors. Here the rapid spin (dictated by tidal locking) and low magnetic field (typical of field white dwarfs) arise naturally without the need to invoke magnetic field decay. Henrichs and van den Heuvel (1983) suggest instead that the eventual coalescence of the two neutron stars in a system such as PSR 1913 + 16 might also lead to a rapidly rotating neutron star. The Alpar et al. suggestion associates millisecond pulsars with low-mass X-ray binaries. The latter are 100 times more common in globular clusters than in field stars. Kulkarni and Narayan (1988) point out that during the spin-up phase, the systems should be bright low-mass X-ray binaries, and the existing population of such objects is one to two orders of magnitude too small to be millisecond pulsar progenitors. White and Stella (1988) find that spin-up is not quantitatively consistent with existing millisecond binaries. Hamilton et al. (1985) had searched for candidate millisecond pulsars in such clusters, finding one likely candidate, which turned out to indeed harbor a millisecond pulsar (several more have been seen since; sec. 10). Alpar and Shaham (1985) argue that quasi-periodic oscillations seen in the bright galactic bulge source GX5-1 (van der Klis et al. 1985) might be from a millisecond pulsar spun up by accretion.

The original millisecond pulsar, PSR 1937+214, has a period of only 1.558 ms, about 20 times shorter than that of the previous record holder (in the Crab Nebula). Previously the entire range of pulsar periods from fastest to slowest was only about a factor of 100 in a sample of over 330 pulsars, so it was quite a surprise on empirical grounds alone to find a pulsar with so different a period. On the other hand, there was also an observational selection effect in that such short periods are instrumentally more difficult to detect and, given the paucity of pulsars with periods near that of the Crab, were not anticipated.

PSR 1937 + 214 was already a known radio source, 4C 21.53, or more precisely one member in a complex of radio sources. As such, it was one of several already singled out (see listing in Rickard et al. 1983) as having pulsar-like properties: interstellar scintillation (therefore compact) and a steeply falling radio intensity at higher frequencies. Erickson (1983) and Purvis (1983) specifically single out this source as being a likely pulsar. (The Erickson paper, for example, was submitted to *Astrophysical Journal Letters* on 2 August and accepted on 9 September, the pulsar was confirmed 25 September and the discovery paper was submitted to *Nature* on 22 November and published 16 December, while the Letter appeared postdiscovery on 1 January of the following year!)

Following its discovery, PSR 1937+214 proved to be quite different from the other very fast pulsars (the Crab and Vela being the usual comparison objects) but not too different from the run-of-the-mill pulsars, other than being so fast. Specifically, it is not a strong X-ray source (fairly stringent upper limits are known at this time: Becker and Helfand 1983), it is not embedded in a strong synchrotron nebula, the electromagnetic power output is mainly in the radio, and it is decelerating extremely slowly, not rapidly. Indeed, the slowing-down rate of $dP/dt \approx 1.2 \times 10^{-19}$ (Backer et al. 1983; Ashworth et al. 1983) corresponds to a formal lifetime (starting from zero period) of 4×10^8 years. Given our prejudice that pulsars are formed in supernova events, the lack of a detectable remnant nebula suggests an age of more that about 10^4 years, so this object could have formed anywhere in between that and the above rough upper bound. Heiles et al. (1983) put the distance at 5 kpc.

Other than being so rapid and an excellent clock (Davis et al. 1985; Hankins et al. 1987), PSR 1937 + 214 is not very exceptional. The radio luminosity of about 3×10^{30} ergs/s is even consistent with at least one scaling prediction (Michel and Dessler 1983), and the pulse and polarization properties are no more or less individual than those found among pulsars in general. However, the mass and high densities normally associated with pulsars are much more strongly constrained by the fact that it can spin so fast and still stay together. PSR 1937+214 has a mass of probably at least 0.7 $M_\odot$ (Datta and Ray 1983), a mean density of at least about 2×10^{14} g/cm^3 (Cowsik et al. 1983; Ray and Chitre 1983; Harding 1983), and a radius of less than 16 km (Harding 1983), results entirely consistent with expectations from neutron star theory (see, for example, Shapiro et al. 1983; Friedman 1983). However, these objects are spinning at very close to breakup for neutron star models (e.g., Friedman et al. 1984, 1988). Discoveries of significantly more rapid rotators would be something of an embarrassment. Usov (1983) has argued on the basis of the standard model that significant gamma-ray fluxes might be expected from this object. After some initial excitement that optical emission from this object or a companion might have been observed (Djorgovski 1982), further efforts failed to confirm the association (Middleditch et al. 1983a; Djorgovski and Spinrad 1983). The tentative identification with a red star would have been difficult to understand, even for a pulsar. The low value of $\dot{P}$ would rule out (excluding bizarre coincidental cancellation) orbital motion in the vicinity of an ordinary star.

Soon afterwards, a second pulsar (PSR 1953+29) of the "millisecond class" was discovered (Boriakoff et al. 1983). Although appreciably slower (6.133 ms) than PSR 1937+214, it has the important and exceptional property of being in a binary system with a period of about 120 days, thereby combining two exceptional properties into a single system. Unlike the original binary pulsar, PSR 1913 + 16, which has approximately twin neutron stars both with masses near the Chandrasekhar limit of 1.41 $M_\odot$, PSR 1953 + 29 has a companion with a mass of at most only 0.3 $M_\odot$. It was immediately suggested (Joss and Rappaport 1983; Paczyński 1983; Savonije 1983; Helfand et al. 1983; Alpar et al. make the same suggestion in connection with PSR 1937+214) that this object was originally a binary X-ray source. In some ways this confirmation created its own problems: why was there no longer a companion about PSR 1937+214 to begin with? Ruderman and Shaham (1983, 1985) anticipate this issue when they suggest that the companion may have been destroyed. The eclipsing binary millisecond pulsar PSR 1957+20 has a companion of only 0.02 $M_\odot$, which compounds the mystery. That this pulsar could "evaporate" its companion (Kluźniak et al. 1988a; van den Heuvel and van Paradijs 1988) is not excluded but also not unambiguously supported by the data (sec. 10.c below). One idea is that mass loss was previously driven self-consistently and matter lost from the companion was accreted by the neutron star, which in turn irradiated the companion to maintain the mass loss at something comparable to the Eddington limit (see sec. 10.4). At some point the companion would become too weak a source of matter and the process would abruptly halt. The much weaker pulsar radiation would then be required to finish off the companion. Rawley et al. (1988) found a very small proper motion for PSR 1937+214, consistent with its low galactic latitude, which puts limits on formation by disruption of a binary system. They cite white dwarf/neutron star coalescence (van den Heuvel and Bonsema 1984) as a possible alternative. Only the isolated millisecond pulsars are listed in table 1.3; even more have been identified in binary systems (table 1.6). An association with binaries could serve to support the spin-up hypothesis or indicate that these objects are formed in type I rather than type II supernovae (sec. 8).

TABLE 1.3—**Short List of Millisecond Pulsar Properties**

Property	1937 + 214	1821 − 24
Period (ms)	1.55780644887275	3.0543144932
Period derivative (10^{-20})	10.51054	155.
Interpulse	Yes	Yes (not near 180°)
L_{radio} (ergs/sec)	1.1×10^{26}	1.3×10^{27}
DM (pc/cm^3)	71.0440	120.
Distance (kpc)	5.	5.8
Magnetic field (10^8 gauss)	4.	20.
Comments	No X-rays accidental associated H II	In a globular cluster
Aliases	4C 21.53W	M28 pulsar

b. *Very Long Period Pulsars*

Given the discovery of the millisecond pulsars, a similar bias may exist at very long periods; typical searches have been conducted in a period range of about 0.01 to 100 seconds. Again, there are theoretical arguments against very long period pulsars based on precisely the same scaling law: if the slowing-down torque scales as $B^2 \Omega^3$, if all pulsars have a "standard" field near 10^{12} gauss, and if pulsars get to be slow pulsars only by slowing down from significantly higher spin rates (1 s, say), then the electromagnetic torques will not spin them down to periods longer than about 5 s within the age of the galaxy (currently estimated to be 10 to 20×10^9 years). The same reasoning would require them to be very dim, hard-to-detect objects as well. In fact, there are a significant number of known neutron stars with even much longer periods: the pulsating X-ray sources. These objects have periods that overlap those of the radio pulsars (0.069 s for A0538–66): up to 835 s for 4U 0352 + 30. Since they are in binary systems (see sec. 10.1), their evolutionary history is potentially much more complicated than that of single pulsars. Consequently it remains unclear whether very long period radio pulsars exist. To spin down to periods of much more than a minute in a Hubble time (estimated age of the Universe) would require that these pulsars had very strong ($\geq 10^{14}$ gauss) fields. It is interesting that the pulsating binary X-ray sources have periods much *longer* in general than those for the radio pulsars, despite the episodes of rapid spin-up that they are seen to undergo. None to date have been caught in the final stages of spinning up a millisecond pulsar.

4. Slowing-Down Rates

All radio pulsars are found either to be slowing down on the long term or (in a few cases) to have time derivatives too small to be determined. The normal measure of this slowing down is $\dot{P}$, a dimensionless quantity typically of the order of 10^{-15}, although often quoted in the literature in units of s/s. Given that pulsars have periods around 1 s and that there are about 3×10^7 s in a year, it is clear that the characteristic time scales are then of the order of a few million years.

For the phenomenological slowing-down relationship

$$\dot{\Omega} = -K\Omega^n \tag{8}$$

defines the "braking index" n (=3 for magnetic dipole radiation), and integrating gives

$$T_{\text{age}} = -(n-1)^{-1}(1 - x^{n-1})(\Omega/\dot{\Omega}), \tag{9}$$

where the coupling constant K was removed using equation 8. The quantity $x \equiv \Omega/\Omega_0$, where Ω_0 is the spin rate at birth. If x is small and $n = 3$, we obtain the usual formula to estimate the pulsar "age," $T = P/2\dot{P}$. This is the time it would take an initially rapid pulsar to slow down to its present period, assuming the magnetic torque scaling. The estimate is insensitive to the initial spin period because the pulsar would slow down so rapidly at first. Consequently the initial spin period of a pulsar could even have been comparable to the present value without greatly changing the pulsar "age." For the Crab pulsar, this age is 1240 years while the historical record gives 934 years, which is close but not on the mark. Setting $x = 0$ gives $n = 3.7$ whereas n is observed instead to be about 2.5 (at present). Thus n could have changed and indeed gravitation radiation would give $n = 5$ (but a much smaller coupling constant K, perhaps zero, because the pulsar would have to be out-of-round to produce such radiation). Alternatively, we have a nonzero value of x. We conclude from $x = 0.50$ ($n = 3$) or $x = 0.58$ ($n = 2.5$) that the pulsar had an original period of between 16 and 19 ms. These ages are therefore ballpark estimates, and if the pulsar magnetic field were to have evolved, for example, they could be seriously misleading (see sec. 1.6.2).

A popular representation of the data is to plot $\dot{P}$ versus P as shown in figure 1.6. The theoretical slowing-down model would cause the pulsars to evolve along the gently sloping line from left to right. A steeply rising line fitted along the lower right of the "V" formed by the data is sometimes called the "death line" (because pulsars, if evolving from upper left to lower right, are no longer seen, presumably because they cross this imaginary line). Such pulsars are also becoming quite dim, so observational selection must certainly enter to some extent. Many pulsars approach this "death line" null, and an alternative suggestion is that pulsars turn off not so much by getting dimmer but simply by being in a null longer and longer, at some point for good. Manchester et al. (1981), however, argue that the nulling pulsars tend systematically to be slower, whereas to explain the cutoff in pulse periods, many pulsars must turn off at fairly short periods. Almost every facet of pulsar statistics is open to several interpretations.

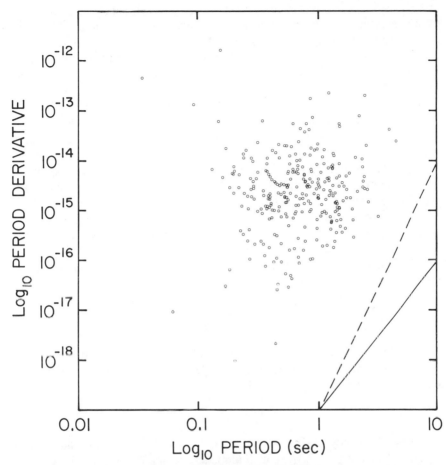

FIG. 1.6. Period derivative versus period for 365 pulsars. No millisecond pulsars plotted. Note scattered appearance in form of a "V." Crab pulsar is in upper left-hand corner; an $n = 3$ evolutionary history would carry it on a trajectory passing slightly above the core of the distribution. Given that pulsars are observed to slow down, they must evolve to the right, and, judging from a handful of pulsars, presumably down). The dashed line shows the slope of an empirical cutoff or "death line" with $\dot{P}/P^5 = $ constant, which can be interpreted as a fixed magnetic field (about 2 gauss) at the light-cylinder; the latter is sometimes taken as empirical evidence for emission at the light-cylinder (there is no corresponding theory, however). The solid line is the slope if pulsars are simply luminosity selected, assuming that the radio luminosity scales proportional to the spin-down luminosity. Both lines would be drawn further to the upper right, but there seems to be no very satisfactory criterion for fitting lines to the *edges* of data.

5. Radio Luminosities

If the rotating magnetized neutron star hypothesis is correct, then slow pulsars should have less energy to give up to radio-frequency emission. It should be realized, however, that the radio output accounts for only about 10^{-5} of the total power output. Thus the radio luminosity, bright as it is, seems to be a "dirt effect" insofar as the overall energy budget of a pulsar is concerned. The lower limit at present to (inferred) absolute radio luminosities seems to be about 10^{25} ergs/s whereas the brightest rarely exceed 10^{31} ergs/s. Although this dynamic range spans 6 orders of magnitude, it is much less than 4 times the 3 orders of magnitude range in periods (i.e., the Ω^4 power output for magnetic dipole radiation from equation 1.3.1), showing again that observational selection may be important on the one hand and that the theory may be oversimplified on the other (there seems to be a falloff in radio efficiency for the very energetic Crab and Vela pulsars, which, albeit fairly bright, are extremely inefficient at producing radio emission).

Let us define some of the characteristic ways of rating the energy given off by a pulsar, because several similar looking but different representations are possible. In general, radiant energy from a distant source falls on the Earth (or, more specifically, on a radio telescope) at a certain rate, which can then be converted to watts per unit area of the receiving surface. The power input is usually filtered so that the telescope is sensitive only within a certain band of frequencies, Δf, centered at some frequency f. One therefore measures the radiation per unit area per unit bandwidth at frequency f. The standard unit is one *jansky*, where

$$1 \text{ Jy} \equiv 10^{-26} \text{ watts/m}^2 \text{ Hz} \tag{10}$$

with Hz (hertz) being cycles per second. According to a standard relationship given by Manchester and Taylor (1977), a pulsar at 1 kpc with a luminosity of 1 mJy will have a total luminosity of

$$3.4 \times 10^{25} \text{ ergs/s,}$$

where the frequency has been assumed to be 400 MHz and the radiation to be emitted in a $10°$ cone. Although the latter assumption may not be correct, it is generally built into the definition of pulsar luminosities so we will adopt it for consistency. Thus, for example, S_{400} is the flux in janskys at 400 MHz in typical notation. Clearly one would like to know S_f at all f, to know the spectrum of the source.

a. *Spectra*

Because there are only certain "windows" allocated for radio astronomy that are not used by FM, TV, telecommunications satellites, etc., and because there are practical limitations on how much time one can allot to study just one of the many radio objects, it is rare for S_f to be measured at more than one or two frequencies, except in the case of certain well-studied objects. In general, the spectra are found to be steeply falling functions of frequency, which often makes it difficult to observe a pulsar at higher radio frequencies unless it is an intrinsically strong source. At very low frequencies, however, the spectra are often observed to turn over below about 100 MHz (Malofeev and Shitov 1981).

To estimate the energy output, then, one must assume that the integrated energy flux is of the order of

$$F \approx fS_f.$$ (11)

To find the absolute radio luminosity one can multiply the flux density in mJy at 400 MHz by the distance in kpc estimated from the DM and use the conversion factor 1 mJy kpc^2 = 3.4×10^{25} ergs/s, which assumes a 10° wide beam (i.e., a factor of order 5 to correct for radiation *not* beamed elsewhere). Often luminosities are quoted in the literature in the mJy kpc^2 units. Another practical consideration is scintillation, which causes the pulsar signals to fade in and out as distant clouds of interstellar electrons drift through the line of sight (see, for example, Wolszczan 1983). If the pace of such scintillation is slow, the flux estimated at a given observing run may not be representative, and a considerable effort to study and remove such scintillation effects may be required. Observing time priorities may not aways make such studies feasible.

Let us review the likely radiation geometry. First, a 1 kpc sphere is not necessarily uniformly illuminated by a pulsar. In fact it cannot be; otherwise the pulsar wouldn't be pulsing. Because we only observe one point on the sphere, we need some hypothesis or theory to "fill in" the rest. The rotating neutron star hypothesis allows us to trace the one point around some (unfortunately unknown) axis to give us a circle (not necessarily a great circle) on the surface of the celestial sphere. Owing to the pulsed nature, the flux density varies widely along this circle and one must exercise a bit of care because there is a huge difference between the *average* flux density and the *peak* flux density (typically, a factor of around 30). In the above numerology, we have assumed that S_f represents the *average* flux density. Unfortunately, there is yet a different way in which the data can be presented, and that is in terms of the *integrated* flux density, namely the power received per pulse. If one is not careful, the distinction between the units Jy m^{-2} Hz^{-1} and Jy m^{-2} can be overlooked. Furthermore, most pulsars have periods of about 1 s, and consequently their integrated flux densities and their average flux densities are roughly the same. For rapid pulsars a large error can be made, so it is important to know which flux measure is being discussed.

It was immediately recognized that these large radio intensities from compact objects posed serious theoretical problems. One would not have expected to find an intense source of low-frequency radiation that was also pulsed with very short periods, but pulsars are just such objects. The reason is that many sources of radiation are limited to what a blackbody at some effective temperature might radiate. Blackbody radiation at low frequencies has a differential surface emissivity that is well represented by the Rayleigh-Jeans formula

$$\frac{dJ_0}{d\omega} = kT\omega^2/(2\pi c)^2 \, (\hbar\omega \ll kT).$$ (12)

Notice that, unlike the total blackbody radiation which increases as T^4, the low-frequency radiation in any fixed-frequency interval $d\omega$ increases only linearly with T. The number of degrees of freedom for waves in a fixed volume is fixed ($\approx \omega^3$), and only the average thermal energy ($\approx T$) increases the energy density at fixed frequency. At the same time, it is physically difficult for a source to pulse once a second and yet be much larger than 1 light-second. It is widely taken as a rule of thumb that a source cannot have a pulse period (P) that is short compared to the light travel time across the source. More realistically, the limiting velocity is the longitudinal wave propagation time across the object, so the former criterion is conservative indeed. The radius of the object is therefore assumed to be

$$R \leq cP/2, \tag{13}$$

and the total integrated luminosity up to some observation frequency ω_0 is then

$$L \leq L_0 = J_0 4\pi R^2 = kT\omega_0^3 P^2/12\pi. \tag{14}$$

As noted above, pulsars are typically at distances of hundreds of light-years or more and therefore must have radio luminosities of the order of 10^{28} ergs/s. The observation at 81.5 MHz ($\omega_0 = 5.12 \times 10^8$ rad/s) of such luminosity from a pulsar with a period of 1 s then gives a limiting luminosity

$$L_0 = kT(6 \times 10^{24})\,\text{ergs/s}, \tag{15}$$

which requires $kT \approx 10^4$ ergs, hence a formal temperature of about 10^{20} K. And this temperature is certainly a minimum one; if the radiating surface were as small as the neutron star itself, we would require 10^{29} K. One is reluctant to accept such high temperatures as the actual kinetic temperatures of these sources. Instead, it must be assumed that they are somehow radiating coherently. Two popular mechanisms for producing coherent radiation are (1) population inversion (e.g., masers and lasers) and (2) particle bunching. We will discuss these mechanisms in more detail in sec. 2.5. One must also avoid reabsorption of the radiation. Maser mechanisms only work if amplification exceeds reabsorption, although one must still worry about where the absorbed energy would go. Bunching mechanisms work best in inhomogeneous media where there is a paucity of neighboring particles to provide reabsorption (e.g., radiation from current-carrying flux tubes into vacuum regions surrounding them).

6. High-Frequency Luminosities

A handful of pulsars emit detectably in other than the radio bands, as given in table 1.4. The physics of these emissions is discussed in more detail in sec. 2.7.

The first pulsar in the LMC, PSR 0529–66, was reported by McCulloch et al. (1983) and possibly identified with a very nearby (8 arcsecond) X-ray source (Long et al. 1981), but no recent articles have appeared on this potentially interesting object. The Vela pulsar is a marginal optical source (Wallace et

TABLE **1.4—Pulsars with Both Radio and High-Frequency Emission**

Pulsar	Other Names	P	$\dot{P}$	DM	Distance	Comments
0531+21	Tau X-1	.033	4.22×10^{-13}	57	2000	$n \approx 2.51$ optical to gamma-rays
0529−66	First LMC PSR	.976	?	125	55,000	X-rays (?)
0833−45	Vela pulsar	.089	1.24×10^{-13}	69	500	25th magnitude optical gamma-rays
1509−58	MSH15−52 G320.4−1.2 RCW 89	.150	15.4×10^{-13}	235	4200	Slowing-down age 1500 yr vs. nebula age approx 10^4 $n \approx 2.83$ no glitches to date (>600 d) $L_x \approx 5 \times 10^{34}$ ergs/s
0540−69.3	LMC pulsar	.050	4.79×10^{-13}		55,000	Crab-like remnant 22.7th magnitude optical $n \approx 3.6 \pm 0.8$

al. 1977; Peterson et al. 1978), and it is curious that no new studies using high-sensitivity charge-coupled devices (CCDs) have been reported. The Crab pulsar has long been prototypical of an optical pulsar, with PSR 0540–69.3 apparently being a quite Crab-like object (Chanan et al. 1984) complete with optical pulsations (Middleditch and Pennypacker 1985). Harnden and Seward (1984) have excellent pictures of the X-ray emission from the Crab pulsar and surrounding nebulosity at different pulsar phases. Manchester et al. (1982) reported PSR 1509–58 in MSH 15–52, Weisskopf et al. (1983) first obtained a period derivative from the X-ray emissions, and later Manchester et al. (1985) established that $n = 2.83$. There is a possibly significant discrepancy regarding the expansion age of the associated nebula, which van den Bergh and Kamper (1984) determine to be >5000 years compared to the short characteristic age for the pulsar of only 1500 years.

7. Polarization

The radio pulses are often highly polarized, usually linearly, but sometimes with important amounts of circular polarization as well. It is difficult to summarize the polarization characteristics; however, there are some general trends.

(1) The polarization of individual subpulses is often quite high, approaching 100%. These high polarizations are sometimes averaged out in the integrated (time-averaged) pulse shape, but not necessarily, e.g., PSR 1929+10.

(2) Intensity often anticorrelates with polarization, at least in the integrated pulse profiles, and the wings of a pulse are more strongly polarized than its core.

(3) Many pulsars display a "swing" in the position angle of the linear polarization, corresponding to a nearly constant rate of rotation during the strong part of the pulse, with some slackening at the wings to give an "S"-shaped variation of position angle with phase.

(4) Also observed are orthogonal mode changes wherein the polarization abruptly changes by 90° rather than rotating smoothly as above. A simple model for this effect would have two separate (i.e., uncorrelated, overlapping) sources of emission with orthogonal polarization; then whichever source happened to be strongest would completely determine the net polarization (see Cheng and Ruderman 1979).

(5) A change in sense of *circular* polarization is frequently observed in the middle of a pulse (Rankin 1983a). For example, Biggs et al. (1985) report that PSR 0826-34, which has an unusually long duty cycle, shows such a circular polar sign change (ibid., figure 6), and very wide ($\approx 60°$) regions of one sign of polarization.

The observation of significant levels of *both* circular and linear polarization in pulsars seems significant. If the source were entirely linearly polarized, one could not easily filter it to give circular polarization: a modest difference in propagation velocity would simply rotate the axis of linear polarization (Faraday rotation) while a large enough difference that the two senses of circular polarization arrived at slightly different times would introduce such huge Faraday rotations that the linear polarization would be wiped out entirely. The alternative of a purely circularly polarized source seems implausible given that the strong polarization is commonly linear; what would give such a precise phase shift to convert circular to linear in so many pulsars? Moreover, the linear and circular intensity patterns for a given pulsar seem rather insensitive to the observing frequency, whereas most propagation effects would be very frequency dependent because they depend on the proximity to zeros or poles in the dispension relation. These circumstances all seem to point to a polarization that is intrinsic and little modified by propagation effects.

Some caution is urged in interpreting "average" properties such as the integrated pulse shape; often a seemingly simple pulse can be shown to be a superposition of several independent components (Krishnamohan and Downs 1983). Also, some position angle "swings" can result from orthogonal mode changes that jump around in the pulse longitude of individual pulses (Stinebring et al. 1984a,b; see figures 7 and 8 for PSR 0834+06). Backer and Rankin (1980) list polarization properties for a number of pulsars, as did Lyne et al. (1971). Furthermore, there can be significant instrumental cross-coupling between the linear and circular polarization measurements (Stinebring et al. 1984a). Older published pulse profiles showing polarization changes with longitude may be somewhat deceptive for either reason. The quality of data that are only collected incidentally will naturally be inferior to that collected for the express purpose of elucidating, say, polarization behavior.

a. *Stokes Parameters*

Radiation observed at a given frequency (and within some reasonable bandwidth) can be parameterized by four quantities in general. As is often the case, there are various possible combinations of these parameters. Although the textbook choice is usually the Stokes parameters, these are not the parameters usually quoted. Radiation is conveniently divided into a polarized and an unpolarized component, the latter having no properties (at a given frequency) other than just the intensity. Introductory textbooks concentrate on perfectly monochromatic light, which literally must be 100% polarized. For light to become unpolarized at a certain frequency, it must be broken up into successive or overlapping wave trains, each with a different polarization so that the average is zero. The breaking into wave trains of finite length destroys the pure monochromatic nature of the light as well, so unpolarized ("natural") light is actually quite complicated. The remaining polarized component is, in the most general case, elliptically polarized, which in turn can be separated into linear and circular polarization. The circular polarization has an intensity and sign of polarization (left or right), while the linear has an intensity and axis or position angle. Thus we have the three intensities of the unpolarized, linearly polarized, and circularly polarized components together with an angle and a sign, or four numerical values (the sign can be incorporated into one of these numbers, such as the intensity of the circular polarization, with the understanding that "negative" intensity is just the other sense of polarization). These are the quantities typically published on pulsar characteristics. They are not the Stokes parameters.

The Stokes parameters are given (Born and Wolf 1986) as s_0, s_1, s_2, and s_3, and usually defined in terms of wave amplitudes and phases, which is precise but usually takes some puzzling over to interpret in the above terms. Stokes parameters have dimensions of *intensity* but combine more or less like *amplitudes*. Radio astronomers often use I, Q, U, and V, and the notation I, M, C, and S also exists. The first parameter, s_0, is just the total intensity. The last parameter, s_3, is the intensity of the circularly polarized component, carrying the sign in question ($+$ for right-handed, $-$ for left-handed) *if there is no linear polarization*. We will return to this restriction momentarily. It might be supposed that the remaining two intensities are something like the intensity of linear polarization in one axis and the intensity in the orthogonal axis, when in fact the intensity of linear polarization is given by $(s_1^2 + s_2^2)^{1/2}$ *if there is no circular polarization*. Note that this definition is what one would expect if s_1 and s_2 were amplitudes and not intensities. But the way s_1 is defined is that for purely linearly polarized light, $s_1 = +s_0$ for polarization along a conveniently defined principal axis (the x-axis or north or whatever), and $s_1 = -s_0$ for polarization *orthogonal* to that axis. This is the same convention as that used for circularly polarized light: "+" is for one polarization sense and "−" is for the orthogonal sense, even though we are dealing with quantities having the dimensions of intensities and not amplitudes. We do not really have light with negative intensity! For either of the above cases, $s_2 = 0$. If, instead, the polarization is at 45°, then $s_2 = \pm s_0$ depending on which way the axis is rotated relative to the principal axis. Mathematically this simply corresponds to

$$s_1 = s_0 \cos 2\psi, \tag{16}$$

and

$$s_2 = s_0 \sin 2\psi, \tag{17}$$

where ψ is the position angle.

The powerful thing about the Stokes parameters is that, with these conventions, if one has two or more beams combined (e.g., if one is receiving signals from several places on a pulsar), the Stokes parameters of the resultant beam are just the sum of those of the individual beams. For example, suppose we have equal contributions of linear and orthogonal polarization from two sources. Then one source might be described with $s_0 = 1$, $s_1 = +1$, $s_2 = 0$ (ignoring circular), and the other with $s_0 = 1$, $s_1 = -1$, $s_2 = 0$. The total is then simply $s_0 = 2$, $s_1 = 0$, $s_2 = 0$, which correctly shows that the effect of adding two orthogonal linear polarizations of equal intensity is total depolarization. If the orthogonal polarization "jumps" in pulsars are caused by such an effect, they should be accompanied by a percentage polarization that goes to zero, which in fact seems generally to be the case. The same thing will happen when one adds opposite circular polarizations. (This analysis is not correct if there is *phase coherence* between the two sources such as would occur if a single source were split by multipath propagation—as in interstellar scintillation—which then gives interference effects and one adds amplitudes rather than Stokes parameters.)

Returning to the issue of intensities, the total intensity of the beam is simply s_0 of which the unpolarized intensity is given by $s_0 - (s_1^2 + s_2^2 + s_3^2)^{1/2}$, with $(s_1^2 + s_2^2 + s_3^2)^{1/2}$ the polarized intensity. Clearly, then, this intensity is not the sum of s_3 for the circular and $(s_1^2 + s_2^2)^{1/2}$ for the linear. We have already seen that s_1 and s_2 are not "orthogonal" in the conventional sense, and s_1 is not even the intensity that would necessarily be registered with a polarizing filter aligned with the principal axis. In the same way, the linear and circular are not, strictly speaking, independent modes of polarization, but rather correspond to two orthogonal modes and a phase. If two orthogonal linear polarizations are in phase, the electric vectors simply sum at $45°$, giving, say, $s_1 = +1$ for $s_0 = 1$. If now the relative phases are shifted, one goes to circular polarization with, depending on the sense of the shift, $s_3 = +1$, and then to the orthogonal linear with $s_1 = -1$, and finally through the orthogonal circular polarization $s_3 = -1$ before repeating all over again. It should be clear from this pattern that s_1 and s_3 are out-of-phase harmonic functions of the phase shift and therefore

$$s_1 = s_0 \cos \phi \tag{18}$$

and

$$s_3 = s_0 \sin \phi \tag{19}$$

when the linear component is along the principal axis or orthogonal to it. It follows, then, that the total polarized intensity is $(s_1^2 + s_3^2)^{1/2}$ for this example, or $(s_1^2 + s_2^2 + s_3^2)^{1/2}$ in general. What is generally quoted in the literature for the separate linear and circular "intensities" is in fact just $(s_1^2 + s_2^2)^{1/2}$ and s_3, but it is evident that one does not simply add these intensities to get the total. For most pulsars such fine points are unimportant, but can lead to noticeable discrepancies for some, such as the Vela pulsar, which is almost completely linearly polarized except near the peak, where there is some significant circular polarization. The linear plus circular polarization intensities as plotted can then be *greater* than the total (e.g., Manchester and Taylor 1977, figure 2-8)!

Optical polarization studies (e.g., of the Crab Nebula) are most easily done with Polaroid filters, in which case any circularly polarized light (of which there is little if any) passes as unpolarized. If intensities are recorded at $0°$, $60°$, and $120°$ (I_0, I_1, I_2, respectively), it is straightforward to show that

$$s_0 = \frac{1}{3}(I_0 + I_1 + I_2), \tag{20}$$

$$s_1 = \frac{2}{3}(2I_0 - I_1 - I_2), \tag{21}$$

and

$$s_2 = \frac{2}{\sqrt{3}}(I_1 - I_2). \tag{22}$$

The polarized intensity is given from

$$s_p^2 \equiv s_1^2 + s_2^2 = \frac{16}{9}(I_0^2 + I_1^2 + I_2^2 - I_0I_1 - I_0I_2 - I_1I_2), \tag{23}$$

and the unpolarized (plus circular) is given from

$$s_u = \frac{1}{3}(I_0 + I_1 + I_2) - \frac{1}{2}s_p. \tag{24}$$

The position angle (from wherever the $0°$ position was defined, usually north) is given by inverting equations 16 and 17. If observing time is not an issue, a simpler choice of filter positions is $0°$, $45°$, $90°$, and $135°$ insofar as data reduction goes.

For radio observations one does not have "filters" but rather antennae sensitive to either one sign of linear polarization or one sign of circular polarization. It is then necessary to observe with two antennae of orthogonal polarization and also observe the phase of the signal.

8. Binary Pulsars

For a relatively long period (≈ 8 years), pulsars were observed to be single objects, and numerous attempts failed to detect any evidence of companions. It had even been suggested that the glitches might be caused by the passage of small orbiting objects (e.g., planets) in very eccentric orbits near the pulsars (Michel 1970a; Hills 1970) but such passages would not explain the much larger relative changes in $\dot{P}$ which accompany the glitches. The discovery of the so-called Hulse-Taylor binary pulsar broke this drought, surely one of the more important advances in the field. At this writing, a handful of binary pulsars are known, about 1% of the total known pulsars. Ordinary stars, by contrast, are commonly found in binary systems. Thus, either pulsars are formed from a class of objects that are not in binary systems themselves or the formation of the pulsar leads to the unbinding or disintegration of the companion (the latter scenario could form a disk of matter about the neutron star).

a. General Relativistic Effects

The first binary pulsar, PSR 1913 + 16, was discovered by Hulse and Taylor (1975; see also Taylor et al. 1976; Fowler et al. 1979). This system is spectacularly important because it shows the predicted behavior for a system radiating gravitational waves as expected from general relativity (Wagoner 1975; Taylor et al. 1979). (See Cooperstein and Lim 1985 for a discussion of the sometimes controversial radiation formula for gravitational waves.) Unlike Newtonian binary systems, the masses of a system exhibiting general relativistic effects can be solved for both stars, which in this case turn out to be consistent with a pair of neutron stars of mass 1.4 $M_\odot$ (Weisberg and Taylor 1984). Moreover, this system is beginning to set interesting limits on a possible variation of the gravitational constant, presently set at about $\dot{G}/G \leq 3 \times 10^{-11}$/year (Damour et al. 1988). Since 1975, several other binary pulsars have been detected: PSR 0820 + 02 (Manchester et al. 1980; Manchester et al. 1983), PSR 0655 + 64 (Damashek et al. 1982; Lyne 1984 gives a determination of $i \geq 60°$ from scintillation studies), PSR 1953 + 29 (which is also a millisecond pulsar: Boriakoff et al. 1983, 1984; Chadwick et al. 1985a), PSR 2303 + 46 (Stokes et al. 1985), and PSR 1855 + 09 (a second millisecond pulsar in a binary: Stokes et al. 1986). The mass function is the quantity, in $M_\odot$,

$$f_1 \equiv (M_c \sin i)^3/(M_p + M_c)^2 = \Omega_b^2(a_p \sin i)^3/GM_\odot, \qquad (25)$$

where M_p is the pulsar mass, M_c is the companion mass, i is the inclination of the orbital plane, and $\Omega_b \equiv 2\pi/P_b$ is the binary orbital rate (P_b being the orbital period). What is actually measured is the line-of-sight velocity variation of the pulsar through changes in the apparent pulse rate, which, together with the period of this variation (i.e., the orbital period), gives the amplitude of the line-of-sight motion, by definition $a_p \sin i$. Note that for $a_p \sin i$ equal to one astronomical unit and P_b equal to one year, the mass function is unity (by definition). If these quantities are instead in centimeters and seconds, respectively, the conversion coefficient $(1/GM_\odot)$ is just 2.975×10^{-25}. Small values of f_1 indicate either light companions or orbits seen close to face-on. A binary neutron star system seen edge-on would have $f_1 \approx 0.25$.

TABLE 1.5—Some Binary Pulsars (nonmillisecond)

Property	1913 + 16	0820 + 02	0655 + 64	2303 + 46	1831 − 00
Pulse period (s)	0.0590	0.8648	0.1956	1.0664	0.5209
$\dot{P}$ (10^{-20})	864.	10,390.	67.7	56,930	1430
Orbital period	7^h45^m	1232.47^d	24^h41^m	12.34^d	1.81^d
$a \sin i$ (10^6 km)	0.700	48.6	1.24	9.81	0.217
Mass function ($M_\odot$)	0.1322	0.00301	0.0712	0.2463	0.00012
Eccentricity	0.6171	0.011859	<0.00002	0.65838	<0.004
DM	167	23.6	8.73	60.9	88.3
Companion (type)	Neutron star	White dwarf	White dwarf	≥1.2	≥0.06

So far, all determinations seem consistent with a pulsar mass of about 1.4 $M_\odot$. Taylor and Weisberg (1982) review the observational and theoretical situation. Kulkarni (1986) identified the companions of PSR 0655 + 64 and PSR 0820 + 02 as white dwarfs. Kristian and Young (reported by Blandford and De-Campi 1981) had identified a faint star ($\approx 23^m$) at the position of PSR 0820 + 02 and suspected it was an old (or reddened) white dwarf. In particular, the companion of PSR 0655 + 64 turned out to be an *old* cool white dwarf with an estimated age of 2×10^9 years, consistent with the spin-down age. It is clear from table 1.5 (see also table 1.6) that of the eight pulsars listed so far, PSR 1913 + 16 has a very likely neutron star companion and PSR 2303 + 46 could have such a companion. The rest would have to be almost face-on systems for the companion masses to be of the order of 1.41 $M_\odot$.

That about 1% of all pulsars might be binary was predicted (Guseinov and Novruzova 1974). (See also Trimble and Rees 1971a,b; Shvartsman 1971; Bisnovatyi-Kogan and Komberg 1974; Barker and O'Connell 1975, 1976; and Hari Dass and Radhakrishnan 1975.) Since the magnetosphere is not implicated in gravitational radiation, binary pulsars are presently of interest not only as probes of gravitational theory (sec. 2.7.5) but also because of the questions of stellar evolution they pose (sec. 10, below).

Why one might have a neutron star companion that is not itself a pulsar is something of a puzzle because a massive star progenitor evolves so rapidly (order of a million years; sec. 10) that formation of the second pulsar would be expected before the first could spin down.

9. The Pulsar Object (Crab Pulsar)

A natural question is, "What does a pulsar look like?" Photographic plates of the best radio positions of pulsars show, at best, blank fields (those cluttered with candidates are, of course, less useful, not more). There is generally nothing to be seen. Pulsars are evidently so faint optically as to be well below the limit at which a star can be discerned on a photographic plate. That limit is not a firm number; but it is about 25th magnitude, and CCD technology should allow

TABLE 1.6—Short List of Binary Millisecond Pulsar Properties

Property	1957 + 20	1953 + 29	1855 + 09	1620 − 26
Period (ms)	1.60740171	6.13316488729	5.362100452367	11.07575
Period derivative (10^{-20})	1.2	2.95	1.7	
Interpulse	Yes	Yes	Yes	No
L_{radio} (ergs/s)	10^{27}	6.8×10^{25}	1.7×10^{25}	2×10^{27}
DM (pc/cm^3)	29.128	104.58	13.3	63.
Distance (kpc)	1.	3.5	0.4	2.
Magnetic field (10^8 gauss)	1.	4.	3.	
Orbital period	9.17^h	117.35^d	12.33^d	195^d
Mass function ($M_\odot$)	0.0000052	0.00269	0.0052	0.007
$a \sin i$ (10^6 km)[a]	0.027	9.39	48.6	19.2
Eccentricity	<0.001	0.0003	0.00002	
Companion mass	0.02	≥0.2	≥0.2	≥0.2
Comments	Eclipsing 20.5^m variable	42% duty cycle		In globular cluster
Aliases		2CG065 + 00(COS B)		M4 pulsar

a) $a \sin i$ = projected semimajor axis.

the limits to be pushed further down should interest in this issue revive. The Sun at 10 pc would be about a 5th magnitude star, so a 25th magnitude star at that distance would be 20 magnitudes, or 10^8 times fainter than the Sun (5 magnitudes $\equiv$ 100 times). At 100 pc it would still have to be 10^6 times fainter, and at 1000 pc it would be 10^4 times fainter than the Sun to appear as a 25th magnitude star. The exception sorely tests the rule. It is again PSR 0531 + 21 in the Crab Nebula which is in fact a visible object and appears to be a 16th magnitude star (at 2000 pc). Moreover, the optical emission is pulsed at the same rate as the radio, 33 ms. As a visible star, the Crab pulsar has a peculiar, featureless spectrum and was nominated many years ago by Minkowski (1942) as the likely stellar remnant from the supernova explosion that formed the Crab Nebula (see also Baade 1942). This explosion was evidently observed and recorded by the Chinese, who date it at A.D. 1054, giving an independent determination of the nebula's age. The Vela pulsar has also been detected at optical wavelengths, at a very weak level of optical emission, close to the "plate limit." The *companions* of some binary pulsars have been detected as in the case of the binary eclipsing pulsar PSR 1957 + 20. The spectrum of this pulsar seems featureless, possibly owing to violent oscillations of the photosphere in response to the pulsar electromagnetic emissions.

It is sobering to realize that if a spacecraft could be sent to the nearest pulsar and if the trajectory were similar to that of the *Voyager* spacecraft past Jupiter, essentially the same magnetic field strengths would be registered (sec. 7.2). The entire pulsar magnetosphere out to the light-cylinder distance would be the size of Jupiter, but the pulsar itself would be the size of the smallest known satellites of Jupiter. It is difficult for a space probe to survive even a flyby of Jupiter itself owing to the trapped radiation about the planet, presumably trivial compared to that of a pulsar with a total energy output roughly comparable to the Sun and at such close proximity.

10. Pulsar Formation in Supernovae

How and why are supernovae thought to form pulsars? As noted already, there are obvious associations of pulsars with some supernova remnants (Crab, Vela, etc.), while other remnants have *something* in their centers, and yet others show nothing (Seward 1985). It is surprising that, given that they are such unusual events, there are at least two distinct types of supernova. The two types have similar energy outputs and similar occurrence rates, but significantly different spectra (basically the spectrum of type II exhibits hydrogen lines while type I is hydrogen deficient). Indeed, there are yet further spectral classifications, but it is not clear whether each type requires a physically different model or is just a complication (e.g., owing to the presence of a companion star in the system) of one of the two basic models.

With unfortunate inscrutability, the two types are called "type I" and "type II" supernovae. (For reviews, see Trimble 1982, 1983, and Woosley and Weaver 1986.)

Type I (the hydrogen-deficient supernova) is often modeled as a contact binary system of two white dwarfs (hydrogen-deficient stars, in general) which are transferring matter from one to the other, thereby pushing the one toward the Chandrasekhar limit (e.g., Sutherland and Wheeler 1984).

Type II is modeled as a massive star which has reached the point where all the hydrogen and much of the helium in its core has been "burned" into heavier nuclei, again leading to a core ready to slip over the Chandrasekhar mass limit. Such stars are modeled as being somewhat more massive than the Sun (3 times or more). Owing to their greater masses, the central pressure and hence temperature of these objects must be higher. Consequently their cores evolve much more rapidly because they progressively exhaust internal sources of energy to maintain that temperature, burning first hydrogen to helium, then helium to carbon, oxygen, etc. (once such a core burns to iron, no further exothermic nuclear reactions are left because heavier isotopes are energetically uphill, eventually reaching the point at which they spontaneously fission or alpha-decay). As a result, the central region condenses until the density becomes so great that electron degeneracy pressure, not thermal motion, supports what has now become essentially a white dwarf surrounded by a massive "atmosphere," i.e., the rest of the star. At the surface of this core, however, heat must continue to be produced, and the core steadily increases in mass until it approaches the Chandrasekhar limit, and at some point it simply collapses. Electron capture further reduces the stabilizing pressure, and the collapse becomes irreversible. At this moment the local temperature shoots up and a vast amount of energy is released impulsively, ejecting the rest of the star while most of the core collapses to form a neutron star (see, for example, Arnett 1969). The theoretical details and mechanisms of energy release, energy transfer, etc., are still strongly debated, but such fine-tuning is beyond our purposes here.

Although the stellar collapse releases a large amount of energy and disrupts the star, the long-term luminosity from both types of supernova is usually attributed to radioactive decay of Ni^{56} to Co^{56} (6.1 day half-life) soon after the explosion, and subsequent decay of the cobalt (78.8 day half-life) over a year or two. Without this subsequent reheating of the supernova ejecta, the expanding cloud would rapidly cool adiabatically and supernovae would not actually appear so "super." This continued heating requires a vast amount of a single radioactive isotope (approaching a significant fraction of an entire solar mass!) to be created in the event. The source of so much of a single isotope is argued to be the reactions induced by the shock heating of the otherwise stable isotopes formed in the core by the nuclear burning. These shocked nuclei accrete, exothermically, increasing numbers of alpha particles until addition of another alpha particle becomes endothermic, which indeed is at Ni^{56} ($=14\alpha$). Such nuclei are, however, proton rich and positron decay (or electron capture) to the stable isotopes with smaller Coulomb energies. Although the beta decays themselves are a source of energy to heat the ejecta, most of the long-term powering would come from beta decay to excited states of Fe^{56}, which then emit gamma-rays in cascading to the ground state. (These gamma-rays are sometimes inaccurately called "cobalt lines.") It is interesting in this respect that the relative abundance of iron isotopes on Earth is 92% Fe^{56} and only 0.3% Fe^{58}, even though the latter is thermodynamically the more stable isotope.

An alternative model is that of a newly formed pulsar that delivers energy to the supernova remnant and keeps it hot (Ostriker and Gunn 1969c; Ostriker 1987). Although pulsar powering must at some point become important (Michel et al. 1987), observations of the iron gamma-ray lines and the exponential decay of the light curve of SN 1987A have so far favored the radioactive decay model.

The observed decay rates of type I supernovae are faster (50-60 day half-lives) than the cobalt decay rate. This discrepancy is argued to result from the declining optical depth of the remnant as it expands, which permits the gamma-rays following cobalt decay to escape instead of being converted into visible radiation. Type II supernovae typically decay *less* slowly than cobalt, possibly because energy is trapped in the much more massive envelope and subsequently released. The very long term behavior (which should eventually lead to decay at or faster than cobalt) has not been followed in detail but seems consistent with cobalt decay (Kirshner and Uomoto 1986). Being rare, supernovae are mainly seen in distant (≈ 10 Mpc) galaxies, where they are difficult to resolve very long past maximum light. Supernova 1987A in the LMC, however, has followed almost exactly the simple radioactive decay curve for cobalt.

On the average, pulsars seem to last much longer than the remnants (several million years versus tens of thousands of years), and the remaining known pulsars are not necessarily expected to be surrounded by a detectable remnant. It is still somewhat puzzling why more known remnants do not have young observable pulsars in them; however, most supernova remnants are at distances of several kpc, near the upper limit of the distance at which pulsars can be detected (unless they are very bright, like the Crab pulsar). Moreover, the remnants themselves are usually bright radio sources against which detection of a central pulsar can be difficult. The Crab pulsar was originally detected because it emits an occasional giant pulse which can even outshine the nebula. Braun et al. (1989) list the young pulsars associated with known supernova remnants, giving useful tabulations of both, and conclude that some remnants would have to dissipate more rapidly (order of 20,000 years) than is usually thought to be the case.

The theoretically favored progenitor of a pulsar is the type II supernova because the type I is thought to lead to disintegration of the white dwarf rather than collapse to a neutron star. Because massive stars (the expected type II progenitor) evolve so rapidly (order of 10^6 years), they are absent from populations of old stars (order of the age of the galaxy, order of 10^{10} years) such as those found in the galactic halo and in globular clusters. These ideas are consistent with observations which show hydrogen-deficient (type I) supernovae in elliptical galaxies. The apparent association of millisecond pulsars with just such a population leads to interesting speculations on their origin.

a. *LMC Supernova (SN 1987A): Type II Model*

The recent supernova in the LMC was coincident with a neutrino burst (Hirata et al. 1987; Bionta et al. 1987) and consequently strongly confirmed the overall theory of type II supernovae and the formation of neutron stars. Without neutron star formation, there would be no important source of neutrinos of sufficient energy (10s of MeV) to activate terrestrial detectors for the several seconds the burst lasted. Collapse to a black hole or disintegration following collapse would give neutrino sources lasting only milliseconds (see Burrows 1988 for a comprehensive analysis). Unfortunately, no sensitive gravitational wave detectors were on line at the time, so nothing was learned about possible gravitational wave emission from a rapidly rotating deformed neutron star. The presupernova object appears to have been the *blue* giant SK–69 202 whereas conventional theory had predicted a red supergiant. Arnett (1987) attributes the difference to the lower metallicity in the LMC, which alters the core evolution, leading to a somewhat different and lower-luminosity supernova event than expected and observed for galactic objects. Woosley et al. (1987) show that the stellar radius is a very sensitive function of core luminosity, varying by a factor of 10 over a rather small change in core luminosity (hence the stellar color can range from red to blue). The amazing fidelity with which the decay of the supernova has followed the cobalt decay rate since the outburst raises puzzling questions as to why such a characteristic decay was not noticed in other type II supernovae. Michel et al. (1987) suggest that pulsar powering is one cause of more complex light curves. If so, the radioactive decay should eventually drop to a level that would reveal such powering. Early detection of a pulsar seems ruled out owing to obscuration by the ejecta, but eventual detection (or even nondetection) should provide valuable information on pulsar origins (ibid.).

The reported detection of an 0.508 ms (!) optical pulsar in the remnant of SN 1987A (Middleditch et al. 1989; Kristian et al. 1989) has caused considerable controversy. It is possible to find equations of state for neutron stars that would permit such spin rates, but the star must be largely supported by centrifugal forces, which requires it to be on the large side and have a *soft* equation of state. But pulsars as a class rotate too slowly for centrifugal forces to be significant. Consequently pulsars with such equations of state would not be as massive as observed (Friedman et al. 1989; Suzuki and Sato 1989; Goldman 1989), with the possibility of one very narrow range of choices. The observation has not as yet been repeated, which is not in itself surprising given the large amount of debris that would surround a neutron star (Michel et al. 1987). Such a detection would almost have to be through a temporary clearing considering the expected amount of debris. If confirmed, this pulsar would be designated PSR 0535–69. Still, there are a large number of unsatisfactory aspects from a theoretical point of view. The pulsar is not yet a significant energy source for the nebula, and the pulsed luminosity in the optical would suggest a magnetic field of order 10^9 gauss. However, Bouchet et al. (1989) report a leveling of the bolometric luminosity at around 10^{38} ergs/s. Such an object would be a millisecond pulsar, and not a typical field pulsar (see sec. 2.4.3.g), which is exactly the inverse of the expected progenitors (see below). In other words, what little is understood about how pulsars are born would be stood on its head, and one would have to concede that both millisecond and ordinary pulsars can be formed in type II supernovae. There were also periodic variations in the period reported, which would imply planetary objects (Jupiter scale masses) in roughly 1 $R_\odot$ orbit; how such an object could get to this proximity to begin with or survive a type II event (which ejects so much mass that orbiting objects would be unbound from the remnant) is a problem too, although the eclipsing binary PSR 1957 + 20 arguably provides a parallel. The possibility that the pulsations are real but represent radial oscillations of a neutron star would remove these objections (Wang et al. 1989; Kluźniak et al. 1989; Tsuruta 1989); such oscillations could be commonplace and short-lived, and would not provide a reliable estimate of magnetic field strength. Thus, even the typical pulsars could have masqueraded as millisecond pulsars after birth without otherwise distorting the standard notions of pulsar evolution. Unfortunately, if the oscillations are sufficiently short-lived, they might never be seen again, leaving SN 1987A with yet another irreproducible result (in addition to reported neighboring spots of emission indirectly detected with speckle photometry and then not found again; see discussion by Felten et al. 1989).

b. *Pulsars in Globular Clusters: Type I Model*

A steep-spectrum polarized radio source in the globular cluster M28 (NGC 6626) was a strong candidate for an unresolved millisecond pulsar (Mahoney and Erickson 1985; Erickson et al. 1987) following a search by Hamilton et al. (1985). It was confirmed as a 3.05 ms pulsar by Lyne et al. (1987). This discovery suggests that a few percent of type I supernovae (of which there are several subcategories, and which are associated with old stellar populations) might form neutron stars as well (Michel 1987b; Chanmugam and Brecher 1987; Grindlay and Bailyn 1988; Epstein et al. 1988). The search for pulsars in globular clusters was stimulated by the suggestion (Alpar et al. 1982) that the millisecond pulsars represent old pulsars whose magnetic fields have largely decayed away and which have subsequently been spun up by accretion from a companion. This view has been amplified (Romani et al. 1987) for the case of M28. The tendency for millisecond pulsars to be in binary systems is broadly consistent with that model. The millisecond pulsar (PSR 1821–24) in the globular cluster M28 again seems consistent with the above scenario of "recycling" an old pulsar. But this model *requires* magnetic field decay, the importance of which in pulsars is open to question. Since then, a *binary* "millisecond" pulsar (actually, at 11.076 ms, much closer to the Crab in period than to the original millisecond pulsar) has been discovered in the globular cluster M4 (Lyne et al. 1988a; McKenna and Lyne 1988). With a period of about 191 days and a likely companion with a mass of about 0.3 $M_\odot$, such a white dwarf companion is essentially a massive planet, far too small to transfer matter and too distant to be affected by the pulsar. Given the small eccentricity ($e = 0.025316$) and the fact that tidal effects are close to negligible in such tiny objects, this must be a rather old system that has never been seriously perturbed, although McKenna and Lyne (1988) argue that the nonzero eccentricity might result from close stellar encounters. Also reported have been three pulsars in M15, a 110 ms pulsar (Wolszczan et al. 1989), a 56 ms pulsar (Anderson et al. 1989a), a 30 ms binary pulsar (Anderson et al. 1989b), and a *pair* of binary millisecond pulsars in 47 Tuc (Ables et al. 1988), summarized in table 1.7. In addition, a 10 ms pulsar with a dispersion measure of about 30.5 and a single pulse with a 15% duty cycle has recently been discovered in M13 (NGC 6205) by Anderson et al. (1989c).

Note that the extremely small mass function for PSR 0021-72A must result from its being seen almost face-on (Ables et al. 1989). For a neutron star with even as light a companion as that of the eclipser (PSR 1957 + 20, see sec. c below), 0.02 $M_\odot$, the Earth must be within $1°$ of the orbital axis, the a priori probability of which is only about 10^{-4} *per binary*. The simplest hypothesis would be that the pulsar rotation axis in such a small system (0.06 $R_\odot$) is parallel to the orbital axis, in which case we would be seeing one rotational pole.

The large *negative* period derivative for the M15 pulsar is attributed to acceleration in the dense cluster by an amount in excess of 5×10^{-6} cm s^{-2}. This acceleration would also be given if the pulsar had been left in a relatively loose binary system ($\approx$300 AU) about a cluster star following a three-body interaction.

The pulsar in M28 is taken as evidence for the spin-up scenario. It is argued that close encounters act to separate the companion from the pulsar and account for the lone M28 object. Given that these systems have sizes of order 1 $R_\odot$, these must be close encounters indeed. The problem of how millisecond pulsars might be single objects (if they are indeed spun up by accretion from a companion) has even stimulated speculation that the companions could be destroyed by interaction with the pulsar. This could only be possible if the companion were extremely close to the pulsar (order of 1 $R_\odot$ or less); otherwise there would not be enough energy intercepted even to lift matter out of the companion's gravitation well within the lifetime of the galaxy.

If the M28 pulsar was formed from a massive star (presumably a type II supernova), it would have to have been formed shortly after the birth of the globular cluster itself and hence be nearly as old or older than the galaxy itself (assuming star formation in globular clusters quickly died out, as is generally believed). One candidate progenitor for a type I supernova is a contact white dwarf binary system (Arnett et al. 1985; Cameron and Iben 1986) wherein mass transfer pushes one of the two over the Chandrasekhar limit. The central region of a globular cluster is an ideal site for making binaries because "heating" from binary formation is probably required to stave off core collapse in these objects. The center of M28 has a central density of about $5 \times 10^4 M_\odot$/pc^3

TABLE 1.7—Properties of some Globular Cluster Pulsars

Property	1821−24	1620−26	2127+11A	2127+11B	2127+11C	0021−72A	0021−72B
P (ms)	3.05431	11.0757	110.6647	56.	30.	4.478953	6.127
$\dot{P}$ (10^{-19})	16	8.2	−200(!!)	−	−	−	−
Interpulse	Yes	No	No	−	−	−	−
DM	120	62.87	67.25	67.2	67.2	65.	65.
Distance (kpc)	5.8	−	9.7	9.7	9.7	−	−
B (10^9 gauss)	−	3.	−	−	−	−	−
Binary?	No	Yes	Possibly	No	Yes	Yes	Yes
Binary period	−	191 d	−	−	Undet.	1924.3 s	7-95 d
$a \sin i$ (l-s)	−	64.8	−	−	−	0.002	−
Mass function	−	0.008	−	−	−	1.16×10^{-8}	−
Eccentricity	−	0.25316	−	−	−	0.33	−
Cluster	M28[a]	M4[b]	M15[c]	M15	M15	47 Tuc[d]	47 Tuc

a) NGC 6626.
b) NGC 6121.
c) NGC 7078.
d) NGC 104.

and a core relaxation time (Madore 1980) of only 5×10^7 years. Given the plethora of supernova classifications, the theoretical uncertainties, and the apparent inevitability of the formation of close binary white dwarfs in globular clusters, one wonders if it is certain that pulsars are no longer formed in these old (population II) systems.

Suppose that a neutron star is formed when one of the two white dwarfs implodes: what are the expected consequences? First, the orbital period of two white dwarfs with masses sufficient to exceed the Chandrasekhar limit of one of them during mass transfer (order of 1 $M_\odot$ each, say) and of radius 10^3 km that have evolved into near contact (prior to the onset of mass transfer) is a few seconds. Tidal dissipation at such close proximity should quickly force the two into synchronous rotation. At some distance, mass transfer would commence because white dwarfs have an inverse mass-radius relationship; consequently the contributing dwarf swells and the recipient dwarf shrinks, which feeds back to promote mass transfer. Subsequent implosion of the accreting white dwarf into a neutron star of 10 km radius would then result in a neutron star with a period of a few tenths of 1 ms. The excess angular momentum could easily be shed by gravitational waves (owing to the triaxial shape induced in objects that spin faster than a critical value) or by spinning off a disk. Thus the white dwarf contact binary model would automatically produce a millisecond pulsar. Indeed, the origin of the idea that pulsars are generally born as rapid rotators is itself obscure, given that little is known a priori about the rotation rates of the cores of massive stars (the presumed progenitors) at the time of implosion. The millisecond rotation rate was simply the maximum possible value, not an expected value from any theory. For the above model, however, the expectation of a millisecond period would be automatic.

White dwarfs in the general field of isolated stars probably have fields more typically of order 10^5 gauss, which would produce a pulsar magnetic field of order 10^9 gauss in flux-conserving collapse (what one observes is that a few are magnetized to larger fields). The expectation that these white dwarfs are the ones that will be paired in the cores of globular clusters leads naturally to the expectation that millisecond pulsars with low magnetic fields would mainly be formed. The occasional formation of a strongly magnetized millisecond pulsar by such a mechanism would make little difference to pulsar statistics because such a pulsar would slow down so rapidly that, unless detected shortly after the supernova event, it would quickly become a relatively common pulsar.

Type I and II supernovae occur at similar rates, despite their utterly different evolutionary scenarios. One per hundred years in a galaxy of 10^{11} stars corresponds to one per 10^8 years for a globular cluster of 10^5 stars, which corresponds to the typical spin-down ages of millisecond pulsars. Thus a given globular cluster would typically have had one type I event recently enough to leave a readily visible millisecond pulsar (if formed with 100% efficiency). The finding of one millisecond pulsar among the 12 globular clusters originally surveyed corresponds to 1 in 12 odds of forming a pulsar in a type I event, broadly consistent with the absence of pulsars in the historical supernovae and the paucity of such objects in general. The true odds could well be substantially lower given the expectation that a larger proportion of close binaries are formed in the dense cores of globular clusters. At the root of such low odds would naturally be the requirements for special circumstances if a neutron star is to be left behind (initial total masses $\geq 1.4 M_\odot$, etc.). The properties of globular clusters and their cores have recently been surveyed by Chernoff and Djorgovski (1989).

Because the orbital velocity of a 1 s white dwarf binary is not dissimilar to the velocity of supernova ejecta (both $\approx 10^4$ km/s), the gravitational impulse from the ejected matter passing the companion white dwarf is neither "sudden" (which would leave the companion in an elliptical orbit but with the periapsis at a small value of order 10^3 km) nor "slow" (which would cause the orbit to spiral out and leave it nearly circular). Consequently, both the eccentricity and the periapsis will be increased, depending sensitively on the fractional mass loss and the ratio of the orbital velocity to ejecta velocity. A white dwarf pushed over the Chandrasekhar limit and imploding to a neutron star could well eject relatively little matter, which would improve its chances of retaining a binary system. In contrast, for type II supernovae most of the star is expected to be ejected and any binary companion would almost certainly be lost, consistent with the observation that the majority of known pulsars (98.5%) are solitary objects.

Thus a simple direct way of making millisecond pulsars with low magnetic fields is to form them from contact binary white dwarf systems of sufficient total mass to form a neutron star (Michel 1987b; Isern et al. 1987; Chanmugam and Brecher 1987; Epstein et al. 1988; Grindlay and Bailyn 1988). Whether the system remains binary may depend on finer details. The idea of recycling old pulsars (which served the important role of calling attention to pulsars in old stellar populations) is dependent on rapid magnetic field decay, a phenomenon which may well not even exist.

c. *Eclipsing Pulsar*

Fruchter et al. (1988b) discovered the extraordinary *eclipsing* binary pulsar PSR 1957 + 20, which is in eclipse over 8% of its orbit. Because the orbital inclination must be nearly 90° to the line of sight to have eclipses, the mass of the companion can be estimated from the mass function:

$$\frac{(m_c \, \sin i)^3}{(m_p + m_c)^2} = 5.2 \times 10^{-6} M_\odot. \tag{26}$$

So for a pulsar mass $(m_p) \approx 1.4 M_\odot$, the companion mass (m_c) is only about $0.02 \, M_\odot$! The basic properties of this system are outlined in table 1.8.

The companion mass is quite unusual, huge compared to known planets, yet less than the roughly $0.07 \, M_\odot$ expected for it to be a true star with nuclear

TABLE 1.8—Properties of the Eclipsing Binary Pulsar

Period (ms)	1.60740171
Period derivative	1.2×10^{-20}
Interpulse	Yes (broad)
L_{radio}	10^{27} ergs/s
DM	29.128
Distance	≈ 1 kpc
Magnetic field	1.3×10^8 gauss
Binary period	33001.9 s
Orbital radius	1.7×10^{11} cm
Roche lobe (L_1)	2.8×10^{10} cm toward pulsar from companion
Companion radius (He dwarf)	3×10^9 cm
Mass function	$5.2 \times 10^{-6} \, M_\odot$
Companion mass	$0.022 \, M_\odot$
Pulsar mass	$1.4 \, M_\odot$ (assumed)
Eccentricity	< 0.001

burning ignited at the center. Even if this object filled its Roche lobe, it would be too small by a factor of 2 to give the observed eclipse. The Roche lobe is the locus of equipotentials, in the corotating frame and including the effective centrifugal potential, that links to the inner Lagrangian point at which a test particle would co-orbit at a fixed distance between the companion and primary. This test-particle orbit is actually unstable to falling toward the pulsar. Thus a star overfilling its Roche lobe is expected to simply "drain" its outer shell over this point, with the matter falling toward the other star (see, for example, Blanco and McCuskey 1961). In other words, there is no physical object that could be this large this close to the pulsar. Yet the eclipses are larger!

It was immediately proposed (Phinney et al. 1988) that this light companion represents erosion by the pulsar wind, which then operates to "evaporate" the companions of those millisecond pulsars that lack companions. Such a huge mass loss, if taking place, is not intercepted by the radio signals (too little time delay). One can at least do a plausibility assessment, taking advantage of the fact that the lighter a white dwarf is, the *larger* it is, with $R \approx M^{-1/3}$. Accordingly, erosion will be rapid at the final phases of evaporation and slow at the early phases. If we take the mass loss to be proportional to the area of the companion (i.e., to the energy from the pulsar that could be deposited on the companion), we find that $M_0^{5/3} - M^{5/3} = T +$ constant. From this we can ask what the a priori probability is of stumbling upon a pulsar in the final instants of evaporating its companion. If for the sake of argument we take an initial mass $M_0 \approx 0.5 \, M_\odot$ and adopt the estimates (Fruchter et al. 1988) that the companion will evaporate in 10^8 years, we can immediately determine that the system must have been evaporating for 2×10^{10} years and is nearly as old as the galaxy. If it has such a small period derivative, it will not have enough power output to evaporate the star: a mass loss of 10^{16} g/s requires about 10^{31}

ergs/s to be deposited just to accelerate the matter to escape velocity from the companion, while the companion intercepts only about 10^{-4} of the pulsar power output, hence requiring a pulsar power output of 10^{35} ergs/s or a period derivative of 10^{-18}. That period derivative would give the pulsar an age of only about 30 million years. In fact, the measured value is $\dot{P} \approx 10^{-20}$. The apparent conclusion is, then, that the companion was never much more massive than it is now. Tavani et al. (1989), however, suggest that the mass loss could be self-excited when heating from matter falling into the pulsar generated X-rays that in turn drove the mass off the companion. Such a process could, at some point, simply turn off for good, leaving a light companion and leaving the issue of the pulsar evaporating the companion moot. Another hypothesis is that the companion was almost totally disrupted by the supernova explosion that formed PSR 1957 + 20, and with slightly different parameters the companion can be completely obliterated. But then, of course, there is no field decay followed by accretion spin-up. Such a view would be consistent with the notion that the millisecond binaries are formed in type I supernovae (Michel 1987b). Fryxell and Arnett (1981) argue, however, that a supernova explosion should have little effect on a companion star (see also Cheng 1974).

The eclipses of PSR 1957 + 20 by its companion seem quite diagnostic, as shown in figure 1.7. For small orbital eccentricity (measured to be less than 0.001) and an eclipse extending from 0.21 to 0.29 of phase (zero phase is the ascending node and unit phase is a full revolution), one has to first-order a symmetric eclipse centered on inferior conjunction (i.e., just as if the pulsar were occulted by the body of an orbiting sphere 4×10^{10} cm in radius). This degree of symmetry almost immediately rules out eclipses of the pulsar by a cometary structure around the companion, because, owing to the intrinsic curvature of such cometary tails, the flanks would be intercepted at different distances and at different phase angles on entry and exit from eclipse, as discussed in more detail below. Such curvature can be minimized near the companion if the material is accelerated quickly, but then any tail would be swept back too sharply to account for the large eclipse angles. Moreover, there is a measurable time delay of up to at least 400 μs that trails off over another 15° following the eclipse, but essentially *no* time delay preceding the eclipse. The entire event therefore subtends about 45° of the orbit (12% of phase). It is tempting to suppose that this time delay is directly related to the eclipse, but it is too small in magnitude to produce an eclipse by having this plasma refract the radio pulses away from the observer. Because refraction by plasma is *away* from high plasma concentrations, eclipses can occur, and because refraction by neutral gas is toward high gas concentration, no complete interruption can occur: the *Voyager* spacecraft was tracked continuously behind Uranus, for example, owing to the latter effect. Given that the system radius is about 5.8 light-seconds, the average index of refraction along a line of sight that reaches to within 4×10^{10} cm of the companion (the impact parameter at eclipse) differs from unity by less than 1 part in 10^{-4}, which is much too small to produce an eclipse with such a large

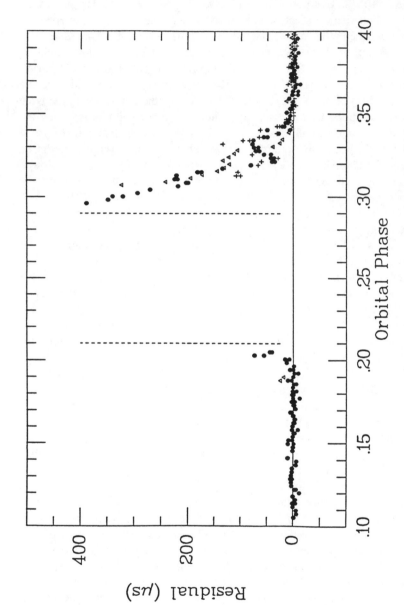

FIG. 1.7. Time delay versus phase in PSR 1957 + 20. Eclipse is indicated by dashed lines, essentially symmetric about 0.25. Essentially all the time delay follows eclipse (from Fruchter et al. 1988b). Reprinted by permission from *Nature*, 333, 237 (figure 1). Copyright © 1988 Macmillan Magazines Ltd.

angular extent. Simply bending the beam through 10° would introduce about 90 ms of delay, two orders of magnitude larger than observed. The fact that there is *some* time delay asymmetrically associated with the eclipse shows that there is indeed a wake blown back from the system, probably plasma eroded off the companion by interaction with the large-amplitude 622 Hz pulsar electromagnetic radiation (*not* just by the pulsar radio-frequency radiation). The column density of the latter is about 5×10^{16} electrons/cm^2 from the time delay (assuming zero refraction).

At this proximity to the pulsar mass, tidal damping should have removed any orbital eccentricity and spin so that the companion should now be in synchronous orbit. As a result, the same hemisphere constantly faces the 622 MHz large-amplitude (20 gauss!) electromagnetic waves from the pulsar. In that case the wind luminosity is about $L_{\text{total}} \approx 10^{35}$ ergs/s, and, if the companion absorbed and reradiated all of the energy it geometrically intercepted, its surface temperature on the illuminated side would be about 6000 K, giving a total luminosity of about 0.02 $L_\odot$ corresponding to about a 21^{m} object at 1 kpc, the distance suggested by the dispersion measure. This "star" would be very unusual in that it should be a deeply modulated variable at the orbital period of 9.2 hours; also, any spectral lines would very likely be washed out by Doppler broadening from the plasma motions driven by the pulsar radiation. Such expectations seem broadly consistent with observation (Fruchter et al. 1988a; van Paradijs et al. 1988; Kulkarni et al. 1988b; Djorgovski and Evans 1988), which shows at least three magnitudes of variation in the candidate, a surface temperature of around 6000 K, and a size somewhat larger than the 3×10^9 cm estimate for the size of the companion, which is, however, dependent on the interstellar absorption assumed.

To summarize the essential observations: (1) The eclipse is highly symmetric. (2) The time delays are quite asymmetric with little time delay before eclipse and a long drawn-out time delay starting at about 500 μs and falling off in an irregular manner that seems to be bracketed between models for plasma receding at (a) constant velocity and (b) constant acceleration. (3) The observed time delay corresponds to about 5×10^{16} electrons/cm^2, which converts to a mass-loss rate after being multiplied by some characteristic length transverse to the orbit (reasonably, the eclipse size of 4×10^{10} cm) and some characteristic velocity (reasonably, the escape velocity of 3×10^7 cm/s). The latter then gives roughly 10^{12} g/s, which is unimportant by four orders of magnitude insofar as the evaporation of the companion taking place in an interesting length of time. Three models have been proposed (see discussion in Michel 1989).

(1) The Interacting Wind from Companion Model. This seems the obvious first choice, given that matter driven off the companion clearly gives the time delay (Phinney et al. 1988; Wasserman and Cordes 1988). Unfortunately there are two problems with the model. First, as noted above, the amount of observed plasma is insignificant for evaporation and is too small to give refractive eclipses. The time delay is less than 1 ms, whereas the system size

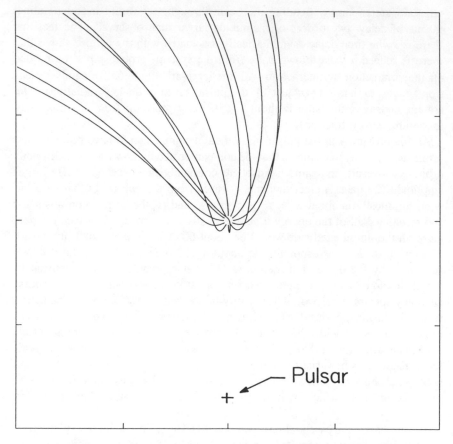

FIG. 1.8. Cometary trajectories of escaping gas. Single-particle trajectories seen in the coorbital frame of the companion. Here the velocity was 600 km/s and the radial acceleration was 12 times the pulsar gravitational attraction (chosen to give roughly the scales required to account for the observed eclipse extent). From Michel (1989). Reprinted by permission from *Nature*, 337, 236 (figure 1). Copyright © 1989 Macmillan Magazines Ltd.

is thousands of light-milliseconds across; thus the index of refraction differs from unity by little more than that of air. So what causes the eclipse? Phinney et al. suggest that it is the leading head of a cometary structure, as shown in their figure 1. A similar sketch in my figure 1.8 shows the particle trajectories for interaction with an outflowing wind which is then accelerated back radially. Note the cometary shape and lack of a sharp trailing edge. Postpublication arguments have been circulated by various authors that some sort of "collimation" might salvage the model, perhaps if the pulsar wind behaved as a $\gamma = 4/3$ gas while the companion wind behaved as a $\gamma = 5/3$ gas and thereby the tail expanded less rapidly downstream. These arguments have not yet been supported by quantitative simulations.

(2) The Magnetosphere Model. Symmetry of eclipses can be had if the

pulsar wind is halted at a magnetosphere surrounding the companion. One basically needs a magnetic field of about 20 gauss at the magnetosphere to balance the 20 gauss large-amplitude 622 Hz radiation from the pulsar. For a degenerate helium companion of $\approx 3 \times 10^9$ cm, the magnetosphere extends out about 13 radii and corresponds to a surface magnetic field of about 10^5 gauss, which seem entirely reasonable if the companion is the remnant of a former white dwarf or white dwarf-like stellar core. Evidence for a larger companion size rests on assuming a nominal interstellar absorption (Djorgovski and Evans 1988); however, the paucity of H_β to H_α observed in the surrounding nebulosity by Kulkarni and Hester (1988) may point to larger interstellar absorption and a smaller, more evolved core. This nebula has itself the form of a comet and is probably an interaction front with the pulsar wind, distorted by the motion of the system through the local interstellar medium. The advantage of the magnetosphere model over the wind model is that the plasma density inside the magnetosphere—on the closed magnetic field lines—can be orders of magnitude larger than in a magnetospheric tail blown back by the wind interaction. The latter could give the cometary-like time delay while the dense magnetosphere would give a symmetric eclipse. The problem with this model is the lack of any appreciable Faraday delay. At small magnetic fields there is Faraday rotation of any linear polarization, while at high densities there is essentially depolarization of any linear component and a differential time delay for the circularized components of a factor $4\omega_c/\omega$ of the total time delay, where ω_c is the cyclotron frequency and ω is the observation frequency. For each sign of circular polarization we obtain

$$\delta t_\pm = \delta t \left(1 \pm \frac{2\omega_c}{\omega}\right), \qquad (27)$$

where δt is the mean time delay. Because this differential delay has an upper limit of about 2 μs, the magnetic field in the tail must be less than about 0.2 gauss in some path-average of the line-of-sight component (Fruchter et al. 1989). There is no good theory for calculating the magnetic field in a magnetospheric tail, but on general principles one would guess it to be comparable to that at the magnetopause, within an order of magnitude. Thus the low limit to the tail field seems an important problem. However, a low plasma density implies a large (relativistic) Alfvén velocity, which means that the tail is not probed until it has expanded extensively, in which case the small limit on B need not reflect values near the magnetopause.

(3) The Occulting Disk Model. A disk of particles around the companion could cause eclipses by scattering the radio waves. This idea is fairly simple and unforced. If the pulsar wind actually interacts directly with the companion instead of with a wind or magnetosphere, then absorption of the wind energy close to the surface is consistent with the observed light curve (which varies essentially like the phases of the Moon or Venus). At the terminator, however, the radiation pressure will drive plasma off the companion (accounting for the

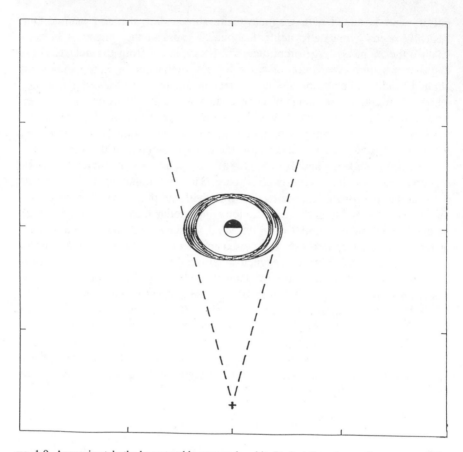

FIG. 1.9. Approximately the largest stable retrograde orbit. Dashed lines show eclipse extent, while meandering orbit shows roughly the largest orbit that will persist in the system for a significant length of time. These orbits are beginning to be chaotic, and a sharp threshold between stable and unstable orbits probably does not exist.

time delay) and some neutral gas will be entrained briefly and swept back in this flow, leading to a "fountain" of gas ejected behind the companion which tends to wrap around it in a retrograde sense (the extreme limit of a cometary flow). If the companion is evolved, the carbon in the gas phase that should be present can condense to form highly refractory particles. Thus it is perfectly possible that the companion is surrounded by carbon particles. These particles could be anything from soot balls to diamonds (which form if hydrogen is present in the gas phase, in which case the companion might be named "Lucy"). If one calculates the furthest stable retrograde orbit, it turns out to be almost exactly the eclipse size shown in figure 1.9. The natural model is therefore one in which a large number of large particles ($\approx$10 cm to effectively scatter the radio emission) have grown owing to repeated passage through the gas fountain, with the particulate cloud limited in size by orbital stability. The difficulty here is that the cloud cannot be too thick or it would also obscure the companion, and it cannot be too thin or it would not occult the pulsar.

If we overlook their intrinsic difficulties, what could the various models tell us about the pulsar wind? The wind model requires a magnetohydrodynamic wind from the pulsar, namely one with enough particles to cause thermalization of the Poynting flux of the wind. Otherwise there is no "shock" and the pulsar wind simply reflects like a flashlight beam. Additionally, a large flux of energetic radiation from the pulsar must accompany the wind and drive the wind off the companion. Indeed, Eichler and Levinson (1988) find it hard to come up with a mechanism with which to drive a sufficiently strong wind. The magnetosphere model probably requires an electromagnetic (EM) wind that is dominated by the Poynting flux. Reflection in this case would arguably minimize the intimate interaction between the two flows that leads to tail formation and thereby permits a weakly magnetized tail. The occulting disk model also would point to an EM wind so that the interaction would be pushed down onto the companion surface (to form the neutral fountain) and so that direct interaction of the wind with the orbiting particles would be minimized.

1.6 Pulsar Properties and Statistics
(Exceptional Properties)

1. Statistics

One would naturally hope to learn something about pulsars from their statistical distribution, once a large enough data base is built up. For most pulsars one can measure their
- period (P)
- radio luminosity
- pulse width (duty cycle)
- slowing-down rate ($\dot{P}$)
- subpulse multiplicity (if any)
- radio-frequency spectral features (spectral index, cutoff frequency, etc.)
as well as other interesting morphological features, many of which, however, seem to be exhibited by only a few pulsars. Many of these data have been tabulated for 330 pulsars (Manchester and Taylor 1981), but PSR 0904 + 77 is now thought to be spurious (Dewey et al. 1985). In general, nothing correlates very incisively with anything else. Certainly, there are general trends (e.g., short period with large slowing-down rate, at least before the millisecond pulsars were discovered!) that seem physically reasonable, but nothing like a Hertzsprung-Russell diagram. Classification schemes have been proposed (Taylor and Huguenin 1971; Huguenin et al. 1971; Backer 1976; Roberts 1976; Klyakotko 1977; Kochhar 1977; and authors cited below). One of the more clear-cut distinctions is whether the pulsars null, a phenomenon in which the pulsar becomes undetectable for relatively long stretches of time (up to 10 to 100 pulse periods) before abruptly reappearing. About 30% of all pulsars null. Another distinctive phenomenon is drifting wherein the time-averaged pulse

is actually composed of one or more narrower subpulses that are seen within the pulse "window" at successively earlier times. About 5% of all pulsars are drifters. A drifter that also nulls (PSR 0809 + 74: Unwin et al. 1978) has the interesting property that after a null the drifting subpulse reappears at the same position it occupied just before the null. Thus the sections containing nulls could be removed from the record and the drifting would appear to be uninterrupted. Such a phenomenon among drifters that null would suggest a close relationship between the two phenomena. Manchester and Taylor (1977) further divide the pulsars into S (simple) and C (complex) according to their average pulse shapes. So far, the record for complexity is held by PSR 1237 + 25, which has five distinct subpulses evident in its overall pulse. (It is interesting that one does not see all five subpulses in any single pulse from PSR 1237 + 25; they are apparent in the average waveform, figure 1.10, but are not "illuminated" simultaneously.) Other pulsars with complex pulses are PSR 1857–26, PSR 1737 + 13, and a few mentioned by Rankin et al. (1988b). Later Backer (1976) subdivided the C classification into D (double), T (triple), and M (multiple). However, these classifications are not entirely frequency independent. Rankin (1983a) accordingly subdivided the S classification into those that became more complex at low frequencies ("conal") and those that became more complex at high frequencies ("core"), with the core components having steeper spectra than the conal ones. In this model it was literally assumed that the pulsar emission pattern physically consists of a narrow core of emission surrounded by a circular cone of emission (see sec. 3.2). Lyne and Manchester (1988) in an exhaustive review of pulse morphology effectively reclassify pulsars into those dominated by a cone (82 examples), by a cone within their core (32 + 4 uncertain cases), by a core (10 + 20 uncertain), and by a partial cone (32 + 18 uncertain).

2. Period Variations

There are three obvious variations to be expected of a rotating magnetized object: (1) precession or nutation of the rotation axis, (2) variations in the magnetic field, such as those seen in both the Earth's field (major changes in millions of years) and the Sun's field (the 11 year cycle), and (3) evolutionary changes in the size and shape of the solid body, which would change its angular momentum.

a. *Glitches*

It is not hard to find variations in pulsars. The total average intensity of pulsar emission can vary systematically over months or years. There can also be variation from pulse to pulse, there can be subpulse drift, and there can be variation of the microstructure within the pulses. Then there are the timing "residuals," the difference between the actual pulse arrival times and that predicted for a perfect, albeit slowing-down clock. Few of the pulsar data connect any of the theoretically expected variations with observation, with the possible exception

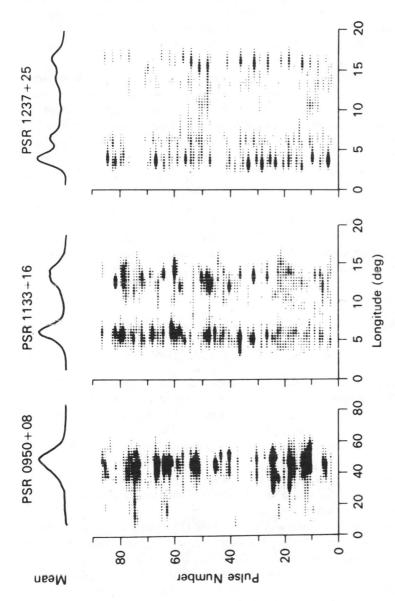

FIG. 1.10. Integrated pulse profiles versus individual profiles. Note that the at least five distinct subpulses exhibited by PSR 1237 + 25, not all of which are necessarily active at the same time, require three nested "hollow cones" (sec. 3.2), each of which is only partially lit up at any one time. From *Pulsars*, by R. N. Manchester and J. H. Taylor. Copyright © 1977 by W. H. Freeman and Company. Reprinted by permission.

of variation (3), which is exemplified by the starquake hypothesis. Starquakes were originally proposed to explain an abrupt drop in the period of the Vela pulsar (a "glitch"). The Crab pulsar has numerous microglitches (compared to Vela), although it experienced a major change in 1975, as was noticed later in the timing data (see Lohsen 1981; Demiański and Prószyński 1983). Because these had long been known as the fastest pulsars, it was thought that glitching was related to spin rate; however, other fast rotators discovered later such as the millisecond pulsar seem virtually glitch-free.

The problem with quake models is basically the expectation that the solid object—neutron star—in a pulsar ought to be very nearly spherical. This means that precession or nutation has little to act upon (a rotating sphere cannot nutate) and there is not enough potential moment-of-inertia change to explain the magnitude-plus-frequency of observed glitches in pulsars, unless we have been lucky in the sense of having seen a string of closely spaced glitches, which are actually rare on the average. This difficulty is particularly acute for the glitches of the Vela pulsar. The glitches typically correspond to a spin-up of $\Delta\Omega/\Omega \approx 2 \times 10^{-6}$ and are spaced a few years apart (see figure 1.11). This pulsar has a characteristic age $(P/\dot{P})$ of 22,700 years, which means that it would be expected to produce about 7000 more glitches. The equality of the relative moment-of-inertia change, $\Delta I/I$, with the frequency change then implies a total change in momentum of inertia of

$$\Sigma\Delta I \approx (2 \times 10^{-6} \text{ per event}) \times (7 \times 10^3 \text{ events}) = 1.4 \times 10^{-2}. \quad (1)$$

The centrifugal distortion from being a perfect sphere is easily calculated to be

$$\delta \equiv (r_E - r_p)/r_p \approx \Omega^2 a^3/2\,GM, \quad (2)$$

where r_E and r_p are the equatorial and polar radii, and a is the average. For the Vela spin rate and nominal pulsar parameters, we then get $\delta = 1.35 \times 10^{-5}$. The maximum moment-of-inertia change (2δ) is therefore a factor of 500 too small to maintain the observed glitch rate.

Magnetic field changes (item 2) are unpromising because the entire theory relies on the existence of a quasi-permanent magnetic field, and one does not see any systematic change in mean pulse shapes associated with glitches. Hence there is no evidence that this presumably vital factor waxes and wanes. Some observed glitches are listed in table 1.9.

Note the two glitches, one massive, observed (Lyne 1987) in what should be a rather old pulsar, PSR 0355 + 54, with $P/\dot{P} = 1.2 \times 10^6$ years, as well as two "typical" (?) glitches in another old pulsar, PSR 0525 +21 (Downs 1982), with $P/\dot{P} = 3.0 \times 10^6$ years. The detailed parameters observed by Lyne are given in table 1.10, where the errors are of the order of 1 or 2 in last places.

Note that the large values of the time constant $\dot{P}/\ddot{P}$ are systematically much smaller than the formal "age" given by $P/\dot{P}$, which immediately signals the presence of some influence on the timing behavior other than simple system-

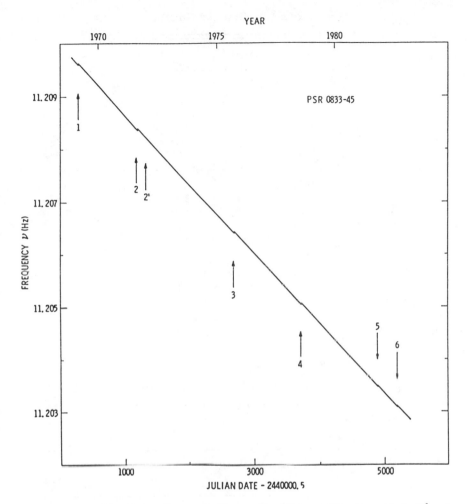

FIG. 1.11. Vela pulsar glitches. For $n = 3$ this curve should be a section of a parabola ($P^2 \approx t$) rather than a straight line; however, the deviation would amount to only about 1 part in 10^{-3}, too small to be seen on this scale and obviously small compared to the variations introduced by the glitches. From J. M. Cordes, G. S. Downs, and J. Krause-Polstorff, 1986, *Ap. J.*, 330, 847.

atic slowing-down. Otherwise these two characteristic times would be of the same order and from them one would be able to determine the braking index. (Sometimes a large "braking index"—see next section—is quoted on the basis of such figures, but calculating a value for n from the values in table 1.10 seems meaningless insofar as any long-term behavior is concerned.) Indeed, the significant entries in table 1.10 are arguably not the tiny period changes but the comparatively *huge* changes in the slowing-down rates. If pulsar theory is even vaguely correct, the slowing-down rate is determined mainly by the magnetic field of the pulsar. If it is, a tiny change in period (e.g., from a change in

TABLE 1.9—Observed Glitches

Observation	Interval (days)	$\Delta\Omega/\Omega$	$\Delta\dot{\Omega}/\dot{\Omega}$
Crab pulsar (0531 + 21) ($P = 0.0331$ s):			
1		5.99×10^{-9}	3.36×10^{-5}
2	670	2.99×10^{-9}	1.26×10^{-5}
3	1285	38.5×10^{-9}	19.82×10^{-5}
4	476	1.04×10^{-9}	0.75×10^{-5}
5	745	2.5×10^{-9}	1.53×10^{-5}
Vela pulsar (0833 − 45) ($P = 0.0892$ s):			
1		2.34×10^{-6}	1.02×10^{-2}
2	900	1.96×10^{-6}	1.63×10^{-2}
3	1500	2.02×10^{-6}	1.09×10^{-2}
4	1000	3.06×10^{-6}	2.81×10^{-2}
PSR 1641 − 45 ($P = 0.4550$ s):			
1		2.0×10^{-7}	2×10^{-3}
PSR 1325 − 43 ($P = 0.5327$ s):			
1		1.0×10^{-7}	
PSR 0355 + 54 ($P = 0.1564$ s):			
1		5.56×10^{-9}	0.18×10^{-2}
2	354 to 425	$4376. \times 10^{-9}$	$\approx 10 \times 10^{-2}$
PSR 2224 + 65 ($P = 0.6825$ s):			
1		1.7×10^{-6}	$< 0.6 \times 10^{-3}$
PSR 1508 + 55 ($P = 0.7397$ s):			
1		2×10^{-10}	6×10^{-3}
PSR 0525 + 21 ($P = 3.745$ s):			
1		1.3×10^{-9}	4.6×10^{-3}
2	1750	0.3×10^{-9}	$< 0.1 \times 10^{-3}$

Adopted largely from Alpar et al. (1981).

moment of inertia or magnetic field strength) would itself have only a comparable relative change in slowing-down rate. Instead, the slowing-down rates are profoundly changed and evolve extremely rapidly. The simplest model is one involving relatively weak coupling between two components of the moment of inertia, such as a core and a crust. In this regard, the magnetic field presents something of a challenge. The short-term time constant would then be the time it takes for the two to come into sync. Even these tiny changes of period imply

TABLE 1.10—Fitted Parameters for PSR 0355 + 54 (Lyne 1987)

Event	Epoch (MJD)	P (s)	$\dot{P}$ (10^{-15})	$\ddot{P}$ $(10^{-24}\ s^{-1})$
0	43874	0.1563809205335	4.387778	− 0.000291
1	46250	0.156381820445	4.3916	− 0.25
2	46504 (earliest)	0.15638123391	4.66	− 61.
2	46504 (latest)	0.15638123467	4.412	

that core and crust will have rotated some hundreds of times relative to one another between events. If they have, the two could hardly be coupled by a strong 10^{12} gauss magnetic field. Even if the field were aligned with the rotation axis, huge currents would circulate between the core and crust and bring the two into corotation. Such coupling would rule out the sort of quasi-stable differential rotation that crosses some threshold and leads to rapid synchronization (presumably the glitch). Moreover, the naive assumption that the slowing-down torques are exerted only on the crust needs careful examination; this assumption would be most plausible if the magnetic field only threaded the crust. If indeed the core and crust are magnetically decoupled, with the main pulsar field localized in the crust itself, we have a system wherein it would be very plausible for the magnetic field to be *generated* by the differential rotation. If anything, the opposite assumption tends to be made (to understand the supposed *decay* of pulsar magnetic fields to an asymptotic value), namely that the core retains its magnetization while the crust loses magnetization, thereby reducing the magnetic moment of the star. Why the entire core and crust should have been uniformly magnetized to begin with, the implication of such models, is equally unclear.

It is possible that some distortion of the shape of a neutron star from spherical might be "fossil" left over from an earlier rapidly rotating phase. However, it is not expected that the crust could withstand the extreme stresses required to freeze in the huge distortions required. Observation of a glitch in such a slow (0.4551 s) pulsar as PSR 1641–45 (Manchester et al. 1978) also poses severe constraints on any theory. A careful analysis by Cordes et al. (1986) of 12 years of timing data (Downs 1981; Downs and Reichley 1983) shows that the situation is much more complicated. On the large glitches ("macrojumps") are superimposed an order of magnitude more small glitches ("microjumps"). Unlike the macrojumps, which always involve an increase in frequency together with an increased deceleration (i.e., recovery from this increase), the microjumps seem to have all four possible combinations of increases or decreases. Cheng et al. (1988) suggest that minor vortex unpinnings might play a role in such "restless" behavior, although that would in a naive picture suggest that the microjumps are simply scaled-down macrojumps. Hamilton et al. (1989) have finally caught a Vela glitch during an observing session. Greenstein (1979) suggests that thermal perturbations might give glitch-like behavior. Rankin et al. (1988) show that large changes in intensity and dispersion measure take place in the Crab pulsar, possibly related to the glitches, but only circumstantially. We will discuss this issue in a bit more detail in sec. 1.7.3.

b. *Braking Index*

In the any long-term slowing-down model one expects (sec. 1.5)

$$\dot{\Omega} = K\Omega^n, \tag{3}$$

where n is the deceleration or braking index and can be determined from observation since

$$n = \frac{\Omega \ddot{\Omega}}{\dot{\Omega}^2} \tag{4}$$

provided that K is indeed a constant. From equation 1.3.5, magnetic dipole radiation carries energy away at a rate Ω^4 and the slowing-down goes as $\Omega\dot{\Omega}$, so one would expect to find $n = 3$.

The glitches and other irregularities raise serious questions on how pulsars actually slow down. It has proven very difficult to determine the rate of change of slowing-down. Both P and $\dot{P}$ can be determined to high precision, but $\ddot{P}$ is frequently masked by apparently erratic components that are comparable to or larger than any secular term. The discovery of PSR 1509–58 with its large $\dot{P}$ but no evidence of glitches (yet) promises to change this situation. Manchester et al. (1985) have determined $n = 2.83 \pm 0.03$ for this pulsar, intriguingly close to the value of 3.0 expected by theory, but also significantly discrepant. Middleditch et al. (1987) have determined that PSR 0540–69 has $n = 3.6 \pm 0.8$ on the basis of three period determinations over a 5 year span. However, Manchester and Peterson (1989) find a braking index of 2.01 ± 0.02 in a series of repeated measurements over two years which were close enough together to determine phases without ambiguities of possibly missing pulses. The difficulty with such sparse data is that glitches could easily be missed. These authors conclude that no glitch larger than 1 part in 10^7 fell in the preceding interval, on the assumption that the true value of n would otherwise have been 3. One puzzle is that extrapolating the periods back does not match the Middleditch et al. observations, requiring the postulation of an intervening glitch. An early determination for the Crab pulsar, which is a challenge considering that this object exhibits large and small glitches along with "restless" timing noise on almost all levels, gives $n = 2.515$ (Groth 1975a,b), while for the Vela pulsar the erratic components dominate and no sensible value for n has been inferred. Statistical analysis of many pulsars gives $n = 4 \pm 1$ (Ellison 1975), but such analysis necessarily assumes that all pulsars evolve over a large range of periods; this assumption is not established and has been questioned for other reasons. Demiański and Prószyński (1983) have studied a huge glitch that was observed

TABLE 1.11—Measured Braking Indices

Pulsar	n	Comment
0531 + 21	2.515	Serious contamination from glitches
0540 − 69	3.6 ± 0.8	Unknown if glitches within observing span
	2.01 ± 0.02	No significant glitch within observing span
1509 − 58	2.83	Glitches not yet observed

in the Crab pulsar in February of 1975. They found a change in braking index associated with the glitch of $n = 2.53 \rightarrow 2.49$, with $\dddot{\Omega}$ changing by $-0.7 \rightarrow +3.7 \times 10^{-30}$ s^{-4}. With a longer data run, Lyne et al. (1988b) fitted across the glitch to determine $n = 2.509 \pm 0.001$ and $\dddot{\Omega} = -0.615 \times 10^{-30}$, which puts a self-consistency check on the deceleration parameter. Blandford and Romani (1988) accordingly point out that such measurements may be able to detect magnetic field variations. Table 1.11 summarizes recent determinations of n.

It was suggested that planets might be a source of timing irregularities (Hills 1970; Michel 1970c; Treves 1971a). But long-term observation has not revealed the implied underlying periodicities, and such a model would give comparable time scales to changes in both P and $\dot{P}$, which is not what is observed. Precession has been suggested (Vila 1969) but discounted (Axford et al. 1970) and would have the same difficulty. Note that the determination of the braking index is predicated on the phenomenological slowing-down equation. Consequently the defining equation (eq. 4) is meaningless if other terms are present (glitches, restless behavior, etc.) unless the effects of these short-term components can be isolated in a separate term and removed. Unfortunately n is sometimes taken to be *defined* by equation 4, which is sensitive to short-term behavior, in which case n no longer has any meaning insofar as equation 3 is concerned, which is intended to represent long-term trends.

If we set $n = 3$ but allow K to vary with time, we find

$$n = 3 + \frac{\dot{K}}{K} \frac{\Omega}{\dot{\Omega}}, \tag{5}$$

so these values of n less than 3 can be attributed to an increase in K with time ($\dot{\Omega}$ being negative). In the simplest models, K is proportional to $B^2 \sin^2 \alpha / I$. Although I could decrease naturally as a neutron star contracted, the rate would have to be alarmingly large, comparable to the spin-down time itself. However, α would decline with time if alignment took place as expected from vacuum torques, which has the wrong sign to account for the observed discrepancy. Formally, the difference in n is due to an increasing magnetic field, but not because the magnetic field is known to increase with time. Rather, the magnetic field is *defined* to be proportional to $(\dot{\Omega}/\Omega^3)^{1/2} = \Omega^{(n-3)/2}$, and this field, *as formally defined*, must increase in time. For the Crab pulsar, observation would require $B \approx \Omega^{-1/4}$, which is actually a rather small effect: the field would only triple by the time the Crab became as slow as its neighbor PSR 0525 + 21, which is in fact (but perhaps coincidentally) the inferred magnetic field of this pulsar. Clearly n is very sensitive to any secular changes in magnetic field, regardless of whether that is why the deceleration parameter differs from the expected value of 3 (or larger). If pulsars indeed "turn on," as suggested by Blandford et al. (1983), one could be seeing the last stages of that turn-on in the observed fast pulsars. One cannot determine meaningful values of n for slow (1 s) pulsars because for any secular decay the time constants $\Omega/\dot{\Omega}$ and $\dot{\Omega}/\ddot{\Omega}$ are comparable (i.e, have a ratio 3). Because $\Omega/\dot{\Omega}$ is typically on the order of 10^{15} s, one needs to measure the period to 1 part in 10^7 over 3 years to determine the period derivative, which is quite feasible given that one obtains 10^8 pulses during that period. But to obtain information about the second derivative, one would need to determine the period derivative to 1 part in 10^7, which would require determining the period to 1 part in 10^{14}. Even if one can determine the phase to 1 part in 10^3, giving a pulse period to 10^{11} places, one is still three orders of magnitude short. Thus a minimum of 30 years' observation would be required to determine n to one significant place. Extremely large values of n (both positive and negative) have been reported (Gullahorn and Rankin 1982) but are probably meaningless in the above context of *long-term* evolution and simply represent short-term timing noise observed over a short time (Ghosh 1984).

3. Radio-Frequency Luminosity

Although the Crab and Vela pulsars are pulsing in optical light (Vela is barely detectable), the vast majority of pulsars have only been observed at radio frequencies. The known radio luminosities range from around 10^{29} ergs/s for the Crab pulsar to about 10^{25} ergs/s for the weakest known (Dewey et al. 1985). Accurate determination of the luminosity of a pulsar is complicated by interstellar scintillation, wherein inhomogeneities in the interstellar distribution of electrons permit radio waves to reach the observer along many paths of slightly differing length and thereby to interfere with one another. An example of intensity as a function of frequency and time for the millisecond pulsar is given in figure 1.12. Indeed, the long-term variations in pulsar intensities observed by Huguenin et al. (1973) have been similarly interpreted as being due to a longer time-constant variation in the interstellar medium (Sieber 1982; Shapirovskaya and Sieber 1984) rather than to intrinsic variations in the radio emission. These patterns thus move over the antenna like clouds and must be averaged out to produce accurate estimates of the intensity (hence spectrum). Even more perplexing problems are introduced in the case of the eclipsing binary pulsar, where one would like to identify any frequency dependence on the eclipse size, and one must correct for scintillation at the same time.

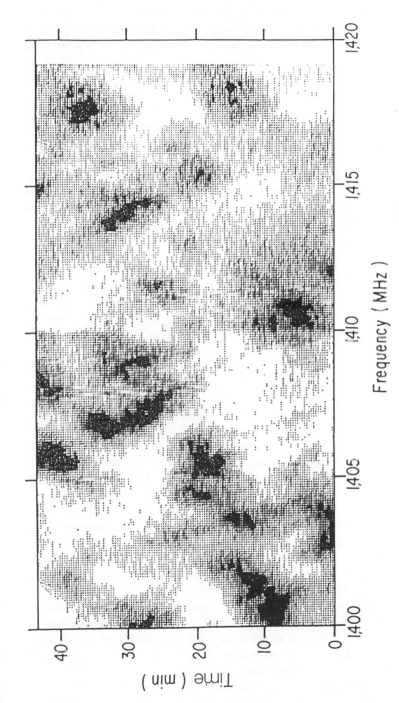

FIG. 1.12. Pulsar scintillation. From Backer et al. (1982). Intensity (gray scale) as a function of time and frequency for the millisecond pulsar. Notice the dramatic changes in intensity at any given frequency over relatively long times (half hour). Reprinted by permission from *Nature*, 300, 615 (figure 1). Copyright © 1982 Macmillan Magazines Ltd.

4. Pulse Width

The pulse width tends to be a fairly constant fraction of the total period; in other words the "duty cycle" tends to be fixed. A duty cycle of 1 in 30 would be a serviceable generalization, corresponding to radiation through 12° of "longitude." It would be less model dependent to use the word "phase," but the practice is to use "longitude." The exceptions (longer pulse widths, up to about 90°) are not yet well correlated with other properties, although there seems to be a movement toward associating complex—hence wide—pulse profiles with aging pulsars. An extreme case is PSR 0826–34, which has a double-humped pulse about 145° wide (Durdin et al. 1979). Actually, the pulse seems to be "on" at almost all phases, although the pulsar itself nulls and is only active about 20% of the time. Manchester and Taylor (1977) plot period versus pulse width (their figure 2-3) and obtain essentially a linear correlation (i.e., constant duty cycle) with the usual scatter about that correlation line.

a. *Micropulses*

Micropulses are ultrashort intensity variations within individual pulses with time scales ranging from about 1 to 1000 ms (Cordes 1979b). In general, autocorrelation of pulses shows a sharp spike at short times (micropulses), followed by a shoulder at longer times (subpulses), followed by a second shoulder out to the pulse period (more complicated behavior, of course, is sometimes seen). These micropulses are evidently fundamental features of pulsar pulses, and presumably a sound theory of pulsar radiation would account for their existence. Models for micropulses basically appeal to spatial or temporal variations. In the former, an intrinsically nonuniform beam is swept past the observer. Benford (1977) appeals to a filamentary beam aligned with polar cap field lines. Ferguson (1981b) proposes that bunches of particles are located at periodically spaced emission regions, possibly concentrated and excited by long-wavelength standing waves above the polar caps. Modulation of the output by changes in the surface temperature (Cheng 1981) is an example of the temporal models. Harding and Tademaru (1981) argue that the pulses are modulated by propagation through sheared regions of the magnetosphere. Michel (1978a) argued that radiation reaction from one departing bunch could briefly inhibit the emission of particles, leading to a subsequent burst of particle emission. Chian and Kennel (1983) propose that nonlinearities arising from relativistic particle mass variation excite a self-modulational instability of strong electromagnetic waves in a pair plasma.

Again, the distinction between temporal and spatial may be somewhat artificial considering the high degree of coherency required, which may be difficult both to create over a large region and to sustain locally for a long period. Boriakoff (1983) finds that micropulses are broad band (i.e., arrive simultaneously independent of frequency), which suggests that one may be seeing fluctuations in the basic emission mechanism rather than interference patterns from local propagation effects (which is somewhat surprising given that the latter effect is commonly found with coherent sources such as lasers). Stinebring and Cordes (1981) find that the micropulses are on the average time-symmetric.

b. *Interpulses*

The typical view is that interpulses result from favorable viewing angles that allow one to see radiation from both magnetic poles of a pulsar. This would happen if the magnetic dipole moment were nearly oblique and the observer were close to the equatorial plane. Manchester and Taylor (1977, p. 201) argue that the interpulses may in fact be the two edges of a very wide beam, and in their compilation of data on 330 pulsars (1981) they quote pulse widths on this assumption (their widths W_{50} and W_{10} for 0531 + 21 are consequently half the pulsar period: W_{50} is the time the pulse takes to climb through 50% of maximum intensity and fall through 50%, etc.). If they are right, the rapid swing in polarization (as seen in the optical: Smith et al. 1988) cannot be due to rotation of the magnetic pole close to the line of sight, as in the hollow cone model (sec. 2.4). If one is seeing both poles (or, less restrictively, *activity* associated with the two poles), the angular constraints are interesting in the case of the Crab pulsar and the millisecond pulsar because in both cases the swing in polarization is in the *same* sense for the main pulse and the interpulse. This would require both poles to pass to the same side of the line of sight, as would happen if the dipole were perfectly orthogonal to the spin axis and the observer were somewhat above (or below) the spin plane. As the condition for orthogonality is relaxed, it becomes increasingly difficult for the observer to see both poles. A list of pulsars with interpulses, from Gil (1986) and Taylor and Stinebring (1986), is given in table 1.12. The R values are the rotation rates of polarization position angle with longitude (degrees/degree) from Narayan and Vivekanand (1982). In general, the location of the interpulse in pulse phase is insensitive to observing frequency (e.g., PSR 1937+214: Cordes and Stinebring 1984), but there too one finds exceptions. Hankins and Fowler (1986) give a shorter list but include frequency dependence. Weisberg et al. (1981) have studied the unusual interpulses of PSR 0940 + 16 and PSR 1530 + 27, which are not at all symmetrically spaced between successive main pulses.

c. *Giant Pulses*

The Crab pulsar exhibits giant pulses, as already discussed, with an exponential probability of occurrence distribution extending up to pulses at least 1000 times the normal size. It is interesting that the fraction of radio power in these largest pulses approaches the *average* fraction exhibited by typical pulsars ($\approx 10^{-5}$). The peculiar "wide" pulsar PSR 0826-34 also exhibits extremely intense pulses up to two orders of magnitude, which possibly puts it in the same class. The origin of these giant pulses is obscure.

d. *Frequency Dependence*

The separation of components has weak frequency dependence, and sometimes this weak dependence appears to have "kinks" in it (Manchester and Taylor

TABLE 1.12—Some Pulsars with Interpulses

Pulsar	Period	$\delta\phi^a$	R_L/a	R_{main}	R_{Ip}
0531 + 21	.033	215	157	6.0	3.1
0823 + 26	.531	180	2500	14.3	− 4.6
0826 − 34	1.84	150 (?)	9000		
0940 + 16	1.087	90	5200		
0950 + 08	.253	155	1200	2.0	0.67
1055 − 52	.192	205	900	1.6	2.1/0.66
1530 + 27	1.125	52	5400		
1822 − 09	.77	167	3600		
1848 + 04	.285	150	1360		
1855 + 09	.0053	143	25		
1929 + 10	.226	176	1000	1.48	− 0.58
1937 + 214	.0015	174	8		
1944 + 17	.440	$177 - 193^b$	2100		

a) $\delta\phi$ is the longitude angle from main pulse to interpulse.
b) Frequency dependent, 318 MHz to 2380 MHz.

1977, figure 2-4; Rankin 1983b). But the changes in position rarely amount to more than a few degrees. Possibly one is seeing to different depths in the emission region and therefore to slightly different emission directions. Cordes (1979b) has proposed that one can identify a radius-to-frequency dependence (which would immediately suggest emission at significantly different heights above the surface; see also Rickett and Cordes 1981).

e. *Multiple Components*

Pulsars display a variety of pulse structure, up to about five components (Rankin et al. 1988b). As discussed in more detail later (sec. 3.2), the popular explanation for multiplicity of components is that emission is in the form of a hollow cone (or an auroral zone). Then a one-component pulsar would correspond to just the edge of the cone swept past the observer. Two components would be seen when the entire cone swept by, with a distribution of pulsars observed ranging from poorly resolved close pairs to clearly separated components. Given that electrical currents must flow both into the neutron star and out if there is to be radio-frequency emission, it is perfectly plausible that there is also current flow of the opposite sense down the center of this hypothetical cone. Emission could be obtained from both, giving three components, as are observed in some pulsars. To explain five, it was originally supposed that a second annulus existed. From a theoretical standpoint, it has been difficult to think of any reasonably unforced reasons for this multiplicity of cones, and even the underlying cone idea has problems (ch. 3). A natural explanation for some degree of multiplicity is simply substructure in a current arc, owing to the tendency of currents to collapse into localized filaments. Some pulsars with a large number

TABLE 1.13—**Pulsars with 5 Components (plus some likely candidates)**

Pulsar	Period
1237 + 25	1.382
1737 + 13	0.803
1831 − 04	0.290
1857 − 26	0.612
1905 + 39	1.256
0621 − 04	1.039
1738 − 08	2.043
1910 + 20	2.233
2210 + 27	1.203

of possible components are listed in table 1.13. Note that the number of components is somewhat poorly defined, particularly when they appear to overlap and merge.

It is clear from table 1.13 that the pulsars with complex pulse shapes are noticeably slow, given that the peak of the period distribution is found at about 0.5 s (the median is at about 0.7 s). This result is somewhat counterintuitive; ohmic decay of currents in the crust should act to simplify any magnetic field inhomogeneities, for example, and the standard estimates of the entire size of the auroral zone indicate that it should be smaller and therefore have less room for structure. Something is causing the pulses to become more complex.

5. Pulse Variations

Except for the pulse-to-pulse variations common among pulsars, which could be likened to "flickering," the integrated pulse profiles are relatively stable. A few pulsars display exceptional behavior in the form of mode changes, nulling, and so forth.

a. *Orthogonal Mode Changes*

Some pulsars are known to change linear polarization abruptly by close to 90°. Careful analysis of some cases shows that the pulsar is actually emitting pulses that are sometimes of one polarization and sometimes of the other, with the relative fraction changing with time (Stinebring et al. 1984a; Gil 1987). As a result, the percentage linear polarization drops to zero at the jump(s) in position angle.

b. *Nulling*

Nulling is an extraordinary property whereby the successive pulses do not simply vary widely in amplitude but rather disappear entirely, as first reported by Backer (1970). Often such nulls consist of only one or two pulses, but the "train" of unseen pulses can be very long, 10 to 100 pulses in length.

From a theoretical point of view, nulling seems significant. Because only about 10^{-5} of the spin-down energy appears in the radio, a simple explanation would be that whatever generates this small output simply falters for a while (with the spin-down presumably continuing). However, from the point of view of coherent emission in some sort of discharge, it is hard to understand the long time constants. The highly fluctuating and unsteady nature of the emission can be attributed to short time constants in whatever mechanism produces the coherence (e.g., bunching). For most theories, any such time constant must be extremely short ($\ll$ the pulsar period) if bunching is to occur before the particles are ejected from the neighborhood of the star and their concentration is reduced by expansion. For the emission to be "off" for many periods therefore seems statistically implausible. Alternatively, one might suppose that the discharge itself turns off. But given the presumed short time constant, why would the discharge turn back on? Such a short time constant is exemplified in gamma-ray/pair production discharge theories (e.g., Ruderman and Sutherland 1975). Rather than the microphysics faltering (e.g., bunching), the global physics might falter (e.g., current flow interruption), and the latter could have a longer time constant. Early models essentially had no global physics. Rather, the theory tended to assume that the mere fact of neutron star rotation and magnetization automatically led to pulsar action locally near the star, with the global physics being one of accommodation of the surroundings to this localized action, a "tail-wagging-the-dog" view still in evidence. Nulling seems inconsistent with that view. Ritchings (1976) suggested that nulling is the process by which pulsars "turn off." Namely they simply null for progressively longer times until no longer seen. It is not clear that a mechanism to turn off pulsars is actually required by the data (sec. 2.4), but it could nevertheless be that something happens to pulsars as they age. Table 1.14 lists some pulsars known to null.

c. *Drifting*

The drifting subpulse phenomenon was originally noticed by Drake and Craft (1968), and the present-day nomenclature describing the drifting in terms of subsidiary periods P_2 and P_3 is due to Sutton et al. (1970). In this terminology, the ordinary period becomes P_1. Some drifters are shown in figure 1.13. Here P_3 is the time (or number of pulses) that the values would have to be shifted so that the bands would again lie one upon the other, about seven periods for PSR 0031–07 as shown and a somewhat ambiguous number for PSR 2016+28. The other period corresponds to the shift in longitude needed to make successive bands line up and is essentially a measure of the slope of the bands (equivalent to the drift rate).

The usual view of the drifting mechanism is that subpulses drift across a "window" (modulated by the viewing angle of the observer relative to the magnetic field structure). Unfortunately this view is not innocuous, because *it suggests that the subpulse continues to exist outside of the viewing window!* Thus in Ruderman and Sutherland (1975) we have the concept of a full

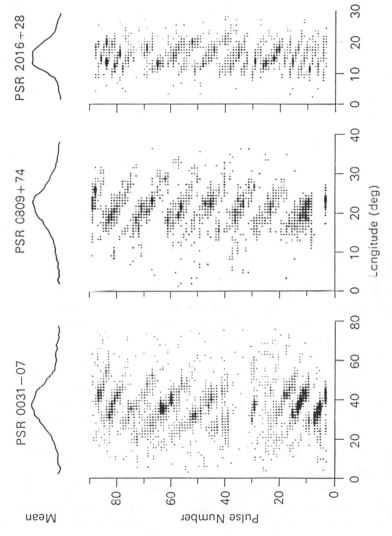

FIG. 1.13. Drifting subpulses. Successive pulse tracings give bands of intensity, showing that some activity appears to move through the pulse "window." From *Pulsars*, by R. N. Manchester and J. H. Taylor. Copyright © 1977 by W. H. Freeman and Company. Reprinted by permission.

TABLE 1.14—Some Pulsars That Null

Pulsar	P	$\dot{P}$
0031 − 07	0.9430	0.4083
0525 + 21	3.745	40.06
0628 − 28	1.244	0.107
0809 + 74	1.292	0.1676
0818 − 13	1.238	2.106
0834 + 06	1.274	6.799
1112 + 50	1.656	2.493
1133 + 16	1.188	3.733
1237 + 25	1.382	0.9595
1857 − 26	0.6122	0.16
1944 + 17	0.4406	0.0240
2045 − 16	1.916	10.96
2111 + 46	1.015	0.7195
2154 + 40	1.525	3.417
2319 + 60	2.256	7.037

series of unseen subpulses marching around the polar caps. It would be interesting if this view could be tested somehow. It is far from obvious that one even sees the "same" subpulse during successive pulses, any more than one is seeing the "same" spokes on the stagecoach wheels, the ones in a western film that seem to revolve in the wrong direction! The "window" idea is contrary to the possibility that the subpulse is born at one edge and drifts across to be extinguished at the other, without that distinction having yet been observationally tested. Indeed, we will see below that the pulsars that both drift *and* null show a dependence on the "window" location.

The drifting subpulse phenomenon strongly suggests that the subpulse represents a localized discharge (as opposed to numerous discrete discharges going on and off, only one of which happens to be beamed at us at any given instant).

Ruderman and Sutherland attribute drifting to sparks circling the polar caps with period P_3, which they estimate at $5.6B_{12}/P$(s) owing to $\mathbf{E} \times \mathbf{B}$ drift. The actual P_3 observed would then be P_3/n, where n is the number of discharges surrounding the polar caps (imagined to be equally spaced, in some sort of standing wave pattern).

d. *Nulling and Drifting*

Nulling was noted in PSR 0809+74 by Cole (1970). An early analysis by Unwin et al. (1978) implied that the drifting *ceased* during the nulls. A later analysis by Lyne and Ashworth (1983) with higher signal-to-noise ratio revealed a more complex behavior, namely that the rate of drifting discontinuously decreased at the time of the null, but then recovered exponentially as the length of the null grew. (This behavior is vaguely reminiscent of glitching, where a discontinuous

TABLE 1.15—Some Pulsars That Both Null and Drift

Pulsar	P (s)	$\dot{P}$ ($\times 10^{15}$)	L_{radio}/L_{total} ($\times 10^5$)
0031 − 07	0.943	0.41	1.0
0628 − 28	1.244	7.11	6.6
0809 + 74	1.292	0.17	2.0
0818 − 13	1.238	2.11	10.2
0834 + 06	1.274	6.80	0.14
1944 + 17	0.441	0.024	5.6

All of these pulsars are at high galactic latitudes except PSR 1944 + 17!

change in period recovers exponentially, but with a much longer time constant.) Such a coherent action suggests that some physical process is involved in the nulling rather than simply some stochastic interruption of the emission mechanism. The question of whether the nulls themselves actually occur randomly is difficult to address; in the case of another pulsar, PSR 0818–13, the nulls appear in quasi-oscillatory groups of two or three (ibid.), so the nulls are not stochastic on short time scales. PSR 0818–13 has a behavior "similar" to that of PSR 0809 + 74, except that there is a change in drift rate before the null and at the null the drift rate *increases* if anything (the statistics are also much poorer, owing to a lack of significantly long nulls in this pulsar). Deich et al. (1986) find *no* memory displayed in the case of PSR 1944 + 17. To complicate matters, PSR 0809 + 74 exhibits no change in pulse amplitude before the null but a significant increase just after the null ($\approx$25%). PSR 1944 + 17 shows a decrease before the null ($\approx$50%) but no change after the null. Apparently, there is a nulling behavior for just about every possibility.

Considerable attention has been paid to the nulling plus drifting phenomenon. As noted above, Ruderman and Sutherland (1975) attribute the drifting to sparks circling the polar caps, while Cheng and Ruderman (1977a,b, 1980) attribute the memory after a null to hot spots left behind on the stellar surface, a model criticized by Filippenko and Radhakrishnan (1982) (spots cool too fast) and Filippenko et al. (1983) (spots give perfect memory in contrast to data). Arons (1983c) suggests nutation about some mean beam direction. Filippenko and Radhakrishnan suggest a transition from rapidly intermittent discharges to a steady discharge. Jones (1983) suggests a competition between thermionic discharges and pair production (see also Jones 1981, 1982).

Note that the nulling pulsars listed in table 1.15 (from Deich et al. 1986; Lyne and Ashworth 1983) show no particular tendency to be less efficient at converting spin-down energy into radio emission than the average pulsar, which is not particularly supportive of the view that the radio emission is at the edge of faltering.

TABLE 1.16—Pulsars Known to Exibit Profile "Mode Changing"

Pulsar	P	$\dot{P}$ (10^{-15})
Observed as drift mode changes:		
0031 − 07	0.943	0.41
0943 + 10	1.098	3.53
Observed as profile changes:		
1612 + 07	1.207	2.36
0329 + 54	0.715	2.05
0355 + 54	0.156	4.39
1822 − 09	0.769	52.3
1917 + 00	1.272	7.68
1926 + 18	1.221	. . .
2319 + 60	2.256	7.04
1237 + 25	1.382	0.96
1737 + 13	0.803	1.45

e. *Mode Changing*

In addition to the orthogonal mode changes *within a given pulse*, there is, in some pulsars, a tendency to make transitions back and forth between rather distinct looking *integrated* pulse profiles (see table 1.16). An example is the five-component PSR 1237 + 25, which alternates between having the central component the brightest and having the two outlying components the bright ones, a distinctive change. Such a phenomenon is reminiscent of laboratory discharges, which often seem to hop between quasi-stable forms, but, of course, an anecdotal comparison is hardly sufficient to establish such a connection. Fowler and Wright (1982) report a complex behavior in pulsars with an interpulse that undergoes mode changing. PSR 2319 + 60 is a pulsar with three distinct modes, drifting subpulses, and also nulls (Wright and Fowler 1981). Morris et al. (1981) observed the first component of PSR 1822–09 drop out, and argue that the interpulse is from the opposite pole (see also Fowler et al. 1981).

6. Pulse Location

The exact location of where the pulses originate is arguably not observable in such a tiny astrophysical system, but a number of attempts have been made to infer where the emission is coming from. The hollow cone model (ch. 3) together with the narrowness of the pulses suggests emission near the surface. Cordes (1979b) has proposed, as noted above, that one can identify a radius-to-frequency dependence, which would immediately suggest emission at significantly different heights above the surface (see also Rickett and Cordes 1981). Krishnamohan and Downs (1983) argue that overlapping subpulse structure indicates that the components originate at different heights (hundreds of kilometers) in the case of the Vela pulsar. Wolszczan and Cordes (1987) using interstellar scintillation to do interferometry on PSR 1237 + 25 deduce that the emission originates within the light-cylinder but well above the surface. Previously, Cordes et al. (1983) concluded that emission must be near (≤ 100 km) the surface. Most theoretical analysis places the emission either very close to the surface (with a neutron star radius or so) or out near the light-cylinder (e.g., Ferguson 1981a). Except for perhaps the disk model (ch. 6), no theoretical models have regions intermediate between the surface and the light-cylinder that might be natural places from which to expect emission.

1.7 Neutron Star Properties

With all due reserve we advance the view that super-nova represents the transition of an ordinary star into a *neutron star* consisting mainly of neutrons (Baade and Zwicky 1934b).

1. Masses

The general expectation is that neutrons stars should have masses close to the Chandrasekhar limit of 1.4 $M_\odot$, because this is the mass at which the core will collapse. Joss and Rappaport (1984a, figure 11) report a tendency for pulsating X-ray binary sources to have a mass near 1.4 $M_\odot$. Moreover, Weisberg and Taylor (1984) find that the binary pulsar PSR 1913 + 16 consists of two objects both with mass near 1.4 $M_\odot$. The latter is possible because relativistic effects permit a full solution to the binary system based on Doppler data from a single member (not possible in the Newtonian limit, where only the projected semimajor axis and mass functions can be deduced).

a. *Maximum Masses*

The major observable properties of nuclear matter are its mean density and compressibility. Physicists report the former in understandable terms ($\rho = 2.67 \times 10^{14}$ g/cm^3) but are fond of giving the latter as the "compressibility modulus," $K_0 \approx 200$ MeV, which is less than transparent. Given that the speed of sound for a bulk medium can be obtained from

$$V_s^2 = dP/d\rho, \tag{1}$$

it follows that multiplying by the nucleon mass gives an energy which is related to the change in density with change in pressure (compressibility). Thus we can guess that $K_0 \approx m dP/d\rho$ but cannot divine the constant of proportionality. The latter turns out to be, for historical reasons, 9 (see Blaizot et al. 1976; Baron et al. 1985). As given in Weinberg (1972, p. 301), the general relativistic hydrodynamics is just given from the usual hydrodynamic equation

$$\frac{dP}{dr} = -g\rho f_1 f_2 f_3 \tag{2}$$

with three correction factors f_i of order unity, which are given (the ordering here is arbitrary) by

$$f_1 = \left(1 - \frac{2GM(r)}{rc^2}\right)^{-1}, \tag{3}$$

$$f_2 = 1 + \frac{4\pi r^3 P}{M(r)c^2}, \tag{4}$$

and

$$f_3 = 1 + \frac{P}{\rho c^2}, \tag{5}$$

with the defining relations

$$M(r) = \int_0^r 4\pi \rho r^2 \, dr \tag{6}$$

and

$$g = GM(r)/r^2. \tag{7}$$

The only thing lacking is the relationship between ρ and P, the equation of state. For solid nuclear matter, this must be of the form

$$\rho = \rho_0 + \frac{P}{V_s^2} + \text{order } (P^2). \tag{8}$$

These equations can then be solved numerically as a simple integration outward from the center, given some assumed central pressure and neglecting the second-order terms. The central pressure will be some incomprehensibly large number in cgs units, so it is convenient to write the pressure instead as a fraction of the local rest-mass energy density, $P = \epsilon \rho_0 c^2$.

For incompressible matter ($V_s \to \infty$), these equations can be solved analytically and one finds that as $P \to \infty$, the radius approaches 8/9 of the Schwarzschild radius ($\equiv GM/c^2$), from which one can deduce that the radius is given from $r_{max}^2 = c^2/3\pi G\rho_0 \to 23.1$ km and $M_{max} = 4\pi r_{max}^3 \rho_0/3 = 6.97$ $M_\odot$. If a finite value for $\beta \equiv V_s/c$ is assumed, then one finds that as ϵ is increased, the mass and radius increase to a maximum radius and then to a maximum mass at a slightly smaller radius; then both decline monotonically, as shown in figure 1.14. The maxima are typically found in the neighborhood of $\epsilon = 1$ (the circled points in figure 1.14), which is not too surprising in retrospect. Thus the mass is a double-valued function of radius, as shown for the $\beta = 0.5$ case. One can see that the maximum radius is found roughly for $\epsilon \approx \beta$, and therefore the compression of the central regions over the uncompressed regions is roughly a factor $1 + 1/\beta$, corresponding to roughly factors

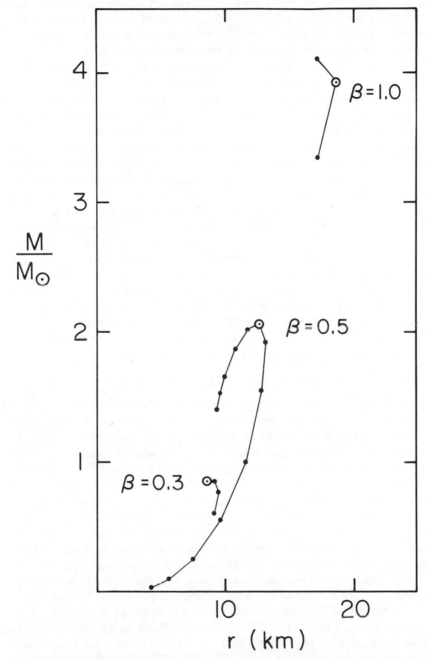

FIG. 1.14. Mass versus radius for neutron star models. Integration of the simple equation of state for central pressure steps of 2, with the central pressure (circled point) equal to $\rho_0 c^2$. Three values of sound velocity in units of c are shown, but only plotted extensively for $\beta = 0.5$.

of 2, 3, and 4 for the values illustrated. One can readily see that the maximum mass is *extremely* sensitive to the velocity of sound in nuclear matter. The more realistic calculations assume an equation of state with an increasing velocity of sound (the neglected higher-order terms in equation 8), and are therefore not directly comparable to the above simple illustrative ones. Baron et al. (1985), for example, use $P = (K_0\rho_0/9\gamma)(\mu^\gamma - 1)$, where $\mu = \rho/\rho_0$, and γ is assigned values between 2 and 3. As a result, the velocity of sound is rather small initially ($\beta \approx 0.15\mu^{(\gamma-1)/2}$) but increases rapidly. For this reason the central densities calculated are significantly larger than above. Baron et al. found that supernova explosions were favored by the *softer* equations of state (smaller K_0), which creates something of a bind because the maximum neutron star masses are correspondingly reduced. Their maximum masses are approximated by $M_{max} = 1.07 \, (\gamma - 1)(K_0/230\text{MeV})^{1/(2\gamma-2)}$; thus for $\gamma = 2.5$ one requires $K_0 \geq 150$ MeV to get a 1.4 $M_\odot$ neutron star, and for $\gamma = 3$ one requires $K_0 \geq 42$ MeV. Brown (1988) discusses the present experimental situation concerning the compressibility modulus, and a good review of neutron star masses can be found in Burrows (1987). A number of corrections are required in such work, such as correcting for the neutron richness of the matter, which is believed to reduce the density and K_0 below the values for ordinary nuclear matter. See also Prakash et al. (1988), who find typically radii of 10 km, maximum masses of about 1.7 $M_\odot$, moments of inertia of about 1.3×10^{45} g cm^2, and again large central compression factors (μ) of order 10 (i.e., the equation of state must be fairly significantly extrapolated).

b. *Rotation and Other Corrections*

Friedman and Ipser (1987) and Shapiro and Teukolsky (1983) have examined the degree to which the maximum neutron star masses are affected by rotation, and conclude that the effect amounts to perhaps a 25% increase over the nonrotating case. The normal modes and stability of rapidly rotating neutron stars have been examined in more detail by Ipser and Lindblom (1989). A number of workers (Shapiro and Lightman 1976; Glatzel et al. 1981; Tohline 1984) have suggested, however, that rotationally sustained neutron stars of large radius ("fizzlers") could be sustained. If they could, the slowdown of a rotating neutron star could conceptually trigger collapse to a black hole. Pulsars, however, seem comfortably less massive than the nonrotating mass limits (2 to 3 $M_\odot$, depending on the model for nuclear matter at very high densities). The maximum mass of neutron stars is argued to be around 2.0 $M_\odot$ for what are regarded as realistic equations of state (Arnett and Bowers 1977) and about 2.7 $M_\odot$ for the stiffest. An upper limit of 3.0 $M_\odot$ is derived from relativity and causality (Rhoades and Ruffini 1974; Chitre and Hartle 1976). Observed values of the compressibility of nuclear matter are consistent with neutron star masses of at least 1.4 $M_\odot$ (Glendenning 1986). Shang-Hui et al. (1981) summarize previous estimates and give 1.7 $M_\odot$ as the maximum neutron star mass.

Fan and Jiping (1982) suggest, in complete contradiction to the above considerations, that pulsars have a steep power-law spectrum of masses extending up the Chandrasekhar limit in order to explain the scattering in the $P/\dot{P}$ distribution. But it seems likely that such scattering results naturally from the distribution of pulsar magnetic fields (sec. 2.4).

2. Surface Properties

The surface characteristics of a neutron star are challenging to predict, although there seems to be a general uncritical assumption circulating that the surface is solid iron. However, the high temperatures at formation, photodisintegration of the nuclei, incomplete reactions, etc., open the door for "contamination" by other entities. Helium, for example, is highly mobile and rather stable, and quite possibly a significant amount of it will evolve out of the crust to form a surface layer (Rosen and Cameron 1972). Similarly, hydrogen, albeit more reactive, could be present. Thus the surface could be a hot, dense liquid metal (Epstein 1988) with a film of helium floating on top. Even whether the crust is solid or a fluid is difficult to predict. Recent calculations (Jones 1985; Neuhauser et al. 1987) suggest that the iron atoms (or whatever) do not bind together and consequently the surface may be a fluid.

What role these trace constituents (insofar as the bulk star goes) might play in the pulsar phenomenon depends on whether positive current carriers are required for the pulsar action; they could be depleted if the charge carriers needed for pulsar action all originated at the surface and were ejected from the system (Michel 1975d). The cooling of neutron stars is rather slow, and even old neutron stars should be about as hot as the Sun. Neutron stars should probably be imagined as tiny hot stars rather than as dark shiny steel balls.

a. Atoms in Strong Magnetic Fields

It was once thought that a strong surface magnetic field caused ions to be so strongly bound that they could not be removed by field emission, but this seems no longer to be supported by calculation.

The entire issue of what structure atoms would have in strong magnetic fields is an interesting pure theoretical question in its own right. And a considerable literature has already been devoted to the states of the atoms in such fields, either from a variational approach (Cohen et al. 1970; Newton 1971; Barbieri 1971; Kadomtsev and Kudryavtsev 1971; Mueller et al. 1971; Smith et al. 1972; Callaway 1972; Wilson 1974; Banerjee at al. 1974; Mueller et al. 1975; Rau et al. 1975; Brandi 1975; Rau and Spruch 1976; Angelie et al. 1980; Hylton and Rau 1980; Wadehra 1983 [but see Herold et al. 1984]) or by numerically integrating the Schrödinger equation (Canuto and Kelley 1972; Simola and Virtamo 1978; Wunner 1980; Wunner and Ruder 1980, 1981; Wunner et al. 1981; Le Guillou and Zinn-Justin 1983). The more difficult problem of calculating the lattice binding energy has been addressed by Ruderman (1971), Witten (1974), Chen et al. (1974), Jones (1985), and Neuhauser et al. (1987).

It was thought at first (Ruderman 1969) that the ions would be so strongly

bound in linear chains that they could not be pulled from the neutron star surface by likely electric fields (whereas the electrons would presumably be freely
available). These early estimates now seem to have been too large (Hillebrandt
and Müller 1976) and have since been revised downward (Flowers et al. 1977;
see also Constantinescu and Moruzzi 1978 and Endean 1973). These later estimates give a binding energy of about 2.6 keV for iron ions in a 10^{12} gauss
magnetic field (the density at the neutron star surface is estimated to be 10^4
to 10^5 g/cc under these circumstances: Ruderman 1979). Alternatively, it has
been suggested that the helium that might be available from the pulsar crust
(Michel 1975d; Burdyuzha 1977) would significantly enhance the availability
of weakly bound ions. The uncertainty is how ideal a real neutron star might
be. See Glasser (1975) for a discussion of the effect of the field on the conduction electrons. It is interesting to note that the contemporary explanation
for X-ray bursts (Woosley and Taam 1976; Maraschi and Cavaliere 1977; Joss
1977, 1978; Lamb and Lamb 1978; Taam and Picklum 1978) postulates, to
the contrary, that sizable amounts (about 10^{21} g) of helium can accumulate
on the neutron star surface. Such quantities are just those necessary to supply
ions to a pulsar wind over the pulsar lifetime (Michel 1975d). However, if all
pulsars are born at high spin rates (where they spend a negligible fraction of
their life), these light ion sources would be depleted (Ruderman 1981). Recent
work by Jones (1985) and Neuhauser et al. (1987) may have rendered much of
the above irrelevant. They find that heavy ions are not bound at all! Only the
lightest atom (helium) is bound, and then only by about 25 eV, which is not
enough to condense even at the 10^6 K pulsar surface temperatures. The theoretical situation may continue to oscillate back and forth on the sign of the effect
(binding versus no binding) because the calculations involve small differences
between large numbers, but the small difference itself seems too small to be of
the significance once thought.

The general issue of quantum electrodynamics in strong electric fields is
discussed by Melrose and Parle (1983a,b,c).

b. *Temperature and Cooling*

The nucleons in a neutron star must be degenerate, and therefore they must have
rather low heat capacity and high heat conductivity. These factors, combined
with the small size of the object, point to rapid cooling at first, typically on the
order of a thousand years (see Tsuruta et al. 1972; Tsuruta 1974, 1975; Maxwell
1979; Tsuruta 1980; Richardson 1980; Glen and Sutherland 1980; Nomoto and
Tsuruta 1981; Van Riper and Lamb 1981; Tsuruta 1986). Nevertheless, the
internal temperature can be significantly higher than the surface temperature
(Gudmundsson et al. 1983). Van Riper (1988) has done an extensive analysis of
the magnetic corrections and thermal conductivities near the surface of a neutron
star. Eichler and Cheng (1989) have noted that impulsive energy injection to
the neutron star either is promptly radiated away (in less than an hour or so) or
is stored in the interior and requires thousands of years to diffuse out. Fujumoto

et al. (1987) examine whether cooling might be detected during an outburst from an accreting neutron star.

3. Internal Structure

Beyond the straightforward theoretical expectations of how the composition of a neutron star should vary with depth (see Baym and Pethick 1975, 1979), almost all the theoretical activity has centered on explaining the timing variations observed in some pulsars. Basically there are thought to be five regions: (1) the surface, (2) an outer crust (or ocean, depending on one's persuasion), (3) an inner crust where free neutrons are coexisting with the "normal" matter, (4) the liquid or superliquid neutron core (plus an equilibrium concentration of electrons and protons), and (5) finally a possibly distinct core of exotica (pion concentrates, quark or hyperon "soup," etc. [see, however, Van Riper and Lamb 1981]).

Observation of period variations, and in particular the attribution of the "glitch" in the Vela pulsar to a starquake (Baym et al. 1969b), has stimulated an enormous literature on this subject. For the most part, it is assumed that the pulsars function more or less independently of what is happening in their interiors, and observation seems to support that view since no dramatic changes in radio output or character (pulse shape, etc.) seem to be associated with period variations. It is not even certain that the period variations have anything to do with the pulsar interior, but the view that they do has been extensively promoted. Roberts and Sturrock (1973) and Pustil'nik (1977) did propose that mass loading and ejection in the magnetosphere might be involved, and Scargle (1969) and Scargle and Harlan (1970) reported activity in the *wisps* near the Crab pulsar following the first reported glitch. The physical nature of these wisps is unclear, and similar structures have been detected about the Vela pulsar (Ögelman et al. 1989). Photographic plates taken decades apart show essentially the same wisp structure whereas the light crossing times are only of the order of a month, so what keeps this supposedly relativistic plasma in place over long time periods is presently unsolved. In any event, the amount of matter associated with the space charge that should form in the pulsar magnetosphere is far too tiny to be kinematically significant. Disk models are, of course, a complete break from this view. The literature is too vast and the subject sufficiently distant from magnetospheric issues for us to treat them in detail here, so we will just sketch the theoretical situation. Oppenheimer and Volkoff (1939) first treated the structure of the neutron star (see also Landau 1932 and Baade and Zwicky 1934a,b). The problem is sensitive to detail, and as we have seen, the maximum possible mass of a neutron star is quite sensitive to the equation of state (Oppenheimer and Snyder 1939). The general relativistic corrections are important, because the central pressure of a star of incompressible matter goes to infinity when the Schwarzschild radius reaches 8/9 of the stellar radius (Møller 1952; Weinberg 1972). At much lower masses, however, the ability of the nuclear matter to hold up a star vanishes as the constituent particles become relativistic, and so the behavior is delicately sensitive to fine details of the equation of state (see Prakash et al. 1988). Observationally, the binary X-ray sources are providing neutron star masses near 1.4 $M_\odot$, so supernova remnant neutron stars may tend to be standardized objects, probably because their masses are limited by the maximum possible for the presupernova core rather than by what their own maximum mass might be. At the moment, it appears that a neutron star mass versus radius relationship would tell us more about nuclear physics than vice versa. Except possibly during the formation event, thermal pressure is not expected to play an important role in neutron star structure unless the temperature is up to around 10^{12} K, and neutrino cooling is expected to carry away such thermal energies rapidly.

a. *Vortex Unpinning*

The original starquake hypothesis has evolved considerably since its inception, with recent works (Alpar et al. 1981) downgrading starquakes and focusing more on "vortex unpinning," namely the decoupling of the core and crust of the neutron star to temporarily change its apparent moment of inertia. Such is the basic picture originally forwarded by Greenstein and Cameron (1969). Other mechanisms for glitches have been proposed, such as changes in the magnetosphere (Scargle and Pacini 1971; Ozernoi and Usov 1972; Roberts and Sturrock 1973), a change of state in the crust (Bisnovatyi-Kogan 1970), or evaporation of material from a planet (Rees et al. 1971). Cordes and Greenstein (1981) have carefully reviewed the timing noise processes; they find the starquake and vortex unpinning hypotheses less attractive than others, such as luminosity fluctuations. Alpar and Ho (1983) argue that the statistics are consistent with either core-quakes or vortex unpinning models.

The overall picture is not too satisfying because the observed glitches vary greatly in magnitude and in recovery properties, and one must argue that different neutron stars have quite different internal properties. Also, one must ask about the role of a strong magnetic field, which would be expected to strongly couple a crust and core. For example, Lodenquai (1984) finds that both short-term and long-term decay components are inconsistent with two-component (crust/core) models of glitches. The strongest support for models with a differentially rotating core and crust is simply a lack of alternatives, which has not historically proven to be an entirely reliable basis for acceptance.

4. Magnetic Field Origin

For some time there was no "direct" evidence that pulsars were either highly magnetized or neutron stars, or that neutron stars even existed as tiny objects with radii of about 10 km. The power balance in the Crab Nebula (Finzi and Wolf 1969; Gold 1969a), however, points to a collapsed object with the correct moment of inertia, which therefore requires Ma^2 to be of the correct magnitude. Originally it was thought that thermal neutron stars were the most plausible candidates for astrophysical X-ray sources. However, early rocket-borne X-ray detectors using a lunar occultation to provide sufficient spatial resolution failed to observe a point source in the Crab Nebula. Actually the point source was there, but just below the sensitivity of the experiments. Subsequent confirmed X-ray objects proved to be just about anything but thermal radiation from a neutron star. Now, however, some of the X-ray burst sources seem to be neutron stars heated impulsively. The subsequent cooling of the neutron star by thermal radiation seems consistent with a radius $a \approx 10$ km (Hoffman et al. 1977; Swank et al. 1977; van Paradijs 1978), providing some support for the present models of neutron star structure. Finally, an X-ray emission line in Her X-1 (Trümper et al. 1978) has been interpreted as electron cyclotron emission from electrons in a 10^{12} gauss field (see also Wheaton et al. 1979). But Nagel (1981) interprets this "line" as a neighboring absorption feature. Observationally, there is some indication that second harmonics of the electron cyclotron frequency have been seen (Kirk and Trümper 1982), but see Gruber and Primini (1982). A cleaner pair of emission lines are reported at 11.5 and 23 keV in the source 4U 0115+63 (White et al. 1983). Rose et al. (1979) previously reported just a broad (± 3.8 keV) iron line at 6.8 keV in this source, although one can see suggestions of the above two lines in their published spectra. More recently, Murakami et al. (1988) report such a pair at 20 and 40 keV in absorption in a gamma-ray burst source. In any event, it now seems to be uncritically accepted that neutron stars typically have 10^{12} gauss fields. There is evidence that gamma-ray burst sources are neutron stars; these sources often display both a 50 keV absorption feature and a 400 keV emission line. The latter is attributed to redshifted positron annihilation near the surface of the neutron star, and the former to the above cyclotron resonance seen in absorption. Melia (1988c) argues that this structure may come from local reprocessing.

It is evident that pulsar theory leans heavily on the expectation that neutron stars are highly magnetized. Why should they be? Pacini (1967) originally suggested that simple flux-conserving compression would result in fields of the order of 10^{10} to 10^{14} gauss. If the internal magnetic flux in a star is conserved by high conductivity, then the surface magnetic field scales as $1/a^2$, where a is the radius. Thus, if the Sun collapsed to neutron star dimensions (roughly a factor of 7×10^4 in radius), the general 1 gauss solar fields would escalate to about 5×10^9 gauss (Ostriker and Gunn 1969a; Imoto and Kanai 1971). This value is shy of 10^{12} gauss, but the Sun is not a candidate supernova progenitor. The implication then is that the presupernova core, with a radius of, say, 10^8 cm, will have a "surface" field (actually deep within a massive giant star) of 10^8 gauss. Note that the magnetic moment, which is proportional to Ba^3, actually decreases in such a flux-conserving collapse, and would vanish for collapse to a point (nonrotating black holes are predicted to be unmagnetized: Anderson and Cohen 1970). One can formulate an empirical "magnetic Bode's law" which shows a rough proportionality between spin angular momentum and magnetic moment (Blackett 1949; Hill and Michel 1975; Ahluwalia and Wu 1978; see also Greenstein 1972). Such a phenomenological proportionality does not even have to be "extrapolated" to pulsars, since the planet Jupiter has both a magnetic moment and an angular moment only slightly less than those typically attributed to pulsars. Such empiricism, however, has not yet proven to show what underlying physical mechanisms might be important.

a. *Cowling's Theorem*

The standard argument for *planetary* magnetism is that it has to do with circulation in the core, presumably driven at least in part by differential rotation. Unfortunately the simplest (axisymmetric) models are excluded by Cowling's theorem (Cowling 1934, 1957). The heart of this theorem is the observation that an axisymmetric magnetic field requires an azimuthal current, and these currents can only decay in the absence of azimuthal electromotive forces. There are obvious exceptions such as ferromagnetism, but the interiors of stars and planets typically exceed the Curie point. Moreover, the well-studied fields of the Earth and Sun are known to vary drastically, so the very persistence of ferromagnetism would be a problem. (The possibility of spontaneous magnetization of the electrons in neutron stars had been discounted by Canuto and Chiu 1968 but put forward later by Canuto et al. 1969; Lee et al. 1969; Canuto et al. 1970.) Stevenson (1983) gives an exhaustive review of the progress to date. The usual assumption is that complicated circulation patterns cause the magnetic field lines to be stretched and folded, leading to amplification. Complicated circulation models, however, do not explain why Saturn has such a nicely aligned dipole magnetic moment (until that observation, the inclined dipoles of the Earth and Jupiter appeared not inconsistent with underlying complexity).

The usual formulation is based on the induction equation, which can be rewritten slightly (see, for example, Levy 1976) to give

$$\frac{\partial \mathbf{B}}{\partial t} = \nabla \times (\mathbf{v} \times \mathbf{B}) + \eta \nabla^2 \mathbf{B}, \tag{9}$$

where η is the magnetic diffusivity representing ohmic decay of conduction currents, and the first term is the induction electric field associated with bulk motions,

$$\mathbf{E} = -\mathbf{v} \times \mathbf{B}. \tag{10}$$

Various workers have challenged the theorem on the grounds that the system could be nearly or apparently axisymmetric (Todeschunk et al. 1981; Braginskii 1964; Parker 1955; Roberts and Stix 1971). Others have accepted it and resorted to intricate diagrammatic arguments in hopes of obtaining the desired field amplification (e.g., Inglis 1955).

b. *Self-Excited Dynamos*

Despite the formal difficulties, it is straightforward to imagine a system in which magnetization appears spontaneously (Bullard 1978), as shown in figure 1.15.

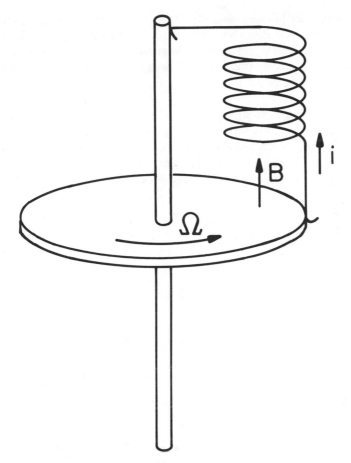

FIG. 1.15. The self-excited Faraday disk. A disk rotating through a magnetic field produces an EMF and drives current through a stationary shunt, here replaced by a solenoidal coil which in turn provides the magnetic field in the first place.

Not only is such an example instructive and illustrative that dynamos actually work, but it provides a counterexample to the Cowling argument, as shown in figure 1.16. One can imagine that the currents are returned through a cylindrical "can" with the spiraling of the current flow owing to the Hall effect induced by the stray magnetic field lines penetrating the surface of the can. Given that this system is manifestly axisymmetric, what happened in Cowling's theorem? Basically, it is incomplete because the conductivity is assumed to be a scalar when in general it is a tensor.

With the above considerations, one can construct a model that could simulate an astrophysical dynamo: a conducting core rotating at a different angular velocity than that of a Hall-conducting shell, as shown in figure 1.17. In such a model two magnetic field components will be generated, an internal toroidal field B_T

and the hoped-for poloidal field B_P. Differential rotation drives the currents that directly produce B_T, which in turn causes the Hall currents that produce B_P. This is just the magnetic field configuration usually proposed on qualitative grounds, namely a strong toroidal field accompanied by a weak poloidal field corresponding to the lines of force that are "wound up" in the core by differential rotation.

The apparent difficulty with such a model, intriguing though it is, is that the Hall currents are typically very small compared to the Pederson current (see sec. 2.6), which would mean that B_P should be correspondingly small compared to B_T. The direct way to get large Hall currents is to have a collision frequency comparable to the cyclotron frequency, which seems out of the question for

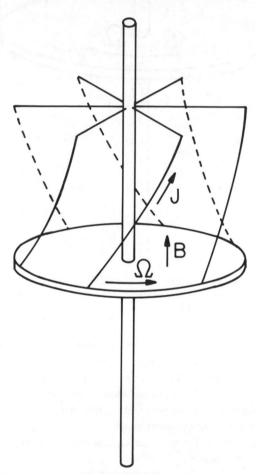

FIG. 1.16. The axisymmetric self-excited Faraday disk. By restyling the solenoid and the sliding contact at the disk edge, the self-exciting system can be made manifestly axisymmetric, in contradistinction to the theorem against such dynamos. It is necessary to spiral the wires returning the current to the periphery, but this spiraling is not a violation of axisymmetry.

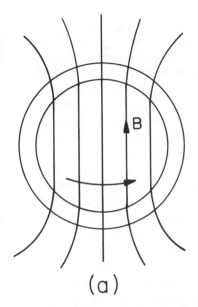

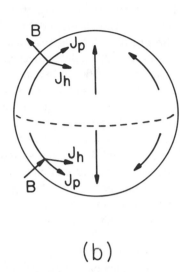

(a)   (b)

FIG. 1.17. A "planetary" self-exciting Faraday sphere. (a) Cross-sectional view showing the rotating core with uniform magnetization, which induces an electric field that causes the current to flow in the surrounding shell. (b) Current flow in the shell showing that the Pederson currents and magnetic field lines crossing the shell conspire to drive the Hall currents in one azimuthal sense, which can thereby provide the source of the magnetization in the first place.

neutron stars. The magnetic field contributes about $B_0^2 a^3/2\mu_0 \approx 4 \times 10^{40}$ ergs of energy, or about 10^{-14} of the rest-mass energy of a neutron star, so toroidal fields approaching 10^{19} gauss could be invoked. Such fields would, of course, have a first-order effect on the maximum mass and structure of a neutron star and essentially invalidate existing estimates. On the other hand, the current carried to form a poloidal magnetic field of 10^{12} gauss is astonishingly small compared to what would be present if, say, the electrons stood still and the ions rotated:

$$B_{\text{limit}} \approx \mu_0 \left(\frac{e}{m}\right) M \frac{\Omega}{a} \approx 2.4 \times 10^{33} \text{ gauss!} \tag{11}$$

It follows that the currents correspond to a difference in rotation rates of only 1 part in 2.4×10^{21} for a 1 s pulsar, so about 10^4 Hubble times would be required for the ions to lap the electrons once. One wonders what small effects might be neglected that could act at this level.

As mentioned earlier, ferromagnetism has been examined as a possible magnetic field source (Silverstein 1969; O'Connell and Roussel 1972; Pfarr 1972; Schmid-Burgk 1973), but seems unattractive. Differential rotation between, say, superfluid protons and normal electrons has also been proposed (Sedrakyan 1970a,b; Sedrakyan and Shakhabasyan 1972; Sedrakyan et al. 1975, 1977; Woodward 1978).

Blandford et al. (1983) have expanded on the idea (Urpin and Yakovlev 1980; Urpin et al. 1986) that thermoelectric phenomena might generate the pulsar magnetic field, under rather special conditions (i.e., deep molten oceans). The growth rates and resultant magnetic field strengths predicted seem quite interesting, although the mechanism as described would result in a patchwork of small-scale ($\approx$500 m) but strong magnetic field regions more or less randomly oriented on the surface. Such magnetization itself would lack the large-scale global organization apparently required to explain the brief pulses observed from pulsars. The physical problem is that a high multipolarity magnetic field falls off so rapidly with radial distance that there is essentially no radiation reaction even in a rapidly rotating system; the 10^{12} gauss surface magnetic fields were inferred on the assumption of a dipolar field. For a quadrupolar magnetic field, one would then require 10^{19} gauss, with values required for higher multipoles to be important even more difficult to believe. These authors suggest that the regions somehow align to give a dipolar field ("long range order"), and surely this model would become particularly interesting should a mechanism for such alignment be identified. The model has been taken by some as an indication of "delayed turn-on" of pulsars long after their formation (to account for the poor pulsar correlation with supernova remnants). It should be noted that to "order" the magnetic patches to give one large dipole is energetically uphill because a much larger volume of space is filled with dipole magnetic field lines than by any other multipole. This very fact has been used by Flowers and Ruderman (1977) in a model for *decay* of the dipole field (to higher multipoles). The possibility that the magnetic field picture eventually decays is discussed in more detail below (sec. 6). Some observational support for delayed turn-on comes from the discrepancy in the ages of PSR 1509−58 and the MSH 15−52 nebula (Helfand 1983), but delayed turn-on could result from other effects than growth of the magnetic field.

c. *Critical Magnetic Field*

It is interesting that the inferred magnitudes of the pulsar magnetic fields are close to the so-called critical magnetic field

$$B_{\text{crit}} = m_e^2 c^2 / e\hbar = 4.4 \times 10^{13} \text{ gauss.} \tag{12}$$

Physically, the cyclotron frequency of an electron in this magnetic field would correspond to an energy $\hbar\omega_c = m_e c^2$. Alternatively, an electrostatic field of strength cB_{crit} would be so strong that the electrostatic potential difference over an electron Compton wavelength would equal $m_e c^2 / e$ and the vacuum could spontaneously break down into pairs to discharge this field (Lerche and Schramm 1977), although Zaumen (1976) estimates that magnetic fields up to 10^{20} gauss could exist.

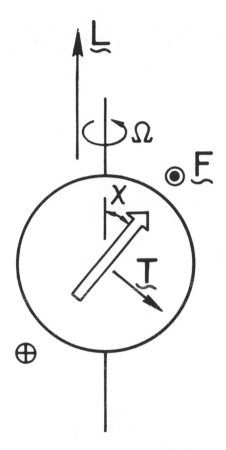

FIG. 1.18. Alignment torques on a pulsar. Sweeping back of field lines by retardation effects gives a force component opposite to rotational velocity and concentrated near the magnetic pole. Resultant torque vector (**T**) has component opposite to the angular momentum vector, which produces the slowing down (**L**), and toward the nearest pole, which produces the alignment. From F. C. Michel, 1982, *Rev. Mod. Phys.*, 54, 1 (figure 24); F. C. Michel and H. C. Goldwire, Jr., 1970, *Ap. Letters*, 5, 21 (figure 1). Reprinted by permission of Gordon and Breach Science Publishers Inc.

5. Alignment of Spin and Magnetic Moment

It was recognized by several groups simultaneously (Michel and Goldwire 1970; Davis and Goldstein 1970; see also Mestel 1968) that the torque on an oblique rotator acts not only to brake the rotation but also to align the magnetic moment (which is assumed to be "frozen" into the neutron star) with the rotation axis (which is not). How this alignment torque comes about is illustrated in a simple heuristic way in figure 1.18, although it can be calculated directly from the near zone fields (Soper 1972; Imoto and Kanai 1972).

To calculate the instantaneous torques (z is along the spin axis, x is orthogonal to z in the plane containing the spin vector and magnetic moment [figure 1.18], and y is orthogonal to both), we can convert from polar coordinates to these Cartesian coordinates, which gives the surface integrals

$$T_x = -\int r(T_{r\theta} \sin\phi + T_{r\phi} \cos\theta \cos\phi) \, dS, \tag{13}$$

$$T_y = \int r(T_{r\theta} \cos\phi - T_{r\phi} \cos\theta \sin\phi) \, dS, \tag{14}$$

and

$$T_x = \int r(T_{r\phi} \sin\theta) \, dS, \tag{15}$$

where the stresses are given using the exact Deutsch (1955) expressions (but neglecting the higher-order electrostatic terms) with

$$T_{r\theta} \equiv B_r B_\theta \tag{16}$$

and

$$T_{r\phi} \equiv B_r B_\phi. \tag{17}$$

The result is just

$$T_x = T \sin\xi \cos\xi \tag{18}$$

and

$$T_z = -T \sin^2\xi, \tag{19}$$

with

$$T = 2\pi a^6 B^2 \Omega^3 / 3\mu_0 c^3. \tag{20}$$

The torque along the z-axis is, of course, the standard result (ξ = inclination angle).

The equations of motion are

$$I(d\Omega/dt) = T_z \tag{21}$$

and

$$I\Omega(d\xi/dt) = -T_x. \tag{22}$$

There is then the constant of motion

$$\Omega \cos\xi = \Omega_0 \cos\xi_0, \tag{23}$$

and if slow (≈ 1 s) pulsars evolved from fast ($\approx 10^{-3}$ s) pulsars, they would have to have been formed as almost exactly orthogonal rotators. Note that this conclusion follows even if the magnetic field changes with time. The braking index is given by

$$n \equiv \Omega\ddot{\Omega}/\dot{\Omega}^2 = 3 + 2\cot^2\xi. \tag{24}$$

The remaining integral of the motion is just

$$\Omega^2 = \Omega_0^2 \cos^2\xi_0 / (1 - \sin^2\xi_0 e^{-t/t_c}), \tag{25}$$

and the spin-down power output then scales as

$$L_{\text{spin-down}} = L_0 e^{-t/t_c} \frac{\cos^4\xi_0}{(1 - \sin^2\xi_0 e^{-t/t_c})^2}, \tag{26}$$

where $t_c = (\Omega_0/2\dot{\Omega}_0)\tan^2\xi_0 \approx 1/B^2\Omega_0^2\cos^2\xi_0$, and, except for a shift in scale, the power output falls exponentially.

Alignment and exponential magnetic field decay would be virtually indistinguishable effects. However, there is an important distinction in time scale. If alignment takes place, it should be rapid when the pulsar has just formed and is spinning down rapidly. Thus the time scale for alignment is short and only becomes long when the alignment is almost complete. Consequently, long alignment times cannot readily be assigned to pulsars that have large angles ξ unless they were formed with low spin rates ("injected": Candy and Blair 1983).

Alternatively, we can assume that T_z is independent of angle. Here a wind torque from the aligned dipole component would be the physical justification. In effect, this is what is done when pulsar magnetic fields are quoted on the basis of the period and period derivative. We would then get

$$\Omega\cos\xi/\sin\xi = \text{constant}. \tag{27}$$

In both cases the luminosity goes to zero with alignment! Either the dipole torque goes to zero despite a finite asymptotic spin rate, or the spin rate itself goes to zero with alignment. Functionally, we have for pure dipole torques, the time evolutions

$$L \approx e^{-2t}; \; \xi \approx e^{-t}; \; \Omega \to \text{constant}, \tag{28}$$

while for the wind plus dipole torques

$$L \approx t^{-2}; \; \xi \approx t^{-1/2}; \; \Omega \approx t^{-1/2}. \tag{29}$$

Aligning mechanisms have also been identified in neutrino emission (Tennakone 1972) and generalized polar wander in analogy with the Earth (Macy 1974). The electromagnetic alignment idea has been refined (Chau et al. 1971) and applied to observation (Henriksen 1970; Chau and Henriksen 1970; Jones 1975, 1976a,b, 1977a). As noted above, the braking index is given by

$n = 3 + 2\cot^2 \xi$, where ξ is the angle between spin axis and magnetic moment (figure 1.18). As already noted, the determination of n for the Crab (Groth 1975a,b) gives 2.515 ± 0.005 and for PSR 1509–58 it is 2.83 ± 0.03 (Manchester et al. 1985), impossible values in the model.

Goldreich (1970) pointed out, however, that for certain assumptions regarding the shape of the neutron star, alignment need not take place. It is easy to understand this argument if one assumes instead that alignment has taken place and looks for a contradiction. One contradiction would be to have the body axes at an angle to the (presumed) aligned spin and moment axes. There would now be no alignment torque, but at the same time the body axes would have to precess, producing a nonalignment. The two most obvious sources of a nonspherical body shape, centrifugal forces and deformation due to the internal magnetic field stresses, produce nonspherical distortions that are, however, aligned. Thus, as Goldreich (1970) points out, it is necessary to suppose some hysteresis wherein the star is still distorted along an earlier spin axis direction. Precession due to triaxiality of the moment of inertia has also been suggested from time to time to explain features of pulsar observations (see Hog and Lohsen 1970; Burns 1970: Chiuderi and Occhionero 1970; Chau and Henriksen 1971; Chau and Srulovicz 1971; Avakyan et al. 1972; and Pines and Shaham 1974). Good and Ng (1985) point out that there are actually two different torques: the alignment torque discussed above (basically caused by displacement currents) and a possible counteralignment torque given by conduction currents if they flow with axial symmetry into the magnetic polar caps (sec. 3.2.3b). This is a point of considerable importance, given that the distribution of conduction currents remains an open question. Broadly speaking, there seem to be no forceful observational data that point to either alignment or precession, which is not to say that these physically plausible effects do not take place. Indeed, Weisberg et al. (1989) report detection of geodetic precession in PSR 1913 + 16.

6. Evolution of the Magnetic Field

The fossil magnetic field model suggests eventual decay of the field, and Ostriker and Gunn (1969b,e) argued for such decay on empirical grounds to explain the paucity of long-period pulsars, although this view is not without its counterarguments (Pacini 1969; Holt and Ramaty 1970; Setti and Woltjer 1970; Chanmugam and Gabriel 1971; Heintzmann and Grewing 1972; Kundt 1986). Other causes of field variation, such as evolution of the internal source field, have also been examined (Vandakurov 1972; Chanmugam 1973; O'Connell 1975; Jones 1976a,b,c; Flowers and Ruderman 1977; Chanmugam 1978). Unfortunately, none of the observed pulsar variations can confidently be attributed to magnetic variations, and their overall evolutionary pattern still remains somewhat obscure. It has been repeatedly suggested that pulsar turn-off at long periods is the result of magnetic field decay (see Gunn and Ostriker 1970; Fujimura and Kennel 1980); however, the theoretical estimates consistently give conductivities that are orders of magnitude too large (Baym et al. 1969a; Ewart et al.

1975). We discuss the statistical evidence for magnetic field decay in sec. 2.4.

The idea of turn-off has proven attractive to many astrophysicists, particularly the observers, but it is far from being an established fact. In some analyses, turn-off is not *just* attributed to magnetic field decay but additionally to the pulsars evolving across a "death line" in the P-$\dot{P}$ diagram. Thus field decay must be supplemented with a second mechanism, all simply to explain one weak possibility (see sec. 2.4). The assumption of exponential decay has recently been shaken by the observation of an *old* white dwarf companion in a binary pulsar system, PSR 0655 + 64 (Kulkarni 1986). The companion is reasonably magnetized at $\approx 10^{10}$ gauss, yet the inferred age of the white dwarf from cooling calculations is of the order of 2×10^9 years, in reasonable accord with the $P/2\dot{P}$ age of 3.6×10^9 years. Evidently the magnetic field of this object has not exponentially decayed away. To salvage the magnetic field decay idea, van den Heuvel et al. (1986) and Taam and van den Heuvel (1986) argue for a "standard" decay rate to 3×10^9 gauss, rather than to zero, followed by a very slow decay. If the entire neutron star started out uniformly magnetized and the crustal magnetic field decayed away, the core size would then necessarily be only 14% of the total radius, completely inconsistent with neutron star models which give large superfluid cores ($\geq 80\%$ of radius; Shapiro and Teukolsky 1983, p. 251). Moreover, there is no clear reason to expect the crust to be uniformly magnetized in the first place. Wendell (1988) suggests that magnetic flux tubes float up from the core and decay in the crust, but that some flux remains in the core to give a bimodal decay behavior. Some theorists remain unconvinced that rapid magnetic field decay actually takes place (e.g., Kundt 1981a, 1986; Lamb 1981; Arons 1986, private communication; Michel 1986; Sang and Chanmugam 1987; Cheng 1989). Indeed, this idea originally stemmed from the observation that few pulsars have periods longer than 3 s. Because pulsars are all slowing down, one would expect an accumulation of slow pulsars, and it was suggested early on that these slow pulsars fade from view owing to the decay of their magnetic fields, as discussed, for example, by Manchester and Taylor (1977). However, they also get fainter and more difficult to discover (sec. 2.4).

Plausible as magnetic field decay might seem, a number of important cautions should be mentioned:

(1) Although it is widely believed, for the theoretical reasons already outlined, that the slowing-down torques scale essentially as period to the fourth power, it has not been possible to confirm this exponent from the data. How much confidence can then be invested in detection of a first-order correction to the slowing down (magnetic field decay) when the zeroth-order slowing down itself cannot be accurately calibrated?

(2) The rapid drop in total spin-down luminosity itself gives a characteristic age of a few million years and a limiting period of a few seconds simply because pulsars become too faint to be seen, assuming that the radio luminosity simply parallels the spin-down luminosity on the average.

(3) If a mechanism *were* required to explain the lack of long-period pulsars,

a number of alternative mechanisms have been identified that would do as well as magnetic field decay. For example, the surface temperature of a neutron star is estimated to fall precipitously after a few million years, just to name one factor having essentially the correct characteristic time.

(4) Theoretical estimates of neutron star conductivity have systematically been too high to give such a rapid decay, except possibly in the outermost crust. The core is even thought to be superconducting.

(5) The binary pulsating X-ray sources are thought to be neutron stars accreting from a binary companion, and some show line spectra suggestive of cyclotron emission in strong pulsar-like magnetic fields. (Lewin and van Paradijs 1986 argue from the presumption of magnetic field decay that these objects must be relatively recently formed [$\approx 10^7$ years].)

(6) The idea that a celestial body is magnetized by flux trapping at formation with subsequent ohmic decay does not work anywhere that it can be tested. It is obviously incorrect for the Earth and the Sun. To the modest degree that planetary magnetism is understood at all, one would even expect rotating neutron stars with superfluid cores to be generators of magnetic field, not sinks.

If the magnetic field decay idea were correct, evolutionary scenarios for many astrophysical objects would be highly constrained. Gamma-ray burst sources are thought to be old neutron stars, for example. If theory were obligated to assume them to be weakly magnetized when in fact they had strong and important magnetic fields, we would get nowhere.

Jones (1988) has recently reexamined and summarized the various mechanisms and concluded that radial Hall drift of magnetic flux tubes could transport magnetic flux out of the core to the crust where it would decay more rapidly. He estimates, qualitatively [B in units of 10^{12} gauss, t in units of million years (My)],

$$\frac{dB}{dt} = -\frac{B^2}{t_H} - \frac{B}{t_D},$$
(30)

where $t_H = 12$ My and the ohmic decay constant itself is estimated to be $t_D \geq\ = 500$ My. In other words, no one can get direct ohmic decay to operate rapidly enough to satisfy the expectation of decay on a time scale of a few million years. The Hall drift is the nonlinear term, from which we can immediately estimate a characteristic time $t_H \approx (140)^{1/2} \approx 12$ My, which Jones holds to be not inconsistent with the statistical expectations. However, this equation can be easily solved to show that $B \approx B_0(t_H/t)$, which means that the evolutionary time scale will increase to 120 My by the time B decreases to just 10^{11} gauss, which is nowhere small enough a field and a vastly too long time constant. The other point is that, in appealing to second-order effects, one should also include effects that could *increase* the field such as dynamo action from differential rotation, so the conservative conclusion is that direct ohmic decay is just too long on all estimates to date, and there are a number of effects of both signs that could well act on a shorter time scale (just as they do for planets in the solar system).

The one new fact supporting magnetic field decay (beyond the debatable analysis of $P/\dot{P}$ distributions) is the observation of an association of millisecond pulsars with binary systems. Clearly a pulsar in a binary system is a tempting candidate for accretion and spin-up (Nomoto 1981; van den Heuvel 1981; Blandford and DeCampli 1981), and by definition millisecond pulsars must have weak fields, so these cases are certainly consistent with field decay followed by spin-up. On the other hand, it is equally curious that the original millisecond pulsar has no companion, which requires postulating that the companion conveniently not only became a supernova but also recoiled or had sufficient mass loss that the binary system became unbound.

If the magnetic field indeed decays, the torque law becomes

$$\Omega\dot{\Omega} = -Ke^{-2t/\tau_D}\Omega^4 \tag{31}$$

and integrating directly as before gives

$$\left(\frac{1}{\Omega^2} - \frac{1}{\Omega(0)^2}\right) = \tau_D K(1 - e^{-2t/\tau_D}). \tag{32}$$

The constant K can be solved from equation 32, defining the characteristic spin-down time $\tau_C \equiv \Omega/2\dot{\Omega}$ to give

$$K = (\tau_D/2\tau_C)e^{\,|2t/\tau_D}\Omega^{-2}, \tag{33}$$

in which case the exponential factors cancel out in the relation

$$\frac{1}{\Omega(\infty)^2} - \frac{1}{\Omega^2} = \frac{\tau_D}{2\Omega^2\tau_C} \tag{34}$$

or

$$\Omega(\infty) = \Omega(1 + \tau_D/2\tau_C)^{-1/2}, \tag{35}$$

in which case, any pulsar with a characteristic age near a magnetic field decay time would already have spun down to essentially its asymptotic spin rate. The constant K can also be isolated by differentiating equation 32 again to give

$$\frac{\ddot{\Omega}}{\dot{\Omega}} - \frac{3\dot{\Omega}}{\Omega} = -\frac{2}{\tau_D}, \tag{36}$$

which then gives the breaking parameter $n = 3 + \tau_C/\tau_D$. Note that n is always larger than 3 but not significantly different until the characteristic spin-down period is large, which is also the criterion for not being able to determine n! In more physical terms, the characteristic of pulsars older than τ_D is that they should decline in luminosity while the period remained roughly constant.

If, as is often supposed, the pulsar was initially very fast ($\Omega(0) \ll \Omega$), the relationship between the true age t and the characteristic spin-down time is

$$t = (\tau_d/2)\ln(1 + \tau_C/\tau_D),\tag{37}$$

so for $\tau_C = \tau_D$, the true age is 0.55 the decay (or characteristic) age, and for a characteristic age even 1000 times the decay age, the true age is 3.45 the decay age. Such considerations ignore the very powerful luminosity selection, which would act to render pulsars with ages much larger than τ_D unobservable.

2

General Analysis of What a Pulsar Is

2.1 Introduction to the Models

We now begin our discussion of models for the pulsar object. From the sketch given in chapter 1 we have attempted to justify the assumptions that (1) pulsars are rotating neutron stars, (2) the rotation period is the pulsation period, and (3) the coupling to the radiation field is by a strong magnetic field in the neutron star. No alternatives have yet been suggested that are not grossly inconsistent with either the data or known physics. It is important to note that pulsars are numerous in the sense that they are being born at about the same rate as their presumed progenitors, the supernovae. The neutron star is thus assumed to be the remnant of the explosion. The other two properties, rotation and magnetization, seem to be ubiquitous, hence plausible, properties of all astrophysical bodies. The theories therefore bifurcate according to whether these properties alone are taken to be sufficient for pulsar action ("vacuum models"), or whether yet additional factors are deemed essential. Because pulsars are so populous, any supplementary requirements must have high a priori probabilities. For example, there might be additional matter near the pulsar, such as a disk, that is essential to pulsar action. Or interaction with the interstellar medium may play an essential role. Both of these have high a priori probabilities. In contrast, as a straw man example, a theory requiring the neutron star to be orbited by a mini-black hole would have a low a priori probability. As mentioned in sec. 1.2.a, oscillating and orbiting models have fallen into disfavor, and with them the idea of extraneous matter near pulsars. Careful timing of the pulse arrivals has ruled out the presence of significant companions (e.g., planets: Lamb and Lamb 1976), which could neither be close to the pulsar, because tidal forces would disrupt them, nor far from it, because then the motion of the neutron star relative to the barycenter would be large and therefore detectable as a cyclic arrival-time shift in the pulses. For example, the Crab pulsar has a pulse only 1 ms or so wide and the precise location of a long string of such pulses can be determined to, say, 1% of that value. Consequently a periodic displacement of the pulsar by only 3 km would be detectable. A planet one-hundredth the mass of the Earth placed as close as the Roche limit (about 1.7×10^6 km) could thus be detectable, and more distant objects would have to be proportionately less massive not to be seen. In the same way, planetary perturbations within the solar system also show up (e.g., Mulholland 1971; Cowling 1983). Planetary companions would have to be placed at huge distances to be undetected so that the long period would be absorbed into a nearly constant period derivative, but then there would be no reason to suppose they play a role in the pulsar action. Furthermore it was plausibly suggested (Ostriker and Gunn 1971) that intense radiation from a rapid newly born pulsar might even create the supernova explosion by accumulating in the collapse cavity, hence driving away and exciting the envelope of the preexisting star. Except for the millisecond pulsars, which are commonly found in binary systems, few of the more typical pulsars have companions (about 1%). Pulsars also seem to have much higher space velocity than the normal stars that populate the galactic disk. Taken together, these various ideas and observations suggest that the supernova explosion was so violent and the resultant pulsar so energetic, possibly even recoiling from the event, that the neutron star would be stripped clean. The popularity of the vacuum models (vacuum in the above sense; it seems necessary that conduction currents flow to and from the neutron star to explain the pulsar action) therefore seems well founded. Nevertheless, it has recently been proposed that material does remain about the neutron star in the form of a disk and that this disk is essential for pulsar action (Michel and Dessler 1981a). We will therefore discuss first the vacuum models and later the more controversial models. Table 2.1 lists some classic papers.

Deutsch (1955) early on considered rotating magnetized *stars* and noted that for rapidly rotating, highly magnetized cases one obtained substantial rotationally induced electric fields capable of accelerating ambient particles to substantial energies. A modern view would suggest that so much plasma surrounds ordinary stars as to short out any such fields, but the work (which includes calculation of the full **E** and **B** field structure from the surface to infinity) is of particular interest to the case of neutron stars. Deutsch thought that the accelerating potentials might produce cosmic rays, an idea persistent to the present.

TABLE 2.1—Classic Papers Related to Pulsar Theory

Author(s)	Date	Model
Deutsch	1955	Acceleration by rotating magnetized stars
Pacini	1967	Prediscovery prediction
Gold	1968	Circulating particle bunch
Ostriker and Gunn	1969	Oblique rotator and magnetic field estimate
Goldreich and Julian	1969	Aligned rotator and plasma source
Sturrock	1970	Aligned rotator and pair production
Ruderman and Sutherland	1975	Extension of Sturrock model (gap, sparks)

Pacini (1967, 1968) anticipated that a rotating neutron star could power the Crab Nebula, before the discovery of pulsars, and he assumed that a strong magnetic field of the neutron star would be required to couple the rotational energy out of the star. That the star itself would be a source of *pulsed* radio emission was not specifically suggested, although some degree of rotational modulation would have been a natural expectation.

Gold (1968) suggested that a bunch of electrons (10^{22} within a 1 m sphere) corotated with the pulsar, trapped in the pulsar's equatorial magnetic field and located out almost at the aforementioned light-cylinder (i.e., where corotation would be at the speed of light, $R_L = c/\Omega$). Why such a bunch would be so located, or be stable, or be replaced if lost, was left open. In fact, radiation reaction would cause the bunch either to be ejected or to retreat from the light-cylinder, even if it could be formed somehow in the first place. Moreover, particles do not actually try to drift at the speed of light when placed in crossed **E** and **B** fields where $E = cB$ and do not move in the $\mathbf{E} \times \mathbf{B}$ direction, as can be shown by elementary calculations. Observationally, the model would have had difficulties in accounting for special phenomena such as nulling and the complex variation of polarization patterns in such a model, because the radiation would be essentially that of a terrestrial synchrotron and hence would have fixed polarization properties (see also Good 1969). Nevertheless the idea of bunching as a mechanism for coherence seemed sound and influenced future work.

Ostriker and Gunn (1969a) (or Gunn and Ostriker 1969) introduced the basic idea of intense magnetic fields (10^{12} gauss) by equating the energy loss from pulsars to electromagnetic waves emitted by a rotating oblique magnetic dipole, again not a pulsar model per se. They also looked at the acceleration of particles to high energies in these large-amplitude electromagnetic waves (see sec. 5.2). Most observers expect that pulsars must as a class be oblique rotators, so such acceleration remains an interesting mechanism.

Goldreich and Julian (1969) proposed the simpler case of a rotator with an aligned magnetic dipole moment (which would not, of course, radiate electro-magnetic waves) and showed that electrostatic forces would pull plasma off the pulsar surface to fill the magnetosphere (see also Michel 1969a). This space-charge density estimate appears frequently, even in quite different models (because it is simply the dimensionally scaled charge density), and is sometimes called the Goldreich-Julian density. Their model concentrates on the physics of the aligned rotator and is not a model for pulsar action per se. Instead, they note that the emitted plasma must settle into corotation with the star and therefore the magnetosphere beyond the light-cylinder distance must be spun away to form a wind. This wind zone clearly places the interesting dynamics in the polar cap regions (locus of the magnetic field lines that reach out to R_L or beyond) and therefore makes these regions the logical place from which to expect radio emission. Goldreich (1969) supposed that the oblique rotator had the same physics as an aligned rotator (specifically, radio-emitting polar caps, although that was not actually demonstrated in the model) and that the obliquity served only to sweep this beam around to give *pulsed* radio emission, a view widely adopted by subsequent theorists. This supposition led to the so-called hollow cone model (emission concentrated at the edge— "auroral zone"— of the magnetic polar caps), which has enjoyed a long popularity with observers in interpreting their data. This short paper, published in a somewhat difficult-to-find journal, repeats the Deutsch idea for particle acceleration. The term "light-*cylinder*" is variously interpreted in theoretical papers, with some literally assuming that there is a cylindrical surface at which the magnetospheric properties change, and others simply taking it as a characteristic distance in the equatorial plane at which there is a transition between closed and open magnetic field lines but at which there is no particular significance out of the plane (in which case the term "cylinder" is irrelevant).

Sturrock (1970, 1971a) introduced the first "modern" pulsar model: particles continuously ejected from the magnetic poles (an idea already implicit in the Goldreich-Julian model) at a controlled rate (*space-charge limited flow*) with electrons radiating gamma-rays because they must follow curved field lines (emitting *curvature radiation*, which is just synchrotron radiation except the particle motion is curved because the particles follow curved magnetic field lines rather than because they execute orbital motion *about* these field lines). These gamma-rays in turn, by producing pairs in the same strong magnetic field, would produce yet more radiating particles, leading to a *pair production cascade*. Coherence is attributed to bunching in the counterstreaming electron-positron plasma. The idea that pair production is essential to pulsar action has also enjoyed a continuing popularity with theorists.

Ruderman and Sutherland (1975) elaborated and improved upon the Sturrock model (the original way of handling the space-charge limited flow is now known to be oversimplified: see Michel 1974c, Fawley et al. 1977, and sec. 2.3.6, below). The Ruderman-Sutherland reformulation has received much more attention than did Sturrock's model in its time, possibly because the former addresses some of the observational puzzles that arose in the interim (e.g., drifting subpulses are attributed to localized discharges, "sparks," that drift systematically about the polar caps).

All of these theories are internally inconsistent in the sense that they do not clearly show how current closure took place. In the Goldreich-Julian model the polar caps are found deep within an enveloping space-charge-separated plasma of one sign, as we will discuss in chapter 4. Thus the normal expectation would be loss of just that one sign of charge. However, the argument that corotation leads to loss of plasma in the form of a wind seemed so secure that the general attitude was that the system must work, and that the current closure therefore worked out somehow; a return current into the auroral zone was the standard expectation. An alternative suggestion was that the pair production could neutralize the outflowing plasma. Consequently, later models simply *assumed* current closure was guaranteed and typically concentrated on what activity might take place as a consequence of the currents into the polar cap regions that would account for the radio emission, which was mainly the type of input the observers wanted so that they could interpret their data (e.g., Ruderman and Sutherland 1975). Attempts to understand the global physics were largely regarded as minor exercises useful to tidy up the theory.

Instead, a number of people eventually stumbled on what now seems to be the correct answer, namely that the magnetosphere does *not* need to rotate with the star. In fact, an aligned model would not necessarily have any plasma at the light-cylinder distance to be spun away in the first place. In other words, the current closure problem is solved in an unanticipated way: *there need be no current!*

A number of responses have been forthcoming. The idea that pairs not only are responsible for coherent radio emission but *also* neutralize the wind has taken on new urgency in some quarters. Such an exact neutralization is in fact difficult to get to work, and would not help much because with zero net current there would be no source of electrical power. Each electron-positron pair would be produced and confined to essentially one and the same magnetic field line, so the current density is zero if both escape. A more radical departure has been the suggestion that a pulsar is not an isolated neutron star but has a disk of perhaps planetary mass orbiting it. Such a system would certainly be electrodynamically active, given two conductors in differential motion threaded by a common magnetic field, but the requirement for a disk bothers many. Another view is that the currents only circulate locally from the star out to the light-cylinder distance and back, with the driving mechanism radiation reaction at the light-cylinder. Such a model essentially makes a virtue of what was a flaw in the Gold model. Unfortunately, the baseline models suggest that there need be no plasma at the light-cylinder to begin with, so it is not clear how the flow could ever be set up. Even the proponents of this model have been unable to show that radiation reaction would lead to the required circulation to and from the star. It has also been suggested that the currents are closed by particles from the interstellar medium. Such a source is very difficult to imagine, given the powerful outgoing winds which should result if in fact the pulsar *is* active, and which would act to expel any external plasma. (Quantitative discussion follows in later chapters.)

Despite these continuing uncertainties, a powerful set of tools has been arrayed with which to understand the physics of the pulsar magnetosphere. The above models can now be analyzed in sufficient detail to see if they merit serious consideration. Thus the nature of pulsars is steadily being tracked down, thanks to the improving theoretical modeling and the continuing refinement of the observational techniques.

2.2 Observational Constraints

The central problem of pulsar theory is to explain how the mechanism for such intense coherent radio emission could arise spontaneously (hence what mechanism it might be). The next problem is to decipher the large number of observational data to see how pulsars differ from one another and to organize these data. Before either of these tasks can be performed, however, it is necessary to account for the overall gross energetics. The Crab pulsar has become a prototype, and it is often felt that if this pulsar can be explained, all pulsars can be explained. There are those who worry, however, that any one specific object could have very unusual morphological properties and hence need not really be representative. In fact, some models for the Crab pulsar cannot be scaled down to work for the more ordinary pulsars.

One needs to explain the slowing down of pulsars, which translates into accounting for the loss of rotational energy in the rotating neutron star model. Indeed, slowing down is the basis for estimating the magnetic field strength. It would be nice to account for the excitation of the nebula around the Crab pulsar, the energy output from which is, within uncertainties, comparable to the rotational energy output from the pulsar. Moreover, the nebula contains electrons too energetic to have been left over from the supernova explosion and a magnetic field too strong to be simply the adiabatically expanded remnant of any conceivable internal stellar field. The pulsar is therefore the most plausible source of the excitation, the particles, and the magnetic field. Unfortunately, the plausible connections are not infrequently the wrong ones. For example, the sign of the magnetic moment determines whether electrons or positive particles tend to flow out along field lines from the polar caps. Since there is no a priori reason for one sign or the other sign, can this dichotomy show itself in two distinguishable pulsar families? This interesting question promises to be with us for some time. The observed pulsars do not seem to fall neatly into two distinct groups, and perhaps pulsars of one of the two groups do not pulse significantly (or this sign makes no difference). Indeed, the millisecond pulsars may be a distinct class, but this distinction is not yet readily attributed to whether electrons or ions are the polar cap species. A selected subset of pulsar theories is listed in chronical but not necessarily logical order in table 2.2.

TABLE 2.2—Rotating Neutron Star Models

Author	Date	Mechanism	Pulses?	Stable?	Current closure?
Gold	1969	Trapped bunch	Yes	No	Yes: no current
Gunn and Ostriker	1969	Dipole torque	No	Yes	Only displacement current
Goldreich and Julian	1969	Polar cap	No	?	Claimed
Sturrock	1970	Pair/polar cap	No	No	?
Ruderman and Sutherland	1975	Pair/polar cap	No	?	Assumed
Michel	1980	Particle trapping	No	Yes	Not required: not pulsar
Michel and Dessler	1981	Kepler disk	Yes	?	Yes
Arons	1983	Slot gap	Yes	?	Assumed
Beskin et al.	1983	Pair/polar cap	Yes	?	Assumed

1. Energy Output

Of all the issues involved, this one has been flogged to death. The choice of a magnetic field of order 10^{12} gauss together with rotation at the pulse rate and a neutron star radius of about 10 km immediately guarantees that the energy loss will scale to the correct order of magnitude. Any model with such a starting point will "be consistent with the data" (i.e., should get sufficient power output).

2. Radio Emission

That the radio emission must be coherent if the source is to be as small as a neutron star or even the light-cylinder is a severe constraint. Unless one has a global model of where the currents flow and what their characteristics are (i.e., just electrons, or comoving pairs, or counterflowing pairs, etc.) one simply makes a guess and goes from there. About as far as any model has gone to date is to point to some instability with a promising growth rate (if that), and assume that it can do the job. The steep decline of radio flux with increasing frequency is broadly consistent with emission from particle bunches, but other mechanisms are not excluded.

3. Sharp Pulses

Concentration of the radio emission into narrow pulses is typically attributed to relativistic beaming, which requires the particles (almost uniformly assumed to be electrons) to have Lorentz factors of at least order 10^2. If anything, it is difficult to understand why the Lorentz factors around such highly charged objects would be so *low*, given that the available electrostatic potential drops should be huge. The tendency, regardless of the radiation mechanism, is to suppose that the radiation is directed parallel to magnetic field lines near the surface (for emission models that act near the surface, of course).

4. All the Rest

The uncertainties mount rapidly even in accounting for the ubiquitous pulsar properties. For the more specialized and individual phenomena, the existing models must either ignore them or require ad hoc embellishments. For phenomena that apply to only a restricted set of objects, it may indeed be that special circumstances are involved. One hope is that some observation will be so cleanly diagnostic that the underlying radiation mechanism or current flow or whatever will be revealed. Given the steadily expanding data base and the discovery of new pulsar systems such as the eclipsing binary pulsar, this may indeed be a realistic hope. The present situation with regard to the so-called standard model for pulsars (sec. 4.3) is summarized in table 2.3.

2.3 Physical Constraints

Regardless of how pulsars actually work, a number of acknowledged physical constraints must be met. Here theory does us the service of restricting the possible models (unusual in much of astrophysics). The theory may, of course, be

TABLE 2.3—Theory versus Expectation in Pulsar Models

Consideration	"Standard" Model	"Actual" Pulsar[a]
Rotation = period	Yes	Yes
Strong magnetic field	Yes	Yes
Alignment	Yes	No
Coherent radio emission	Not demonstrated	Yes
Pair production	No[b]	Yes
Ions for surface	Yes	No consensus

a) According to conventional wisdom, namely a consensus based on a lack of better information.
b) Pair production is sometimes invoked, but not in any self-consistent model as yet.

incomplete, but that should be signaled by an inability to construct a satisfactory model. Unfortunately it is commonly assumed that we have in hand the essential physics and, if not, what is required is some exotic feature such as cosmic strings or quark matter. More often we simply overlook effects because they are in an unfamiliar setting or because we are hesitant to move more than one step beyond observational confirmation. For example, it has been argued for two decades that neutrinos are copiously produced by a newly formed neutron star in a type II superova, yet only belatedly has it been realized, given confirmation from SN 1987A, that these neutrinos can have important nucleosynthetic roles in their interaction with the ejecta!

1. Rotationally Induced Electric Fields

A surprising number of people are perplexed that rotating a permanent magnet also charges it differentially. There seems, for example, to be no $\partial \mathbf{B}/dt$ term to provide the induction. But we must remember that a pure $\mathbf{B}$ field in one reference frame is a mixed $\mathbf{E}$ and $\mathbf{B}$ field in another moving relative to the first. Thus a conductor moving relative to a pure $\mathbf{B}$ field sees this induced $\mathbf{E}$ field and must shield it with induced surface charges. A conducting sphere therefore becomes an electric dipole when in motion. From a single-particle point of view, a particle can be in uniform motion in a magnetic field only if it also sees an electric field such that the $\mathbf{E} \times \mathbf{B}/B^2$ drift velocity matches precisely the particle velocity. Indeed, failure to match induces the currents that lead to the surface charging until the match is attained. The same picture carries over to rotation without creating anguish over accelerated frames of reference, etc. Rotation simply requires a tiny electric field imbalance to balance the centrifugal forces, but this imbalance is typically ignored. In general there is an electric field required to keep the electrons and ions in a quasi-neutral configuration despite the fact that they experience quite different pressure gradient, gravitation, and centrifugal forces, but because this electric field acts to balance these forces, it drops out of the drift equation when the magnetic field is put into the problem.

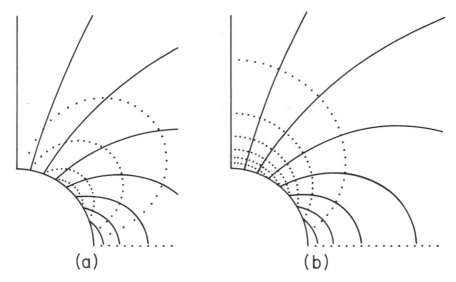

(a)　　　　　　　　　(b)

FIG. 2.1. Magnetic and electric field lines about an aligned rotator. Solid lines are the dipole magnetic field lines, while the dotted lines are the electrostatic field lines, (a) for the vacuum case and (b) for the Goldreich-Julian case. From F. C. Michel, 1982, *Rev. Mod. Phys.*, 54, 1 (figure 4).

Thus the induced electric field causing the electrons and ions to corotate through the magnetic field is precisely the same for each.

As should be familiar by now, the central idea for pulsar action invokes the above rotationally induced electric field and the responses of charged particles in the vicinity to the combined electric and magnetic fields. A conducting star rotating through its own (dipole) magnetic field creates an induction (quadrupole) electric field in a vacuum which would initially give

$$E_{||} \equiv (\mathbf{E} \cdot \mathbf{B})/B = a\Omega B_0 (a/r)^4 \cos^3 \theta. \tag{1}$$

However, equation 1 gives a force greatly exceeding gravity on any charged particle near the pulsar surface (see table 2.4, which summarizes the numerical estimates in this section). Thus the electrostatic forces are expected to ultimately create a space charge such that $\mathbf{E} \cdot \mathbf{B}/B^2 \to 0$, which in turn suggests that the magnetic field lines become equipotentials. The magnetic (solid) and electric (dotted) field lines for a rotator in a vacuum are shown in figure 2.1a, while figure 2.1b shows the modified electric field lines if the magnetic lines become equipotentials and the appropriate space charge surrounds the rotator. If the magnetic field lines are equipotentials, then it immediately follows that the plasma motion (which is just the $\mathbf{E} \times \mathbf{B}/B^2$ drift velocity, neglecting any motion along field lines) corresponds to *rigid corotation* of the plasma with the pulsar. This result is easily demonstrated as shown in figure 2.2. Here two magnetic field lines are selected differing in electrostatic potential by $\Delta\Phi$ and confining a

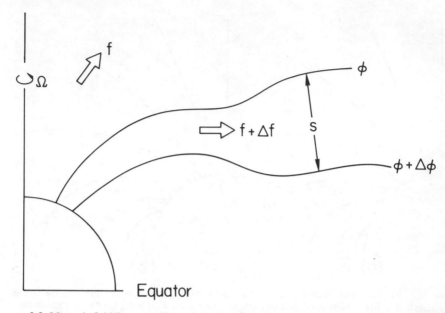

FIG. 2.2. Magnetic field lines as equipotentials. The field lines can be labeled, given axial symmetry, by the total magnetic flux (f) that would be enclosed by rotating the field line about the axis. The flux Δf between two field lines of potential difference $\Delta \Phi$ is therefore geometrically in a fixed ratio along each field line. From F. C. Michel, 1982, *Rev. Mod. Phys.*, 54, 1 (figure 5).

magnetic flux Δf between their respective surfaces of revolution (often termed "magnetic shells"). If the perpendicular distance between the two is s (at some arbitrary point), then it follows that

$$E \approx \Delta \Phi / s \qquad (2)$$

and

$$B \approx \Delta f / sr \sin \theta. \qquad (3)$$

Thus the azimuthal drift velocity is just

$$V_\phi = E/B \approx r \sin \theta \qquad (4)$$

and the constant of proportionality is determined at the surface by the condition that any free conduction electrons move with the surface material; thus we obtain the important results

$$V_\phi = \Omega r \sin \theta \text{ (rigid corotation)} \qquad (5)$$

and, moreover,

$$\Delta \Phi = \Omega \Delta f. \qquad (6)$$

(Note that the magnetic vector potential is just $A_\phi = f/r \sin \theta$.) These results are sometimes termed the "Ferraro law of isorotation" (see Ferraro and Plumpton 1961). Equation 6 looks innocuous enough, but it has proven to be a difficult boundary condition to satisfy, as we will detail. Rigid corotation must fail, even for massless particles, if $V_\phi \geq c$ or at an axial distance

$$\rho = r \sin \theta \geq c/\Omega, \tag{7}$$

which defines the *light-cylinder* distance (Gold 1968). The most elementary and fundamental requirement of any electrodynamical system is that it conserve current. In a steady state system, then, the net current to or from an object must be zero. It turns out to be quite difficult, however, to theoretically model a rotating magnetized star, particularly if it has an aligned dipolar field, to give a closed current flow. The basic problem is that, while electrons (say) could easily be ejected to infinity along polar field lines, the positive particles are instead injected on the strong closed field lines near the star. Evidently, then, the system would become positively charged. One would expect this positive charge to modulate the emission over the polar caps; however, once the stellar charge became large enough to do that, electrons would be unable to escape the system (they could be ejected from the surface, but not with sufficient energy to escape the system).

There is growing support, in fact, for the view that the current closure problem has no solution because the system simply ends up trapping both electrons and ions (or positrons). Let us consider, for example, what might take place near the light-cylinder. If $E \geq cB$, it simply follows that the particles would cross equipotential surfaces. The canonical assumption of the standard model is that this does not happen (a point which we will return to) and therefore the field lines are obliged to acquire an azimuthal component with distance, which adds to B^2 without changing $\mathbf{E} \times \mathbf{B}$ and thereby keeps V_ϕ less than c. If this did not happen, the strong currents from particles crossing field lines would presumably act to create such a component. Such a magnetic field component requires currents to flow in the meridional plane and therefore to flow parallel to the magnetic field lines in that plane. However, these currents cannot flow on closed magnetic field lines (e.g., they cannot flow out of one hemisphere, follow a closed dipolar field line back to the other hemisphere, and finally flow through the star to the original hemisphere) in the absence of an electromotive force to counter ohmic dissipation in the solid body of the pulsar. It would then seem to follow that currents flow only on open field lines, and that all field lines that cross the light-cylinder are open. We mention below some counterarguments.

It is traditional to separate radiation fields from static fields, even though the distinction is not always sharp. For example, one calculates the structure of the hydrogen atom using just the static Coulomb field and then one calculates the lifetimes of excited states by treating the coupling to the radiation field as a perturbation. A few pulsar models are, in effect, based on the argument that such separation is unphysical. In the model of Mestel et al. (1979), for example, it is proposed that a closed current can flow out from the neutron star and return to it. To do this, the particles are assumed to cross magnetic field lines near the light-cylinder as a result of radiation reaction there. If the coupling to the radiation field could be "turned off" (e.g., by reducing e/m of the particles), this mechanism would be ruled out. Jackson (1981a,b) has also proposed a model with closed currents flowing as a consequence of radiation, but not necessarily radiating just at the light-cylinder (see also Jackson 1976a,b and Rylov 1978). Such models pose an interesting challenge to physical intuition. Is it possible to find examples of such radiational "bootstraps," namely dynamic systems that function and radiate only because they radiate? The resultant structure of the standard model, with a corotation zone of closed field lines and a magnetized stellar wind flowing out of the polar caps, is shown schematically in figure 2.3. Mestel (1966) had already given a quite similar qualitative picture.

Note that the power output can be calculated from two different points of view. Consider a Faraday disk rotating in a magnetic field **B** that is being shunted (e.g., figure 1.16). If the current is i, then it flows radially along with the radially induced electric field $E = \Omega r B$ and therefore one is expending power at a rate $\delta P = i\delta\Phi = iE\delta r = i\Omega Br\delta r$. Equivalently, a $\mathbf{J} \times \mathbf{B}$ force is exerted on the disk with $F = Bi\delta r$ corresponding to a torque $T = rF$ and a power output $\delta P = \Omega T = i\Omega Br\delta r$. Several important results can now be obtained without more detailed analysis.

2. Magnetized Stellar Wind

It is clear from figure 2.3 that, as a first approximation, we can take the corotation zone boundary to be dipolar, in which case the magnetic flux may be estimated from

$$f = f_0\rho^2/r^3 \tag{8}$$

and for the first field line to the light-cylinder, the axial distance is $\rho = c/\Omega = r$; thus

$$f_E = \Omega f_0/c \tag{9}$$

is the magnetic flux escaping to infinity from each polar cap. The polar magnetic field strength at the pulsar is

$$B_p = 2f_0/a^3, \tag{10}$$

where $r = a$ is the radius of the pulsar. If we look down from a pole at the field lines, we see that they must spiral (to keep $V_\phi \leq c$ as shown in figure 2.4), since the plasma must flow radially outward at velocity slightly less than c. Since the magnetic flux f_E is trapped between consecutive spirals, it is clear

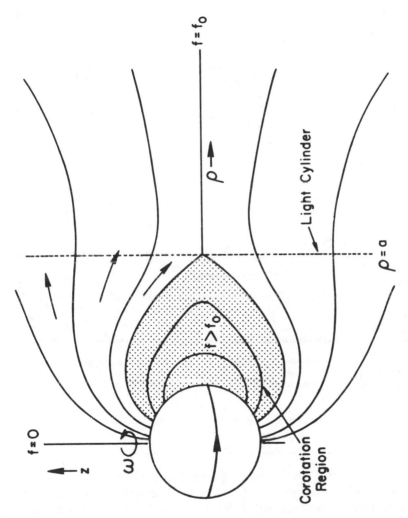

FIG. 2.3. Hypothetical aligned rotator magnetic fields. Dashed vertical line locates the light-cylinder. The field line f_0 is the "last open field line." Shaded region contains the closed field lines. From F. C. Michel, 1974b, *Ap. J.*, 187, 585 (figure 1).

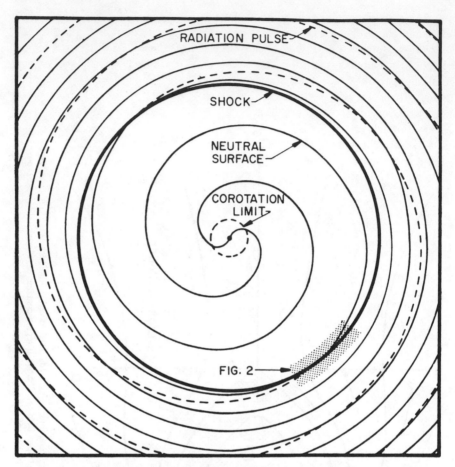

FIG. 2.4. Spiral asymptotic field structure projected on equatorial plane, showing spiraling of the magnetic field lines (from Michel and Tucker 1969). Reprinted by permission from *Nature*, 223, 277 (figure 1). Copyright © 1969 Macmillan Magazines Ltd.

that the magnetic field becomes azimuthal and

$$B \to B_\phi \approx \Omega f_E / cr = \Omega^2 B_p a^3 / 2c^2 r, \tag{11}$$

where cr/Ω is roughly the area through which the flux must pass (we are neglecting, of course, factors of order unity which describe the distortion of the corotation region and the detailed distribution of the flux in the meridional plane). Thus, for example, the pulsar in the Crab Nebula should contribute a nebular field of about

$$B_{\text{neb}} \approx 2 \times 10^{-5} \text{ gauss} \tag{12}$$

for $B_p = 10^{12}$ gauss, $a = 10^4$ m, $\Omega = 200$ radians/s, and $r = 1$ lt-yr $= 10^{16}$

m. The magnetic field could be greatly enhanced by shock and magnetohydro-dynamic compression as the wind interacts with the nebular material (Michel 1969a; Piddington 1969; Rees 1971b; Pacini and Salvati 1973; Rees and Gunn 1974). The nebular magnetic field can also be estimated from the spectrum and intensity of the continuum synchrotron radiation from the nebula; the total energy (particles plus field) is a minimum for $B \approx 10^{-3}$ gauss and rapidly rises for either very much weaker or very much stronger fields (Burbidge 1956). This method of analysis has become a standard one for estimating the energy stored in a synchrotron source. Moreover, the above analysis for the nebula concludes that the radiating electrons must have an energy of about 10^{11} eV, which is also that obtained below (eq. 29), which reinforces confidence in the general approach.

3. Torque on the Pulsar

We have implicitly ignored the energy carried by the particles; thus the total energy-loss rate is at least that in the Poynting flux away from the pulsar, or

$$ P = \mu_0^{-1} \int (\mathbf{E} \times \mathbf{B}) \cdot d\mathbf{S} \approx cB_\phi^2 r\pi r^2 / \mu_0 = \pi \Omega^4 B_p^2 a^6 / \mu_0 c^3, \qquad (13) $$

which is, apart from a factor of 2/3, also the electromagnetic power that would be radiated by a magnetic dipole moment $B_p a^3$ rotating orthogonal to the moment axis. Separate calculations (sec. 9.2) indicate that the particles may in fact carry relatively little energy. Thus one result is that the net torque is largely independent of the magnetic moment/spin orientation and is proportional to Ω^3. For an orthogonal magnetic moment the outflow is in the form of large-amplitude electromagnetic waves while for the aligned case it is supposedly in the form of a stellar wind. (The same scaling argument follows for a stellar wind as for the dipole radiation case, since in either case the asymptotic flux to infinity would be those field lines crossing the light-cylinder; however, we will see that there are some problems with existing pictures of wind production.) The expected slowing-down behavior is then

$$ \dot{\Omega} \approx \Omega^n, \text{ with } n = 3. \qquad (14) $$

Kaplan et al. (1974) have argued against this view, citing effects from turbulent, ultrarelativistic plasma near the pulsar and obtaining $n = 3.4$.

4. Nebular Excitation (Crab Nebula)

In the frame of reference of the (essentially stationary) nebular material, the pulsar wind contains an electrostatic field

$$ E \approx cB, \qquad (15) $$

and consequently dissipative currents will be excited which act to stop the flow (and accelerate the nebular material). As a result, a substantial fraction (if not all) of the pulsar wind energy should be dissipated in the nebula (see sec. 2). In the case of the Crab pulsar, there is in fact good agreement between the total luminousity of the nebula itself and the inferred total luminosity of the pulsar ($I\Omega\dot{\Omega}$, where I is the moment of inertia of the neutron star), both being about 10^{38} ergs/s. Energy may be deposited to accelerate the expansion of the nebula (Trimble and Rees 1970). The total pulsed luminosity of the pulsar, largely in X-rays and gamma-rays, is about 10% of this grand total, whereas the pulsed radio luminosity is only about 10^{-8} of the grand total. The radio emission per se is consequently only a minor factor in the overall energy budget of this and most other pulsars (10^{-5} is more representative). The nebula is thought to be a hollow shell with the synchrotron-emitting region confined within it (Clark et al. 1983; Michel et al. 1990).

5. Particle Injection Rate

If the field lines are equipotentials, the plasma corotates with

$$\mathbf{V} = \mathbf{\Omega} \times \mathbf{r}. \tag{16}$$

Thus we can immediately write

$$\mathbf{E} = -\mathbf{V} \times \mathbf{B}, \tag{17}$$

and

$$q = \epsilon_0 \nabla \cdot \mathbf{E} = 2\epsilon_0 \mathbf{\Omega} \cdot \mathbf{B} \tag{18}$$

is the space charge, corresponding to a number density of

$$n = q/e, \tag{19}$$

where e is the elementary charge of the plasma particles. Because the particle rest-mass energies are small compared to the electrostatic potential differences, the particles should flow at essentially c in response to these strong electromagnetic forces; thus the particle flux is nc and this flux flows along the open field lines. A dipole magnetic field line satisfies

$$f = \sin^2 \theta / r = \text{constant}; \tag{20}$$

thus the polar cap radius $\rho = a \sin \theta$ at $r = a$ is intercepted by the same field line that extends to the light-cylinder, $R_L = c/\Omega$, if $\rho^2 = \Omega a^3/c$. Thus the *polar cap* area is just

$$A = \pi\rho^2 = \pi\Omega a^3/c. \tag{21}$$

It is here that a very important and controversial assumption is made, namely that all particles having access to field lines beyond the light-cylinder distance are centrifugally expelled. Given that the particles are exposed to such huge electric fields, it is natural to assume that they move essentially at the velocity of light. With these two assumptions, the two polar caps would contribute a particle flux

$$\dot{N} = 2ncA = 4\pi\epsilon_0\Omega^2 B_p a^3/e \tag{22}$$

corresponding to 1.3×10^{34} particles/s from the Crab pulsar (here B_p is taken to be 4×10^{12} gauss, as required to give the correct total Crab luminosity). Just as for the magnetic field strength estimate itself, the particle-loss rates are dimensionally scaled to give the correct order of magnitude. They are not presently observed quantities, and factor of 2 uncertainties are more than likely. This particle-loss rate is an upper limit in the sense that fewer particles could escape or they could escape at a lower velocity, and it is a lower limit in that a somewhat arbitrary number of electron/positron pairs could be added to the flow. The latter uncertainty is of course limited by energetics because it takes about 1 MeV ($= 1.6 \times 10^{-6}$ ergs) to create the pair. An even greater problem is encountered if the positive charge carriers are ions, because it takes about 100 MeV just to lift an ion out of the gravitational well of the neutron star. For a run-of-the-mill pulsar with 10^{30} ergs/s of spin-down luminosity, the particle flux would then be limited to 10^{34}, which is comfortably above equation 22 (given the smaller Ω; see table 3.1) but limits the possible Lorentz factors in the wind to 10^3; depending on the model, this may or may not be a problem. If the ions were singly charged iron, the energy required would be multiplied by another factor of 50.

The much larger flux from the Crab of 10^{41} electrons/s that is frequently quoted (Shklovsky 1968, 1970a, 1977) is obtained by assuming that the energy input into the nebula is entirely in the form of energetic electrons (or electrons and positrons). In the discussion here, to the exact contrary, most of the energy would be in the Poynting flux of the magnetized stellar wind (Michel 1975d). In this latter case the energetic electrons could be produced by local reacceleration, energy being transferred from the electromagnetic field to the particles as a result of the wind stagnation upon interaction with the surrounding nebular shell. The Shklovsky interpretation is also difficult to reconcile with the pairs being produced near the pulsar, because the nebula is also strongly magnetized and that field also presumably comes from the pulsar. Thus, to get acceptable energetics, one needs something close to equipartition of energy between the magnetic fields and particles in the outer nebula. Equipartition is plausible if the energy is locally being transferred from the Poynting flux back to the particles. But if the particles are created and energized at the pulsar, how would they "know" what numbers and energies would be required for them to end up in rough equipartition with the remnant magnetic field a light-year away? A number of quantities pertinent to the pulsar problem are summarized in table 2.4.

TABLE 2.4—**Physical Parameters of the Standard Model**

Quantity (at surface)	Estimator	Crab	Typical
Magnetic field (B)	—[a]	4×10^{12} gauss	10^{12} gauss
Rotation rate (Ω)	—[a]	200 radians/s	6 radians/s
Radius (a)	—[a]	10^6 cm	10^6 cm
Mass (M)	—[a]	$1.4\,M_{\odot}$	$1.4\,M_{\odot}$
Moment of inertia (I)	—[a]	10^{45} g cm^2	10^{45} g cm^2
Electric field (E)	ΩaB	8×10^{12} V/cm	6×10^{10} V/cm
Polar cap area	$\pi\Omega a^3/c$	2×10^{10} cm^2	6×8^{10} cm^2
Polar cap radius	$(\Omega a^3/c)^{1/2}$	8×10^4 cm	1.4×10^4 cm
Gravitational/electric force	$m_e g/eE$	10^{-14}	10^{-12}
Pole to equator potential	$\Omega a^2 B/2$	4×10^{18} V	3×10^{16} V
Potential across polar cap	$\Omega^2 a^3 B/2c$	3×10^{16} V	6×10^{12} V
Electron cyclotron frequency (ω_c)	eB/m_e	7×10^{19} radians/s	1.8×10^{19} radians/s
Electron concentration (n_e)	$2\epsilon_0\Omega B/e$	9×10^{12}/cc	7×10^{10}/cc
Electron plasma frequency[b] (ω^p)	$(e^2 n_e/\epsilon_0 m_e)^{1/2}$	1.7×10^{11} radians/s	1.7×10^{11} radians/s
Alfvén velocity[c]	$(2\omega_c i/\Omega)^{1/2}c$	$4 \times 10^7 c$ (Fe)	$10^7 c$ (Fe)
Particle flux[d] ($\dot{N}$)	$4\pi\epsilon_0\Omega Ba^3/e$	1.1×10^{34}/s	2.5×10^{30}/s
Slowing-down rate ($\Omega/\dot{\Omega}$)	$3I\mu_0 c^3/8\pi B^2 a^6 \Omega^2$	1340 yr	2×10^7 yr

a) Input assumption or observation.
b) An equivalent expression is $\omega_p^2 = 2\Omega\omega_c$; hence $\omega_p \ll \omega_c$.
c) Here we use the ion cyclotron frequency; singly ionized iron ions give the lowest value.
d) A simple equivalent is (Michel 1978b) $L_T/i^2 = 15$ watts/amp^2, where L_T is the spin-down luminosity and i is the net current ($\equiv e\dot{N}$).

6. Electron Energies: Space-Charge Limited Flow

As we have just seen, a plausible estimate can be derived for the injection rate from the pulsar. Given this flux, we can estimate the particle energies. For a vacuum tube diode, there is a fixed current that can flow in response to a fixed plate potential, regardless of how hot the filament is. In the same way, the current flowing from a pulsar is directly related to the accelerating potential; the assumption that plasma be freely available from the surface does not imply that unlimited currents can flow. We can simply invert this fact here and use the current (eq. 22, essentially) to calculate the accelerating potential. In other words, the field lines are equipotentials only if sufficient space-charge plasma is present, and if not, an appropriate acceleration potential develops to provide it. Knowing the loss rate then allows one to estimate the required potential and hence the particle energy.

a. *Vacuum Tube Analogy*

Sturrock (1971a) using exactly the above analogy overestimated the electron energy. If in a one-dimensional geometry one has a current J_0 of relativistic electrons, the space-charge density is just J_0/c and the accelerating potential is then

$$\Phi = J_0 h^2 / 2c\epsilon_0, \tag{23}$$

where h is the linear distance (here height above the polar cap surface). One now estimates h to be the radius of the polar cap, using equation 21. The argument is that geometric divergence limits the validity of equation 23 to this distance, which is just an approximate way of solving the actual three-dimensional electrostatics (another "astrophysical" approximation).

b. Differential Space Charge

The above method overestimates the potential (Michel 1974c) because the "current" J_0 was estimated in the first place by multiplying the space-charge density by c. But this density is already consistent with the condition $\mathbf{E} \cdot \mathbf{B} = 0$, hence with zero accelerating potential (Tademaru 1974). Thus the accelerating potential arises only from deviations from this charge density, not from the charge density itself. The basic consequence is that the potential does not increase quadratically, but only linearly. In fact, the differential space charge is almost all confined to a thin sheath (Buckley 1976) at the surface within which the particles are accelerated to relativistic velocities. Fawley et al. (1977) have carefully reviewed and reconfirmed this (Michel 1974c) result. The derivation is straightforward using the Ansatz that it is only a charge difference from the space-charge density (now q_0) that leads to an accelerating potential. We have then, in one dimension, Poisson's equation for the accelerating component of the electric field:

$$\frac{d^2\Phi}{dx^2} = (q_0 - q)/\epsilon_0. \tag{24}$$

If we write the particle energy as $e\Phi = (\gamma - 1)mc^2$ and q_0 as $2\epsilon_0 \Omega B$, we have a natural length scale

$$\lambda = (mc^2 / re\Omega B)^{1/2}, \tag{25}$$

which is less than about 1 mm using Crab pulsar parameters. Electrons become relativistic in traversing this small distance. The resultant equation reads, writing $d\gamma/dx$ as p and $d^2\gamma/dx^2$ as $\frac{1}{2} dp^2/d\gamma$,

$$\lambda^2 \frac{dp^2}{d\gamma} = \frac{q}{q_0} - 1. \tag{26}$$

If we set $q \to q_0$ at large distances, the particles having attained an asymptotic Lorentz factor γ_0, the right-hand side vanishes, giving asymptotically, using

$q/q_0 = \beta_0/\beta,$

$$(\lambda p)^2 = \int_1^{\gamma_0} (\beta_0/\beta - 1)\, d\gamma = 1 - \frac{1}{\gamma_0} \approx 1. \tag{27}$$

What has happened is that the acceleration of electrons from rest to relativistic velocities always leaves behind a charge layer wherein $|q| > |q_0|$, and the electric field from this charge layer is not canceled out because $q \to q_0$. In three dimensions this electric field is not constant but vanishes as $r \to \infty$, while in one dimension one must limit the validity of equation 27 somehow, and a reasonable estimate is to take, as before, a height of about one polar cap radius

$$h_0 \approx (a^3 \Omega/c)^{1/2}, \tag{28}$$

which is about 10^4 cm for the Crab pulsar. Then the maximum Lorentz factor can be estimated from equation 27 by writing $p \approx \gamma_0/h_0$ and solving for γ_0,

$$\gamma_0 = h_0/\lambda \approx 10^5, \tag{29}$$

which completes the argument and gives an estimate of the particle injection energy.

c. Effect of Pair Production

Here we show that equation 29 will be an upper limit if pair production is included. We add pair production to the model by assuming that at some height h_2 pairs are formed by *magnetic pair production* in the strong pulsar magnetic field by hard gamma-rays produced owing to *curvature radiation* from the relativistic electrons as they follow the strong magnetic field lines (*synchrotron radiation*, in contrast, is caused by electrons circling the field lines). This process produces relatively low energy positrons, some of which will be halted and accelerated back toward the surface to form a downward flux of relativistic positrons of density

$$\delta q = -q/2\gamma_2, \tag{30}$$

where γ_2 is an undetermined constant to this point. We will see below that the ejected particles end up with a Lorentz factor of order γ_2. If one imagines that one has "copious" pair production as Ruderman and Sutherland (1975) discuss (i.e., $\delta q \gg q_0$), one finds an impossible situation if any significant fraction of these particles return to the surface, since the space charge would now have the wrong sign above the surface, a problem already recognized by Sturrock (1971a). On the other hand, if one increases q_0 to keep the space charge negative, the charge density above h_2 is now not only larger than q_0, but much larger and we cannot asymptotically approach q_0 beyond the acceleration region as before. Thus, for a steady state solution, one can only tolerate a small downward positron flux (assuming throughout that the primaries are upward moving electrons); hence we must have $\gamma_2 \gg 1$, and indeed we find that the system quite naturally achieves such a condition. In equation 27 we assumed that at $\gamma = \gamma_0$ the asymptotic condition had been reached. With pair production at height h_2, we have a charge density $q - \delta q$ below this altitude (primary electrons minus secondary positrons) but a charge density $q + \delta q$ above (primary plus secondary electrons). It therefore follows that we can apply the "asymptotic" condition simply by setting $q + \delta q = q_0$ (hence there is no acceleration above h_2), which in turn requires the charge density in the accelerating region to be

$$q - \delta q = q_0 + 2\delta q = q_0 \left(1 - \frac{1}{\gamma_2}\right). \tag{31}$$

As a consequence, we get a mixture of the first two theoretical treatments, with some of the accelerating field due to the surface sheath and the rest due to a small, fixed deviation from the space-charge density in the region below h_2 from the down-flowing secondaries, namely

$$\lambda^2 p^2 = 1 - \frac{\gamma}{\gamma_2}, \quad \gamma < \gamma_2 \quad (h < h_2). \tag{32}$$

Integrating equation 32 ($p \equiv d\gamma/dx$) gives

$$\gamma_2 = h_2/2\lambda. \tag{33}$$

Now, however, the relationship between h_2 and γ_2 is determined by the detailed physics of the pair production, as shown in figure 2.5. The dashed line is the curve $\gamma = h/\lambda$, namely the acceleration just from the negative current sheet, while the curved line indicates the moderating effects of the intervening excess positrons. Notice that we are avoiding actually trying to determine from first principles where the transition takes place; rather, we will simply deduce what happens regardless of the location. If $h_2 \gg h_0$, the pairs are unimportant since they all escape, and the particles are accelerated to γ_0 as before. If $h_2 \ll h_0$, pair production is the controlling factor and particles are accelerated only to $\gamma \approx \gamma_2$; hence pair production reduces, if anything, the particle energy. The above is a heuristic version of the work by Arons and Scharlemann (1979).

In summary, Sturrock's (1971a) way of handling space charge is to assume that the particle sees progressively more space charge between itself and the star as it departs; hence $\Phi \approx h^2$. With the $\mathbf{E} \cdot \mathbf{B} = 0$ correction, the particle sees only a fixed sheath of space charge covering the stellar surface; hence $\Phi \approx h$. If pair production is included, one simply imposes an additional limitation owing to a tiny excess volume charge from down-flowing charges of sign opposite to the sheath; namely Φ will increase only to the point where pair production is initiated and is then terminated ("poisoned" in the view of Arons and Scharlemann 1979).

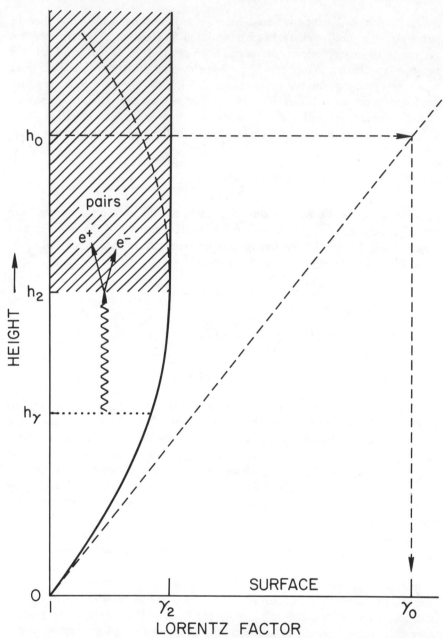

FIG. 2.5. Space-charge flow limitation by pair production. Electrons gain energy $(\gamma - 1)mc^2$ as they accelerate to height h (solid curve). At height h_γ the curvature radiation first produces photons, which are energetic enough to be converted into pairs at h_2. Because only a very small downward flux of positrons can be tolerated, the accelerating field must essentially vanish quite close to h_2 in order that positrons produced at and above h_2 are not returned. Consequently the electron energy never reaches the value γ_0 it would have attained if acceleration all the way to h_0 had been possible, as illustrated. For $h_0 > h_2$, pair production becomes unimportant in limiting γ, and for $h_0 < h_\gamma$, there is little or no pair production at all. From F. C. Michel, 1982, *Rev. Mod. Phys.*, 54, 1 (figure 8).

d. *Distributed Space Charge*

We have emphasized the role of a thin polar cap surface charge whose main effects are confined within a small distance comparable to the size of the polar cap regions (typically a few hundred meters). Arons and Scharlemann (1979) develop the role of a quite different, widely distributed component deviating from the space-charge density which results from the curvature of the magnetic field lines. A relativistic flow of space charge along magnetic field lines is only able to satisfy the static space charge required by the force-free condition $\mathbf{E} \cdot \mathbf{B} = 0$ for field lines that are straight. For field lines that curve away from the rotation axis, the case for all the field lines in a perfectly aligned rotator, the flowing space charge exceeds more and more that required for the magnetosphere to be force-free. The natural conclusion is that the resultant electric field acts to halt the flow in this case. For field lines of the opposite curvature, one expects an accelerating field, and in this way it is proposed that the particles end up crossing a net potential comparable to that across the polar cap rather than the very small potential implied by local conditions near the polar caps. Larger potential drops imply larger particle energies, thereby enhancing the possible importance of pair production. The above consideration is important, particularly because it emphasizes the global nature of the magnetospheric physics, despite the natural hope of breaking the problem down into manageable elements. It is therefore essential that the full global problem be solved. Indeed, solutions with no flow of space charge from an aligned rotator exist (sec. 4.2.2).

e. *High-Current-Density Solutions*

It might appear by now that there is no way to get large currents to flow into the polar cap regions. If no way exists, any model that appeals to concentrated current flows would be in trouble, yet such flows seem physically plausible. Indeed they are, for one simply requires a counterstreaming current flow, the usual case for electric arcs in the laboratory. The problem with producing such a counterstreaming with positron production is that the counterstreaming is essentially concentrated near the surface, resulting in reduction in the space charge near the surface and amplification further from the surface, as discussed above. However, if there were a separate source of particles (or if the positrons were produced by some other mechanism far from the surface), then the entire beam could be neutralized with the result that the current density could be quite large without there being a large concomitant charge density. Such flows would require either a source of new particles or some "target" on which to convert positrons. A nearby disk of matter could, for example, serve either function.

7. Pair Production

Ever since Sturrock's work, people have been convinced that pair production plays an essential role in pulsar action (see, for example, Arons 1979, 1983a). It would be surprising if pair production did not take place in the vicinity of pulsars, but the conclusion that it is essential remains model dependent. We can now calculate one condition for pair production to be important, because (figure 2.5) the electrons must create hard enough photons at h_γ to produce pairs at h_2. The electron Lorentz factor at h_γ is just

$$\gamma_\gamma = \gamma_2(1 - \Delta^2/h_2^2), \tag{34}$$

where $\Delta \equiv h_2 - h_\gamma$, and such an electron will radiate photons with energy (in units of $m_e c^2$) up to

$$\gamma_p = 3\xi\gamma_\gamma^3\lambda_c/\rho_c, \tag{35}$$

where $\lambda_c(\equiv hc/2\pi m_e c^2 = 4 \times 10^{-11}$ cm) is the electron Compton wavelength and ρ_c is the field line curvature. The factor ξ is of order unity and parameterizes a slight uncertainty over what constitutes a "significant" flux of photons, because the synchrotron spectrum still extends somewhat beyond the critical frequency. In principle, we could calculate ξ self-consistently, but an exact value is not essential here (ξ is most probably between 1 and 3). Magnetic opacity has the property (see sec. 2.7) that absorption goes almost discontinuously from zero to "infinity" at the critical condition

$$\gamma_p B_p \sin \zeta = B_{\text{crit}} = 4.4 \times 10^{12} \text{ gauss}, \tag{36}$$

where B_{crit} was introduced in sec. 1.7.4. Here B_p is the polar cap magnetic field and ζ is the angle between the photon and the local field orientation. Photons are created with $\zeta \approx 1/\gamma_p$ and hence would never be reabsorbed in a typical pulsar field were it not for the field line curvature ρ_c, which permits ζ to grow with distance as the photon propagates. As a consequence, after going the distance Δ they satisfy

$$\sin \zeta \approx \Delta/\rho_c \tag{37}$$

and equation 32 becomes

$$y(1 - y^2)^3 = \rho^2/l^2, \tag{38}$$

where $y = \Delta/h_2$ and the numerical constants (having units of distance squared) are packaged into l^2. Using equation 33 to eliminate γ_2 then gives

$$l^2 = 3\xi B_p\lambda_c h_2^4/8\lambda^3 B_{\text{crit}}. \tag{39}$$

Since the left-hand side of equation 38 has a maximum value of 0.238... at $y = 7^{-1/2}$ we have a condition on the curvature, namely that if

$$\rho_c \leq l/2, \text{ pair production obtains}, \tag{40}$$

$$\rho_c \geq l/2, \text{ no pair production obtains}$$

(we have approximated the square root of the maximum, 0.48787..., by 1/2).

Because $h_2 < h_0$ if pair production is to be important, we can replace h_2 with h_0 (hence with the polar cap radius) to obtain a liberal condition for pair production (i.e., pair production might still not be important even if the condition is met). Inserting Crab pulsar parameters into equation 39 then gives $l \approx 4 \times 10^2$ m, which would require tightly curved magnetic field lines (i.e., highly multipolar, corresponding to magnetic "spots," etc.) even for the Crab pulsar. As noted above, Sturrock (1971a) originally contemplated much more energetic particles as a result of his excessive estimate of the space-charge limit to the flow.

This analysis points to the unimportance of pair production because the accelerating potential has been assumed to be limited to the differential acceleration of equation 29. Not all investigators adopt this limitation. One route is to suppose that the unfavorable curvature of the field lines permits larger accelerating potentials (Arons and Scharlemann 1979; Arons 1981a), at least in fast pulsars like the Crab. For the more typical pulsars, nondipolar magnetic fields would be additionally required (Barnard and Arons 1982). In an alternative suggestion put forward by Ruderman and Sutherland the discharge is driven by breakdown for cases in which ions are what would be pulled from the polar caps. Assuming the ions to be tightly bound (an assumption that has become questionable of late), the potential drop would then self-consistently rise to whatever would be required to produce pairs. In its simplest form, this model would appear to have no space-charge limitation here (see figure 2.6), because one has equal numbers of upward-moving positrons and downward-moving electrons. However, this cancellation is not exact at the edges of the gap (just above the surface and just below h_2) and we have already seen that the acceleration sheath at the surface alone can exercise an important limitation (see Cheng and Ruderman 1977b).

8. Radiation Reaction

Because electrons can radiate copiously, it is sometimes supposed that radiation reaction will be the fundamental limiting process, robbing energy as fast as it can be delivered to the particles. This does not seem to be the case for the parameters under consideration. The radiation loss due to curvature radiation is

$$mc^2 d\gamma/dx = e^2\gamma^4/6\pi\epsilon_0\rho_c^2, \tag{41}$$

whereas the input rate is just mc^2/λ. Thus, equating the two gives an asymptotic radiation-reaction limited Lorentz factor (γ_r)

$$\gamma_r^4 = 6\pi\epsilon_0\rho_c^2 mc^2/\lambda e^2, \tag{42}$$

and for a conservative estimate ($\rho_c \approx 10^6$ cm, the neutron star radius), one obtains

$$\gamma_r \approx 3 \times 10^6. \tag{43}$$

This value is significantly higher than the space-charge limited flow value (i.e.,

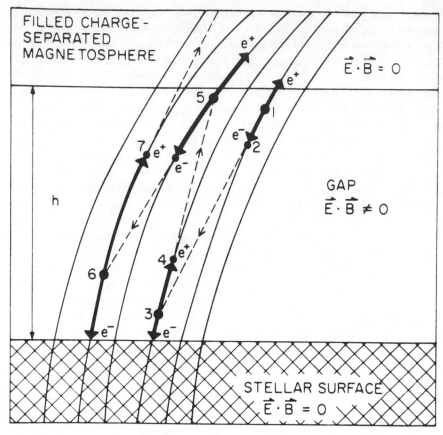

FIG. 2.6. Nature of a pair production discharge. All the potential drop must appear in the gap (*h*); otherwise the system would be flooded by downward-accelerated electrons. Above the gap the energetic primaries would continue to radiate and produce yet more pairs, resulting in a relatively dense pair plasma. The gap width would be maintained so that pair production within the gap would be kept just as threshold (otherwise the average number of particles there would exponentiate). Pair production takes place at points 1, 3, 5, ··· and gamma radiation at points 2, 4, 6, ···. Because the process requires curved field lines, it automatically "marches" toward the least-curved field line, suggesting that it would either extinguish itself or exhibit a relaxation type of oscillation (note event 6 is numbered inconsistently). From A. F. Cheng and M. A. Ruderman, 1977b, *Ap. J.*, 214, 598 (figure 1).

radiation reaction is unimportant for the selected parameters). If we replace the left-hand side of equation 41 with the full corotationally induced electric field in equation 1, we still find a limiting Lorentz factor of 7×10^7. Space-charge limitation and pair production are already more effective mechanisms at limiting the particle energy.

9. Frozen-in-Flux Concept

The field lines are taken to be equipotentials, an assumption worthy of comment. Within the context of the aligned rotator, where the system acts to null out any parallel electric field at the surface by emitting particles, *one* plausible end result is the Goldreich-Julian type of inner magnetosphere where one has a static space-charge distribution with $E_\parallel \equiv 0$ (see, however, sec. 4.2.2).

The frozen-in-flux concept (FIF) actually embraces two assumptions: (1) that the magnetic field lines are equipotentials, and (2) that the plasma only experiences $\mathbf{E} \times \mathbf{B}$ drift. When these two conditions are met, the magnetic flux through any arbitrary closed loop is constant even though the loop, each point of which is taken to be fixed in the local rest frame of the plasma, is distorted and displaced as the plasma circulates.

FIF is frequently assumed in treatments of the pulsar magnetosphere, and we have no specific criticism of that assumption. However, a few cautionary notes should be made.

a. *Equipotential Field Lines*

For the magnetic field lines from the stellar surface to be equipotentials everywhere is not possible; some potential drop is necessary to accelerate plasma from the surface. In the space-charge limit, this potential drop is tiny compared to the cross-polar-cap potential. Thus FIF seems a reasonable approximation. In the free-flow pair production theories (e.g., Ruderman and Sutherland 1975) the cross-polar-cap potential is also taken as the accelerating potential and FIF is violated everywhere except perhaps at large distances. In both cases, the particles are highly relativistic and consequently unresponsive to the presence or absence of perturbing parallel electric fields, which now have negligible effect on the motion and do not create a neutralizing space charge (unlike the quasi-neutral plasma approximation that quasi-*stationary* plasma particles are available to respond to stray electric fields).

Finally, it is not obvious that plasma fills the system, in which case vacuum gaps can open the circuit along the magnetic field lines. Such gaps are impossible in a quasi-neutral plasma but are the norm in nonneutral plasmas such as may surround neutron stars (sec. 4.2.2). Indeed, there must be some resistive or open circuit somewhere in such a system (Pilipp 1974).

b. $\mathbf{E} \times \mathbf{B}$ *Drift*

The assumption of $\mathbf{E} \times \mathbf{B}$ drift requires $E < cB$, and the latter condition is expected to be valid near the star (where $E \approx \Omega a B$ and $\Omega a \ll c$) but not necessarily at large distances. Indeed, the wind theories uniformly give $E \rightarrow cB$.

10. Relativistic Bending of Light

The gravitational bending of light near a neutron star can be substantial. For light rays going almost vertically away from the surface, the bending is negligible, but for rays making substantial angles to the surface, the bending can be pronounced. As long as the neutron star radius is larger than $3m$ ($m \equiv GM_{\text{neutron star}}/c^2 \approx 2.09$ km), or 6.3 km, all radiation emitted from the surface will escape to infinity. Because this limiting radius is smaller than those quoted for neutron stars (10 km being canonical), the issue is then the bending

TABLE 2.5—Light Bending for 1.41 $M_\odot$ and 10 km Radius

ϕ_0	ϕ_∞	Ratio
10	10.05	1.00
20	20.38	1.02
30	31.32	1.04
40	43.20	1.08
45	49.61	1.10
50	56.40	1.13
55	63.63	1.16
60	71.36	1.19
65	79.65	1.23
70	88.57	1.26
75	98.22	1.31
80	108.85	1.36
85	122.63	1.44
89	180.14	2.02

of emissions near the surface rather than whether these emissions can escape. It makes quite a difference for near-surface emission models whether the emissions are directed upward or at substantial angles to the surface. In the latter case there can be a sizable widening of the beam observed at Earth from the width it had at the origin (Pechenick et al. 1983).

The trajectory of a light ray emitted at angle ϕ_0 ends up making an angle ϕ_∞ at large distances. The light ray trajectory is given in a simple form in terms of inverse distance $u \equiv 1/r$ and polar angle ϕ, as (see Shapiro and Teukolsky 1983, p. 350)

$$(du/d\phi)^2 = k^2 - u^2 + 2mu^3, \tag{44}$$

where k^2 is determined from the boundary condition that the photon starts from the surface $u_0 = 1/a$ at an angle ϕ_0 which corresponds to $(du/d\phi)_0 = u_0 \cot(\phi_0)$. The asymptotic angle is then given by integrating the change in ϕ in going from $u = u_0$ to $u = 0$. Hence

$$\phi_\infty = \int_0^{u_0} du(k^2 - u^2 + 2mu^3)^{-1/2}. \tag{45}$$

For small angles the only important term is k^2, and the relativistic corrections are of the order of $2mu_0 \sin^2 \phi_0$, which is obviously small at small angles. At large angles, however, the corrections can become large. The relation between ϕ_0 and ϕ_∞ for a given range of angles is shown in table 2.5.

We can see working backwards that a 10° beam would have to be about 7° wide if emitted near 45° at the surface and only 5° wide if emitted near 70°. Note the large excursion at 89°. The photon capture cross section is $3\sqrt{3}m$, or 10.9 km, which is essentially the neutron star radius, so neutron stars are themselves "black holes" insofar as incident light on them is concerned. The importance of such corrections depends on the detailed model under consideration. Lindblom (1984) gives rather generous bounds on the gravitational redshifts that are possible (see also Riffert and Mészáros 1978). Chen and Shaham (1989) have extended these analyses to include the effects of rotation by using the Kerr metric.

2.4 Pulsar Statistics

1. The Pulsar "Garden"

Imagine visiting a friend who has a nice garden in the back, filled with flowers of various kinds. You notice that the white flowers seem to be taller than the rest. The friend agrees and wonders why. You suggest that perhaps the white flowers can grow larger because they don't have to devote energy to manufacturing pigments. Thus is born a "quasi-fact" (flowers grow taller if they are white), which has now been explained by a "theory" (white flowers are more efficient without pigment). Pulsar lore abounds with such arguments. But it is difficult even for a string of random numbers to appear exactly random; some degree of accidental correlation is statistically required. The flowers growing in a garden are even less likely to be random or to be representative of all flowers everywhere. In science, a curious thing happens when theory supports such quasi-facts. If the theoretical explanation is attractive enough, the "facts" are accepted even if the observational support is weak or nonexistent. This is the inverse of what is usually preached as part of the "scientific method," wherein one is supposed to be suspicious of facile explanations above all else. Which is not to say that all quasi-facts or their attendant theories are necessarily wrong. The problem is that the wrong ones detour workers into dead-end directions, often at just the time an open mind is called for. The current state of pulsar research suggests that the time for an open mind has come. Unfortunately once a quasi-fact has been *published*, it attains a constituency that nurtures and maintains it. In pulsar theory, the first plausible idea published has repeatedly become the accepted idea. Often both the theorist and observer are bound together in a symbiotic relationship; the observer is gratified that his results can be reduced to quotable results, and the theorist is gratified to have a theory that the observers say is correct. Entire symposia might be held on topics such as "Origin of Pigmentation: Reverse Evolution?"

A full list of pulsar quasi-facts would be boring to read, but a few can be listed for provocation: (1) all pulsars are made by type II supernovae, (2) pulsars radiate into a circular ("hollow cone") beam which sweeps by some observers and not others, (3) pulsars disappear because their magnetic fields decay away, (4) pair production guarantees pulsar action, (5) a rotating magnetized neutron star with a large magnetic field and fast spin is all that is required for pulsar action, (6) millisecond pulsars are old dead pulsars (hence weak fields according to quasi-fact 3) that have been spun back up by accretion from a companion,

(7) there is a "death line" corresponding to the values of P and $\dot{P}$ at which pulsars turn off. My guess is that at least one of these propositions is true. Many observers believe firmly in all! But none of the propositions are unambiguously supported by observational data. Instead they seem plausible ideas. To the degree that the observational data do not unambiguously *exclude* them, they are believed. Alternatives to the above? Some pulsars are made when accreting white dwarfs are pushed over the Chandrasekhar mass limit; pulsars radiate into a fan beam; pulsars will "disappear" anyway owing to a simple Ω^4 luminosity law (their accumulation at long periods does not overcome their declining brightness, as we will demonstrate); electron/positron pairs could well be generated but only incidentally; perhaps there are orbiting disks of debris around some or all active pulsars; millisecond pulsars could be formed in type I supernovae; and pulsars "die" simply when their luminosity is below the detection threshold. It is of course statistically unlikely that all of these alternatives are correct.

To reiterate the relationship between observation and fact, let us distinguish between three quite different entities: (1) observable parameters, (2) evolutionary parameters, and (3) theoretical parameters. The three are so intimately intermingled that one often cannot make sense of which is which in some discussions of the data.

The *observational parameters*, of course, are those that can be observed. Ideally one would like to observe every single neutron star in the galaxy (perhaps 10^8 of them) and tabulate the period, period derivative, position in space, radio luminosity, beam shape, size, mass, etc., of each. We have no such list and probably never will. Instead, we have a list of 450 or so nearby radio-bright pulsars, obviously a highly selected sample from the conceptually complete list. The vast majority of the neutron stars are presumably slowly rotating radio-dim objects. Whether they are "off" is a moot point; even the planet Jupiter functions as a weak pulsar (sec. 7.2).

Next come the *evolutionary parameters*, required but unknown because our list of pulsars is essentially a snapshot of an evolving population (otherwise the period derivatives would be zero). The plausible operational assumption is that the general distribution of properties in this list is stationary, which cannot literally be true because the list is constantly growing as the sky is searched more thoroughly and with greater sensitivity. Given a stationary distribution, *all* the evolutionary parameters must be inferred. That is to say, there is no practical way to actually observe how individual pulsars move through the P-$\dot{P}$ diagram (figure 1.6), or even how they flow through it as an ensemble. Even if the list were complete and unbiased, which it is not, we would have this same difficulty. Consequently, one must make models with certain evolutionary parameters.

To understand the $P - \dot{P}$ distribution we would need (1) the pulsar "birth" rate, (2) the distribution of "start" values of period and period derivative, (3) the time evolution of the spin-down rate, (4) the time evolution of the radio luminosity, (5) the spatial velocity acquired at birth, and (6) the spatial distribution of progenitors. Each of these factors must be represented by one or more adjustable parameters. In actuality these are neither adjustable parameters nor *observable* parameters when one is limited to a snapshot. Fortunately, we do better than just having a snapshot because we do have the period *derivatives*; otherwise there would be no objective way to assess a time scale. Given any distribution, the various parameters can be modeled, but the question is whether those values mean anything (values can always be adopted). These parameters are apparently "orthogonal" in the sense that one could imagine any one parameter being changed without directly influencing the others (e.g., the velocity in (4) could, as a *parameter*, be changed without requiring the others to be changed; a complete theory would not allow such fiddling, but that is a different level of analysis).

The major unadvertized difficulty with determining the evolutionary parameters is that they are *not* independent when they must be determined from a stationary data set. As a trivial example, consider the birthrate. This parameter can be halved if the evolutionary rates are also halved. The resulting model would give the same distribution, but the period derivatives would now be changed by a factor of 2 *on some average*. That does not mean that the period derivative for any given pulsar would be noticeably incorrect, because a statistical model does not (cannot) track the actual objects. But the biased sample also biases the averages. In the full (nonexistent) sample, we expect the period derivatives of all the pulsars in the galaxy to range from values somewhat larger than in our finite subset to a vast number of very tiny values, essentially all of which are missing from our subset. So unless we know precisely how the selection effects reshape the actual observed set of period derivatives, we do not know how to define the appropriate average!

It is important to note that pulsars are no longer discovered by wiggles on a chart recorder. Rather, the data (which at the sensitivity limit are probably indistinguishable from noise) are fed into a computer program which searches through steps in dispersion measure (necessary to undisperse the signal to give a potentially detectable pulse) and steps in period to see if a statistically significant signal exists. What biases might unintentionally exist in these complicated programs are difficult to characterize. Generally, observers are gratified if known pulsars are redetected in a new survey, but given that the program itself may well be a clone of the one used in the original survey, it is not necessarily certain that missed pulsars are not still missed. Consequently the lower luminosity limit for pulsar detection must be somewhat fuzzy and indistinct.

The finite sample chops off the rare pulsars with large values of period derivative while luminosity selection chops off those with small values. Simply performing some average (linear, geometric, or whatever) and matching it to some model allows one to quote a value for, say, the birthrate. But without solving the selection problem, one does not know how close this model value might be to the actual value. The statistical uncertainties (if quoted) can be insignificant compared to these systematic errors. One can model the selection effects themselves, but to do this we need (4) the time evolution of the radio luminosity to understand how pulsars are selected out on the basis of their apparent luminosity, and (5) the spatial velocity acquired at birth and (6) the spatial distribution of progenitors to understand to what degree the local population density is diminished by pulsars drifting away from the galactic plane. The luminosity function could be obtained immediately from a full list of all pulsars, but of course in the luminosity-selected list it must be inferred indirectly. Items (5) and (6) are almost hopelessly intercorrelated with one another and with (4). A pulsar with a spatial velocity of 100 km/s perpendicular to the galactic plane and a lifetime of 10^6 years will glide upward a distance of 100 pc (neglecting the deceleration of gravity toward the plane). Thus, even if pulsars were "shot" out of the galaxy entirely, they would still appear to be confined to the galactic plane because they would fade out before getting to extreme heights above the galactic plane (z). One can see even from the abbreviated list above that the number of adjustable parameters is becoming rather large (order of 20 or more). If we imagine fitting a least-squares average for a *single* parameter to a set of data, we get the average plus a spread which represents some confidence value (e.g., 95%). If we fit two parameters to a set of data, we get instead an error ellipse which is in general canted at some angle instead of being aligned with one or the other parameter axis. Consequently the uncertainties in each parameter are generally larger than would be suggested by fixing the second parameter at its average and asking what range of parameters satisfies the confidence level (see figure 2.7) because the parameters are correlated. For three parameters we have an error ellipsoid, and for N parameters an N-dimensional hyperellipsoid of difficult-to-determine size, shape, and orientation.

In practice, workers often assume that the observational parameters remain orthogonal and try to determine each independently from the data. Consider, for example, the evolution of the radio luminosity. It is here that the lack of a clear theoretical model becomes critical. In practice, the dependence of radio luminosity is either guessed at or obtained from a least-squares fit to the original biased data set itself! The latter gives a "phenomenological" luminosity law $L_{radio} \approx L_{total}^{1/3}$ in rounded exponents (Lyne et al. 1975; Vivekanand and Narayan 1981; Prószyński and Przybycień 1984). If one looks at the data, this is a very weak correlation even in the available data set, dominated by the relative inefficiencies of the Crab and Vela pulsars, which are arguably not representative of pulsars as a class (see figure 2.13). In any case, this luminosity "law" is then used to assess the selection effects on the data set to correct for missing members and yield up other parameters such as the pulsar birthrate. But now what observational selection factors have also been swept up into the least-squares fit in the first place? Note that such a "law" necessarily has a characteristic luminosity below which the radio luminosity *exceeds* the spin-down luminosity. Adding in magnetic field decay avoids this built-in unphysical consequence.

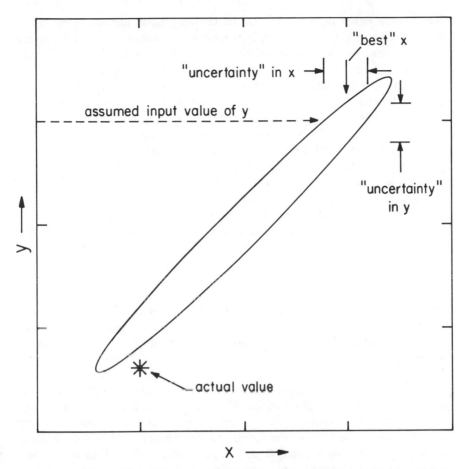

FIG. 2.7. Error ellipse. Effect of having two correlated parameters; least-squares fit gives elliptical contours of constant probability (3σ, say). It is quite possible to adopt an input value for one parameter (y here), find a best fit to the other within apparently narrow limits, and then confirm that the first guess was also rather "close." Within statistical uncertainties, the actual value could be located as shown, not only far from the presumed solution but also far outside of the perceived uncertainties. Although uncertainty contours can be directly calculated in the process of making a least-squares fit, it is rare to find such discussions in the literature.

Alternatively, Gunn and Ostriker (1970) suggested $L_{\text{radio}} \approx B^2$, (i.e., *no* period dependence!), which is physically absurd but innocuous in their model because they also assumed that the magnetic field B decayed away exponentially anyway; no one seriously believes that the radio luminosity could be independent of the period, for the physical reasons outlined at the outset. In both cases, magnetic field decay is required by the assumed insensitivity of the radio luminosity to spin rate. An alternative assumption is that the radio luminosity is a *fixed* fraction of the spin-down luminosity (Beskin et al. 1984), which is certainly different from the above phenomenological luminosity law, but actually rather representative of the *nearby* pulsars neglecting the exceptional Crab and Vela pulsars, and is at least not implausible from a theoretical point of view. We will see below that it works rather well, as does taking the magnetic field to be fixed at 10^{12} gauss, at least for some applications.

Ultimately one would like a first principles *theoretical analysis* which would give (1) functional variation of the radio luminosity, (2) functional variation of the period derivative, (3) functional variation of the magnetic field, (4) evolution of progenitor(s), and (5) dynamics of the formation event, to name a few. Notice that these theoretical parameters are only indirectly related to the observational parameters; for example, the time evolution of the radio luminosity where time is an explicit variable depends to some extent on at least all of the first three, which have time as an implicit variable. Thus the fact that there is considerable confidence in the functional form of (2) from the dipole magnetic field model is insufficient, given the uncertainty in the form of (1), and it seems amazing that one would expect much of anything to be said about (3) alone, given the limited biased data set.

2. Monte Carlo Simulations

The literature contains many estimates of pulsar birthrates, for example, and it is variously claimed that these rates are consistent or inconsistent with supernova rates. To understand the considerations involved in this issue and others, let us simply simulate some expected pulsar populations by restricting ourselves to a limited number of options. It might naively be thought that the problem is to obtain the number of observed pulsars as an equilibrium of trial creation and death rates. Such a simple program is in fact easy to implement, but it may have unsatisfactory features. For example, the resultant distribution of pulsar periods may be completely inconsistent with observation, with the oldest pulsar having a period of only a second! At this point what started off as an objective attempt to model pulsars easily gets sidetracked into a rather subjective one when additional effects are included in the model (beaming, magnetic field decay, distribution in z-height, etc., etc.) as the period distribution and other aspects of the data are brought in to be reproduced. As more and more effects are included, more parameters appear that are highly intercorrelated. In the end, then, the pulsar birthrate derived from the model has numerous implicit "ifs" associated with it, as do the values assigned to all the other parameters. Even what constitutes a "good fit" is quite subjective. The easiest way to demonstrate the issues involved is simply to show the results from a few such simulations. These simulations are *not* intended to be best fits at all (although they don't do too badly). Rather, they characterize the types of results one gets and illustrate the problems in identifying unique causes.

a. *Baseline Model: Dipole Spin-Down*

The baseline simulation will be the purest possible case: (1) assume pure magnetic dipole torque with no alignment starting with *zero* period, (2) assume

$L_{radio}/L_{spin\text{-}down} = 10^{-5}$, (3) assume constant magnetic field, (4) assume pulsars stay where they are formed, (5) assume pulsars are created randomly in space within a certain fiducial volume, and (6) create pulsars at a constant rate until no additional observable ones are formed. The only idiosyncratic choice we will make is in the shape of the fiducial volume, which we will take to have a range in galactic (z) height of 0.3 kpc, a y range (tangent to the spiral arms) of 3 kpc, and a radial x range outward from the galactic center of 10 kpc. (Actually this distance was taken to be 30 kpc, but a dispersion measure limit of 300 was imposed as well, so effectively the extent was only 10 kpc.) The latter ratio is simply to crudely compensate for the high sensitivity but low galactic longitude range of Arecibo compared to the other radio telescopes, while the first is just a typical estimate for the galactic disk thickness.

Let us do some simple estimates. If pulsars decay by pure magnetic dipole radiation without change in orientation, we have

$$P(t) = P_0(t/t_0)^{1/2}, \tag{1}$$

where t is the time since formation, with t_0 the present epoch and P_0 the typical period. For simplicity, we take $P = 0$ at birth although this cannot be strictly true, so in the simulations we only count pulsars beginning *after* one birth cycle and ignore the infinitely brilliant newborn ones. The period derivative is then

$$\dot{P}(t) = \frac{P_0}{2t_0}(t_0/t)^{1/2} = \frac{P_0^2}{2t_0 P(t)}. \tag{2}$$

Conservation of pulsar "flux" in time gives

$$\frac{dN}{dP}\frac{dP}{dt} = \text{constant}, \tag{3}$$

and because $\dot{P} \approx P^{-1}$, we can integrate to obtain

$$N(P) \approx P^2, \tag{4}$$

which can also be obtained by noting that the number of pulsars is assumed to accumulate linearly with time, hence from equation 1 as P^2. An important selection effect is luminosity, and the fixed efficiency assumption gives us $L_{radio} \approx P^{-4}$ while the received flux density falls as $S \approx L/d^2$. We can estimate the mean distance between pulsars (d) as $d \approx N(P)^{-1/2}$, so $S \approx P^{-3}$. What this means is that observable pulsars do *not*, as has been casually supposed, "accumulate" at long periods. Quite the opposite happens. Old pulsars simply are too dim, regardless of their increasing concentration in the galaxy, to ever be seen, even fortuitously. As we will see, a steady state is reached rather quickly and pulsar creation could have been happening essentially forever without producing additional observable pulsars. We can normalize the above parameters to represent the observed pulsars by taking $P_0 = 1$ s and $\dot{P}_0 = 10^{-15}$. For an injection rate of one new pulsar per 10^{11} s (3×10^3 years) one only needs to create 10^5 pulsars before saturation is complete and no new observable pulsars are added to the list. Of these pulsars, only the apparently bright ones will be seen. Moreover, large dispersion measures effectively weaken the pulsar signal by spreading it over a longer time. So we will select a luminosity limit of 10^{25} ergs/s kpc^2 and a DM limit of 300, roughly consistent with the limits on the Manchester and Taylor (1981) data set (the few large dispersion measure cases they have are of little statistical importance). We take the DM to be 30×the distance in kpc, regardless of direction. There is argued to be a plausible dependence of electron concentration on z and radial distance from the galactic center, but these would introduce at least two additional adjustable parameters.

Even for a bare-bones simulation, we are forced to make a number of nontrivial assumptions, but they can always be dropped or modified later. We will make four simulations, the above baseline model, baseline plus space velocity, baseline plus space velocity *and* magnetic field decay, and baseline plus space velocity and Gaussian beaming (but no decay). In all of these, pulsars are added at an assumed birthrate until no further observable pulsars are made, which is usually only a small fraction of the galactic age. The typical S-shaped saturation curves are shown in figure 2.8. These are all the observable pulsars ever created (after the last pulsar is formed, one can create a thousandfold more without creating a singe additional observable pulsar). The various assumptions are summarized in table 2.6. Note that the entire pulsar population is a recent population extending over the past 100 million years (only about 1% of the age of the galaxy) with half of the pulsars less than 4 million years old (from spin-down alone). Consequently the pulsar data may be sensitive to recent local events.

In all of these models, the effective area seen is 30 kpc^2, the efficiency of producing radio luminosity is fixed at 10^{-5}, the moment of inertia is 10^{45} g-cm^2, there are no alignment effects, and all begin as rapid rotators (there is no "injection" of slow pulsars). No millisecond pulsars are produced, of course, because they would slow down too rapidly if formed with typical fields (the parameter $\dot{P}_0$ in table 2.6 and in eq. 3). Recall that, although the pulsars are born with zero period, they are not listed until after one gestation period.

b. *Baseline plus Space Velocity (increases z height)*

As the next step in simulating reality, we will give the pulsars random z velocities in the range 0 to 200 km/s while reducing their formation region to 0.1 kpc (from 0.3, above) because we are now letting them drift to large z values rather than producing them at those values. This drift causes the pulsars to be depleted locally, because locally we see mainly old-but-near pulsars. These pulsars will tend to be depleted by drifting away to large z and becoming unobservably dim.

c. *Baseline plus Magnetic Field Decay*

To include magnetic field decay, we use (sec. 1.7.6)

$$P\dot{P} = Ke^{-2t/t_d} \equiv P_0\dot{P}_0 e^{-2t/t_d}, \tag{5}$$

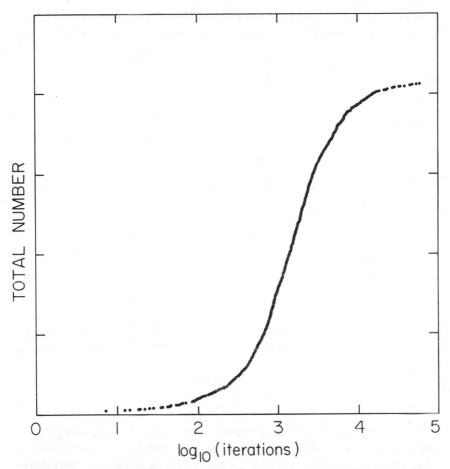

FIG. 2.8. Number of observable pulsars versus time. Example from the baseline simulation, showing how the number of observable pulsars saturates with time (i.e., the number of iterations times the birth rate, which is here 1 per 10^{11} $s \approx 3 \times 10^3$ years). The cutoff is harder than it appears (the statistics leave a deceptive looking positive slope); iterating for 10 times longer did not add a single pulsar.

and integrating gives

$$P^2 = P_0 \dot{P}_0 t_d (1 - e^{-2t/t_d}). \tag{6}$$

Note that we cannot normalize $\dot{P}_0 = 10^{-15}$ at $P_0 = 1$ s because $P_\infty = (P_0 \dot{P}_0 t_d)^{1/2}$, and if we used decay times of 3×10^6 years $\approx 10^{14}$ s, we would have no pulsars with periods longer that 0.3 s! Accordingly, in this third simulation we must increase $\dot{P}_0$ by 100 (thereby increasing the initial magnetic field by an order of magnitude). Now, however, the pulsars are spinning down so fast that they *still* disappear at short periods, and it is necessary to increase the birthrate by 10 to get a comparable number of survivors.

TABLE 2.6—Simulation Parameters

Parameters	Model			
	Baseline	Base $+z$	Base $+z+B$	Base $+z+$ Beam
Birthrate^{-1} (years)	3200^a	3200	320	640
$\dot{P}_0$ (see note c)	10^{-15}	10^{-15}	10^{-13}	10^{-15}
Oldest ($\log_{10}$ years)	8.25	8.00	7.54	7.92
Total number observable	818	722	762	864
P_{max} (s)	3.34	2.50	3.16	2.30
Closest (kpc)	0.05	0.17	0.08	0.06
z_{max} (kpc)	0.3	1.51	0.25	1.35
Total (mostly unseen) created	10^5	10^5	10^5	5×10^5
Birthrates^b ($\times 10^{-5}$/yr/kpc^2)	0.51	0.72	5.8	2.4

 a) Actually 10^{11} s.
 b) Best fit to existing data (see sec. f below on birthrates).
 c) Defined in equations 3 and 4.

d. *Baseline plus Beaming (but no field decay)*

The effect of a Gaussian beaming can be simulated by forming the pulsar at a random angle to the line of sight with a random magnetic-inclination angle and an effective luminosity multiplied by the beaming factor

$$\exp\left[-\left(\frac{\delta\theta}{\sigma}\right)^2\right],\qquad (7)$$

where $\delta\theta$ is the difference in the above two angles and σ is the beaming factor, taken to be 1/9 radian. Whether the random inclination is a good assumption is an open question, but the random orientation seems consistent with the random orientation of binary stars (Gillet 1988). It is then necessary to increase the birthrate by a factor of 5 to obtain roughly the same total number of pulsars (no attempt was made to get precisely the same number for each of the four simulations).

The goal of these simulations is to illustrate side by side the consequences of each effect as it is included, particularly because they are not at all independent. Although the pulsars are created randomly, the same pulsar is alway created at the same place because the "random" number generator always starts with the same "seed" number. This nonrandom aspect has the advantage that the effect of each assumption can be seen on any given pulsar. Thus rerunning a simulation always gave the same total number of pulsars, so one could immediately see the effect of changing a parameter.

e. *Analytic Model*

In addition to the numerical simulations (for which the reader must take our word, although we give a pseudocode listing at the end of this section), it turns out to be quite revealing to give an analytic model as well. The reason for this is that there is an unintentional built-in scale length to the observations!

Consider one pertinent question, "How many pulsars are observable?" We can rewrite equation 4,

$$N(> L) = L^{-1/2}. \tag{8}$$

The practical limit of flux that can be detected (S_0) gives us a limit on the luminosity of pulsars that can be detected

$$L < S_0 r^2. \tag{9}$$

Let us take N to represent the volume density of pulsars. The density of *observable* pulsars will then be, using the limit in equation 9,

$$N(< r) \approx r^{-1}, \tag{10}$$

and the total will be

$$n \approx \int N \, dV = \int N r \, dr \approx r. \tag{11}$$

The number of pulsars is linearly proportional to how far we search, assuming a disk population and a given limiting S_0.

Why don't we see countless pulsars then? This is a version of Olbers's paradox for pulsars. The answer is crucial to understanding many features of the pulsar data and very simple but easily overlooked. The galaxy is finite, but that is not the essential consideration limiting equation 11. Instead, the distance constraint is the increasing dispersion measure of more distant pulsars. Imagine that the galaxy were filled uniformly with a lot of dust. We might still see the nearby stars, but the more distant ones would be smeared into fuzzy spots and ultimately lost into a background haze. In the same way, at any level of observational technique, pulsar searches will be to some limiting flux density and some limiting dispersion. Thus there is a characteristic distance (not very precisely defined, but roughly determined) at which pulsars become rare. We will show from the existing data that this distance (r_0) is about 4-5 kpc, which certainly does not encompass the entire galaxy.

What is important is that observing pulsars is *not*, as might naturally be assumed, entirely limited by their apparent brightness. It is also limited by their distance compared to r_0. Young pulsars would, at our fixed efficiency, be very bright and perhaps be detectable in neighboring galaxies! A population of objects confined to an infinite disk would have no natural scale length. Thus their properties would have to scale with power-law exponents, as illustrated in equation 8 and following. But such power-law scalings are invariably divergent, and equation 4, for example, suggests that there should be hordes of long-period pulsars (which some have apparently taken literally in advocating a "cutoff" in pulsar periods).

Newly formed pulsars could, as noted, be detected at huge distances. In practice, we detect pulsars out to about r_0. What this means is that the pulsars fall into two luminosity classes, those with $L > S_0 r_0^2$ (we see every one of these, because luminosity is not a constraint), and those with $L < S_0 r_0^2$ (we only see those which happen to be close and which are therefore highly selected according to luminosity). As a result we have in effect two scaling laws mixed together according to how luminosity selection enters.

For example, the most common period for a pulsar is 0.5 s (figure 2.9). Does that mean that, when pulsars get to periods around 0.5 s, they necessarily begin to function differently? No. What this peak in the period distribution tells us is that a typical pulsar with a period of 0.5 s can just be observed to a distance of r_0! Unfortunately the data are assembled from telescopes of widely different sensitivity and de-dispersion equipment, so necessarily r_0 is even a fuzzier quantity than it would be given a standardized instrument. In the above simulations, r_0 is roughly the geometric mean of the 3 kpc width and 10 kpc length of the fiducial volume, or about 5.5 kpc.

f. *Pseudocode for Simulations*

The simulation programs were utterly trivial as computer programs go, but they are listed below should the reader wish to check the work described below or to run other possible models.

Definitions

x = distance from Earth along Arecibo line of sight
z = distance from galactic plane
y = distance along third axis
d = distance
DM = dispersion measure
t = time
P = period at time t
Ω = spin rate
$\dot{P}$ = period derivative at time t
$\dot{\Omega}$ = spin rate derivative
t_0 = standard epoch
P_0 = period at standard epoch
$\dot{P}_0$ = period derivative at standard epoch
dt = time between pulsar formation
I = moment of inertia = 10^{45} [g cm^2]
eff = radio conversion efficiency
L = absolute radio luminosity
S = relative luminosity (flux density)

Input Values

$P_0 = 1.0$ [s]
$\dot{P}_0 = 1.0 \times 10^{-15}$
$t_0 = P_0/2\dot{P}_0$
$dt = 1.0 \times 10^{11}$ [s]
eff $= 1.0 \times 10^{-5}$
$t = 0$ [start at present]

Program

Do 100,000 times:

$t = t + dt$
$P = P_0(t/t_0)^{1/2}$
$\dot{P} = P_0^2/2t_0 P$
$\Omega = 2\pi/P$
$\dot{\Omega} = \Omega\dot{P}/P$
$n =$ random number [between 0 and 1]
$z = 0.3 \times n$ [kpc]
$n =$ new random number
$y = 3 \times n$
$n =$ new random number
$x = 30 \times n$
$d = (x^2 + y^2 + z^2)^{1/2}$
$L = \text{eff} \times I\Omega\dot{\Omega}$
$S = L/4\pi d^2$
$DM = 30 \times d$ [pc/cm^3]

skip output if:

t larger than age of galaxy
S less than 10^{25}
DM more than 300
P less than 0.001

print:

line number, d, z, L, t, and P

This program was intended to examine a number of issues. However, with the input parameters illustrated, output is never skipped for the age limit or the period limit and numerous pulsars are created that never fall in the dispersion measure limit. For the application in this section, the program would therefore run 3 times faster if $x = 10 \times n$, but the pulsar ages would be equally spaced for about the first half of the output.

For the z drift, the line $z = 0.3 \times n$ was replaced by the three lines $z = 0.1 \times n$, $n =$ new random number, and $z = z + n \times 2 \times 10^7/3.1 \times 10^{21}$ [velocity in cm/s; distance in kpc].

For the magnetic field decay, in addition the lines $t_0 = P_0/2\dot{P}_0$, $dt = 1.0 \times 10^{11}$ [s], and $P = P_0(t/t_0)^{1/2}$, $\dot{P} = P_0^2/2t_0P$, were replaced with $t_d = 1.0 \times 10^{14}$, $K = P_0\dot{P}_0t_d$, $dt = 1.0 \times 10^{10}$, and $expf = e^{-2t/t_d}$, $P = K^{1/2}(1 - expf)^{1/2}$, $\dot{P} = expf \times P_0\dot{P}_0/P$, $t_c = P/(2\dot{P} \times 3155760)$ [years], respectively, and the formal characteristic age t_c was output instead of the actual age t.

For the beaming simulation, one again begins with the baseline model but following the first random number comes the line $\cos \theta_m = n$ [radians] to give a random magnetic axis relative to the spin axis, and following the second random number comes $\cos \theta_o = n$ to give a random observer colatitude. Finally $L = \text{eff} \times \Omega\dot{\Omega}$ is followed by $L = L \times e^{-81 \times (\theta_o - \theta_m)^2}$.

3. Simulation Results

Let us now compare the results of the simulations with the data set of Manchester and Taylor (1981) (330 pulsars, 293 of which have $\dot{P}$ measured).

a. *Period Distribution*

As we have seen above, the expected number of pulsars per unit period interval should be linear with period at small periods, as is indeed given by both the data and the simulations (figure 2.9). This is a very important result, because it is in exact agreement with dipole spin-down. Thus any other effect such as the "injection" of pulsars with long periods must either be confined to periods within the first bin (0 to 0.1 s) or take place beyond the peak ($P > 0.5$ s); otherwise there should be an upward curvature. Just to make a point, the curves have not been labeled. It might be instructive to try to guess which curves correspond to which model or the data before reading further. In any event, one sees the linear rise clearly, followed by different falloffs. Curve A is actually the model mentioned frequently in the literature: z drift plus magnetic field decay. This model does indeed cut off long-period pulsars, but rather brutally. In fact, the production of pulsars has to be increased with the result that the distribution is actually shifted toward the long-period pulsars, which are then cut off exponentially. The net effect is to produce *too many* old pulsars, which is exactly the opposite of the intended effect. Curve B is the baseline model, with a power-law paucity of old pulsars simply because they become too dim to be seen. Curve C is the data. Curve E is the baseline model with beaming. With beaming, the population of old pulsars is impoverished because one only sees them if the beam sweeps close to the observer, while the fast ones are enriched because, being bright, they can be seen further off axis. As a result there is a noticeable curving-over of the distribution as the maximum is approached, which is not noticeable in the actual data. Curve D is just the effect of drift to large z, which also cuts off a few of the older pulsars.

The overall conclusion from this simulation is that *the simplest possible simulation* gives about as good an accord with the data as any of the more "sophisticated" assumptions. In any case, we see that the period distribution

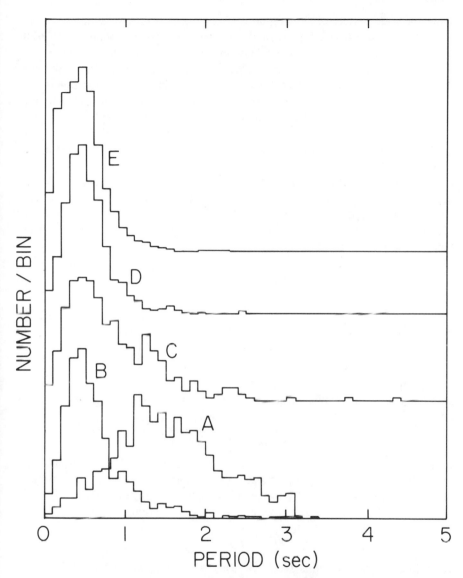

NUMBER / BIN

E

D

C

B

A

PERIOD (sec)

FIG. 2.9. Simulated period distributions. Number of pulsars per 0.1 s period interval. This linear plot clearly shows the linear rise at short periods expected from dipole spin-down (often plotted log-linear, which makes this behavior difficult to see). The identification of the curves is given in the text (try to guess which are the data if you haven't read that far).

at long periods is strongly correlated with all the effects included: beaming, z drift, and magnetic field decay *all* deplete long-period pulsars, so any statement about magnetic field decay implicitly constrains these other assumptions as well.

We have not optimized any given simulation. For example, all simulations give too narrow a period distribution, which arguably arises in the actual data from the wide variety of slowing-down rates seen for pulsars with essentially the same period (presumably the range in magnetic fields). Although the overall shape of the period distribution curves in figure 2.9 would be the same, the maximum would be moved around, which would broaden the distribution. The poor fit for magnetic field decay can of course be improved by increasing the time constant for decay. But this time constant is already comparable to the mean age of the pulsars! Increasing it much more would simply render the distributions insensitive to the decay rate. There is no way the data can tell us that pulsar magnetic fields decay away *after* the pulsars become too dim to be observed.

Note that the most probable pulsar periods in the simulations match those of the data. As noted above, this must certainly (at least in the simulations!) reflect the characteristic distance scale. For short periods, we obtain $dn/dP \approx P$, consistent with differentiating equation 4 and having no distance selection. The long-period pulsars are entirely luminosity selected, giving

$$\frac{dn}{dp} \approx \int_0^r \frac{dN}{dP} r \, dr \tag{12}$$

with $dN/dP \approx p$ as before, but because $L \approx P^{-4}$ and the limiting distance is determined by $L \approx S_0 r^2$, we have $r \approx P^{-2}$, which then gives

$$\frac{dn}{dp} \approx P^{-3}. \tag{13}$$

In other words, the number observable per unit period interval climbs linearly and then falls off as the inverse cube.

b. *The Luminosity Function*

If we knew what the actual luminosities of the pulsars were, we would have an important clue as to what phenomenon rules their energy output. This information is generated automatically in the simulations, and is shown for the totality of pulsars in figure 2.10. The resultant luminosity function is

$$N(>L) \approx L^{-1/2}. \tag{14}$$

As pointed out earlier, this result is easily derived by conserving "flux" as the luminosity evolves: $\frac{dN}{dL} \dot{L} \approx$ constant, and given a radio luminosity proportional to dipole spin-down $L \approx P^{-4}$ with $P \approx t^{1/2}$, we get $L \approx t^{-2}$ and $\dot{L} \approx t^{-3} \approx L^{3/2}$. Thus $\frac{dN}{dL} \approx L^{-3/2}$ and integrating gives $N(>L) \approx L^{-1/2}$. A number of interesting modifications can be noted. First, all the models essentially parallel the baseline model until one gets to the low-luminosity end. There the baseline model cuts off abruptly because we stop iterating! Even if we continue to iterate, the pulsars cannot spin down below a certain period (5 s with our parameters) in a Hubble time (which raises interesting questions of

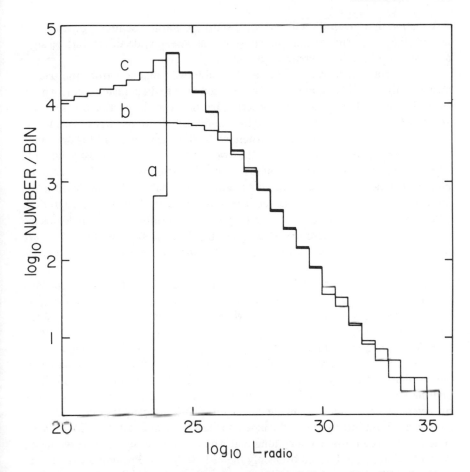

FIG. 2.10. Luminosity function (total). The effect of magnetic field decay (b), and beaming (c) on the baseline total luminosity function (a). The cutoff in (a) is from termination of the simulation program after 10^5 steps; otherwise the luminosity function would continue indefinitely with slope $N(>L) = L^{-1/2}$. Note that even with this abrupt arbitrary termination, the luminosity function for field decay does not turn over and for beaming the drop in number is very minor. If such a drop is real (Manchester and Taylor 1977), it would therefore seem to *rule out* either effect, although it is frequently cited as supporting magnetic field decay.

where the very slow pulsars come from in binary X-ray sources, as discussed in sec. 10.1). It is frequently claimed that there is already a lack of low-luminosity pulsars (as long ago as Manchester and Taylor 1977, and recently in Lyne et al. 1985). The lack of these pulsars cannot be due to magnetic field decay, as can be seen in figure 2.10, where we see that magnetic field decay takes the low-luminosity pulsars and redistributes them to arbitrarily low luminosities with *constant* logarithmic density. As above, if $L \approx e^{-t}$, then $\frac{dN}{dL} \approx \frac{1}{L}$ and therefore $N(>L) \approx$ constant-log L. There is no effect of beaming and z height on the total luminosity distribution. However, if we plot the luminosity distribution in a given direction (i.e., toward the Earth), beaming gives a long tail of otherwise brighter pulsars whose beams sweep wide of the line of sight and therefore have low apparent luminosity.

What is entailed to recover the actual luminosity function from the data is shown in figure 2.11. Again we have been coy about which histogram is which. Basically, they all look rather nondescript, but on closer inspection one sees that at high luminosities the observed and total luminosity functions track one another with a slope of $-1/2$, which is consistent with all the simulations. It is obviously very model dependent how one would invert the observed luminosity function (figure 2.11) to get the true luminosity function (figure 2.10). For example, putative magnetic field decay does not begin to alter the total luminosity function until one is almost out of observable pulsars. Here the data are histogram E while the remaining look-alike curves are A = beaming, B = field decay, $C = z$ drift, and D = baseline. Even given the different simulations, *which are tantamount to complete luminosity-limited surveys*, one can hardly tell one from the other.

Both Manchester and Taylor (1977) and Smith (1977) report total luminosity functions with slope near -1 instead of the $-1/2$ that results from dipole spin-down. These vales were reportedly recovered by correcting for selection effects and relative search depths and areas of coverage, factors which are only roughly included in our simulations by creating the pulsars in a rectangular area. A detailed plot of the number of pulsars with luminosity exceeding a given value versus that luminosity is shown in figure 2.12. Broadly speaking, the data and the baseline simulation track well enough except that there is a noticeable turnover at high luminosities in the data. This high-luminosity region is indeed fit well by a -1 luminosity law. Of course one must remember that the "look-back" time for this region is not very long and we are considering the very youngest pulsars, which were formed over only the last 10^5 years. Presumably the pulsar formation rate is only constant in some average sense, and not necessarily a fixed value in a limited volume of the galaxy. Thus the high-luminosity end of the pulsar distribution would be very sensitive to the natural fluctuating rate of formation events, and it is presently unclear what role that might have played in the observed data.

Again from the analytic model we expect to see all pulsars with $L > S_0 r_0^2$, so the luminosity function at large values should be *the* luminosity function, while that below consists of distance-selected pulsars, which naturally drops off. The simulations and data seem nondiagnostic, with all giving similar lumpish histograms.

Luminosity selection makes it difficult to deduce brightness distributions from the observed luminosities of ordinary stars, which also have a wide range of intrinsic luminosities. This bias can be illustrated by breaking up the pulsars according to distance, with an arbitrary definition of <0.5 kpc for "near" and using logarithmic intervals of 0.5 to 1.0, 1.0 to 2.0, etc., thereafter. The results are shown in figure 2.13.

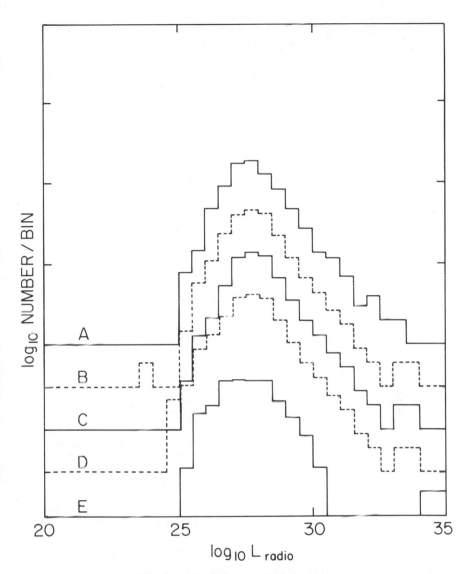

FIG. 2.11. Luminosity function (observed). The various simulations and data are plotted, each curve shifted upwards to help distinguish one from the other. We have added "1" to each bin number before taking the logarithm so that "zero" actually corresponds to zero. See text for identification of curves.

First notice the radio luminosity, which is the agent by which pulsars are usually detected at all. The nearby pulsars are seen to crowd against a limit of 10^{25} ergs/s. In fact, no pulsar has a smaller luminosity in the Manchester and Taylor (1981) catalog. As we go to more distant pulsars, we can see an advancing vertical edge that cuts off ever-brighter pulsars. Returning to the

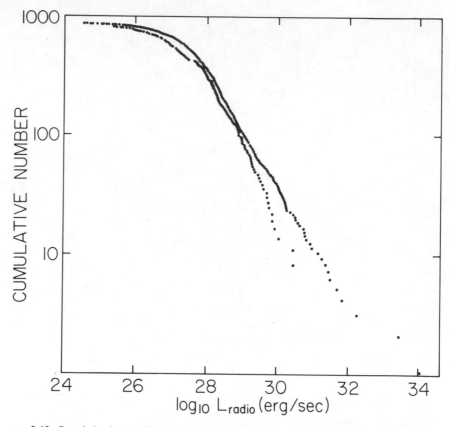

FIG. 2.12. Cumulative luminosities. The upper curve is the total number of pulsars (baseline simulation) with luminosity less than that given. The lower curve is the same plot for the data (293 pulsars), displaced upwards to normalize. Although the two track fairly well, the highest-energy end tends to fall more rapidly in the data.

nearest pulsars, it is not entirely clear from this plot whether the luminosity "limit" at 10^{25} ergs/s is real or statistical. The distribution might appear more balanced perhaps with one or two pulsars to the left in the <0.5 kpc box, but that absence has little apparent statistical weight. Thus the failure to detect very weak pulsars seems a natural feature of the data, and one must wonder a little whether great efforts to identify a "turn-off" mechanism are really required. (There has been no shortage of proposed mechanisms, should such be required; several have even been suggested by this author.) We also see that each distance-grouped sample is cut off to the right as well. Just as for stars, the luminosity function is weighted heavily in favor of dim objects. In the case of pulsars, that weighting must be due in part to their slowing down with age. As we expand the volume, we steadily increase the chances of seeing the rare bright objects. Some of these may not be so distant, and hence their brightness may be overestimated. The brightest members of each group are selected with each step in distance. (The numbers are significantly more complete for the nearby pulsars than for

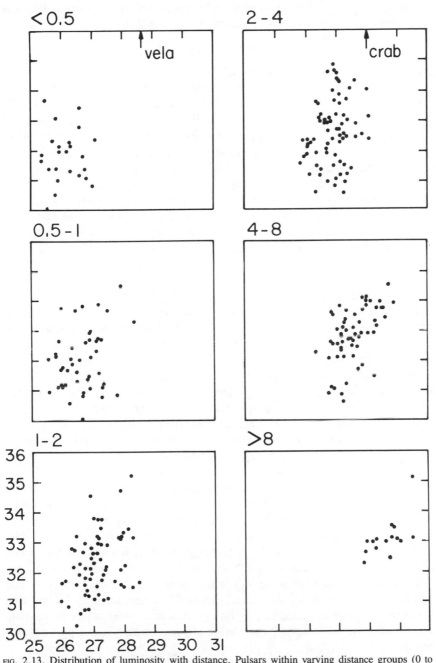

FIG. 2.13. Distribution of luminosity with distance. Pulsars within varying distance groups (0 to 0.5 kpc, 0.5 to 1, etc.) plotted as a function of spin-down luminosity (log scale, vertical) and radio luminosity (as defined by Manchester and Taylor 1977, horizontal scale). Pulsars tend to move on diagonal with $L_{\text{radio}} \approx L_{\text{spin-down}}/10^5$ although there are deviations at the large (and more uncertain) distant scales (see discussion of efficient versus inefficient pulsars). Note that the Crab and Vela pulsars have wildly different luminosities from their distance group companions, which has inspired suggestions that these are not necessarily typical progenitors for the average pulsar.

the distant ones; the latter are dominated by Arecibo observations near 19^h right ascension and become increasingly incomplete with distance.)

c. *Pulsar Efficiency*

Note that drawing a line in figure 2.13 running up at $45°$ broadly typifies a radio luminosity versus total luminosity relationship fixed at 10^{-5}. For the most distant pulsars, we begin to see a tendency for this line to bend over, which is once again suggestive of either the possibility of seeing such pulsars at much larger distances or a selection effect by which "distant" pulsars become more efficient by virtue of their not actually being that distant at all. One problem with the above interpretation is that the distances would have to be grossly incorrect, by a factor of 10, and it is not clear that the path-averaged electron concentration could be that far off the global average, because there are constraints on how discrepant the distances for other pulsars in the same area of the sky might be. There is in fact some support for such a locational correlation. For the "efficient" pulsars, four are all located in the same region: PSR 1929+10, PSR 1916+14, PSR 1915+13, and PSR 1930+22 are all in the galactic plane within about a degree of galactic longitude $l = 48.5$. For the "inefficient" pulsars, another four are similarly grouped: PSR 1323–58, PSR 1323–63, PSR 1309–53, and PSR 1302–64 again are in the plane within about a degree of galactic longitude $l = 306$. There might be a "hole" in the electron distribution through which the pulsars were observed in the one case and a dense cloud in the other. However, we are reluctant to launch yet another quasi-fact.

Another way to choose pulsars at fixed (near) distances is to choose those that also have large galactic latitudes (b). Although one might think that objects uniformly salted throughout a spherical volume would have widely varying distances, the mean distance is 0.8 of the most distant, hence rather less than the factor of about 2 expected from distances determined from dispersion measures. Thus there are 25 pulsars with $d \leq 0.5$ of which 15 have $b \geq = 20°$, as shown in table 2.7.

Note that even in this small sample, presumably not containing any wildly discrepant distances (because both gas clouds and distant pulsars per se should be absent at high galactic latitudes), the apparent efficiency varies by almost two orders of magnitude. If we subdivide this small sample into pulsars with $P < 1$ (average = 0.53) and $P > 1$ (average = 1.42), we find the barest hint of a period-dependent efficiency ($\langle 5.45 \rangle$ for the slow ones and $\langle 5.78 \rangle$ for the fast ones). Since the ratio of periods is just as large as for the "efficient" and "inefficient" pulsars (table 2.10), and can easily be accounted for by less than a factor of 2 distance uncertainty, the "correlation" looks much like a mild version of the same secondary correlation of period with distance plus distance errors.

d. *Galactic Height versus Characteristic Age*

Given the complications from luminosity selection, it would be nice to have at least *one* clean statistical test. One possibility derives from the observation that

TABLE 2.7—Nearby High-Latitude Pulsars ($b > 20°$)

Pulsar	Distance	Period	log Efficiency
0950 + 08	.09	0.25	6.18
1133 + 16	.15	1.18	5.78
0809 + 74	.17	1.29	4.70
0655 + 64	.28	0.20	(binary)
1112 + 50	.32	1.65	5.74
1237 + 25	.33	1.38	4.59
1604 − 00	.36	0.42	5.94
2045 − 16	.38	1.96	5.07
0031 − 07	.39	0.94	4.90
0942 − 13	.42	0.57	5.12
0834 + 06	.43	1.27	5.84
0149 − 16	.44	0.83	6.05
0203 − 40	.47	0.63	6.58
1530 + 27	.49	0.42	5.53
2151 − 56	.50	1.37	6.46
Average	.35	0.79	5.61

pulsars are scattered to larger galactic heights (z) than are the disk population stars, and also large proper motions have been detected in many pulsars (Lyne et al. 1982). Thus, if pulsars are born near the galactic plane and drift outward, there should be a clear correlation between z and the pulsar characteristic age $P/\dot{P}$, regardless of what the distribution of initial velocities is and more or less independent of the evolution in radio luminosity: old pulsars (if they can still be detected) should be at high z. Gunn and Ostriker originally found just such an effect, correlating z and log P rather than $P/\dot{P}$ owing to the relative scarcity of $\dot{P}$ determinations at the time, grouping the data into four period intervals. Manchester and Taylor (1977, p. 163) repeated this analysis with a much larger data set and discovered that pulsars with average ages of about 0.3 My have $z \approx 150$ pc while those with average age of 1.5 My have $z \approx 300$ pc, consistent with a mean space velocity perpendicular to the plane of the order of 260 km/s. Unfortunately, when one examines the yet older pulsar groups, one finds that they have *no greater* mean heights above the plane! Fujimura and Kennel (1980) carefully reexamined this result and plotted the individual pulsars rather than grouped averages. It is clear that the pulsars with large characteristic ages lie near the plane without any significant tendency for larger z heights than that typical of pulsars as a class. Indeed, there seems a smattering of large z pulsars (as plotted by Fujimura and Kennel) with characteristic ages of around 1 My, which seem to suggest a trend in what is otherwise a discouragingly scattered distribution possibly consistent with no correlation at all (in other words, if the coordinates were erased and an independent researcher asked to identify

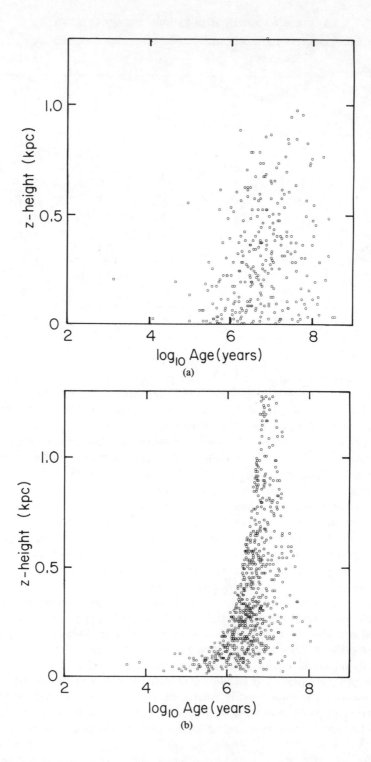

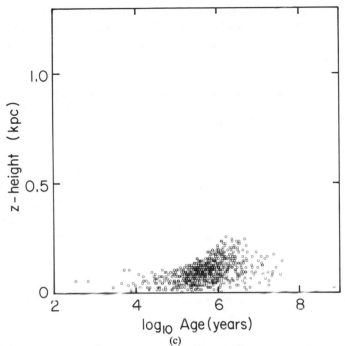

(c)

FIG. 2.14. A plot of z-distribution. In these three figures, (a) is the data, (b) is the baseline simulation (note that old pulsars are at *low* z, and (c) is the magnetic field decay model. The small range in z height for figure 2.14c could be compensated for by increasing the velocities by a factor of 3, but the mean pulsar ages are clearly off by at least an order of magnitude.

any trends in this unknown data set, it is more likely that a Gaussian would be suggested than our anticipated linear correlation). Only nine pulsars have z heights larger than 0.8 kpc and the first pulsar on the list is the Crab pulsar, which is already at 0.2 kpc at essentially "zero" age.

This *lack* of systematic increase in z for very old pulsars is frequently quoted as *supporting* magnetic field decay; the argument goes that their large "ages" are an artifact produced instead by field decay. The z drift simulations are shown in figure 2.14, except for the beaming simulation, which had, as one would expect, no effect on the z drift itself. The artificial curve (b) is the z drift alone while the data are (a) and the magnetic decay is (c). Here the magnetic decay curve doesn't look too bad superficially, particularly because it lacks the sharp edge introduced here by creating z component velocities with a sharp upper limit. But the vertical scale for the magnetic field case is 5 times too small! Moreover, with magnetic field decay there are clearly too many pulsars with ages in the 10^5 to 10^6 year range whereas in the data the bulk are in the 10^6 to 10^7 year range. To compensate, one would again have to lengthen the decay time to the point that its effect on the data became so minor as to be unimportant. The existing data will always be consistent with magnetic field decay times of order 10^7 years simply because decay then makes no difference statistically. Notice that the tendency for old pulsars to be at low galactic latitude is also evidenced in z drift alone; those that drift away get too distant to be seen when they are old.

Although the baseline simulation and data do not seem particularly inconsistent, a number of suggestions have been put forward to "explain" the z distribution: (1) the large characteristic ages (>2 My, say) are not true ages, (2) the large z distribution is not dominantly produced by large spatial velocities, and (3) the high-velocity pulsars are also systematically the rapidly evolving ones, to name a few. Suggestion (2) would be distasteful to observers, (3) has not been explicitly discussed, although it is indirectly involved in a possible velocity versus $P\dot{P}$ correlation (sec. 2.4.4), and (1) is a popular working hypothesis. Manchester and Taylor (1977) and Fujimura and Kennel (1980) suggest that *all* pulsars decay within a lifetime of a few My (e.g., because of magnetic field decay) and that those with long characteristic ages were simply born with large periods and weak magnetic fields (small $\dot{P}$).

There are several problems at present with this idea. (1) None of the *known* pulsars associated with supernova remnants have been born as slow rotators, a short list of only about four, however, so the process could have been seen but so far has not (at low statistical significance). (2) If correct, the actual ages of the pulsars with large characteristic ages are simply *unknown* and therefore they should be excluded from any statistical analysis directed at determining pulsar birthrates or evolutionary processes (which would reduce the 330 pulsars in the Manchester and Taylor 1981 listing to the 48 having $P/\dot{P} \leq 2$ My). The rest of the P-$\dot{P}$ diagram would be irrelevant to evolutionary issues because the evolutionary tracks would only be short arcs. (3) The millisecond pulsar not only has an enormous formal age, but the scenarios involving the spin up of an old pulsar would also date it as old, yet it is found at very low z (a few such cases could of course be dismissed as statistical accidents).

One can see from the simulation that the correlation between z (height above the plane of the galaxy) and τ_c (characteristic age) is not as dramatic as one would have expected. The well-defined envelope is, of course, an artifact of giving all pulsars exactly the same magnetic field and maximum velocity.

e. Number versus Distance

Next we examine a comparison which proves to be insensitive to the models: cumulative number of pulsars versus distance interval, as shown in figure 2.15. As can be seen, the data track the baseline simulation out to almost 2 kpc. In contrast, Manchester and Taylor (1977) perceive incompleteness beginning at 0.5 kpc, on the basis of a log-log plot of the data and the expectation that the cumulative number increases as d^3 for small distances and d^2 for large distances, which would be the case were it not for luminosity selection. In fact, the simulations and the analytic model (and data!) give a *linear* dependence of number on distance. The single simulation that does give a hint of a slight

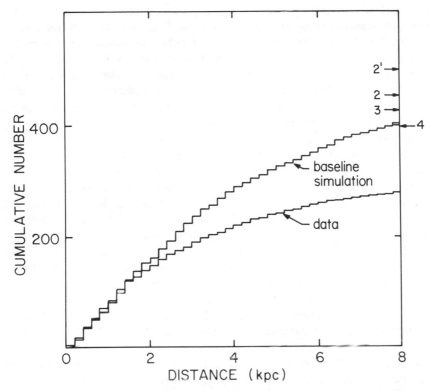

FIG. 2.15. Number versus distance. Except for the z drift, which was slightly "S"-shaped at small distances, all the curves were similar to those shown, with different ("asymptotic") values at 8 kpc when normalized to have the same linear initial slope as the data. Only the asymptotic values are indicated for z drift (2 and $2'$ depending on whether the fit is to the "knee" or to the initial slope), magnetic decay (3), and beaming (4). All suggest an increasing incompleteness in the data beginning at about 2 kpc.

upward curvature is the z-drift case, where the depletion at small distances can be traced to the drift of older (only seen closer) pulsars away from the galactic plane (and Sun). That case is difficult to fit to the data because the data seem quite consistent with a linear dependence of number on distance. Note, as discussed before, that the linear rise of the distance distribution is exactly what is expected. The turnover is largely attributable to having $r_0 \approx 4$ kpc, as can been seen in figure 2.15. Of course, the finite area of the galaxy should also come into play here, given that in the anticenter direction we are probably close to running out of stars at such distances.

f. Pulsar Birthrates

The various number-versus-distance simulations suggest that the Manchester and Taylor (1981) data set is between 55 and 70% complete to a luminosity of 10^{25} ergs/s/kpc (i.e., that there are a total of between 400 and 500 such pulsars out to

8 kpc). If so, our simulation rates are about twice the values necessary, so if we correct to achieve the asymptotic limit in figure 2.15, we obtain the rates given in table 2.6. The baseline model requires a pulsar formation event every 235 years, every 167 years with z drift, every 50 years with beaming added, every 21 years with magnetic field decay instead of beaming, and about every 5 years if *both* beaming and field decay are included. Similar numbers are quoted in other discussions (e.g., Narayan and Vivekanand 1981). The first two are consistent with event rates for supernovae observed in external galaxies. The remaining higher rates require special pleading, such as some sort of "invisible" event besides supernovae that makes pulsars. Neutrino observations should ultimately provide constraints on even that possibility, because we now know that neutrinos from neutron star formation can be seen from as far away as the LMC. The 5 year figure seems unlikely, although it is not at all uncommon for beaming and magnetic field decay (which would impose such rates) to be invoked together in discussions of the data. Haxton and Johnson (1988) have proposed isotopic tests based on the accumulation of rare neutrino-induced reaction products that build up in ore deposits. Such data could constrain the average supernova rate. High supernova rates *do* occur in some galaxies, however.

Kronberg et al. (1981, 1985) and Watson et al. (1984) observe numerous radio and X-ray supernovae in M82 (3.2 Mpc away), some up to 150 times brighter than Cas A (the brightest remnant in our galaxy) and fading as rapidly as 9%/year.

Lyne et al. (1985) argue that there are about 70,000 potentially observable pulsars with luminosities above 0.3 mJy kpc^2 ($=10^{25}$ ergs/s assuming 400 MHz and a $10°$ circular beam width). Their finding that the z distribution has a scale height of about 400 pc is much larger than that ($\approx$70–90 pc) for likely progenitors such as the OB stars. They then infer a magnetic field decay time of 9×10^6 years with a pulsar birthrate of 1 per 230 years if all pulsars are detectable or 1 per 50 years assuming a beaming factor of 5, broadly consistent with plausible supernova rates and with the simulations here. However, the next section, which introduces the effects of having a distribution of magnetic fields, suggests that even these birthrates need to be tripled. The expected rates for stellar collapse (which would have to equal or exceed the pulsar birthrate) are reviewed and refined by Ratnatunga and van den Bergh (1989).

a. *Magnetic Field Distribution*

As emphasized by Ostriker early on, the magnetic fields of pulsars, as interpreted from their spin-down rates, of course, are quite narrowly distributed, as shown in figure 2.16. The width at half-maximum in figure 2.16 corresponds to a factor of 5 in magnetic fields. In our simulations, we have only considered a single magnetic field strength (10^{12} gauss). The temptation is to introduce a distribution of magnetic fields similar to that in figure 2.16 into the simulation, and see what happens. Rerunning the simulations with fixed fields, however, reveals something surprising. For one thing, the total number of observable pulsars *in-*

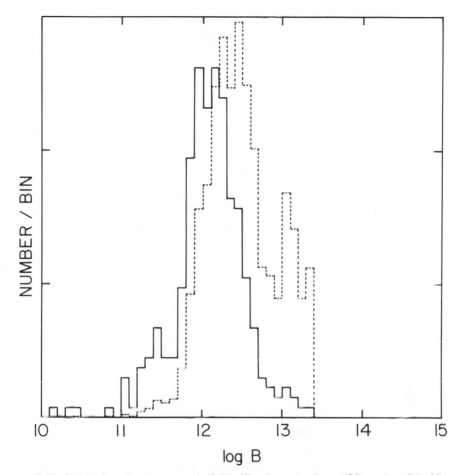

FIG. 2.16. Distribution of pulsar magnetic fields. The observed pulsars (295) are in solid while the corrected distribution taking into account the number versus magnetic field scaling for dipole spin-down is shown dashed. The secondary "peak" is probably just a statistical fluctuation.

creases as the magnetic field is *decreased*. We presumably only observe pulsars to begin with because they have such strong magnetic fields, so it may seem paradoxical that we would observe *more* if only they had weaker fields! But the reason we would is simple: pulsars with lower magnetic fields last longer and therefore have a better chance of being close enough to be observed before they get too dim. The simulations (as well as a simple dimensional analysis) give an inverse correlation between the number of observable pulsars and magnetic field strength. Consequently the histogram of $\log_{10} B$ for observed pulsars must be significantly distorted from the corresponding histogram for the created pulsars. Using this correlation to correct the distribution to the initial input gives the dotted histogram in figure 2.16. The bimodal form results from multiplying points of low statistical significance by roughly 10, so it is probably not signif-

TABLE 2.8—Effect of Magnetic Field on Pulsar Statistics

$\log B_0$	N_{obs}	P_{max}	P_{min}	$\log L_{max}$	$\log L_{min}$
11.0	8296	.003	1.33	36.0	24.1
11.5	2610	.005	1.96	35.0	24.4
12.0	818	.010	3.34	34.0	24.6
12.5	264	.050	7.76	33.0	24.0
13.0	76	.140	8.73	32.0	24.8

icant. Even more surprising, however, is the result that the younger pulsars are actually *brighter* if the field is *weaker*, as shown in table 2.8. This inverse correlation follows because weak-field pulsars decelerate more slowly relative to the strong-field pulsars and therefore have shorter periods. Because $P^2 \approx B^2 t$, $L \approx B^2/P^4 = 1/B^2 t^2$, so for a given birthrate, the weak-field pulsars would be brighter. This assumes that the initial spin rate is arbitrarily large. If instead pulsars are typically born with something like a 0.10 s period, one recovers the more intuitive result that the weaker the magnetic field, the dimmer the pulsar. However, the increase in abundance of weak-field pulsars over strong-field ones still holds even if they all start out with 0.10 s periods. We are talking about effects near 10^{12} gauss, of course, so when B falls to the point that P would still be less than 1 ms (or whatever limit is appropriate) after one birth cycle, we have to introduce the constraint that $P_0 \geq 1$ ms. Thereafter, we do get the usual expectation of fewer and dimmer pulsars with weaker field. A second important consideration is the ages of the pulsars, which begin to exceed the age of the galaxy if the simulation is run to saturation. For magnetic fields below 10^{10} gauss, saturation is not reached within 10^{10} years and we are discussing a class of pulsar that has not reached steady state conditions since the galaxy formed.

The above simulations show yet another cross-correlation with birthrate. If the typical pulsar had a field of 10^{13} gauss, we would have to up the birthrate by a factor of 10, which would even render the baseline model inconsistent with likely supernova rates as progenitors. In fact, we see from the histogram that the more likely input distribution of pulsars must have a typical field of about 3×10^{12} gauss to produce an observed typical field of 10^{12} gauss, which *in itself* roughly triples the birthrate!

It is more sensible to adjust the birthrates to give equivalent numbers of pulsars per simulation, because the maximum and minimum periods are strongly biased by the birthrate. As we can see from table 2.9, the inverse correlation between birthrate and magnetic field extends over a wide range of magnetic fields. If we did not have an upper limit for the spin rate, the correlation would apply to all field strengths, because we are simply renormalizing the same pulsars. The result of keeping the total number of pulsars produced approximately

TABLE 2.9—Effect of Magnetic Field (normalized)

B	τ_b	N	P_{min}	P_{max}	L_{max}	L_{min}	τ_{max}
7.5	6.5	"0"	.0010	.0010	29.8	29.8	10.0
8.0	6.5	864	.0010	.0025	30.6	29.0	10.0
8.5	6.5	992	.0010	.0079	31.5	28.0	10.0
9.0	6.5	629	.0013	.0251	32.2	27.0	10.0
9.5	6.0	727	.0011	.0793	33.4	26.0	10.0
10.0	5.5	812	.0014	.251	34.0	25.0	10.0
10.5	5.0	818	.0025	.593	34.0	24.5	9.7
11.0	4.5	818	.0045	1.05	34.0	24.5	9.2
11.5	4.0	818	.0080	1.88	34.0	24.5	8.7
12.0	3.5	818	.0141	3.34	34.0	24.5	8.2
12.5	3.0	818	.0252	5.94	34.0	24.5	7.7
13.0	2.5	818	.0447	10.5	34.0	24.5	7.2

constant is shown in table 2.9, the same as table 2.8 except that the assumed birthrate (τ_b) and the age of the oldest pulsar produced (τ_{max}) are now included. Notice that the range of luminosities is *independent* of magnetic field for typical pulsar fields! We get the same number of pulsars because they are the same pulsars, just rescaled in magnetic field and birthrate so that their luminosities are unchanged. Furthermore, the minimum *absolute* luminosity in these simulations is almost exactly the minimum listed in the Manchester and Taylor (1981) data set, 10^{25} ergs/s. Thus the observed luminosity function is also independent of the magnetic field in this range. Several interesting selection effects can be seen in table 2.9. For fields below 3×10^{10} gauss (indicated by a line in table 2.9), the total number of pulsars becomes dependent on the age of the galaxy, and therefore the number observable begins to drop off if the product $B\tau_b$ is kept constant. As a result, the range in periods begins to narrow at lower fields. At about the same point, the period limit of 1.0 ms begins to remove the very bright pulsars and the maximum luminosity begins to drop. This decline in maximum luminosity causes the pulsars to become more difficult to detect, and one must wait longer before a pulsar is accidently created close enough to observe, which makes them older as a class and they then run into the galactic age limit. Consequently we have left the birthrates constant below 10^9 gauss. As a result, the number of observable pulsars *jumps* at first from 629 to 992 with the *decrease* in magnetic field. Even then, the general dimming as we go to smaller magnetic fields culls more from view and we find no pulsars whatsoever with fields of 3×10^7 gauss even though as millisecond rotators they would in principle be bright enough to detect. But none are close enough, simply because at the above birthrate not enough of them can be formed within the galactic age.

We therefore can directly see the major selection effects on surface magnetic field:

(1) Pulsars with strong fields ($\geq 3 \times 10^{13}$ gauss) slow down so fast that they quickly disappear and are underrepresented in the observed distribution. Consequently little can be said about the rate of formation of pulsars with very strong magnetic fields because they quickly disappear from view. Only immediately following a supernova would they be seen, as is in fact the case for the Crab pulsar. These pulsars must be the source of the longer-period pulsars on the one hand, yet their ages are proportionately smaller. Thus, for magnetic field decay to be important, it must fall quite precisely within a very narrow band: faster decay would completely distort the pulsar period distribution, and slower decay would have no significant effect.

(2) Pulsars with weak fields ($\leq 10^{11}$ gauss) must be created quite rarely or they would have accumulated and would dominate the observed pulsar population.

(3) Millisecond pulsars must be formed extremely rarely compared to typical pulsars (by about three orders of magnitude or more) or again they would have accumulated excessively.

(4) Pulsars with very weak fields ($< 10^{8}$) gauss could be bright enough to see, but unless the formation rate for such objects increased sharply compared to that for the observed millisecond pulsars, not enough would be formed to be close enough to ever be seen.

(5) One can normalize the characteristic ages of pulsars according to their magnetic fields.

(6) By something of a fluke in choice of units, the limiting *apparent* luminosity of 10^{25} ergs/s kpc^2 results in a limiting *absolute* luminosity of about the same numerical value, 0.3×10^{25} ergs/s. Thus our failure to see dimmer pulsars tells us nothing about the luminosity function at lower absolute luminosities.

a. P versus $\dot{P}$ Distribution

Many statistical analyses of pulsar data concentrate on the $P - \dot{P}$ distribution (figure 1.6). Unfortunately the dominant feature of this distribution is the scatter, which is essentially Gaussian in magnetic field (i.e., number distribution versus $(P\dot{P})^{1/2}$) and Gaussian in total spin-down luminosity (i.e., number distribution versus $\dot{P}/P^3$). What remains is roughly a "V" shape, caused by a straggle of pulsars with small fields, in effect a sort of archipelago leaving only the peaks of the distributions observable as the representative pulsars become rarer. It is sometimes claimed that there are insufficient rapid pulsars to the left to "feed" into this distribution, but the simulations suggest otherwise. The total (spin-down) luminosity distribution is plotted in figure 2.17 on top of the same distribution for the baseline simulation. Notice that there are no free parameters because the same moment of inertia is assumed in both cases. Except for a slight shift in the baseline simulation toward higher spin-down luminosities, the distributions are statistically indistinguishable. Recall that in

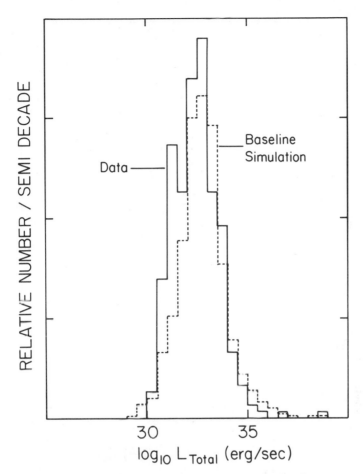

FIG. 2.17. Total luminosity distribution. Absolute comparison between baseline and observed spin-down luminosities (we are plotting $\dot{P}/P^3$ times the same constant). The baseline fits quite well if shifted a semi-decade, although there may be some "missing" pulsars below 10^{30} ergs/s in the data.

the baseline simulation all the pulsars evolve from rapid rotators and disappear entirely owing to a sharp luminosity and DM cutoff. To see if there is a factor involved other than simply luminosity selection such as a "death line" having a $\dot{P}/P^5$ dependence (Manchester and Taylor 1977; see Barnard and Arons 1982 for a possible theoretical justification), we divided the pulsars into two groups according to their formal magnetic field strength (the median is 1.26×10^{12} gauss). These two distributions are shown in figure 2.18. Note that in recent years the "death line" exponent has evolved from 5 to 4, which makes it a line of constant spin-down luminosity! Again, the weak-field pulsars appear to have a slight shift to smaller observed spin-down luminosities. Indeed, it is entirely

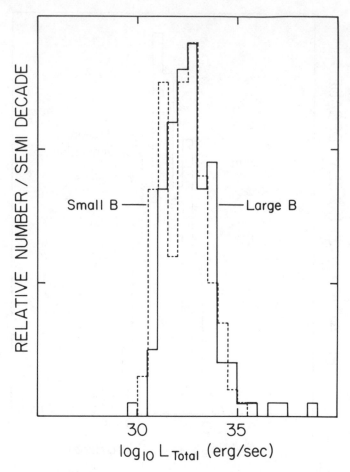

FIG. 2.18. Strong versus weak field distribution. Histogram of luminosity (same as figure 2.17) with the data divided evenly into those pulsars above median magnetic fields and those below. The one semi-decade shift of the weak field distribution is what keeps the baseline simulation from matching quite closely the full data set.

this shift that gives the above shift in the baseline model; the baseline model agrees well with the distribution of the strong-field pulsars.

All in all, the simplest simulation seems to do a credible job of describing all aspects of the data, not necessarily because it is the correct description of how pulsars evolve but rather because the data are so insensitive to the models. The model parameters, however, are extremely sensitive to the data! Thus, to match the data, we require wildly different pulsar birthrates, depending on the model. The remaining features, which have been interpreted as pointing to this or that interesting phenomenon (e.g., death lines, injection, late turn-on, changing beam shapes, field decay, beaming), are either picked out of very minor trends or biased by the properties of only a few pulsars. The two most extreme pulsars seem to be the Crab and Vela pulsars, which have statistically very different properties than their companion pulsars (e.g., figure 2.13).

a. *Time Evolution of the Radio Luminosity*

Let us return to the simple question, "How efficient are pulsars?" Any sound theory should at least be able to account for the observed range of efficiencies (L_{radio}/L_{total}) as well as the "typical" efficiency, and perhaps identify observable properties that should correlate with high or low efficiency. Ideally, we would like to know why a given pulsar is, say, more efficient than average and have some confirmation that the explanation is reasonably accurate. The efficiency can be determined largely from observation, given the radio intensity, dispersion measure, period, and period derivative. It is beginning to look as if neutrons stars may be rather standard objects, in which case the variation in moment of inertia (to give the total power output) may not be large. Distances from the dispersion measure are supposedly good to a factor of about 2. Thus the relative efficiencies are arguably good to within an order of magnitude, say. The beaming is harder to deal with, because a pencil beam could introduce a large variation in apparent radio luminosities depending on how close to the observer it swept, whereas a fan beam would introduce rather less variation. Neither possibility accords with the data. A fan beam would give a "delta" function in efficiencies, while a pencil beam would give the same with an exponential tail. Instead, the efficiencies are a broad Gaussian centered near 10^{-5} and extending at least two orders of magnitude in each direction, with a long tail to very low efficiencies as shown in figure 2.19. An inkling of what is going on can be found by selecting the most efficient pulsars and comparing them with the least efficient ones, as shown in table 2.10.

The weak systematic increase in efficiency with period is intriguing, but the power-law dependence would have to be huge, $L_{radio} \approx P^{10}(!)$, in which case it would hardly show up only as a vague statistical correlation but would be obvious from the raw data. Also suspicious is the correlation with distance. Suppose we are wrong about the pulsars' distances. Obviously, if the "inefficient" pulsars are not as close as thought, they would not be so inefficient, and vice versa. Indeed, as a class, one would expect nearby pulsars to be (old) slow ones, while the "inefficient" ones would instead be nearby *fast* ones, so the correlation with period is likely to be a secondary correlation with the distance errors. However, the usually quoted uncertainties in distance (factor of 2) are insufficient to account for such a large difference in apparent efficiency (factor of 40,000). Alternatively, we do expect to see efficient pulsars at larger distances than the inefficient pulsars. One would expect to see such efficient pulsars at vast distances (were it not for the observational difficulties encountered for large dispersion measures). However, we see from table 2.10 that only two of the pulsars are actually at vast distances.

What these statistics indicate is a built-in problem with correlating poorly

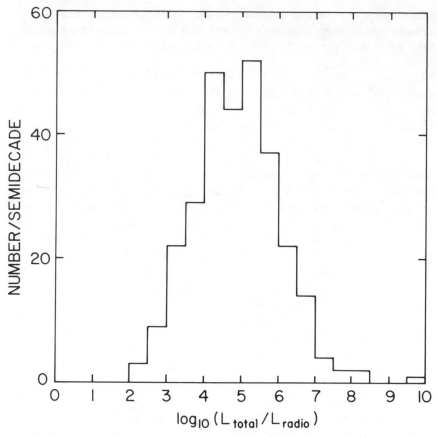

FIG. 2.19. Histogram of efficiency. An average of 10^{-5} seems representative, but the large scatter suggests that something besides magnetic field and spin rate scales the radio emission process. Data from R. N. Manchester and J. H. Taylor, 1981, *Astr. J*, 86, 1953 (table IV).

understood and poorly characterized data: if we look at the extreme objects, *we also select for the ones with the largest errors* in determining the parameter in question.

Exactly the same problems have been encountered with correlating solar and terrestrial data: one is bound to find a correlation between one data set and another if one sorts through enough data sets. The usual test of such correlations is to see if they persist when future data are gathered.

This search for weak correlations is not innocuous because least-squares fits to the data systematically emphasize the extreme data points, which may also be the ones most likely to be affected by errors. Even if the correlation is valid, it may represent a trivial secondary correlation of little direct physical significance (e.g., men tend to be stronger than women but do not live as long; hence being strong is hazardous to one's health?). Moreover, there are at least three distinct "least-squares" fits, depending on whether it is the deviation in the ordinate, the abscissa, or the perpendicular distance from the fit that is minimized.

TABLE 2.10—**Pulsar Efficiencies**

Pulsar	log Efficiency	Period (s)	Distance (kpc)
"Inefficient" pulsars:			
0531 + 21	9.53	0.033	2.00
0833 − 45	8.22	0.089	0.50
1929 + 10	8.16	0.226	0.08
1916 + 14	7.75	1.181	0.76
1001 − 47	7.61	0.307	1.60
0656 + 14	7.24	0.385	0.40
1822 − 09	7.24	0.769	0.56
1702 − 18	7.14	0.299	0.74
1930 + 22	7.07	0.144	7.00
0740 − 28	6.97	0.167	1.50
0611 + 22	6.95	0.335	3.30
1915 + 13	6.91	0.195	2.40
Average	7.30	0.401	1.20
"Efficient" pulsars:			
2111 + 46	2.25	1.015	4.30
1819 − 22	2.35	1.874	4.00
1924 + 14	2.41	1.325	5.80
1323 − 58	2.66	0.478	9.80
0138 + 59	2.67	1.223	3.00
1906 + 09	2.73	0.830	6.20
1323 − 63	2.75	0.793	18.00
1700 − 32	2.85	1.212	3.40
1942 − 00	2.89	1.046	2.20
2319 + 60	2.91	2.257	2.80
1309 − 53	2.93	0.728	4.80
1302 − 64	2.94	0.572	19.00
Average	2.70	1.113	6.94
"Average" pulsars:			
0031 − 07	4.99	0.943	0.39
0254 − 53	5.05	0.448	0.59
0903 − 42	5.03	0.965	1.70
1118 − 79	5.07	2.281	0.95
1503 − 66	5.03	0.356	4.50
1729 − 41	5.07	0.628	6.10
1844 − 04	5.04	0.598	3.70
1913 + 10	5.03	0.404	6.40
1913 + 16	5.02	0.059	5.20
1933 + 15	5.00	0.967	5.30
Average	5.03	0.765	3.48

Let us return to the models for evolution of pulsar luminosity that seem popular in the literature.

(1) The "Gunn and Ostriker" model. Here the radio luminosity is taken to be proportional to the surface magnetic field *independent of rotational period*. This is an odd model to be suggested by two theorists, particularly when they present no physical rationale; there would be no energy source to radiate for a nonrotating neutron star, and yet the neutron star could still be magnetized (hence radiating?)! They chose that scaling as a statistical fit, and they did it on the basis of a handful of pulsars. Perhaps they originally expected to see a large dependence on period and found that the dependence wasn't that dramatic. Because they assumed the fields to decay away, the intrinsic physical absurdity of the model could be finessed because the field would vanish before the neutron star could grind to a stop. Lyne et al. (1985) adopt this model and conclude that you *need* magnetic field decay to fit the data, which of course you do. Curiously, Gunn and Ostriker also emphasize the near constancy of B, which makes it an even more idiosyncratic choice for a dominant *variable*, especially one that decays.

(2) The $L \approx \dot{P}^{1/3} P^{-1}$ "fit to the data." It is true that if you take the data, such a fit is not too bad. Typically an inverse correlation between L_{radio} and P is evidenced in the data (Lyne et al. 1975; Vivekanand and Narayan 1981; Prószyński and Przybycień 1984). The last give $L_{\text{radio}} \approx \dot{P}^{0.36\pm.06} P^{-1.04\pm.15}$. Grotch and Michel (unpublished) get similar results (figure 2.20). They obtain efficiency $= 6 \times 10^{-5} P^{1.6} / \dot{P}^{0.7}$, and since L_{total} scales as $\dot{P}/P^3$, this result corresponds to $L_{\text{radio}} = \dot{P}^{0.3} P^{-1.4}$. At least part of this "fit" follows from data contaminated by distant "efficient" pulsars and nearby "inefficient" ones, as discussed above. But what does the result mean? We are talking about taking the biased data set, fitting it, and using that fit as representative of the *unbiased* input data set! A more reasonable alternative, albeit one that is labor intensive, would be to search for an input luminosity law that, *after* including the selection effects, reproduced the observed data (which almost certainly would fit to a different set of exponents). That is not what has been done. Even then, it is not clear that anything is left to be learned, for if one uses a fit to the data to analyze the data themselves, what does one expect to actually be able to isolate? Recently Chevalier and Emmerling (1986) reported that, on the basis of using the above fit plus assuming magnetic field decay, many slow pulsars must be born (i.e., injection). Phinney and Blandford (1981) consider a dynamic flow analysis for deducing the source function for pulsars from the observational data, essentially a continuum approximation to the above simulations. Like others, they favor magnetic field decay with a time constant of 10^6 years or, as an alternative, a narrowing of the beam with age (which is effectively a form of "anti-injection"). Recently, Narayan and Ostriker (1989) gave statistical fits to various models, some with as many as 16 free parameters (with 301 pulsars as the data base), and reconfirm all of the conventional expectations: magnetic field decay with a My time scale, radio luminosity scaling as total luminosity to the 1/3 power, etc.

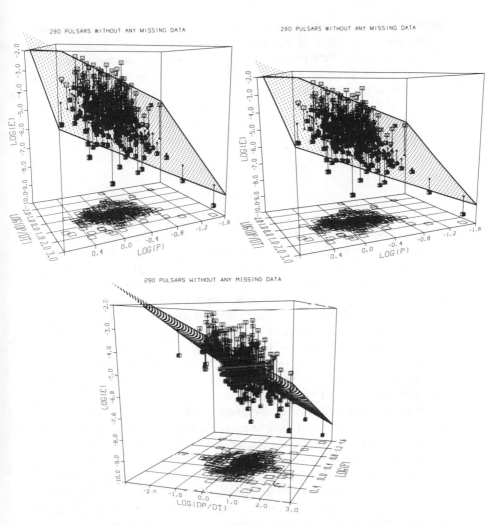

FIG. 2.20. L_{radio} versus P versus $\dot{P}$. Least-squares fit of a plane through the data of Manchester and Taylor (1981), where E is their quoted radio luminosity divided by the total spin-down power (assuming a moment of inertia of 10^{45} g/cm^{-2}). Each figure is rotated by $45°$ about the vertical to give a three-dimensional sense of how the data points (shown as small cubes) relate to the plane. The projection of these points on the horizontal plane gives the usual representation of the data as points (here small squares) in the $P - \dot{P}$ plane.

All of these results follow directly from the assumption of a radio luminosity that declines *slower* than total spin-down luminosity. Because the latter has an Ω^4 dependence, pulsars quickly spin down and then remain at slow spin rates for long times. But by the above assumption, the radio luminosity is comparatively insensitive to spin rate and therefore remains substantial. Under such assumptions, pulsars *would* pile up at long periods, and something else (magnetic field decay) would then be required to turn them off to give agreement with the data. The origin of the slower-than-spin-down scaling seems to be largely the Crab and Vela pulsars, which are very inefficient radiators in the radio part of the spectrum. If this behavior is taken to be representative (see figure 2.13) of the trend with spin rates, it follows (?) that any very slow rotators must be extremely efficient in turn, a conclusion that is not falsifiable because these objects are not to be found (because they have been "turned off" by imposing a second supposition: magnetic field decay).

4. Other Possible Correlations

a. *Velocity versus* $P\dot{P}$

A correlation between pulsar space velocity and the product $P\dot{P}$ was reported by Helfand and Tademaru (1977). Harrison and Tademaru (1975) proposed a mechanism for accelerating pulsars from their radiation provided the effective magnetic dipole was displaced from the center of the star. This effect essentially arises from interference between magnetic dipole and magnetic quadrupole radiation. The first varies as $\cos(\theta)$ while the second as $\sin(\theta)\cos(\theta)$ Thus the two interfere constructively below the plane (say) and destructively above the plane. Consequently the radiation pressure is higher below the plane and the pulsar is thereby accelerated along its axis. In nuclear physics, such effects are observed in the same way, through angular correlation between the direction of emission of the photon from an excited nucleus and the nuclear spin; however, the mixing of M1 and M2 or E1 transitions is possible only through parity-violating interactions. Such correlations are therefore sensitive indicators of parity violation but have not (yet) been used to propel nuclei along their spin axes. As more pulsars are added, the correlation seems to persist (Lyne and Smith 1982; Anderson and Lyne 1983; Cordes 1986 [figure 18, reproduced here as figure 2.21]), although Anderson and Lyne feel the data do not support the above mechanism over, say, asymmetric supernova explosions. The large logarithmic range plus the large scatter within the data set itself is almost as significant as the correlation itself. Such weak correlations pose a problem, because they can very well be accidental or second-order correlations. Indeed, the theory itself gave correlations with *initial* values of $P\dot{P}$ (i.e., at birth), not the *present* (evolved) values, so the prediction itself is second order, and if one were actually looking at the ghost of a previous first-order correlation, one could equally well be looking at the ghost of some selection effect. On the other hand the correlation might be real and could be trying to tell us something. Cheng (1985) attributes this correlation to the activation of pulsars by the accretion of dust particles from the interstellar medium (see sec. 4.4.10.e). The weakness of such correlations can be illustrated by imagining a correlation between the observed values of P and the observed values of $\dot{P}$ (figure 1.6). One only has to look at the figure to see how it is truncated by different selection effects to the point that it looks like a narrow inverted pyramid. The preferred fit would be a line

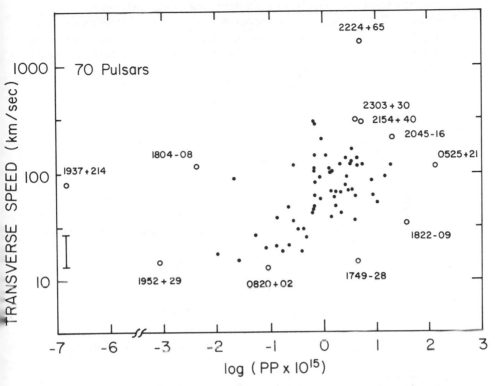

FIG. 2.21. Velocity versus $P\dot{P}$. From J. M. Cordes, 1986, *Ap. J.*, 311, 183 (figure 18), redrawn without lines to guide the eye. Note break in horizontal scale.

slanting down through this distribution, but since one could reflect this distribution in a mirror, it is clear that another possible "best fit" would correlate pulsars having short periods and small slowing-down rates with those having long periods and large slowing-down rates. Such a picture would even have a simple interpretation: pulsars are born with essentially their presently observed periods and "die" before those periods have significantly changed. We are not proposing such an interpretation, but rather cautioning against overinvesting in weak correlations.

The large space velocities that pulsars have are sometimes cited as evidence for off-center supernova explosions (but see Fryxell 1979) or formation in binary systems that are disrupted by the explosion (unlikely for the favored hypothesis of type II supernova origin).

5. Conclusions

Given these various suggestions and possibilities, what is the overall status of pulsar statistics? The conservative conclusions here are that the statistical quality of the data seems in general to be insufficient to confidently show any very fine details. The baseline model is quite similar to the actual data, but not a perfect fit by any means. Beyond that, the parameter space quickly becomes too large to single out any one effect and pronounce it established on the basis of the data (contrary to practice). A wide variety of effects will of course be "consistent" with the data. So far no single parameterization of the data seems to reveal clean diagnostic effect.

2.5 Emission Mechanisms

First we should address some standard incoherent mechanisms of relevance.

1. Synchrotron Radiation

Synchrotron radiation is emitted by a relativistic particle moving in a magnetic field. This radiation has been analyzed extensively by Sokolov and Ternov (1968), and is normally covered to some extent in standard texts on electromagnetism (e.g., Jackson 1975; see also Bekefi 1966). For nonrelativistic particles, one gets *cyclotron* radiation, which is simply a monochromatic wave of radiation at the frequency at which the particle circles, namely the cyclotron frequency,

$$\omega_c(0) = eB/m = 1.76 \times 10^7 B \, \text{s}^{-1} \, (B \text{ in gauss}) \tag{1}$$

(or equivalently $f_c = 2.80$ MHz B), where the zero argument indicates the limit of zero energy, because particles with nonzero energy have a cyclotron frequency declining as

$$\omega_c = eB/m\gamma = \omega_c(0)/\gamma \approx \omega_c(0)(1 - E/mc^2), \tag{2}$$

where γ is the Lorentz factor (it is often approximated that ω_c is constant for nonrelativistic particles; however, there is a possibly significant linear frequency dependence on kinetic energy E even in this limit, as shown above). With increasing energy, harmonics appear in the cyclotron radiation spectrum because to a distant observer in the plane of motion (for example) the sinusoidal particle motion begins to look nonsinusoidal, with the particle appearing to spend more time in the receding part of its orbit than in the approaching part. Since radiation reaction is proportional to the apparent acceleration of the particle, the radiation begins to be enhanced during the approaching part of the orbit and the frequency of this radiation begins to mount (equivalently, the radiation can be viewed as being Doppler blueshifted and bunched by relativistic beaming). In the limit of large γ, the harmonics blend together and one obtains a "magnetic bremsstrahlung" spectrum. This spectrum is *independent* of the particle energy at low energies and rises very slowly (as $\omega^{1/3}$) to a high-frequency turnover of order

$$\omega_H = \gamma^3 \omega_c = \gamma^2 \omega_c(0), \tag{3}$$

where the γ^3 enhancement partially compensates for the γ^{-1} decline in cyclotron frequency, an important distinction when we discuss curvature radiation below. The classical bremsstrahlung spectrum for charged particles hitting a target and stopping in it is a constant intensity (i.e., energy radiated at a rate proportional to the frequency interval) up to photons having the full particle kinetic energy, $\hbar\omega_{max} = (\gamma - 1)mc^2$. The magnetic bremsstrahlung is very similar except for the weakly rising spectrum and the cutoff at ω_H instead of ω_{max}. This radiation is emitted preferentially along the instantaneous velocity vector of the particle and is confined to roughly a cone of opening angle γ^{-1} radians. The energy-loss rate is given by

$$\dot{W} = e^4 B^2 \beta^2 \gamma^2 / 6\pi\epsilon_0 m^2 c. \tag{4}$$

The two powers of particle mass in the denominator limit this mechanism largely to electrons ($\beta \equiv v/c \approx 1$ for relativistic cases). In practical units, the lifetime of a nonrelativistic electron is (Bekefi 1966, p. 181)

$$\tau = 2.58 \times 10^8 / B^2 \text{ s} \,(B \text{ in gauss}), \tag{5}$$

while for relativistic electrons it approaches

$$\tau = 8.86 \times 10^8 / B^2 \gamma \text{ s}, \tag{6}$$

so in 10^{12} magnetic fields the lifetimes are negligible for most purposes. In the case of sufficiently large γ, however, ω_H will exceed ω_{max}, in which case the cutoff is now at this physical limit (Harding and Preece 1987). In the case of extremely strong magnetic fields, the underlying harmonic structure of the synchrotron radiation should not be forgotten, because there will be no radiation below the fundamental frequency, $\omega_c(0)$, which corresponds to photon energies of $\hbar\omega_c(0) = 11.5 \, B_{12}$ keV, where B_{12} is the magnetic field in units of 10^{12} gauss. Thus an energetic electron moving across a magnetic field of pulsar strength would not radiate radio waves or even optical waves, but rather the spectrum would start with hard X-rays and extend upwards to the high-frequency cutoff. Indeed, for $\gamma B \geq 10^{16}$ gauss, the electron would be de-energized before it could make a single gyration. Clearly, then, pulsar radiation is not due to synchrotron radiation per se. However, for energetic electrons moving on weak magnetic fields such as those in the interstellar or intergalactic medium, synchrotron radiation is an important mechanism for generating radio waves.

2. Curvature Radiation

In synchrotron radiation, the particles move in circles (or helices) owing to their having a component of motion across the magnetic field lines. But the calculation of the radiation reaction is independent of *why* the particles move in a circle, and only assumes that the particles move in a circle at a fixed velocity. A convention is therefore to reserve the term "synchrotron radiation" for relativistic particles circling a magnetic field line and to use the term "curvature radiation" for any other case that forces circular motion, usually because the electrons follow curved magnetic field lines.

For circular motion, relativistic particles radiate in a narrow cone about the particle velocity vector, so it is not at all necessary that the particle move in a circle per se; the particle could move along any trajectory, and its radiation would be characterized by the instantaneous curvature of the trajectory where the velocity vector intersected a distant observer.

Insofar as pulsars go, the most often cited reason for particles to move in curved trajectories is that they follow magnetic field lines. Normally, particles are thought of as spiraling along magnetic field lines, but as we have seen, the particles will instantly lose any perpendicular energy component of motion to synchrotron radiation, cascading down to zero free energy in roughly 10 keV steps (or multiples thereof). These states are quantized of course and are often referred to as the *Landau* states for the particle, with energy levels forming the quasi-harmonic oscillator series

$$W_n = mc^2(\gamma^2 + 2nb)^{1/2} \tag{7}$$

where γ is the Lorentz factor for motion *along* the magnetic field lines, b is the magnetic field in units of the critical magnetic field of 4.41×10^{13} gauss (sec. 1.7.3), and $n = l = 0, 1, , \cdots$, the angular momentum quantum number of electrons with spin-down. For electrons with spin-up, $n = l + 1$, so the ground state is for electrons with spin-down relative to the magnetic field direction. For straight field lines, there is nothing to affect the motion along the magnetic field line. Generally field lines are not straight and therefore the particles will lose energy to curvature radiation at the much slower rate

$$\dot{W} = e^2\beta^4\gamma^4 c/6\pi\epsilon_0\rho_c^2, \tag{8}$$

where ρ_c is the radius of curvature. Substituting $\rho_c = \beta\gamma mc/eB$ for motion in a magnetic field returns us to equation 4 for synchrotron radiation.

Consider a particle moving on a closed dipole magnetic field line from a neutron star. The particle will radiate and then hit the star. Particles moving in planetary magnetic fields typically mirror magnetically and are prevented from hitting the planet ("precipitating"). Without perpendicular energy, the particles cannot mirror. Interaction with the intrinsic magnetic moment of a particle will force precipitation at one or the other magnetic pole. Alternatively the field line might be open and lead to infinity, in which case the particle would radiate strongly on the tightly curved portions of the field line and then weakly as the field line straightened out. In both cases, the time the particle has to radiate strongly is limited by the likely field geometry about the neutron star, in contrast to the case for synchrotron radiation, where the particle may circle as many times as one might wish. In both cases, this time is of the order of $\delta t = \rho_c/c$. Consequently we can estimate the total energy lost by a given particle from

$$\delta W \approx \dot{W}\delta t = e^2\gamma^4/6\pi\epsilon_0\rho_c. \tag{9}$$

This energy is distributed in intensity $f(\omega)$ ($\equiv dI/d\omega$) over all frequencies from (almost) zero to ω_H, so

$$\dot{W} = \int_0^{\omega_H} f(\omega) \, d\omega. \tag{10}$$

Because pulsars are typically observed at radio frequencies, the power available at those bands is of an order roughly approximating the integral:

$$\dot{W} = f(\omega_R)\omega_R. \tag{11}$$

With these relationships, the fraction of power radiated by the particle in the radio band is roughly

$$\delta W_R = \delta W \frac{f(\omega_R)\omega_R}{f(\omega_H)\omega_H} = \delta W \left(\frac{\omega_R}{\omega_H}\right)^{4/3} \tag{12}$$

using $f(\omega) \approx \omega^{1/3}$. Because $\omega_H \sim \gamma_3$, the energy dependence drops out of the radio energy emitted per particle (provided of course that $\omega_H \gg \omega_R$). Thus we find

$$\delta W_R = \left(\frac{e^2}{6\pi\epsilon_0\rho_c}\right) \left(\frac{\omega_R\rho_c}{c}\right)^{4/3} \tag{13}$$

Note that the dependence on curvature is only to the 1/3 power. Given these considerations, one can generally estimate the curvature power output independently of the model.

The plausible radii of curvature for magnetic field lines range from about the neutron star radius ($a \approx 10^4$ m) to about the light-cylinder distance ($R_L \approx 10^7$ m), while the curvature near the surface of a pure dipole magnetic field line reaching out to an equatorial distance R_L is given by (Sturrock 1971a)

$$\rho_c = \frac{4}{3}(aR_L)^{1/2}, \tag{14}$$

about an order of magnitude less than R_L. For illustration, we can take $\omega_R \approx 3 \times 10^9$ (≈ 480 MHz) and the smaller value of $\rho_c \approx a \approx 10^4$ m to estimate $\delta W \approx 4 \times 10^{-7}$ eV per particle. Obviously *incoherent* curvature radiation from relativistic electrons (by definition, $E \geq 10^6$ eV) would convert less than 10^{-12} of the particle energy into radio emission, *regardless of the particle energy*, an amount totally insufficient to account for the pulsar radio luminosity, which amounts to 10^{-5} of the total power output on the average.

3. Mechanisms for Coherent Emission: Radio Spectrum

The simplest way to boost the low-frequency luminosity can be seen by noting that the power output scales as the square of the particle charge. Thus, if we had a fictitious electron with 10 times the charge, it would radiate 100 times more curvature radiation. No such particle exists, but as an alternative we can imagine 10 electrons all moving together in a tight bunch. For wavelengths long compared to the bunch size, the radiation from each electron has the same phase and therefore the radiation fields add constructively; thus we get fields 10 times stronger and therefore a power output 100 times larger (i.e., proportional to the field amplitude squared), just as for a single particle. One might worry that the "particle" represented by this bunch would also have an effective mass 10 times larger, but that issue is irrelevant because curvature radiation, unlike synchrotron radiation, is independent of the particle mass. This constructive interference can work only for wavelengths long compared to the particle size. At wavelengths short compared to the bunch size, the phase relations become scrambled, and eventually at wavelengths short compared to the interparticle spacing the bunch simply radiates incoherently with an intensity 10 times the single-particle intensity. Thus we get an amplification of a factor N, where N is the number of particles in the bunch (not a factor of N^2 because we are discussing the output *per particle*). The resultant spectrum (somewhat generic because the spectrum will in general depend on emission angle and shape of the bunch) is shown in figure 2.22.

The extremely high brightness temperatures of pulsar radio emission (sec. 1.5.4) already require a coherent source for pulsar radiation. Coherence on such a large scale came as a surprise. However, large-scale coherence does occur elsewhere in nature. For example, the electromagnetic radiation from a lightning bolt is highly coherent or it would not cause radio interference. A more ominous source of coherent radio waves is the giant electromagnetic pulse (EMP) created by prompt gamma-rays from a high-altitude nuclear explosion (the Compton scattered electrons are all produced nearly in phase to gyrate in the Earth's magnetic field: Broad 1981). Another, less accessible example is maser action in giant molecular clouds. Sound waves, plasma waves, and water waves are intrinsically coherent phenomena. For pulsars, three basic mechanisms have been advanced to explain the high brightness temperatures (see, for example, Ginzburg et al. 1968): (1) particle anisotropy in physical space (e.g., bunches), (2) particle anisotropy in velocity space (e.g., maser-like), and (3) true masers (e.g., population inversion in discrete quantum states).

The first mechanism acts in the case of lightning because the current is bunched, providing coherence by having many charges (electrons) radiating together in phase. Such a mechanism immediately favors low frequencies, as discussed above. The brightness temperature can then be as high as N times the particle energy; $kT_B \approx NE$. Another example, possibly relevant to pulsars, is the radiation in an electron storage ring (Michel 1981). Such rings produce useful quantities of incoherent synchrotron radiation at X-ray wavelengths. However, the electrons are stored as bunches (size of order of a centimeter) and consequently radiate coherently at wavelengths long compared to a centimeter. The coherent power output at such wavelengths can become the dominant power loss and may cause problems in storage ring design. This effect seems to have been observed (Nakazato et al. 1989).

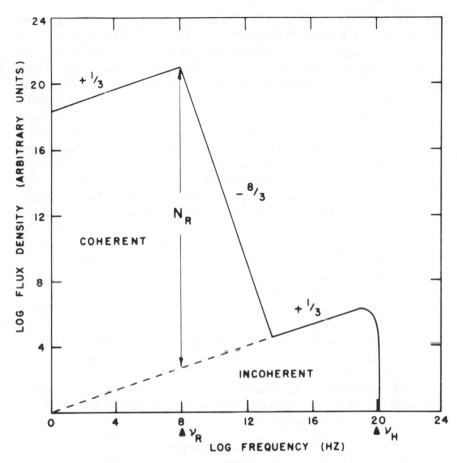

FIG. 2.22. Idealized spectrum from bunched particles. The maximum coherent amplification (N_R, the number of particles in a bunch) is obtained at wavelengths long compared to the bunch size (which then acts as a single particle of a very large charge). No coherence is obtained at wavelengths shorter than the mean spacing between particles, and the behavior between these limits is simply interpolated. The steep drop with increasing frequency is an observed property typical of pulsars. However, the Crab pulsar falls off faster than the $-8/3$ law shown, more nearly a $-10/3$ law. From F. C. Michel, 1978a, *Ap. J.*, 220, 1101 (figure 1); 1982, *Rev. Mod. Phys.*, 54, 1 (figure 30).

The second mechanism appeals to some peculiarity in the velocity distribution. Roughly speaking, if a Maxwellian fit to the *local* velocity distribution requires a *negative* temperature, then one has what amounts to a classical population inversion and the appropriate wave modes grow exponentially at first and continue to grow until the velocity-space distribution is modified to remove the effective population inversion. The simplest example in plasma physics is the two-stream instability wherein, for example, counterstreaming electrons in a uniform positive background excite plasma oscillations (simple longitudinal oscillations about local charge neutrality). These oscillations do not radiate directly, because they do not happen to propagate, owing to the simplicity and symmetry of this example. However, the negative temperature aspect is easy to see in this example of a one-dimensional velocity distribution of two cold counterstreaming (velocity V_0) beams. The velocity drops precipitously for either $V > V_0$ or $V < V_0$, assuming the beams themselves have velocity spreads small compared to V_0. But for $V > V_0$ or $V < V_0$ the population of fast particles is larger than for slow particles; hence temperatures are negative (the only way a Maxwellian can mathematically match such a situation), and hence there is a maser-like exponential growth of waves. A combination of the above two mechanisms is exemplified by the free electron laser. The coherence basically comes from particle bunches (mechanism 1), but the bunching itself is caused by interaction of the electrons with the radiation field. Hence the exponential growth is from feedback, not population inversion. A similar type of mechanism apparently amplifies whistler emissions in the Earth's magnetosphere (see, for example, Helliwell et al. 1980). Goldreich and Keeley (1971) have shown that a uniform beam in a storage ring is unstable to bunching, and have suggested that this mechanism may be active in pulsars (see sec. 8.2; see also Asséo et al. 1983, who confirm this analysis and generalize it).

The third mechanism appeals not to an analogy with population inversion but to a true inversion. Here, instead of atomic states, one has the Landau levels of electrons in a strong magnetic field; each electron is in a quantized Landau orbital. Population inversion could then lead to true maser emission. An early set of theories were based on such ideas (sec. 8.3.1).

Ochelkov and Usov (1980a,b) argue that the radio turnover at low frequencies that is typically seen in pulsars is due to self-absorption rather than being intrinsic to the radiation mechanism.

4. Cherenkov Radiation

Another mechanism for producing coherent radiation is the Cherenkov mechanism, wherein particles (typically relativistic) moving through a medium faster than the phase velocities of electromagnetic waves will radiate these waves in the form of a "shock" cone expanding behind the particle. Because the phase velocity of light is typically small in liquids and solids, the radiation is readily produced by fast electrons in these media (e.g., the blue glow from relativistic beta decay particles—electrons—seen from reactor cores immersed in water).

The radiation rate for Cherenkov radiation is (Jackson 1975)

$$\frac{dW}{dx} = \left(\frac{e^2}{4\pi\epsilon_0 c^2}\right) \int \omega \left(1 - \frac{1}{\beta^2 n(\omega)}\right) d\omega, \tag{15}$$

where the range of integration is, naturally, only over positive arguments (i.e., where the particle velocity exceeds the phase velocity), with n here the index of refraction. The upper limit to this loss rate for radio waves is clearly

$$\frac{dW}{dx} \geq \left(\frac{e\omega_R}{c}\right)^2 \frac{1}{8\pi\epsilon_0}, \tag{16}$$

completely independent of any knowledge of the actual medium or particle energies, which gives a value of 1.4×10^{-7} eV/m. The path length cannot be very long owing to the rapid variation in B and plasma concentration with radial distance, so even taking a path length of 10^4 m (the characteristic scale for variation; the neutron star radius) would not give adequate power, even if the plasma dispersion relations were themselves favorable. The latter is itself unlikely because the large indices of refraction are typically found near poles at some characteristic frequency; the plasma frequency ($\omega_p \approx 10^{10}$) would be in the correct ballpark, but simple plasmas have instead a *zero* in the index of refraction at the plasma frequency. The poles are instead at the cyclotron frequency, which is far beyond radio frequencies. Even if the correct medium were available, it would still be necessary to bunch the particles to boost the coherence to get enough radio power out, in which case the more certain mechanism of curvature radiation would seem the more natural choice.

5. Radiation from a Relativistic Linear Oscillator

If plasma oscillations are excited in a plasma near a neutron star, it is unlikely that they will be mild. Rather, our expectation is that they will be large enough for the particles to become relativistic. A detailed analysis of this radiation can be found in Wagoner (1969). Panofsky and Phillips (1955) give as the energy-loss rate for acceleration parallel to the velocity vector

$$\dot{W} = \frac{e^2}{16\pi^2\epsilon_0 c} \dot{\beta}^2 \frac{\sin^2\theta}{(1 - \beta\cos\theta)^5}, \tag{17}$$

which can be integrated over a solid angle

$$\int \frac{\sin^2\theta}{(1 - \beta\cos\theta)^5} \sin\theta \, d\theta \, d\phi = \frac{8\pi}{3}\gamma^6, \tag{18}$$

which gives the same radiation formula as for synchrotron radiation, except that the γ dependence is here the 6th power instead of the usual 4th power (which is misleading). If the electrons are driven by a large-amplitude electric field oscillating parallel to the ambient magnetic field (e.g., see figure 2.23), we have

$$\gamma\beta = f(t) \equiv f_0\sin\theta t, \tag{19}$$

and we therefore have the general expressions

$$\beta^2 = \frac{f^2}{(1 + f^2)}, \tag{20}$$

$$\gamma^2 = 1 + f^2, \tag{21}$$

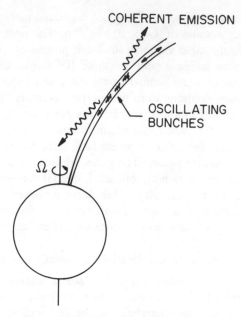

FIG. 2.23. Coherent radiation from large-amplitude oscillations. If the current to the pulsar is in the form of a sheet into the vicinity of the auroral zone, large-amplitude plasma oscillations can radiate into low-density plasma regions on the sides.

and

$$\dot{\beta}^2 = \frac{\dot{f}^2}{(1 + f^2)^3},$$ (22)

which substituted into equations 17 and 18 give

$$\dot{W} = \frac{e^2}{6\pi^2 \epsilon_0 c} \dot{f}^2,$$ (23)

which is exactly the same as the nonrelativistic oscillator! The effective amplitude is therefore the same as if the particle could be driven faster than c, and

$$\dot{W} \approx (\text{constant}) E(t)^2$$ (24)

regardless of the amplitude or form of the driving electric field, $E(t)$.

   The radiation pattern is of course relativistically beamed in a narrow cone along the line of the oscillation, just like synchrotron radiation. Most of the radiation is created (1) when the receding electrons stop and are accelerated toward the observer and (2) when the approaching electrons stop and are accelerated away. However, these two events are seen almost simultaneously by the observer, so the radiation is in the form of a close double pulse emitted every cycle of oscillation, which has a relative spacing of about $(\theta/2)^2$ of the total between pulses. The pulses themselves are compressed by the usual factor of γ. As Wagoner (1969) points out, the spectrum in any given direction looks vaguely blackbody, albeit chopped up owing to interference between the double pulses. The total power spectrum (power per unit frequency) is very much like bremsstrahlung, almost flat and declining slowly rather than exhibiting the $\omega^{1/3}$ rise in intensity in synchrotron radiation, with a cutoff at about a frequency of $\omega_0\gamma^2$. Thus the low-frequency radiation is preferentially emitted at large angles and vice versa, which would greatly enhance emission from near the surface of thin current sheets.

2.6 Propagation

Although the propagation of signals can often be treated separately from the emission of the signal, at high brightness temperature the question of reabsorption and nonlinear effects becomes important.

1. Propagation

The standard approach to the wave propagation properties of plasmas is to linearize the Lorentz and continuity equations, typically by assuming that the plasma particles are at rest in a background magnetic field and by giving all of the perturbations a harmonic structure. Thus the particle velocities are $v(r, t) = v \exp i(\mathbf{k} \cdot \mathbf{x} - \omega t)$, with v a constant (a complex number in order to handle relative phases), etc. Maxwell's equations then become

$$i\mathbf{k} \cdot \mathbf{E} = \rho/\epsilon_0, \tag{1a}$$

$$i\mathbf{k} \times \mathbf{E} = +i\omega\mathbf{B}, \tag{1b}$$

$$i\mathbf{k} \cdot \mathbf{B} = 0, \tag{1c}$$

and

$$i\mathbf{k} \times \mathbf{B} = \mu_0(\mathbf{J} - i\omega\epsilon_0\mathbf{E}). \tag{1d}$$

These algebraic equations are simple to solve. Taking the "gradient" of equation 1d by taking the scalar product with $\mathbf{k}$ eliminates $\mathbf{B}$, and $\mathbf{E}$ can be eliminated using equation 1a; hence

$$\omega\rho - \mathbf{k} \cdot \mathbf{J} = 0, \tag{2}$$

which is just the continuity equation (i.e., conservation of charge). Dimensionally, $\mathbf{J} \approx en\mathbf{v}$ and therefore is generally nonlinear because both the particle concentration, n, and velocities, $\mathbf{v}$, are perturbed in general. This difficulty is solved by simply discarding the nonlinear terms and keeping the leading term, $\mathbf{J} = en_0\mathbf{v}$, where n_0 is the average particle concentration. In the same way, the Lorentz force has the nonlinear term $\mathbf{J} \times \mathbf{B}$, which is linearized by keeping only $\mathbf{J} \times \mathbf{B}_0$, where $\mathbf{B}_0$ is the background field (if any). The usual transversality of light waves is given by equations 1b and 1c, which largely carries over to the vacuum except that nontransverse solutions are possible in the vacuum (the "extraordinary" mode waves and plasma oscillations), and one has the weaker condition that $\mathbf{B}$ vanishes if $\mathbf{k}$ parallels $\mathbf{E}$. The relationship between $\mathbf{E}$ and $\mathbf{J}$ is given from the linearized Lorentz force, namely

$$J_i = \sum \sigma_{ij} E_j, \tag{3}$$

where the sum is over the three coordinates ($j = x, y, z$). The conductivity tensor is (taking $\mathbf{B} = B\mathbf{e}_z$)

$$\sigma_{ij} = \begin{bmatrix} \sigma_P & \sigma_H & 0 \\ -\sigma_H & \sigma_P & 0 \\ 0 & 0 & \sigma_0 \end{bmatrix}, \tag{4}$$

where σ_0 is the so-called zero field conductivity (if $B = 0$, $\sigma_H \equiv 0$, and $\sigma_P = \sigma_0$); σ_P is the Pederson conductivity and the current that flows in the direction of the applied electric field; σ_H is the Hall current that flows orthogonal to both the applied electric field and the ambient magnetic field. It is straightforward to invert the Lorentz force, using

$$m\ddot{x} = m \left[\frac{\partial v_x}{\partial t} + \mathbf{v} \cdot \nabla v_x \right] \approx -i\omega v_x = -\frac{i\omega J_x}{e n_0}, \tag{5}$$

because we assumed $\langle \mathbf{v} \rangle = 0$. Note that the second term, $\mathbf{v} \cdot \nabla$ (the so-called convective derivative), has a simple physical meaning. If we imagine water spilling over a waterfall and examine the fluid velocity at a fixed x below the lip of the fall, the water will always have essentially the same velocity. But the water is accelerating. In this case it is the first partial derivative that is zero (the water always has the same velocity at fixed location) and the second that represents the acceleration (as we change x we find different velocities, even at a fixed instant). We can now eliminate ρ, $\mathbf{J}$, and $\mathbf{B}$ to write

$$\mathbf{k}(\mathbf{k} \cdot \mathbf{E}) - k^2 \mathbf{E} + \frac{\omega^2}{c^2} \sum N_{ij} E_j, \tag{6}$$

where

$$N_{ij} = I_{ij} - \frac{i}{\omega \epsilon_0} \sigma_{ij} \tag{7}$$

(I_{ij} is just the unit tensor). For electrons (all frequencies are positive numbers)

$$\sigma_0 = i\epsilon_0 \frac{\omega_p^2}{\omega}, \tag{8a}$$

$$\sigma_P = i\omega\epsilon_0 \frac{\omega_p^2}{\omega^2 - \omega_c^2}, \tag{8b}$$

and

$$\sigma_H = \epsilon_0 \frac{\omega_p^2 \omega_c}{\omega^2 - \omega_c^2}, \tag{8c}$$

with $\omega_c \equiv eB_0/m$ the electron cyclotron frequency and $\omega_p^2 \equiv e^2 n_0/\epsilon_0 m$ the electron contribution to the plasma frequency. These are the basic starting equations for the theory of linear plasma waves; however, for the case of strong magnetic fields they can be simplified radically.

a. Quasi-Neutral Plasmas

The dielectric coefficients are (just the electron part)

$$N_0 = 1 - \frac{\omega_p^2}{\omega^2} + ..., \tag{9a}$$

$$N_P = 1 - \left(\frac{\omega_p^2}{\omega^2 - \omega_c^2} + ... \right), \tag{9b}$$

$$N_H = \frac{i}{\omega} \left(\frac{\omega_p^2 \omega_c}{\omega^2 - \omega_c^2} - ... \right), \tag{9c}$$

where the ellipses indicate the sign of contribution one would have for the equivalent contributions from positive particles. The dispersion relation is then equation 6, which can be written

$$\sum D_{ij} E_j = 0, \tag{10}$$

where in general we can write $\mathbf{k} \cdot \mathbf{B} \equiv kB \cos \theta$ and define an index of refraction $n = ck/\omega$ to obtain

$$D_{ij} = \begin{bmatrix} -n^2 \cos^2 \theta + N_P & N_H & n^2 \sin \theta \cos \theta \\ N_H^* & -n^2 + N_P & 0 \\ n^2 \sin \theta \cos \theta & 0 & -n^2 \sin^2 \theta + N_0 \end{bmatrix}. \tag{11}$$

Near a pulsar, ω_c is huge, and in fact the corresponding length is much shorter than the distance between plasma particles at any likely density, and at such frequencies the above equations would not be applicable anyway. Consequently

$\omega \ll \omega_c$, the Hall contribution is tiny, and the Pederson contribution becomes

$$N_P \to 1 + \frac{\omega_p^2}{\omega_c^2} \to 1. \tag{12}$$

(For an electron-positron plasma, the Hall contribution vanishes identically.) For all practical purposes, the waves cannot budge the electron across the strong magnetic field lines, and therefore the electrons become invisible. Thus the index of refraction is unity for all waves except waves propagating at an angle θ to the magnetic field with the electric vector in the $\mathbf{k} - \mathbf{B}$ plane (the "ordinary" mode wave). For this wave the index of refraction is

$$n^2 = \frac{\omega^2 - \omega_p^2}{\omega^2 - \omega_p^2 \cos^2 \theta}. \tag{13}$$

a. *Dispersion Measure and Rotation Measure*

The propagation of waves across the thin, weakly magnetized plasma of interstellar space puts us in an entirely different regime. Here we get the important phenomena of frequency-dependent pulse delay (dispersion measure) and rotation of the plane of polarization of the wave (rotation measure: RM). These effects stand in curious contrast to one another. Dispersion measure represents the differences in *group* velocity of waves, which is caused by interstellar electrons and is virtually independent of the magnetic field. Observers quote dispersion measure in *practical* units of electron column depth in units of electrons/cc times pc. Rotation measure on the other hand represents differences in the *phase* velocity caused additionally by the existence of a component of magnetic field along the line of sight. Observers quote rotation measure in *observed* units of radians of rotation per wavelength squared. Fortunately the latter units happen to be almost equal to the units for DM times microgauss. Thus, if a pulsar had a DM of 30 and a RM of 30, the pulsar could be interpreted as being about 1 kpc distant with an interstellar magnetic field along the line of sight that had an average value of about 1 μG (1.232 μG to be numerically exact).

For waves with $\mathbf{k}$ parallel to $\mathbf{B}$, the dispersion relationship can be diagonalized by treating circularly polarized waves instead of linearly polarized waves. Thus we can write

$$N_L \equiv N_P + iN_H \approx 1 - \frac{\omega_0^2}{\omega^2}\left[1 - \frac{\omega_c}{\omega}\right] \tag{14}$$

and

$$N_R \equiv N_P - iN_H \approx 1 - \frac{\omega_0^2}{\omega^2}\left[1 + \frac{\omega_c}{\omega}\right]. \tag{15}$$

The convention here is that electromagnetic vectors for right-handed waves rotate in the same direction (at a fixed position) as electrons do about the ambient magnetic field. Thus these waves can resonate with the electrons if their frequency is the cyclotron frequency. The phase velocity for right-handed waves is just

$$V_{\text{phase}} = \frac{\omega}{k} = cN_R^{-1/2}, \tag{16}$$

and consequently the phase difference is

$$\phi = \phi_R - \phi_L = \frac{\omega L}{c}(N_R^{1/2} - N_L^{1/2}) \approx -\frac{\omega_c\omega_p^2 L}{\omega^2 c} \tag{17}$$

after propagating a distance L. The functional dependence is B from ω_c and n_e from ω_p^2, and accounting for variation with distance gives, because the effect is linear, $\text{RM} \approx \int n_e B dL$. It is only necessary to remember the factor of 1.232 that must be multiplied to convert from radians/m^2 to μG $\times$ electrons/cc $\times$ pc.

The dispersion measure is well known from whistler phenomena, wherein lightning strokes give, in the same way, waves whose group velocities differ with frequency, with the high-frequency components traveling faster than the slow ones. Strictly speaking, whistlers observed on Earth are right-handed waves below the electron cyclotron frequency, but the same effect is obtained regardless of polarization for waves with frequencies well above the plasma frequency. Only the effect is much smaller. For these waves (neglecting ω_c)

$$n^2 = N_0 \tag{18}$$

and the group velocity is

$$V_{\text{group}} \equiv \frac{d\omega}{dk} = \frac{c^2}{V_{\text{phase}}} \approx c\left(1 - \frac{\omega_p^2}{2\omega^2} - \frac{\omega_c\omega_p^2}{\omega^3}\right), \tag{19}$$

so the time difference after going a distance L will be directly proportional to that distance and ω_p^2, which only depends on n_e; hence $\text{DM} \approx \int n_e \, dL$. Note the typically weak sensitivity to cyclotron frequency. The observations are usually quoted in practical units. As discussed in sec. 1.5.10, if the propagation were through even a comparatively short distance of a strong magnetic field, the plane of linear polarization would be rotated through so many cycles and would be so sensitive to the exact path taken that the signal would effectively be depolarized. On the other hand there would then be a contribution from the last term in equation 19 and the pulse would arrive slightly earlier in one sense of circular polarization than the other. For a Gaussian-shaped pulse, the leading edge would then have one sense of circular polarization, which would reverse to the other sense at the trailing edge. Although some pulsars do show such a sense reversal, they also are accompanied by substantial linear polarization, with the exception of the eclipsing pulsar PSR 1957 + 20, which has extremely weak linear polarization if any.

a. Pair Plasmas

For a plasma composed chiefly of electron/positron pairs, the dispersion relation simplifies considerably because the Hall term (eq. 9c) vanishes identically. Consequently a pair plasma has no rotation measure. The off-diagonal elements in equation 11 can then be removed by a rotation of coordinate systems, and we have two pure linearly polarized wave solutions, one polarized orthogonal to the magnetic field and one polarized at an intermediate angle not quite transverse to **k** and **B**. Langdon et al. (1988) discuss shock waves in pair plasmas, while Alsop and Arons (1988) treat solitons. Propagation through ultrarelativistic pairs has been addressed by Arons and Barnard (1986), and possible correlation with pulsar width versus frequency has been considered by Barnard and Arons (1986).

a. Nonneutral Plasmas

Wave propagation in nonneutral plasmas (sec. 4.2.2) presents a special problem because the underlying assumptions for the quasi-neutral plasma fail: the particle velocities cannot be taken to be zero, and the boundaries cannot be taken to be at infinity. Of course for a high-frequency wave, the plasma could be treated locally as a plasma with only one sign of charge contributing to the dielectric constant. But the relative directions and magnitudes of all the components will in principle be constantly changing as the wave propagates, necessitating a WKB approximation at a minimum. For low frequency waves, one must solve eigenvalue problems involving the free boundaries. At present it is a formidable task even to locate where the free boundaries can self-consistently reside (Krause-Polstorff and Michel 1985a,b), much less to find the normal modes involving such boundaries.

2. Propagation in a Moving Medium

If the pulsar radio emission propagates through local plasma, that plasma is probably expanding away from the pulsar; hence any interaction with the pulsar radio-frequency emission is Doppler shifted. To estimate the importance of this effect, consider the propagation of a weak wave of angular frequency ω_0 through a slab of plasma of thickness δr_0 and plasma concentration n_0. The phase change in going through the slab is

$$\delta\phi = \omega_0\delta t = \omega_0 \frac{\delta r_0}{V_{\text{phase}}}, \qquad (20)$$

and we can write for the phase velocity, using $V_{\text{phase}} = c/n$ with $n = (1 - \omega_p^2/\omega_0^2)^{1/2} \approx 1 - \omega_p^2/2\omega_0^2$, giving a phase change

$$\delta\phi = e^2 n_0 \delta r_0 / 2m\epsilon_0\omega_0. \qquad (21)$$

We can now transform to a moving coordinate system, using the fact that the phase change is a relativistic invariant and the usual transforms $n = \gamma n_0$, $\delta r = \delta r_0/\gamma$, and $\omega = \gamma(1 + \beta)\omega_0$, to obtain for the time delay in the wind

$$\delta t = \frac{e^2}{2m\epsilon_0\omega^2} \left(\frac{1+\beta}{1-\beta}\right)^{1/2} \int n\,dr, \tag{22}$$

which is just the standard time delay except for the relativistic correction factor, which, for extreme relativistic flow, is $\approx 2\gamma$. Thus it is more difficult to penetrate plasma that is flowing away because the wave frequency is reduced in the comoving frame. However, the local plasma contribution $\int n\,dr$, using the usual numerology, would correspond to a DM of order 3×10^{-3}. Accordingly, the likely importance of such propagation effects is very model dependent.

3. Large-Amplitude Waves

For large-amplitude waves, the linear assumption in the previous section fails. As the amplitude increases, the current driven in the plasma increases to the point where (1) the particles become relativistic and (2) the perturbation magnetic field becomes comparable to the ambient magnetic field. Then the particle motion fails to be purely oscillatory, and the particle is accelerated rapidly in the direction of propagation of the wave. All of this is simply a description of the bulk radiation pressure exerted by the wave on the plasma. In the linear theory the radiation pressure is discarded in the process of linearization, which permitted oscillatory solutions in the first place. Yet linear stability analyses are commonly relied upon (e.g., Beskin et al. 1987).

What might seem surprising is that pulsar waves are "large-amplitude" waves simply by virtue of their low frequencies. The operative equation is just the electrodynamic acceleration of the particles,

$$m\ddot{x} = eE, \tag{23}$$

and writing $\ddot{x} = i\omega v$ and $E = cB$ gives a limiting field for $v \approx c$ of

$$B \leq m\omega/e. \tag{24}$$

For electrons at the rotational frequency of the Crab pulsar, B is about 1 microgauss! Pulsar winds are already relativistic simply owing to their low frequency, which enables even weak electric fields to exert themselves for comparatively long times.

The second criterion reads

$$B = \mu_0 i, \tag{25}$$

where i is the current density within a sheet one wavelength thick, or

$$i = env\lambda, \tag{26}$$

which gives, substituting as above for v and replacing $\lambda = c/\omega$, not a field limit but a frequency limit

$$\omega^2 > \omega_p^2, \tag{27}$$

which is just the plasma frequency limit on propagation. This tells us (something often ignored in the linearized treatment but physically obvious) that, when a wave is reflected from a plasma because it cannot propagate through it, the wave momentum is exerted on the plasma as an accelerating force. (Students studying plasma physics routinely solve for the reflection of waves by a $T = 0$ plasma that simply sits in place!)

The boundary conditions are not always obvious for such cases. If large-amplitude waves pass through a magnetosphere, for example, the magnetic field lines are anchored to the source to some degree. Depending on the topology, they might be blown back to form a wind or tail, or they might be held in place.

4. Relativistic Plasmas

Space does not permit a deep discussion of relativistic effects, so we will concentrate on a very simple example, namely the effect of a harmonically bound particle driven by a sinusoidal force. This could represent the effect of a perturbation wave field driving particles out of charge neutrality, with the plasma frequency playing the role of the restoring force, or a circularly polarized wave driving the particle's cyclotron motion, etc. The only thing we will add are the special relativistic corrections. This standard equation can be written in nonrelativistic notation as

$$m\dot{v} = -kx + eE\cos(\omega t), \tag{28}$$

and it is useful to eliminate units by noting that the natural frequency is just $\omega_0 = k/m$, in which case we can measure time in units of $1/\omega_0$, distance in dimensionless units $x\omega_0/c \rightarrow x$, and driving force in dimensionless units of $a \equiv eE/\omega_0 mc$. Thus we have the universal equation

$$\dot{\beta} = -x + a\cos(\alpha t), \tag{29}$$

where $\alpha \equiv \omega/\omega_0$ and $\beta = \dot{x}/c$. The relativistic version is simply

$$\frac{d(\gamma\beta)}{dt} = -x + a\cos(\alpha t). \tag{30}$$

A "relativistic plasma frequency" is sometimes defined simply by replacing m with γm in the usual expression; hence $\omega_p^2 \rightarrow \omega_p^2/\gamma$. The force of our analysis here is to what extent that definition means anything. If we examine the unforced ($a = 0$) solutions with large (dimensionless) amplitude, we discover that indeed the oscillation time is effectively reduced. Basically the particles always move with c and therefore the frequency is proportional to the amplitude rather than being independent of amplitude. In the driven solutions, however, the amplitude dependence of the natural frequency means the removal of the usual strong resonance when the driving frequency approaches the natural frequency. What happens is very simple. At resonance, the amplitude grows linearly, just as for the nonrelativistic case, but as it grows, the natural frequency shifts out of resonance. As a result, a phase difference steadily grows between the driving force and the particle motion and the driving force begins to take energy *out* of the motion. The particle amplitude, then, only grows to a certain limit before decaying back to zero and then being reaccelerated anew. The relationship between maximum γ and maximum amplitude x_0 is just

$$\gamma = 1 + x_0^2/2, \tag{31}$$

and numerical simulations show that, for weak driving forces ($a \ll 1$) at (low-amplitude) resonance, $x_0 \approx 2.2a^{1/3}$, which essentially says that the particles can never be driven relativistic by resonant excitation. The natural oscillation is essentially modulated with a harmonic envelope that has a period $T \approx 11/a^{2/3}$ compared to the natural period of 2π. Particles are only driven relativistic if the driving force itself is relativistic, namely the wave has such large amplitude or low frequency that $a/\alpha > 1$ itself. In this case the response becomes "chaotic" because the modulation period becomes comparable to the natural period. However, in that limit the radiation pressure on the particles becomes enormous (this pressure was discarded in the linear approximation when the wave magnetic field was neglected compared to the background magnetic field). For waves propagating across the magnetic field near the pulsar, such neglect would be sensible, but not for waves propagating parallel to the magnetic field.

5. Literature

Since the pulsar magnetosphere, at least in the standard vacuum model, is in relative motion near and beyond the light-cylinder, there is the question of how such motion would modify outgoing pulsed radio waves. Lerche initiated a considerable dialogue on propagation in inhomogeneous, moving media (Lerche 1974a to i; Lee 1974; Lee and Lerche 1974, 1975; Lerche 1975a to d, 1976; also Elitzur 1974 and Harding and Tademaru 1979, 1980; see also Dorman et al. 1973). Ko and Chuang (1978) have disputed some aspects of this work.

Propagation in a strongly magnetized medium has been treated by Ochelkov et al. (1972), Novick et al. (1977), Heintzmann and Schrüfer (1977), Pavlov and Shibanov (1978, 1979), Ventura (1979), Fang and Liu (1976), Cocke and Pacholczyk (1976), and Nagel (1981). The case for streaming plasma has been examined by Elitzur (1974), Heintzmann et al. (1975a,b), Ko (1979), Onishchenko (1975), and Melrose (1979). Evangelidis (1979), Galtsov and Petukhov (1978), and Ignat'ev (1975) have included general relativistic effects.

The issue of propagation is potentially an essential one, but given the difficulty in determining what global model to use to describe a pulsar, the background information necessary to go into such fine-tuning is simply absent. It can be guessed at, but to quote Hewish (1981): "Until these atmospheric [magnetospheric] problems have been solved, it may be rash to consider detailed radiation mechanisms." This caution seems well taken; if, in 10 years say, the correct theory is at hand, 90% or more of the above work will probably be irrelevant because the physical environment postulated will not be that appropriate to actual pulsars. Ideally, such works should be analyzed to illustrate how observational data might be used to constrain pulsar models more tightly. Ruderman (1981) provides an example of such an analysis. Our interest here has been in seeing how far one can go from first principles.

2.7 Other Pulsar Emissions

1. Optical Emission

The Crab, Vela, and LMC pulsars are the three known pulsars visible in the optical. Vela is quite faint, but the Crab corresponds to roughly a 16th magnitude star. The LMC pulsar, at 23rd magnitude, is bright considering its distance. Unlike the radio emission from pulsars, the optical pulse from the Crab is extremely stable in shape and amplitude and contains no lines (Jones et al. 1980; Mahoney et al. 1984). No flickering of the emission (as commonly seen in the radio) has been detected (Hegyi et al. 1969; Jelley and Willstrop 1969; Horowitz et al. 1972; Miller et al. 1975), although some modulation had been predicted (Sturrock et al. 1971). The pulse shape over a wide range of frequencies is shown in figure 2.24. The consensus seems to be that these high-frequency emissions are therefore incoherent and arise from a distinct mechanism (albeit not necessarily from a distinct emission region). As can be seen in figure 2.25, the high-frequency emissions appear to have a distinct spectrum relative to that of the radio. It is marginally possible that the optical could be coherent (Sturrock et al. 1976), but it is extremely difficult to get coherence into the gamma-ray energies whereas spectrally the optical emission extends smoothly up to gamma-ray energies. Tsytovich et al. (1970) suggest a maser mechanism; Elitzur (1979) suggests Compton boosting of the radio photons off relativistic electrons. (For other models involving the Compton effect, see Tsytovich and Chikhachev 1969; Apparao and Hoffman 1970; Arons 1972; Sweeney and Stewart 1974; Stewart 1974, 1975; Bonometto and Scrascia 1974; Shaposhnikov 1976.) Sturrock et al. (1975) suggest that, contrary to the above general view, the optical may in fact be a coherent emission phenomenon (see also Epstein and Petrosian 1973 and Eastlund 1971). Pacini and Salvati (1983) argue that the optical arises from incoherent synchrotron radiation in a pair plasma. Shklovsky (1970b) argues for synchrotron radiation at the light-cylinder.

2. Gamma-rays

The Crab and Vela pulsars radiate detectable fluxes of gamma-rays (Fishman et al. 1969a,b; Tümer et al. 1984). The early reports that PSR 1747–46 and PSR

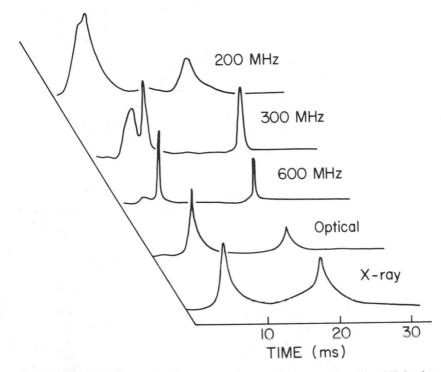

200 MHz

300 MHz

600 MHz

Optical

X-ray

TIME (ms)

FIG. 2.24. The Crab pulse profile from radio to X-ray. The arrival time phase shift has been removed, and the pulse alignment over 11 decades of frequency suggests a simultaneous emission at all frequencies. At frequencies below about 100 MHz (not shown), scattering within the nebula blurs the radio pulse into a sign wave and much of the pulsed amplitude is converted into a powerful steady source (since the spectrum continues to rise). At about 10 MHz, only the steady component is detectable and not the pulsed emission (although one presumes that the pulsar itself is actually emitting sharp pulses at this frequency). From *Pulsars*, F. G. Smith (Cambridge: Cambridge University Press, 1977).

0740–28 are emitting detectable levels of gamma-rays have not been confirmed (Massnou 1980). A number of point sources ("*COS-B*" sources, named after the observing satellite: Swanenburg 1981) that do not seem to be pulsed could conceivably be steady components from as yet unidentified pulsars. The 6.1 ms pulsar 1953+29 might be the source of 10^{12} eV gamma-rays from the source 2CG065+0 (Chadwick et al. 1985a) at the 5σ level, and PSR 1802–23 may be associated with 2CG006–00 (Raubenheimer et al. 1986). These associations require an identification of periods based on a handful of observed gamma-ray events, which of course can be fit by a large number of periods. Thus careful judgments must be made as to whether a close coincidence of one such candidate period with the known pulsar period is statistically significant. In general the data cannot be inverted to *evaluate* the pulsar period without knowing the approximate answer in advance, because each event is separated by a large number of periods.

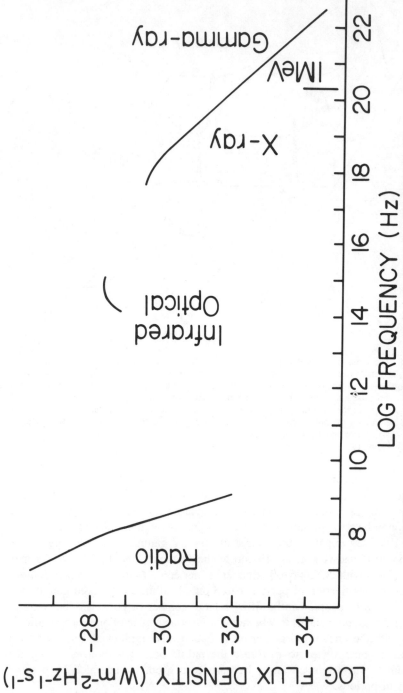

FIG. 2.25. Power spectrum from Crab pulsar. It seems plausible that the optical to gamma-ray emission is from a single mechanism and possibly distinct from that of the radio emission. The total nebular luminosity roughly equals that of the pulsar at the lowest frequency shown, is roughly flat out to the optical, then declines to parallel and roughly equal to the gamma-rays. From *Pulsars*, F. G. Smith (Cambridge: Cambridge University Press, 1977).

Such energetic photons are interesting probes of the pulsar magnetosphere since they will be absorbed in the strong magnetic fields usually proposed. For the Crab pulsar, the gamma-ray pulses seem to be in phase with the radio pulses, which suggests either that these two emission regions are near each other or, if they are widely separated, that there must be a specific geometric constraint imposed. For example, emission near the surface, as is often suggested for the coherent radio emission, would place the gamma-rays in the strongest possible absorbing fields, whereas locating just the gamma-ray source at, say, the light-cylinder would apparently introduce a phase shift of the order of a radian between the gamma-ray pulse and a radio pulse emitted near the surface (Arons 1981c). The Vela pulsar, however, does not seem to show such a phase alignment (Buccheri et al. 1978). The total time delay is given from

$$\delta t = 4 \times 10^{-3} \, \mathrm{DM} \, f_9^{-2} s,$$

where f_9 is the observing frequency in GHz (Manchester and Taylor 1977, p. 103). Thus for the Crab pulsar the delay is of order 0.8 s or about 24 cycles for typical radio frequencies, but negligible at optical frequencies. The role of pair production and photon splitting in a pulsar magnetosphere is largely model independent. For all practical purposes, gamma-rays of a given energy can only escape from beyond a roughly spherical region surrounding the pulsar. Those produced inside are all absorbed. The size of this sphere is not especially sensitive to the precise parameters (owing, paradoxically, to the extreme sensitivity of the absorption coefficients), and one finds that photons much in excess of about 10^6 eV coming from near the surface would be reprocessed via pair production and reradiation in the magnetosphere. The observation of a cutoff energy above these energies would therefore be indicative of where in the magnetosphere the gamma-rays are emitted. Gamma-rays from the Crab and Vela pulsars have now been observed (Kanbach et al. 1977) with energies in excess of 2×10^9 eV, which puts some interesting restrictions on where these photons could have been generated, owing to those propagation effects.

a. Pair Production

The attenuation length for the conversion of an energetic photon into an electron-positron pair while traversing a magnetic field is given by Erber (1966), Erber and Spector (1973), and Tsai and Erber (1974):

$$\kappa = \alpha \omega_c \sin \theta T(\lambda)/2c, \tag{1}$$

where $\lambda \equiv (3/2)(B/B_{\mathrm{crit}})(\hbar\omega/mc^2) \sin \theta$, θ is the angle between the field direction and the photon propagation direction, $B_{\mathrm{crit}} = m^2c^2/e\hbar = 4.41 \times 10^{13}$ gauss, α is the fine structure constant ($\approx 1/137$), and $\omega_c = eB/m$. The factor $T(\lambda)$ is extremely sensitive to λ at small values:

$$T(\lambda) \approx C_p e^{-4/\lambda}, \quad \lambda \ll 1 \tag{2}$$

but only slowly varying at large values:

$$T(\lambda) \approx D_p \lambda^{-1/3}, \quad \lambda \gg 1. \tag{3}$$

$T(\lambda)$ has a maximum value of order unity for λ of order unity (for p = parallel polarization, $C = 0.612$, $D = 1.04$, and $T(\text{max}) = 0.17$, while for p = perpendicular polarization, $C = 0.306$, $D = 0.69$, and $T(\text{max}) = 0.27$). However, much of this detail is irrelevant because for $B = 10^{12}$ gauss, $\omega_c = 1.8 \times 10^{19}$/s and therefore $\kappa = 2 \times 10^6 T(\lambda)$ cm^{-1}. Clearly, then, this attenuation coefficient is huge even considering the small scales associated with the pulsar object ($\approx 10^6$ cm radius neutron star). Consequently the behavior at large λ is largely irrelevant; the photon will have interacted almost immediately to produce an electron-positron pair, which in turn will almost immediately radiate their energy away in the form of new photons. As illustrated in table 4.7, the gamma-rays are quickly reprocessed and reduced in frequency until finally all the photons at the end of the $e\bar{e}$ cascade can freely escape. Consequently $T(\lambda) \approx e^{-4/\lambda}$. Now, however, the attenuation coefficient is exquisitely sensitive to λ, which is going to be small even at the point of most probable absorption. Thus the exponential is going to be large and even a tiny change in λ will produce a large change in T. The magnetosphere therefore absorbs much like the surface of an opaque solid object; there is no attenuation whatsoever and then suddenly complete absorption. The spatial variations of B and θ are largely irrelevant, and they can be replaced by their values at the absorbing "surface." The location of this surface can simply be placed at the point where $\kappa r \approx 1$. Thus

$$1 \approx (r\alpha\omega_c \sin\theta/2c)^{-4/\lambda} \tag{4}$$

or

$$\lambda^{-1} \approx \frac{1}{4} \ln (r\alpha\omega_c \sin\theta/2c) \approx 7.5. \tag{5}$$

This latter step is a "reasonable" approximation as opposed to a "good" approximation; in other words the value 7.5 is probably only good to 10-20%, but that is entirely adequate for our purposes. Thus the fact that r, $\sin\theta$, and ω_c are variables really makes no difference in estimating λ itself; that step is the insensitive one. It then follows that the absorption surface is located where

$$\gamma b \sin\theta \approx 8.8 \times 10^{-2}. \tag{6}$$

Here we write $\gamma = \hbar\omega/mc^2$ and $b = B/B_{\text{crit}}$ to obtain this dimensionless relationship. Essentially the same estimate was given by Sturrock (1971a). Given a specific photon energy, a viewing angle, and a magnetic field model (e.g., dipolar), equation 6 defines a three-dimensional surface surrounding the pulsar. A photon of this energy (γ), if produced inside this surface, would be absorbed and degraded into electron-positron pairs, which in turn would reradiate to produce another set of gamma-rays unless or until the daughter photons were too low in energy to satisfy equation 6 (see sec. 4.5 for further discussion of this cascade process).

Shabad and Usov (1982) have noted that the photons tend to follow the magnetic field lines, which would then act to reduce the effective opacity. The importance of this effect is controversial at the moment (Herold et al. 1985; Shabad and Usov 1985, 1986). In principle, such a process could increase the "survivability" of energetic photons by allowing them to be guided by the magnetic field lines (as are the electrons).

b. Photon Splitting

The only other known process for photon absorption in a vacuum is photon splitting (Adler et al. 1970; Adler 1971), where the incident photon converts in the magnetic field into two outgoing photons. Here to a good approximation

$$\kappa \approx 0.12(b \sin \theta)^6 \gamma^5 \text{ cm}^{-1}. \tag{7}$$

Simply setting $\kappa r \approx 1$ with $r \approx 10^6$ cm gives

$$(b \sin \theta)^6 \gamma^5 \approx 10^{-5}. \tag{8}$$

One can now solve for the photon energy and $b \sin \theta$, for which photon splitting and pair production are comparable, to obtain

$$\gamma_* \approx 5 \times 10^{-2}, \tag{9}$$

$$(b \sin \theta)_* \approx 2. \tag{10}$$

The small value of γ_* alerts one to the fact that we have exceeded the limits of applicability of equation 2, because pair production vanishes for $\gamma < 2$. Thus equation 9 tells us that photon splitting is unimportant if it is in competition with pair production: for any γ of interest (certainly one greater than this tiny value for γ_*) and for any likely value $(b \sin \theta)_*$ (since pulsar magnetic fields are much less than B_{crit}), the absorbing surface will be found at much stronger field strengths ($\approx 10^3 \times$) than the pair production absorption surface. Therefore, photon splitting becomes unimportant because the photons would have already produced pairs. If, on the other hand, for $\gamma < 2$, only photon splitting is operative, from equation 8 we find that, for $\gamma = 2$ and $\sin \theta = 1$, a minimum field of 4×10^{12} gauss is required to split these (10^6 eV) photons. Thus a pulsar such as the Crab would be a marginal candidate for splitting the photons emitted in this critical energy range. (Because only photons with electric vectors perpendicular to **B** are split, there could be spectral ranges of 100% polarization.) However, since the Crab spectrum continues to much higher energies, the expected onset of absorption by pair production is not seen,

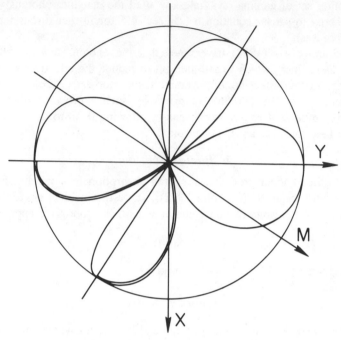

FIG. 2.26. Cloverleaf pattern of gamma-ray absorption. **M** is the magnetic dipole axis, and, for an observer viewing along the x-axis, gamma-rays are visible if viewed almost exactly along the local field lines. Gamma-rays are not visible from the other three dips because they would have to cross absorbing regions to reach the observer, hence are shadowed. After E. Massaro, and M. Salvati, 1979, *Astr. Ap.*, 71, 51 (figure 1).

and therefore looking for photon splitting effects in this pulsar does not seem too promising.

c. *Shape of the Absorbing Surface*

For a fixed gamma-ray energy, equation 6 describes a cloverleaf pattern in a dipole magnetic field (Massaro and Salvati 1979; figure 2.26). The deep dips over the polar caps are suggestive of a gamma-ray beaming mechanism, as pointed out by Salvati and Massaro (1978). However, the modulation largely vanishes if the observer does not happen to be viewing exactly along the polar field line, and no interpulse would be seen unless the alignment were nearly orthogonal with the observer in the equatorial spin plane. There may therefore be statistical difficulties with such a beaming mechanism, although Massaro et al. (1979) point out that the nearly equal-strength interpulse/pulse emissions actually observed for the Crab and Vela could be understood in such a model. Curvature radiation produces photons with $\theta \approx 0$, but, owing to just this same curvature, the photon is soon crossing magnetic field lines at significant values of θ. It is therefore difficult ever to see gamma-rays except those created outside the cloverleaf viewing pattern. The characteristic size of this pattern is just

$r/a \approx (\gamma/2)^{1/3}$ in a 10^{12} gauss field. Thus, for us to observe 2×10^9 eV photons ($\gamma = 4 \times 10^3$) from the Crab Nebula, they must be emitted on the order of 13 stellar radii away. But the light-cylinder is only about 150 radii away. This consideration is one of the motivating considerations in the "outer gap" model of Cheng et al. (1976), whose goal is to find a plausible site for gamma radiation not too close to the star.

d. *Spectrum of the Radiation*

It is difficult to account for the observed gamma-ray spectrum (of the Crab) since it falls as a power law with index ≈ -1 (energy flux per unit energy window) whereas the spectrum from the mechanism usually suggested, curvature radiation (Bertotti et al. 1969a,b; Sturrock 1971a; Treves 1971b; Ozernoi and Usov 1977; Hinata 1977a,b; Hardee 1977; Salvati and Massaro 1978; Harding et al. 1978; Hardee 1979; Ayasli and Ögelman 1980), rises with an index of $+1/3$ and then cuts off exponentially beyond the critical frequency (unless the particles themselves have a power-law distribution). Earlier works suggested synchrotron radiation (i.e., perpendicular motion across field lines) as the source of the gamma radiation (Apparao 1969; Dean and Turner 1971). If the electrons are accelerated across a potential drop, almost all of the radiation comes from the particles with maximum energy, which continue, moreover, to radiate upon leaving the acceleration region. Thus the spectrum would essentially be that expected from a monoenergetic beam. Reasonable fits can be made near the turnover portion of the synchrotron spectrum (Massaro and Salvati 1979; Ayasli and Ögelman 1980; Harding 1981), but the resultant spectra become quite deficient at lower energies. The observed power-law spectrum extends to the optical; a range of 10^9 and "minor" differences in spectral index become quite pronounced.

e. *Compton Boosting*

The gamma-rays can also be produced by inverse Compton scattering of the radio photons by energetic electrons (Cheng and Ruderman 1977c; Schlickeiser 1980) or of the thermal blackbody radiation (Blandford and Scharlemann 1976; Daugherty and Harding 1989; see also Arons 1984). Such processes seem most plausible for energetic pulsars such as the Crab. It has been suggested that photon-photon collisions could *absorb* gamma-rays (Pollack et al. 1971) and that gamma-ray lasing might be observed (Rivlin 1980). Cohen and Mustafa (1987) have reconsidered the idea of direct acceleration of electrons to the huge *vacuum* potentials about a pulsar with subsequent curvature radiation at gamma-ray energies (Cheng et al. 1986a,b), although it is unclear how the issue of radio emission might be addressed by such models.

3. X-rays

Data were taken in hard X-rays of the Crab pulsar's pulsations before the pulsar was discovered, but not analyzed to search for, and find, that periodicity until afterward (Fishman et al. 1969a,b). Since the X-ray emission from the Crab pulsar is apparently part of a general power-law spectrum extending from the optical to gamma-rays, it is not clear that a special explanation of just the X-ray portion is required. Wilson and Fishman (1983) report a narrowing of the Crab pulse profile in the X-ray. X-rays have been attributed to curvature radiation (Ochelkov and Usov 1980a,b), synchrotron radiation (El-Gowhari and Arponen 1972; Aschenbach and Brinkmann 1975), plasma instabilities (Hardee and Rose 1974), and differential magnetic absorption of thermal surface emission (Daishido 1975). Silk (1971) proposed that the diffuse X-ray background came from young pulsars. The Vela pulsar is apparently deficient in X-rays while being seen in gamma-rays. As noted before, an important class of intrinsic X-ray pulsars are not radio pulsars and appear to function by accretion. In fact, it was proposed some time ago that old "dead" radio pulsars might become active X-ray objects owing to accretion (Shvartsman 1970; Ostriker et al. 1970; see also Michel 1972).

Seward and Wang (1988) group pulsars and their associated X-ray emission into pulsars that emit pulsed X-rays: the Crab pulsar (0531+21), PSR 0540–693 in the LMC, and MSH15–52 (1509–58); pulsars that appear as point sources but are weakly pulsed if at all: the Vela pulsar (0833–45), CTB 80 (1951+32), PSR 0656+14, PSR 1055–52, PSR 0950+08, and PSR 1929+10; and pulsars with very nearby (about 2 arcseconds or less) X-ray sources: PSR 1642–03 (possibly coincident), PSR 1700–18, PSR 0355+54, PSR 0031–07, and PSR 1449–64 (see also Helfand 1983). Upper limits of importance can be set on the millisecond pulsar 1937+21 and others such as the relatively energetic PSR 0740–28. Cheng and Helfand (1983) find extended (10") diffuse emission around PSR 1055–52, which Cheng (1983) attributes to a pulsar wind interaction with the relative wind of the interstellar medium owing to rapid pulsar motion. Greenstein and Hartke (1983) argue that the X-rays are from hot poles in PSR 1509–58 and also in PSR 1055–52 (which they take to be aligned although it has interpulses). The observations are shown in table 2.11, together with hydrogen column densities N_H in units of 10^{21} atoms/cm^2. The luminosities are log-10 in ergs/s, and $\dot{P}$ is in units of 10^{-15}.

4. Cosmic Rays

As putative generators of highly relativistic particles, pulsars are natural candidates for sources of cosmic rays (Gunn and Ostriker 1969; Gold 1969b, 1974; Arnett and Schramm 1973; Kennel et al. 1973), and a rather extensive literature now exists. Recent workshops and reviews have summarized the theoretical situation (Osborne and Wolfendale 1975; Cesarsky 1980; Hillas 1984a). Broadly speaking, a number of difficulties must be reconciled in such theories. Naive versions of the standard model give more or less monochromatic particle fluxes, not a power law as observed. It is possible to get the highest-energy cosmic rays ($\approx 10^{21}$ eV) with the extreme pulsar parameters (Michel and Dessler 1981b). Even if they are fast, highly magnetized (and hence presumably very young) pulsars could produce sufficiently energetic particles. These particles must be

TABLE 2.11—Pulsars with Associated X-rays

Pulsar	P	$\dot{P}$	L_{total}	L_x	L_x(neb)	dist (kpc)
0531 + 21	0.033	422.	38.67	35.98	37.38	2.0
0540 − 69	0.050	479.	38.18	36.38	37.02	55.
1509 − 58	0.150	1540.	37.25	34.37	35.27	4.2
0833 − 45	0.089	124.	36.85	32.91	33.77	0.5
1951 + 32	0.039	5.92	36.58	33.80	34.68	3.
0656 + 14	0.385	54.3	34.59	32.49		0.4
1055 − 52	0.197	5.83	34.48	32.50		0.92
0950 + 08	0.253	0.229	32.75	29.39		0.126
1929 + 10	0.227	1.16	33.60	28.45		0.047
1642 − 03	0.388	1.78	33.10	31.93		1.3
1937 + 21	0.001	0.0001	36.04	< 32.48		2.5
0740 − 28	0.167	16.8	35.16	< 32.36		1.5

able to diffuse rapidly through the surrounding supernova remnant so as not to be adiabatically de-energized by expansion of the remnant and not to experience too many collisions (i.e., traverse matter equivalent to 3 g/cm^2 or less of integrated exposure).

One of the difficulties with attributing cosmic rays to pulsars or any other compact object stems from the natural assumption that these objects would largely put out *energetic* particles, which would on the face of it seem to be just what one needs. However, if energetic (e.g., fast) pulsars give out energetic cosmic rays and weak (slow?) pulsars put out low-energy cosmic rays, then the sources of the low-energy particles have little energy and vice versa. But the simplest expectation would then be that there would be a lot of energy in energetic particles and little energy in low-energy particles. The cosmic ray spectrum is just the opposite: it falls rapidly with energy with a spectral index (≈ -2), and in fact the average cosmic ray energy is little different from the lowest-energy cosmic rays (around 10^9 eV; it is difficult to tell for certain because the solar wind strongly modulates the lower energies and also the Sun itself is a source of particles up to this energy). Thus to get cosmic rays from energetic sources that age to become weak sources would require that the energetic sources emit few particles of high energy and then when they become old and feeble they emit copiously! The only mechanism that might change the spectrum later is shock acceleration, but basically shock acceleration simply takes an input spectrum (delta function, say) and smears it to higher energies. In fact, other than this smearing, which leads to a small fraction of energetic particles, the net increase in energy of the original particles is minor compared to the energy boost given to a few particles at the high end of the spectrum.

Smearing an initially rising spectrum with energy would still leave a spectrum rising with energy, so shock modification would not (on the face of it anyway) modify the kind of spectrum naively expected from pulsars or other energetic compact objects to look like the cosmic ray spectrum.

Arons (1981c) points out that the expected production of gamma-rays at the high-energy end of the spectrum would give too much gamma-ray background, at least according to existing theoretical views. Although at least some of the cosmic ray spectrum may contain particles directly accelerated from pulsars, there presently seem to be too many uncertainties surrounding the propagation and modification of any input spectrum to provide definitive constraints on magnetospheric theory.

5. Gravitational Waves

As noted, the Hulse-Taylor binary pulsar provides direct evidence for gravitational waves. Isolated pulsars have also been viewed as possible sources of gravitational waves (Melosh 1969; Chau 1970; Ipser 1971; Ruffini 1971; Bertotti and Anile 1973; Zimmerman 1978) and as probes for indirect detection of such radiation from binary systems (Brecher 1975; Will 1975, 1976; Wheeler 1975; Wagoner 1975; Esposito and Harrison 1975; Eardley 1975; Barker and O'Connell 1975; Nordtvedt 1975; Blandford and Teukolsky 1975, 1976; Will and Eardley 1977; Will 1977; Epstein 1977; Rosen 1978; Schweizer and Straumann 1979). Pulsar timing also probes for a stochastic background of gravitational waves (left over from the big bang, for example: Rosi and Zimmerman 1976; Sazhin 1978; Detweiler 1979). Photoproduction of gravitons has also been investigated (Papini and Valluri 1975). Direct detection with tuned cylinders (Hirakawa et al. 1978; Oide et al. 1979), laser interferometry (Levine and Stebbins 1972; Lu and Gao 1976), and seismic measurements (Dyson 1969a; Wiggins and Press 1969; Mast et al. 1972, 1974) have not yet been successful (see, however, Sadeh 1972). Lunar mascons have also been suggested to serve as detectors (de Sabbata 1970). If pulsars collapse with excess angular momentum, they may spin off a disk and spend some time with a triaxial shape, radiating gravitational radiation until they are slow enough to become spheroidal (e.g., Michel 1987b).

a. *Radiation Rates*

The rate of emission of energy in gravitational radiation can be estimated exactly as for electromagnetic radiation. In the latter case, one has electromagnetic energy stored in space about a magnetized body. If the body is set into rotation, some of this energy is found beyond the light-cylinder (wave zone) distance, becomes causally disconnected from the rotator, and simply flows away. But the electromagnetic fields cannot terminate at this distance, so the system constantly attempts to replace the lost energy. Thus the energy at that distance must be replaced essentially at every rotation. In this case, we can estimate the power output as

$$\dot{W} \approx W\Omega; \ W \approx \int_{R_L}^{\infty} \frac{B^2}{\mu_0} \, dV. \tag{11}$$

Previously, we used the Poynting flux to make this same estimate, but the two are equivalent strategies. We can similarly associate an energy with the gravitational field about the source. The binding of an object of mass M and radius a is of order GM^2/a. If we distribute this energy as the square of the gravitational field (the acceleration, $g = GM/r^2$), then this energy is necessarily of the order of

$$W \approx \int_{R_L}^{\infty} \frac{\mathbf{g} \cdot \mathbf{g}}{4\pi G} \, dV, \tag{12}$$

in direct parallel. However, the monopole moment is spherically symmetric, so rotating a spherical mass has no effect on this static component. The next higher multipole is the dipole moment of the source, but in gravitation the dipole moment is identically zero because the rotation axis passes through the center of mass. Only when we reach the quadrupole moment do we get time-dependent gravitational waves owing to rotation. Thus the part of $\mathbf{g}$ that can be radiated corresponds to $g_2 \approx GMD^2/r^3$, where D is a measure of the departure from sphericity (in a binary system, it is essentially the orbital radius). These higher powers of R_L and hence Ω greatly suppress the radiation rate, which is given by (Wagoner 1975)

$$\dot{W} = \frac{32}{5} \frac{G^4 M_1^2 M_2^2 M}{a^5 c^5} F(\epsilon). \tag{13}$$

The above dimensional argument will not give us the coefficient of 32/5 or the very important factor $F(\epsilon)$ which corrects for the eccentricity (ϵ). Note also that both the individual masses and the total mass appear in this form for the power output. Given the same period, an eccentric binary system is more efficient than a circular one in producing gravitational waves because the size and rate of change of the gravitational quadrupole moment are larger at periapsis and the rates are so sensitive to these parameters that they more than compensate for the weaker emission at apoapsis. Putting it another way, the quadrupole moment has higher harmonics and the Ω^6 radiation rate is enhanced by these higher harmonics. The above function is a bit odd looking,

$$F(\epsilon) = \frac{1}{(1 - \epsilon^2)^{7/2}} \left(1 + \frac{73}{24} \epsilon^2 + \frac{37}{96} \epsilon^4 \right), \tag{14}$$

but for PSR 1913+16 ($\epsilon = 0.617127$) it has a value of 11.64 and therefore importantly increases the radiation rate. The peak energy is radiated at about the 7th harmonic (Zel'dovich and Novikov 1971). Most of the energy is emitted at

TABLE 2.12—Hulse-Taylor Pulsar

Property	Value
Period (ms)	59.029997929
Period derivative	8.62713×10^{-18}
DM	167
Distance (kpc)	5.2
Magnetic field	2.2×10^{10} gauss
ϵ	0.61713
$a_p \sin i$	7.02042×10^{10} cm (almost exactly 1 $R_\odot$)
P_b	27906.98089 s
$\sin i$	0.734
M_p	1.442 ± 0.003
M_c	1.386 ± 0.003
$M_p + M_c$	2.82837 ± 0.00004
$\dot{P}_b$	$-2.427 \pm 0.026 \times 10^{-12}$
Advance of periapsis, $\dot{\omega}$	$4.2266°$/yr

closest approach. As a result, the eccentricity is slowly declining, with (Wagoner 1975)

$$\frac{d\epsilon}{dP} \frac{P}{\epsilon} = \frac{19}{18} \frac{\left(1 + \frac{121}{304} \epsilon^2\right)}{(1 - \epsilon^2)^{5/2} F(\epsilon)} = 0.34. \tag{15}$$

The measured parameters for this pulsar are given in table 2.12. If one substitutes the above values into equation 13, using in addition the fact that $\dot{P}/P = 3\dot{W}/2W$, one obtains $\dot{P}_b = -2.402 \pm 0.002 \times 10^{-12}$, in excellent agreement with the data. In doing such calculations, one should convert all masses into GM because $GM_\odot = 1.3271243999 \times 10^{26}$ cm^3 s^{-2}, with $c = 2.99792458 \times 10^{10}$ cm/s, is much more accurately known than the mass of the Sun itself. Thus, in table 2.12, the total mass of the system is known to much better accuracy in units of $M_\odot$ than is the unit itself (in grams)! At this rate, the two will spiral together in about 200 million years, with uncertain consequences. Note the large value for the advance of the periapsis compared to the famous value for Mercury about the Sun, which is 0.43"/century. The effect is about two orders of magnitude larger per orbit owing to the relative increase in proximity, and the objects orbit about 270 times more frequently. The effect which allows the determination of $\sin i$ is the gravitational time delay. The Hulse-Taylor binary system is inclined by about 47°, so radio signals emitted when the pulsar is near the Earth have relatively little propagation delay compared with signals emitted when the pulsar is furthest from the Earth and must pass close to the companion. The gravitational power output is substantial, corresponding to 7.8×10^{31} ergs/s in gravitational waves, about 10,000 times more than in radio! Although this system nicely exhibits the relativistic effects of gravitation, it is not in any sense a "strong" gravitational system; the gravitational potential is only about $4 \times 10^{-6} c^2$, and to see any nonlinear gravitational effects would require measuring the first-order relativistic effects (the most sensitive is $\dot{\omega}$) to parts of a million or better, which is about the threshold for the existing system. Lengthening the span of observations should improve these values.

Taylor and Weisberg (1989) give an excellent presentation of the data and theoretical situation. The basic theoretical analyses consist of varying approaches to modeling the system. Blandford and Teukolsky (1976) treat a precessing Newtonian system with relativistic effects "patched" on by hand, to avoid building in the assumption of general relativity. Epstein (1977) instead assumes general relativity, building on the two-body solution of Wagoner and Will (1976), and reparameterizes the data to separate effects that strongly correlate using the methods of Haugan (1988). Damour and Deruelle (1986) recast the Epstein-Haugan theory in a form that does not assume general relativity, along the lines of the earlier Blandford and Teukolsky work.

b. Wave Polarization

The nature and energy carried by gravitational waves are straightforward to calculate. One looks for wave solutions to the vacuum field equations exactly as one does for electromagnetism. In a vacuum one has no stress-energy tensor to perturb the curvature of 4-space, so not only is the Einstein tensor zero, but so is the Ricci tensor. The Ricci tensor is composed of second-order derivatives of the metric tensor plus quadratic terms in the Christoffel functions, which are themselves first-order derivatives of the metric tensor. If we look for wave solutions, we expect that we can write a weak field expansion of the metric tensor of the form

$$g_{\alpha\beta} = g_{\alpha\beta}(\text{flat}) + h_{\alpha\beta}, \tag{16}$$

and in Cartesian coordinates the only derivatives are those of the perturbation h, which is a dimensionless function of position and time,

$$h_{\alpha\beta}(x^\sigma) = h_{\alpha\beta} e^{l\omega_\sigma x^\sigma}, \tag{17}$$

and the h's are constant amplitudes in what follows. To first order, only the second derivatives of g (i.e., h) appear in the Ricci tensor, and these are symmetric in α and β, alternating in sign with each differentiation of g; thus we have immediately the algebraic equation (or we can look it up in any standard textbook):

$$\omega_\sigma \omega^\sigma h_{\alpha\beta} + \omega_\alpha \omega_\beta h - \omega_\alpha \omega_\sigma h_\beta^\sigma - \omega_\beta \omega_\sigma h_\alpha^\sigma = 0. \tag{18}$$

To make sense of this, it is useful to consider a wave propagating along one of the principal axes (or to rotate the coordinates so that this becomes the case). Then we will have, say, nonzero ω_t and ω_x but $\omega_y = \omega_z = 0$. Physically, ω_t is just the frequency and ω_x is the wave number, and for waves propagating at the speed of light the two are equal (if we take $c = 1$); thus $\omega_\alpha \omega^\alpha$ is zero but falls out automatically (fortunately).

At one time people tried to argue that gravitational waves can't be real, but are just fictitious "waves" in the coordinate system having no physical significance. Another complaint was the difficulty of conceiving how "geometry" could carry wave energy. The first view is partially valid, and with it we can prove that each of the four terms in equation 18 is separately zero, the first because $\omega_\alpha \omega^\alpha = 0$, the second because h is traceless, and the third and fourth owing to the "gauge" condition $\omega_\alpha h^\alpha_\beta = 0$. Note that all of these are covariant conditions, even though we took a special choice of axes. It is useful to confirm these results, although such a demonstration can easily be skipped over.

We can also put wave-like wiggles into the metric tensor with a coordinate transform of the form

$$x^\alpha \rightarrow x^\alpha + \xi^\alpha e^{i\omega_\sigma \omega^\sigma}, \tag{19}$$

where the ξ^α are (small) constants. To first order, this generates some terms in the metric tensor that look like the h's, namely

$$g_{\alpha\beta} \rightarrow g_{\alpha\beta} + \text{const.} \times (\partial_\alpha \xi_\beta + \partial_\beta \xi_\alpha) + \text{higher-order terms}$$

$$= g_{\alpha\beta} + \text{const.} \times (\omega_\alpha \xi_\beta + \omega_\beta \xi_\alpha), \tag{20}$$

and we can take this to be the *form* of the change due to the coordinate transform without grinding through the details. The coefficients on these terms are irrelevant. We only need to know what terms exist, and we can then identify which of the $h_{\alpha\beta}$ correspond to phony wiggles in the coordinate system and can be ignored. The form is all we will need because we see that we have exactly four degrees of freedom, which can be imposed by choosing the four values of ξ^α. Specifically, we can transform to set h_{tt}, h_{xx}, h_{xy}, and h_{xz} to whatever is convenient. There are 10 amplitudes, however, so the Ricci components in equation 18 are needed to set yet others equal to zero. In this equation, it would be useful first to get rid of the trace h, and we can use h_{tt} for this purpose. In other words, if the trace has some nonzero value, we simply shift the value of h_{tt} to negative that value and the trace is now zero. Or we can live with nonzero values with the knowledge that they have no physical significance. Setting them to zero helps to underscore that. With our choice of coordinates, R_{ty} involves only h_{ty} and h_{xy}, so setting the latter to zero also zeros the first; the same is true if z replaces y. Then we have three equations involving just the three amplitudes h_{tt}, h_{tx}, and h_{xx} that are zero. Also, we have not yet set h_{xx} to zero, so they are all zero. These are sufficient to give the gauge condition (these coordinate transformations are simply gauge transforms), in which case our field equations reduce to

$$\omega_\sigma \omega^\sigma h_{\alpha\beta} = 0. \tag{21}$$

And we have nontrivial solutions only if $\omega_\sigma \omega^\sigma = 0$, which signifies that the wave travels at the speed of light (or that light travels at the speed of gravity). At this point the only surviving amplitudes are

$$
h_{\alpha\beta} = \begin{bmatrix} 0 & 0 & 0 & 0 \\ \cdot & 0 & 0 & 0 \\ \cdot & \cdot & h_{yy} & h_{yz} \\ \cdot & \cdot & \cdot & h_{zz} \end{bmatrix}, \tag{22}
$$

with $h_{yy} = -h_{zz}$ because $h = 0$. So we see that the entire wave amplitude reduces to just two independent amplitudes, h_{zy} and h_{yy}. This result is closely parallel to electromagnetism because the waves are transverse to the direction of propagation and have two independent polarization axes. The action of the wave is just like a tide, squeezing along one axis and expanding along the orthogonal axis.

Although such tidal distortions could in principle be detected, the best efforts to date have not produced any verified results. The expected amplitude of such waves is extremely small, as we will now see.

c. Energy in a Gravitational Wave

Returning to the full gravitational field equations, we have

$$G_{\alpha\beta} = 8\pi G T_{\alpha\beta}, \tag{23}$$

but we have taken the Einstein (hence Ricci) tensor to be zero, so this equation reads zero equals zero for waves, *to first order!* That's because we neglected the quadratic terms in the Christoffel functions, which are first-order derivates of the metric tensor. Now we repair that neglect. Note that we are inverting the usual view of this equation, which is to regard the right-hand side (matter) as the "source" of the left-hand side (gravity). Instead, we are going to evaluate the left-hand side to second order (the above derivation utilized it as zero to first order), and use the fact that the right-hand side represents energy density and flow to calculate these quantities for a gravitational wave! But before we run off and calculate a lot of stuff, let's just think of the form of the Christoffel functions, namely first derivatives of the metric tensor, and the derivatives simply bring down an ω, while the only metric coefficient to be differentiated is, say, h_{yz}. Thus the only nonzero Christoffel functions are of the form ωh_{yz}, and therefore the quadratic term is nothing but constant $\times \omega^2 h_{yz}^2$, the standard form for the energy in a wave, which is proportional to the amplitude squared and the frequency squared. Skipping over the drudgery of obtaining the numerical coefficient (actually there is just a factor of 2 given from the definition of the Einstein tensor and a factor of 2 in averaging over a harmonic function squared) gives an energy flux

$$T^{tx} = \frac{c^3}{32\pi G}\, \omega^2[h_{yz}^2 + h_{yy}^2], \tag{24}$$

where we add the contribution from the other polarization by inspection. Unfortunately, from the point of view of detecting such waves, the coefficient on the front is huge, $c^3/32\pi G = 4.0 \times 10^{36}$ ergs s/ cm^2. Thus, if the Hulse-Taylor binary imploded when the neutron stars eventually made contact and released all of the rest-mass energy in gravitational waves within 1 ms, the resultant wave amplitude would be about 10^{-17} at Earth. In fact, even as close as the wave zone for the gravitational waves from the Hulse-Taylor binary, the mean amplitude of the waves now being emitted is only about 5×10^{-14}. Modern detectors could (in principle) barely detect the gravitational fluctuations at the closest distance at which they could be classified as waves!

d. *Do Gravitational Waves* Actually *Carry Energy?*

As noted above it is sometimes wondered whether the gravitational wave is just a mathematical illusion, a reasonable worry since we saw that most of the amplitudes could be eliminated by a coordinate transform (note that the coordinate transform cannot touch transverse amplitudes like h_{yz} or $h_{yy} = -h_{zz}$ because $\omega_y = \omega_z = 0$). The waves are real, however. Consider a pair of stationary balls with some fixed spacing (neglecting their mutual gravitational attraction). From the geodesic equation, there is no acceleration for objects at rest unless the g_{tt} term is coordinate-dependent, which it is not because we set $h_{tt} = 0$. Consequently the balls remain at *fixed coordinate positions* even when a gravitational wave passes. But the proper distance between the ball is, say, $L = \int(1+h_{yy})^{1/2}\, dy \approx \Delta y(1 + h_{yy}/2)$ if they are separated along the y-axis, and the proper distance between the balls changes while Δy, the coordinate displacement, stays constant. The balls will then move relative to a rigid rod (which will not be compressed significantly by such weak waves), and if they rub against it they will dissipate energy. Thus the waves are real and carry energy that can be transferred into familiar forms (e.g., heat).

e. *Quantum Gravity*

There is admittedly a gulf between theory and observation here. One can calculate the Compton scattering cross section for gravitons on, say, protons. One can estimate the background flux of gravitons necessary to close the Universe. One can then ask how many scatterings have taken place between all these gravitons and at least one of the protons in the Universe. At the present epoch, not even a single scattering will occur in a Hubble time! Bar detectors for gravity waves depend on the coherent interaction of all the particles in the bar, with a coherent wave involving a huge number of gravitons. Experiments able to establish the quantum nature of gravity will clearly be a challenge.

6. Neutrinos

Neutrino fluxes had not been observed from any astrophysical object (barely, if at all, even from the Sun) until the supernova in the LMC showed that neutrinos are indeed created in a type II (hydrogen-rich) supernova (Hirata et al. 1987; Bionta et al. 1987). The general possibility of such detection has been studied by several groups (Sato 1977; Eichler 1978a; Margolis et al. 1978; Eichler and Schramm 1978; and Helfand 1979). The favorite mechanism seems to be energetic particle interaction with surrounding material (Eichler 1978b,c; Shapiro and Silberberg 1979); however, Hara and Sato (1979) believe any such flux to be too weak to be detectable by the Dumand project. Direct neutron production in the pulsar acceleration regions has also been examined by Skobelev (1976) and Loskutov and Skobelev (1976, 1980).

3

Phenomenological Models

3.1 Introduction

In many areas of science the systems are so complex that they are difficult to model faithfully. The systematics of many-electron atoms and their excited levels forms a coherent physical picture even though one cannot yet calculate, from first principles, accurate excitation energies for every given state. In the same way, many have hoped to construct satisfactory models of pulsar magnetospheres without first having to solve the structure of its magnetosphere. Phenomenology and theory go hand in hand, the former fleshing out what are necessarily simplified pictures offered by the former.

Progress has been made in these directions. For instance, the potentially important role of nonneutral plasmas has been discovered (sec. 4.2.2), which changes in essential ways our views of what may surround a rotating magnetized neutron star.

As Tommy Gold is reported to have said, what is important is not so much the data, but rather *what data to ignore!* With 400 plus known pulsars, we have a wealth of data, but unless the data are organized in some coherent manner, they serve more to overwhelm than to inform. Furthermore we risk mixing apples and oranges when we discuss general pulsar properties at the same time that we are trying to introduce constraints to "explain" weak or possibly even nonexistent correlations. For example, it was thought that the fastest pulsars were the ones that evidenced the phenomenon of *glitching* (a sudden albeit tiny increase in period) because they would be the ones most distorted by rotation and therefore liable to *starquakes* after they had slowed down some and had become too oblate for their new slower rotation rate. The two fastest pulsars discovered early on (Crab and Vela) both showed glitching although the slower Vela showed larger glitches than the faster Crab pulsar (table 1.9). Given a plausible model and some generally supporting if sketchy data (which are too frequently not treated with the tentativeness they merit), we soon have another quasi-fact. Often these ideas retard research by discouraging exploration for better models (e.g., by giving referees grounds for rejecting papers advancing different ideas because a "satisfactory understanding" is already at hand). The

200

data should be the arbiter of which of two hypotheses is in better accord with scientific method, not whether the supporters of one outnumber the supporters of the other. The recent discovery of a virtual Crab Nebula/pulsar clone in the LMC (PSR 0540–693) adds to the pool of such objects a pulsar without (to date) *any* glitching behavior (Manchester et al. 1985). Nor do the more recently discovered millisecond pulsars generally display glitching. Nor could the model itself quantitatively give sufficient amplitude of glitching to agree with observation. The idea that pulsars glitch owing to a bulk change in moment of inertia has finally been abandoned, despite its being "self-evident," and glitch research is now directed at differential rotation of inner parts of the neutron star relative to the crust, which are somehow episodically coupled into corotation.

What has further paralyzed pulsar research is the existence of two or more explanations of every separate phenomenon or physical mechanism. Given such a shopping list, it is quite possible to select favorites which are mutually inconsistent. An obvious example is adopting magnetic field decay to explain the lack of long-period pulsars on the one hand and attributing complex pulse profiles to complex magnetic fields on the pulsar surface on the other hand. Both theory and observation point to higher multipoles that are preferentially removed by ohmic dissipation; so unless something subtle is going on, the two propositions are mutually inconsistent.

The role of phenomenology is, if anything, a more ambiguous one where one constructs models that could themselves turn out to be physically baseless but which nevertheless fit the data well, with the hope that the physical basis will eventually appear. Ideally there will be a convergence with time as physical underpinning is found.

3.2 Hollow Cone Model

One of the most influential models remains the model proposed by Radhakrishnan and Cooke (1969), shown in figure 3.1, which is a specific proposal for the radiation from an oblique rotator. It is common to find observational papers couched in terms of this model. Complex pulse shapes are often parameterized in terms of coaxial nests of emitting cones, whereby almost any conceivable pulse shape can be modeled. A single pulse is attributed to grazing a cone of emission, a double pulse to cutting across a cone, a triple pulse to cutting across an outer cone and grazing an inner cone, etc. Drifting subpulses could be regions of enhanced emission circling the surface of the cone. The swing of polarization discussed in sec. 1.5.7 is also naturally explained in this model. The model, or similar proposals, has been refined by a number of authors (Böhm-Vitense 1969; Radhakrishnan 1971; Komesaroff et al. 1971; Manchester et al. 1973; Oster 1975; Oster and Sieber 1976a,b, 1978; Backer et al. 1976; Backer 1976; Oster et al. 1976a,b; Sieber and Oster 1977; Manchester 1978; Cordes 1978; Ochelkov and Usov 1979; Prószyński 1979; Jones 1980a) and disputed by others (Izvekova et al. 1977). The broad pulse of PSR 0826–34 would require a viewing angle very nearly along the spin axis.

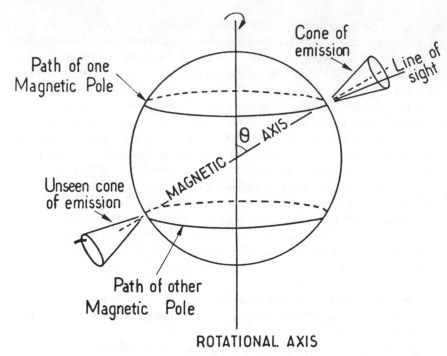

FIG. 3.1. The hollow cone model. Emission is assumed to take place in the form of a cone that in turn rotates with the star, sweeping the pattern past the observer. Interpulses are observed if θ (usually denoted α; see figure 3.2) is near 90° and the observer is near the equatorial plane, in which case both poles will be visible. From V. Radhakrishnan, and D. J. Cooke, 1969, *Ap. Letters*, 3, 225 (figure 1). Reprinted by permission of Gordon and Breach Science Publishers Inc.

1. The Standard Version

The greatest success of the hollow cone model has nothing to do with hollow cones but everything to do with the swing of the linear polarization. If emission is indeed controlled by the orientation of a dipolar stellar magnetic field, then one obtains such a swing regardless of what the actual shape of the emission region is; the presumption that it is in the form of a hollow cone comes not from observation but from the theoretical expectation that the field lines leading out to the light-cylinder are the ones likely to be active.

Following the notation of Manchester and Taylor 1977 (shown in my figure 3.2), ϕ represents the longitude relative to the rotation axis, ζ is the observer's latitude, α is the latitude of the magnetic pole, and ψ is the angle between the projection of the spin axis on the sky and a line from the subobserver point on the star to the magnetic pole. For a pure dipole magnetic field, ψ is essentially the position angle of the magnetic field vector at the surface directly below the observer; moreover, the magnetic field vectors will then also have the same direction everywhere along the observer's line of sight. Assuming ψ to be the position angle of the radiation therefore assumes the radiation to be beamed

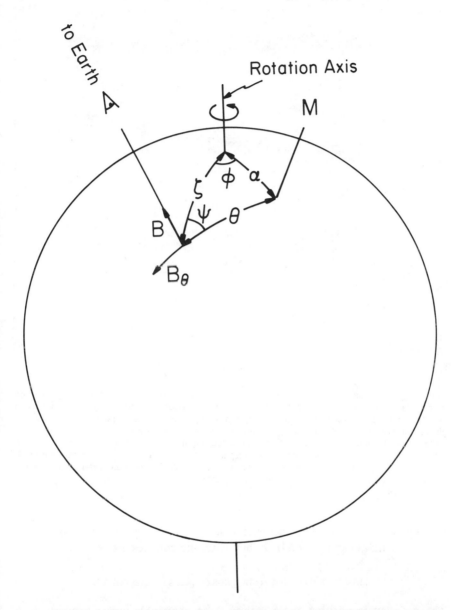

FIG. 3.2. Pulsar coordinates. Definition of the angles discussed in the text. Here M is the magnetic pole. For a pure dipole magnetic field, ψ is the position angle (within a fixed offset) of the magnetic field line and therefore the position angle of the polarization vector if determined by the magnetic field direction. The longitude of the magnetic pole relative to the Earth ϕ is steadily increasing with time while the latitude of the magnetic pole α and latitude of the sub-Earth point ζ are fixed (at least on short time scales). The intensity of the radiation presumably varies with θ, but this angle does not enter in the polarization swing phenomenon; thus a *hollow cone* of emission is irrelevant.

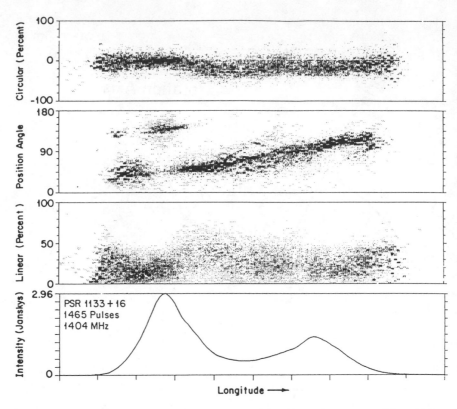

FIG. 3.3. Polarization swing. The position angle of the linear polarized component of the radiation rotates as shown with time (interpreted as increasing neutron star longitude). Longitude is in degrees, and one can see the systematic change in position angle, which appears gradual owing to the compressed scale. Note the orthogonal mode emission which coexists with the dominant sense of linear polarization in the vicinity of the main pulse, but has different probabilities of occurrence with longitude. A time average curve would displace a discontinuous "jump" to orthogonal mode for about a degree of longitude and then a jump back. The shading is an indicator of probability, not of intensity. From D. R. Stinebring, J. M. Cordes, J. M. Rankin, J. M. Weisberg, and V. Boriakoff, 1984a, *Ap. J. Suppl.*, 55, 247 (figure 17).

radially outward along the magnetic field lines, a plausible assumption (but counterexamples can be found). In any case, these angles are related by

$$\tan\psi = \sin\alpha\sin\phi/(\sin\zeta\cos\alpha - \cos\zeta\sin\alpha\cos\phi). \qquad (1)$$

The variation in position angle with ϕ is an observable pulsar property (see figure 3.3), albeit not without complications from phenomena such as orthogonal *mode* changing and differences between the polarization of the integrated pulse profile versus the polarization of the subpulses themselves (the latter tend to be highly polarized while the integrated profile may be somewhat depolarized). The orthogonal mode changing (easily seen in figure 3.3 at the phase of the main pulse) is often handled by simply adding or subtracting $90°$ to make the polarization swings appear continuous. Although the polarization swings are often cited as support for a hollow cone model, there are no hollow cone parameters in this analysis!

To understand pulsar data, it would be very handy to know the viewing geometry. Numerous attempts to invert the polarization swing and determine the two angles α and ζ are in the literature. Unfortunately, this inversion is *not* very robust because only a small range of $\sin \phi$ near zero is typically observed owing to the short duty cycles. Consequently we can expand about $\phi = 0$ to obtain

$$\tan \psi = \sin \alpha \sin \phi / \sin (\zeta - \alpha) + O(\sin^3 \phi), \tag{2}$$

and to this order, there is no distinction between $\tan \psi$ and ψ or $\sin \phi$ and ϕ, so $\sin \alpha / \sin (\zeta - \alpha)$ gives the maximum rotation rate of the position angle with longitude, $R \equiv |d\psi/d\phi|$, a dimensionless quantity typically of the order of 10 to 100. Thus ζ and α cannot be disentangled, although their difference can be estimated to be of order R^{-1}, a number consistent with the duty cycle ($\approx 1/30$), namely the relative width of the pulses themselves, as illustrated in figure 3.1. Note that we cannot even determine the *sign* of $\zeta - \alpha$ because we do not know whether ϕ increases or decreases with time (i.e., which way the pulsar rotates). The sign is nevertheless sometimes quoted (e.g., Manchester and Taylor 1977, figure 10-5) because a least-squares fit will still favor one sign. It is often questionable whether the choice is statistically significant. It is curious that the "S"-shaped polarization swing (see figure 3.3) contains only one piece of information, the maximum rate of swing (R), while one would expect the curvature also to contain some information. But what is mainly established is that the maximum change in ψ is $180°$, which is in itself additional support, consistent with the model. In principle there would be more information for pulsars with wide pulses. The natural coordinates to plot the data are clearly not ψ versus ϕ but $\tan \psi$ versus $\sin \phi$; an "S"-shape in these coordinates determines $\cos \zeta$. If we write $t = \tan \psi$, then $t' \equiv dt/d \sin \phi = R$, and expanding equation 1 gives

$$t''' = -3 \cos \zeta R^2. \tag{3}$$

The minus sign in equation 3 is again an artifact of assuming ϕ to increase with time. However, $\tan \psi$ has a pole at $\phi = \cos^{-1} (\tan \zeta / \tan \alpha)$ only if $\zeta < \alpha$, corresponding to a $\tan \psi$ versus $\sin \phi$ plot sweeping upwards rather than leveling off (in which case there would be no pole and $\zeta > \alpha$). In the case of a leveling off, the maximum occurs at $\cos \phi = \tan \alpha / \tan \zeta$, and for this maximum to be seen in the data (i.e, at reasonably small values of ϕ) would require a very small value of $\zeta - \alpha$, but for small angles, $\cos \phi_{max} = \alpha / \zeta$, so $R = \alpha / (\zeta - \alpha) = \cos \phi_{max} / (1 - \cos \phi_{max})$ and we have only a self-consistent relationship. For the "wide" pulsar PSR 0826–34, $R \approx 1.7$ and $\phi_{max} \approx 50°$,

TABLE 3.1—Position Angle Swing Rates (R values)

Pulsars with "wide" beams	R	Pulsars with "narrow" beams	R
2020 + 28	9.7	1929 + 10	1.5
0833 − 45	5.9	1933 + 16	2.1
1556 − 44	9.1	0540 + 23	1.0
0950 + 08	2.0	0531 + 21	≈ 1
0611 + 22	3.0	1642 − 03	9.1
0329 + 54	28	2016 + 28	5
1508 + 55	19	2021 + 51	3.5
1133 + 16	9.7	1154 − 62	2.0
0823 + 26	14	1604 − 00	1.7
1240 − 64	4.2	1944 + 17	0.7
1237 + 25	59	0628 − 28	4.6
0525 + 21	30	2319 + 60	10
2045 − 16	37	0834 + 06	13
0301 + 19	18	2303 + 30	7.7
1700 − 32	40	1919 + 21	12

from which Biggs et al. (1985) give a "best" fit of $\alpha = 53°$ and $\zeta = 75°$, although 8° and 10°, respectively, are also acceptable, which underscores the difficulty in obtaining reliable absolute angles from polarization data. For most pulsars, the polarization data can only yield R. Even when angles are inferred, it should be kept in mind that the data determining these angles are at the low-intensity wings of pulses and therefore statistically are the weakest.

Note again that the idea of a hollow cone of emission is unrelated to the observation of polarization swings per se. *Any* model of beamed emission from a localized group of magnetic field lines would broadly be expected to have these properties, whether or not they formed a hollow cone. Indeed, Lyne and Manchester (1988) believe that the emission regions are randomly distributed over the polar cap and not concentrated in an auroral zone. The significant observational fact is that the pulse maxima tend to be centered on the most rapid swing rate of the polarization position angle, and this rate tends to be large ($R = 10 - 100$) (see table 3.1, adopted from Narayan and Vivekanand 1982).

Useful as the hollow cone model has been in suggesting organization of the observational data, it contains a number of implicit assumptions restricting theory which have not been confirmed by observation. The assumptions are (1) that the polarization of the emission is polarized parallel or orthogonal to a line from the magnetic pole to the subobserver point on the neutron star, (2) that emission related to a given magnetic field line must be spread out over angles small compared to the total longitude of emission *in both axes perpendicular and parallel to the above line*, and (3) that emission from anywhere within or even outside the auroral zones encircling the magnetic pole is possible (see Rankin 1983a).

Assumption (1) is based on the strict assumption of a pure dipole magnetic field. Obviously, more complex magnetic field structures could give important deviations.

Assumption (2) is more serious because the magnetic field lines bend with altitude above the surface; thus magnetically controlled emission into a narrow cone requires that the range of altitudes over which the emission is important itself be narrow. The most natural place for a *narrow* emission zone would be the stellar surface. Thus the hollow cone model comes close to implying that emission is near the surface without quite insisting on it. Some observers, noting the difficulty of propagating radio-frequency radiation through a plasma with the much larger plasma and cyclotron frequencies required in minimum space-charge models, are willing to put the emission region(s) high above the surface, but still within a narrow height range (e.g., Krishnamohan and Downs 1983).

Assumption (3) is delicate. Permitting emission anywhere in the polar cap is a liberal assumption, but insisting on it is very restrictive. Actual physical models are not necessarily going to permit such general emission. Indeed, the original Goldreich and Julian (1969) model itself suggests emission in an auroral zone about the polar caps, not emission *over* the polar caps.

Assumptions (2) and (3) combine self-consistently, but not necessarily correctly, to exclude emission in a fan beam from a give magnetic field line. Fan beams are excluded because fan beam emission from regions distributed around the polar cap would largely fill the overall beam, giving emission over something like a *steradian* of solid angle, which is definitely not observed. However, a fan beam is the natural emission pattern for curvature radiation, often supposed to be the underlying emission mechanism!

These interactions of the assumptions mean that one cannot easily drop just one or the other of assumptions (2) and (3), but there is no reason not simply to drop both of them. Reluctance to do so makes them appear stronger than they actually are.

Moreover, these assumptions would suggest that there is additional information to be gleaned. For example, because the polarization swings do not depend on the extent of the emission region, one has in principle additional information from combining the width of the emission region with the polarization swing data. This geometry is shown in figure 3.4, which would be consistent with emission from the shadowed region, for example. If one chooses instead a circle or ellipse of emission centered exactly on the magnetic pole, one finds in the data relating polarization swings to duty cycles that in general the polarization swings suggest emission into elongated beams (i.e., fan beams: see Narayan and Vivekanand 1983a; Narayan 1984). But here the overall pattern is argued to be a fan beam, not the pattern of emission from the vicinity of a single magnetic field line. Mapping such a pattern within the usual hollow cone as-

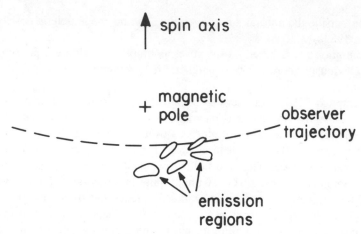

FIG. 3.4. Generalized emission pattern, Emission could be from localized "spots" which might then be regarded as an incomplete "cone."

sumptions forces one to think that the emission pattern on (or near) the surface could come from elongated *polar caps*, which would require such a radical change in theoretical modeling as to call into question the hollow cone model itself. Phinney and Blandford (1981) propose that such elongation may be more common in rapid pulsars than in the slower ones, an alternative to magnetic field decay as an agent to depopulate pulsars of long period. However, there seems presently to be no clear theoretical basis for either assumption nor a clear need for such depopulation. Although the magnetic dipolar configuration should clearly be a significant geometric factor, it is not clear either that the magnetic pole is contained within the emission region, that the emission region encircles the pole, or that the emission region is circular.

It is interesting that the Crab pulsar (see high-precision studies by Smith et al. 1988) and the millisecond pulsars PSR 1937+214 and PSR 1855+09 all have interpulses in which the position angle swings in the *same* direction, which would imply that the observer is viewing first one pole and then the other from the same relative aspect (e.g., from above). Segelstein et al. (1986) display fits for PSR 1855+09 with $\alpha = 30°$ and $\beta = 25.5°°$, but such angles would not permit an interpulse to be seen if anything like a dipole magnetic field obtained, because the opposite magnetic pole would only pass within 124.5° of the observer. If the emission is concentrated in a narrow beam, the two poles must then be nearly in plane (i.e., the rotators are essentially orthogonal). It seems a bit suspicious to presume that all of these interesting objects are sitting with spin axes close to the plane of the sky. Alternatively, the beams could well be fan-shaped, which would free up such geometrical constraints.

2. Other Issues

The following items do not necessarily deal directly with the hollow cone model, but because the model is so popular there is a tendency to try to envision them in such terms.

a. *Degree of Polarization*

Some pulsars are essentially 100% linearly polarized, so it is tempting to suppose that the underlying emission mechanism has this characteristic. For pulsars with lower polarization, the natural supposition would then be that overlap from a more complex distribution of 100% polarized sources leads to depolarization. If one attempts to simulate pulse shapes with any simple model, such as the hollow cone model, this idea fails miserably (Shier 1990). For example, PSR 1039–19 has a double pulse structure with a deep minimum between the two and roughly 50% polarization. Such a signature cannot be due to the two peaks having 100% polarization but overlapping to produce depolarization (obviously the more clearly double the pulse profile, the smaller the overlap). One pulsar with a double pulse was found with high enough polarization to be consistent with the above simple model.

b. *Orthogonal Mode Changing*

Stinebring et al. (1984b) point out that the depolarization discussed above is due largely to orthogonal mode emission, which reduces the integrated degree of polarization (see figure 3.3). But why should orthogonal mode emission so commonly be present? This phenomenon has been a discouragement to emission models based on simple curvature emission, because there is no ambiguity in what polarization one should get. In principle, propagation effects could modify the polarization subsequently, but one would expect such effects to be strongly dependent on frequency and require special viewing angles whereas the phenomena tend to be only weakly dependent and are not uncommon. Cheng and Ruderman (1979) have pointed out that orthogonal mode changing could occur when the radiation of the one polarization decreased in intensity while the radiation with the orthogonal polarization increased in intensity. Adding the two polarizations gives depolarization if the two sources are independent (i.e., uncorrelated in phase), so the polarization appears to jump from one polarization to the other, with a degree of polarization that drops to zero at the "jump" whereas the intensity is maintained without interruption (see discussion of Stokes parameters). In this sense, the "jump" in polarization position angle is an artifact of how the data are presented. Single-pulse polarization studies such as those by Stinebring et al. (1984a,b) show that the two polarizations *overlap*, with some pulses having one polarization and some having the other polarization in the vicinity of the longitude at which the "jump" occurs. This detail is lost when the pulses are averaged. The single-pulse data show that often the *integrated* pulse properties can be seriously misleading, the above being just one example.

If indeed there are two independent sources of polarization, it remains unclear whether this is really *one* source region that can emit in either sense of polarization or two independent source regions with intrinsically different characteristic polarizations. Either view has its problems: if a single source radiates different polarizations in different directions, why is there a pulse-to-pulse overlap; while if two distinct sources give overlapping contributions, why should the two

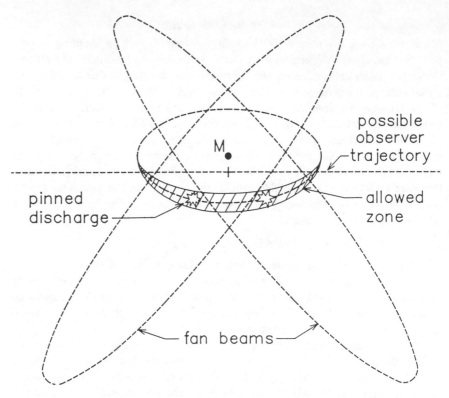

FIG. 3.5. An orthogonal mode-changing mechanism. Here we imagine that two discharges are located along the lower-latitude part of the auroral zone (*M* signifies the magnetic pole). Each discharge produces a fan beam as shown. An observer passing through the two beams as shown will see radiation from one and then the other. Because the two sources are independent, there is no phase relationship between the linearly polarized emissions and when the two intensities are equal but the position angles nearly 90 degrees apart, the percentage polarization will fall to nearly zero and the position angle will switch from being dominated by the first source to being dominated by the second (i.e., appear to "jump" to a new value roughly 90 degrees different). The circular polarizations should also change sign at about this point, as is frequently observed (Manchester and Taylor 1977). Notice that this effect is insensitive to the observer's latitude, although the pulses will be resolved into distinct components if the observer's trajectory crosses wide of the magnetic pole. (In the case of PSR 2020 + 28, the two components associated with the first mode change seem to be of unequal strength and nearly resolved: Manchester and Taylor 1977, p. 29.) From F. C. Michel, 1987d, *Ap. J.*, 322, 822 (figure 1).

polarizations be at 90°? Michel (1987d) suggested a geometrical interpretation of this effect in terms of the distinct sources, assuming that the sources radiate in a fan beam which would allow them to overlap even when they were at distinct loci on the pulsar. Geometrically separated fan beams could also account for complex pulse profiles. If we consider the effect of just two beams, it is clear that the observer will view first one beam and then the other, as shown in figure 3.5. In between, the receding beam is circularly polarized in one sense and the position angle for the linear polarization component has rotated away from the mean, while the approaching beam should have the opposite circular polarization and its position angle should not have reached the mean position angle. The result would be the above pattern of an abrupt jump in position angle accompanied by a change in sense of circular polarization. In such a model, the "orthogonal" change in position angle might just be a selection effect. As long as the two regions are reasonably separated, the polarization jump will be through some large angle of order 90° but not necessarily *exactly* 90° The interpulse of PSR 1855+09 has a jump that is at least 20° off being orthogonal, for example (Segelstein et al. 1986, figure 1). For angles very much different from 90°, the transition would not have the conspicuous minimum in the percentage polarization and one would resolve instead a continuous change. Indeed, pulsar polarizations have statistically significant variations superimposed even on the classic polarization swings (Manchester and Taylor 1977, figure 10-4; Krishnamohan and Downs 1983). Many observers are persuaded, however, that the jumps are indeed usually nearly orthogonal (Cordes, private communication). It is not clear why two distinct, separated sources would have nearly orthogonal polarization, which is then a weak point in this model.

Shier (1990), in attempting to simulate pulsar shapes and polarizations, suggested a different explanation, namely that the orthogonal mode emission is nearly, but not exactly, coincident with the regular mode emission. It would then be the overlap of these two, depending on the viewing geometry, that would lead to the moding and depolarization. Specifically, in a hollow cone picture the regular mode emission would come from a ring about the magnetic pole and the orthogonal mode emission would come from a *concentric* ring just inside (or outside) the first, as illustrated in figure 3.6. This idea is phenomenological of course, but it would explain why regular mode emission is so commonly contaminated by orthogonal mode emission and why nearly 100% polarization is seen but only rarely. It would also be consistent with the observation (Rankin 1983a) that *conal* components (the outriggers in a triple) and *core* components (the central one) tend to have different spectral indices. A simulation based on this model is shown in figure 3.7. Even for such a simple model, there are a large number of parameters to adjust (the two cone opening angles, α and ζ, and the beam width, which need not be circular and which need not be the same shape or have the same intensity for each cone; 6 parameters at a minimum) and the simulation shown happens to do poorly for the degree of depolarization but rather well for the position angle, with orthogonal jumps at about the correct places. A possible physical justification for orthogonal mode emission adjacent to regular mode emission is that the emission takes place some distance above the surface and the orthogonal mode is emission directed downward which has been reflected. Why a reflection process should favor the orthogonal mode is unclear. Alternatively, Gil (1985) has suggested emission from two different altitudes above the neutron star surface.

c. Complex Pulse Profiles

Lacking a generally accepted physical model for pulsar action, observers have fallen back on the hollow cone model and made various phenomenological

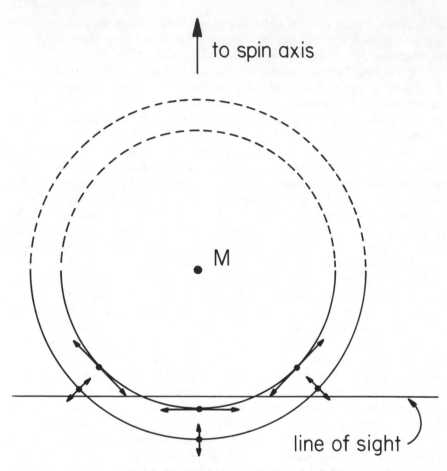

FIG. 3.6. Another orthogonal mode changing mechanism. Emission is split into concentric cones (or arcs: the dashed lines indicate our skepticism that auroral zones close about magnetic pole, *M*), one having "regular" mode emission radial from the magnetic pole and the other having orthogonal mode emission. Order is arbitrary, although a reflection mechanism would suggest that illustrated (Shier 1990).

modifications, such as assuming emission from a "core" as well as the cone itself. Observers have also contemplated two concentric cones with the cones themselves unevenly and sporadically illuminated in order to accommodate the observed pulse shapes (see Rankin 1983a). From a theoretical point of view, something like the Ruderman and Sutherland model would arguably provide *one* illuminated cone, but there seems to be no compelling physical basis for additional concentric structuring. As noted, Gil (1985) has interpreted the two cones as emission from a single set of conical field lines but at two separate altitudes above the surface. At the moment there is no specific theoretical reason to expect such altitude separation, although clearly there should be rapid changes in the pulsar environment with height.

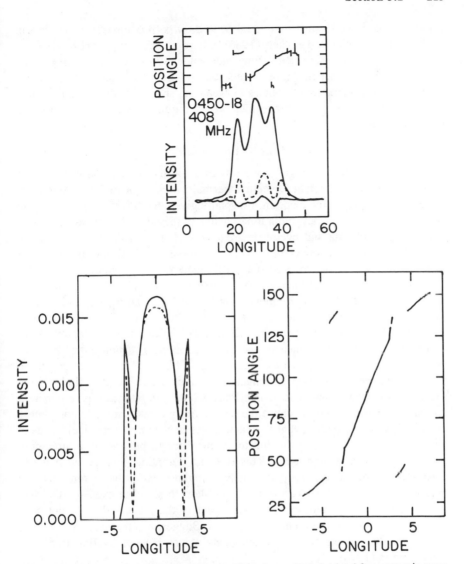

FIG. 3.7. Triple pulse simulation (PSR 0450-18). The pulse profiles simulated for concentric cones have opposite polarization (Shier 1990). This run produced a triple with a polarization swing punctuated by orthogonal mode changes quite similar to the data for PSR 0450-18 (Lyne and Manchester 1988). This simple model cannot give asymmetric pulse shapes, and the choice of parameters did not give adequate depolarization.

Complex pulse profiles could also be interpreted as representing a number of distinct magnetic flux tubes that are preferentially (sec. 8.2.6) illuminated for some reason. The polar caps are presumably favored geometrically for current flow, so if the magnetic field is significantly nondipolar near the surface, a subset of these field lines will themselves be favored. The decomposition of complex pulses into distinct individual *components* (flux tubes?) is manifest in the study by Hankins and Rickett (1986), wherein each component has its own spectral properties, rising and falling independently of the neighboring ones. Emission from distinct flux tubes would be consistent with such data. Rankin (1983a) notes systematic differences between the spectra of the so-called conal and core components, however, which would require special pleading to fit a flux tube model.

d. *Self-Absorption*

The hollow cone model suggests one way to avoid the problem of why there is not extensive reabsorption and eventual thermalization of the radiation as it propagates through the source medium, especially given the difficulty in propagating at all if the low-frequency radiation actually observed had to traverse plasma with the high plasma and cyclotron frequency characteristic of the minimum space-charge electron concentrations. For example, the Crab pulsar radiates at least down to 10^7 Hz whereas its plasma frequency is up at about 3×10^{10} Hz. Thus, even if the density falls as $1/r^3$ as expected, the plasma frequency will exceed the above emission frequency until essentially the light-cylinder. That might argue for a source at the light-cylinder, but then we simply shift to a different set of unpleasant problems. If in fact one has vacuum regions adjacent to plasma regions, which is one of the properties of nonneutral plasmas (sec. 4.2.2), it is possible that the radiation could have been created near these surfaces and then could have propagated into the vacuum regions. Indeed, the excitation of oscillations at such interfaces would automatically give coherent emission into the vacuum because the nonneutral plasma is already completely charge-separated (Michel 1985b), so any waves in such a plasma would already be coherent (as opposed to having to create discrete bunches in a streaming plasma, as is usually supposed). This is a very tempting picture, but it is unclear what the overall topology of such static and nonstatic (the current carriers) plasmas would be, which is a difficult problem to simulate even for the disk model (and which cannot even be addressed in the models that do not self-consistently solve for these current systems). Suggestions that density contrasts play a role have been made on phenomenological grounds by Blandford and Scharlemann (1976).

e. *Drifting Subpulses*

The appearance of two natural frequencies, as exemplified by the drifting subpulse phenomenon, has been a particular problem. Drifting is difficult to interpret confidently owing to aliasing between the two frequencies (the same effect that makes the wagon-wheel spokes seem to go backwards in motion pictures). The conservative assumption is that the drift motion is slow. It could be that the phenomenon is common in pulsars, but only can be observed when slow or when aliased to an apparently slow drift. Ruderman and Sutherland make the attractive suggestion, which is essentially model independent, that there is structure within the emission cone that slowly rotates about the magnetic pole, perhaps in a quantized standing wave pattern. Unfortunately there is no real evidence that there exists such a closed loop of emission to drift about, and indeed we will present physical arguments to the effect that, except for a nearly aligned rotator, the currents are unlikely to flow in such a pattern (sec. 3, below). Michel and Dessler (1981a) speculated that the phenomenon of drifting subpulses represented an interaction between the two natural periods available, the rotation period of the pulsar and the orbital period of the inner disk edge, which were expected to have similar values and thereby alias the subpulse locations. One might think of the discharge as being *pinned* to local features on both the disk and neutron star, which drift apart in longitude until the discharge fails owing to the long path length, therefore reestablishing a new, more proximate point on the disk.

f. *Nulling*

Nulling is particularly vexing, from both a theoretical and an observational point of view. If one really believes that emission is in the form of a hollow cone, perhaps two such cones and also a core of emission, with important variations in emission around the periphery of the cones which vary in time and with perhaps the entire pattern rotating slowly, then one must add to this the possibility of shutting this entire emission system down! At some point such modeling begins to look like an exercise in fitting to epicycles. Theoretically, if a pulsar can actually stop emitting, why should it start back up? This might be related to the absolute electric charge of the system (sec. 4.2.3). Certainly pulsar action would seek one steady state charge, which might not be consistent with the pulsar action per se, with charging or discharging mechanisms that drive the system back into activity once they are off. If nulling pulsars are indeed bi-stable, however, they do not seem to be very regular in this respect.

3. Problems

a. *Inconsistent Current Paths*

The physical basis for the hollow cone model was the "standard" model, where currents *out* of the polar cap regions had somehow to be closed by return currents into regions surrounding the polar cap. In an axisymmetric model, this region necessarily defined a ring on the surface (*auroral zone*) and the extensions of it along the magnetic field lines to form a hollow cone. No one has successfully accounted for this return current, and it may not even exist, given that there are stationary solutions without any currents to infinity (sec. 4.2.3). But if one nevertheless continues to assume the model to work (the old "standard model"), those currents would have to form an auroral zone, leading to the hollow cone model. It is natural to assume that a realistic model would not be axisymmetric and that the magnetic axis would be inclined at some angle to the rotation axis. Consequently the auroral zone was simply tipped but

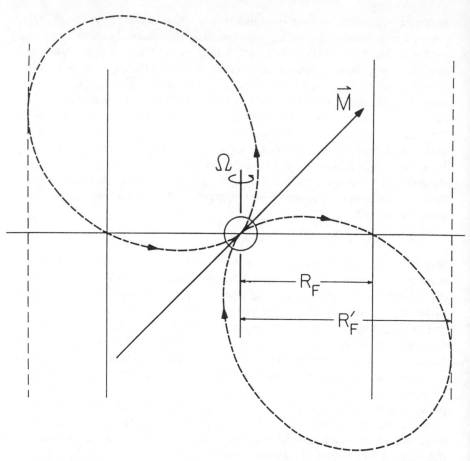

FIG. 3.8. Dipole field lines to light-cylinder for an oblique rotator. The same argument applies to any other fiducial distance that might be chosen for the current-carrying field lines. Note that logical choices for matching to the field lines (R_F to a disk or R_F' to the wind zone) show that the field lines from the low-latitude portions of the auroral zone are much shorter than those from the high-latitude portions, the latter having to arc over the neutron star and terminate on the opposite side from the former. From F. C. Michel, 1987d, *Ap. J.*, 322, 822 (figure 2).

its underlying *axisymmetry* nevertheless retained by fiat in the hollow cone model. The auroral zone is usually taken to be the locus of the footprints of the magnetic field lines that just reach out to the light-cylinder distance, which causes the auroral zone to be distorted from a circle to an ellipse as the magnetic axis is tipped; *that* departure from axisymmetry is sometimes taken into account. In figure 3.8, we illustrate two interrelated difficulties with retaining axisymmetry. If the inclination angle is significant (some pulsars have been interpreted as having the two axes essentially orthogonal), one can immediately see that the distance to the fiducial distance along the field line emerging from the lowest-latitude auroral region is *much* shorter than that from the higher latitude. Certainly, one would expect preferential current flow along the shortest circuit elements. In the disk case, the currents at the disk edge would even have to flow initially in opposite directions (away from the star on the long field lines versus toward it on the short field lines) along the two routes, something of a surprise if a simple potential difference between the two conductors pulls in these currents.

b. *Inconsistent Torques*

The second problem reemphasizes the first even more vigorously. If we look at the torque on the neutron star implied by tipping the old "standard model" current system, we notice something bizarre. Currents on the short field line segments flow across the polar cap magnetic field and produce a $J \times B$ force and a torque that acts to slow down the neutron star rotation (and reduce the inclination angle). *But the currents from the upper half of the auroral zone act to spin up the star!* They also increase the inclination angle, as shown in figure 3.9. Given that the source of the energy to drive the currents is the rotational energy of the star, there is no natural reason to suppose that some of the currents would spin the star back up. It would be entirely possible to have currents from just the lower part of the auroral zone and absolutely physically impossible to have currents from just the upper part. The bottom line is that there is no reason to expect the auroral zone to be fully illuminated, and the real expectation is that any currents would be concentrated near the lowest-latitude parts. This inconsistency in torques was indirectly noted by Good and Ng (1985) when they pointed out that conduction currents into the polar cap give torques that act to decelerate the star but increase the inclination angle whereas the direct retardation torques on a magnetized rotator from magnetic dipole radiation act to decrease the inclination angle (i.e., align the two axes). The difference can now be traced to the effect of the (nonphysical, in our view) current components supposedly spinning up the neutron star. Note that one could not simply reverse the current flowing on the high-latitude field lines because then, in effect, one would have currents flowing out of one polar cap and into the other, which could not happen.

c. *Inconsistent Current Densities*

The problem with charge-separated current flow is that the electrostatic forces are proportional to the current, and the electric current out of a pulsar scales as the space-charge density times the polar cap area times the velocity of light. The immediate consequence is that a flux tube carrying this much current must be as large as the polar cap itself. The same must be true of the putative auroral zones if the return current is itself charge-separated. But even that could not work unless the return current was carried by particles of the same sign. Otherwise the electrostatic fields would grossly violate $E \cdot B \approx 0$ and the whole basis for the model would vanish. The only way to significantly narrow the auroral zones (i.e., to explain pulsars having well-separated double pulses)

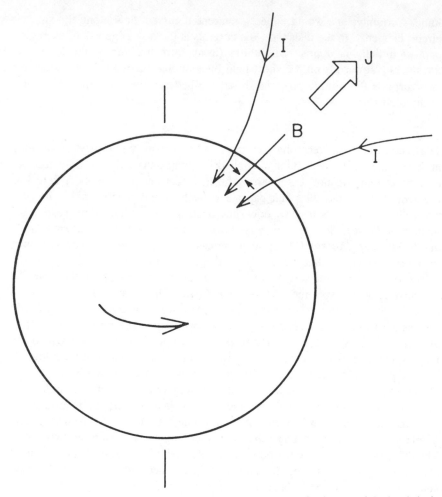

FIG. 3.9. Current flow patterns across the polar caps. The current flowing out of the low-latitude portions of the polar caps act to spin down the star (or, equivalently, the spin energy of the star drives these currents). The currents out of the high-latitude portions necessarily have the opposite effect and (if they were to exist) would require an external source of energy. From F. C. Michel, 1987d, *Ap. J.*, 322, 822 (figure 3).

would then be to have a neutralized beam flow. The same follows even more insistently if one thinks of concentrating the currents into localized emission regions. Adding in electron-positron pairs to the flow is often suggested but does not seem to help because they would not add to the current. It is easy to see that if a primary electron creates a pair, both of which escape, there is *no additional current*. In a quasi-neutral plasma, a current can be carried because the velocities and charge densities of the two species are not directly coupled. In sec. 2.3.6 it was pointed out that the two apparently must escape to make any sense out of the local space-charge effects on acceleration. Counterstreaming is therefore apparently required if the return currents are to be narrow. If one cannot create the counterstreaming by pair production, an alternative would be to attract the charges from some source outside of the system. The interstellar medium would act to discharge any net charge that the pulsar had. However, the huge radiation pressure from the pulsar wind should keep such discharging to an unimportant minimum (sec. 4.5.6). Alternatively, counterstreaming could result from a discharge between two conductors, with charges of one sign from the neutron star and of the other sign from a nearby disk (sec. 6.3). The issue of beam size becomes particularly important for fast pulsars such as the Crab, which should have a large polar cap yet has narrow sharp features in its pulse profile.

d. *Inconsistent Data*

An enormous number of data are available at various levels of detail for a large number of pulsars. Unfortunately, these data are not of uniform quality or are not always available. If, for example, one wished to know how common the phenomenon of orthogonal mode changing was, one would be hard-pressed to find out because no observer has happened to address this particular issue. There is no systematic assessment of orthogonal mode changing on a pulsar by pulsar basis. Although the complaint is commonplace that theoreticians "don't pay enough attention to the data," there is no real consensus on what the important data are. And to track down just one issue as simple as the above would require sifting through an enormous mass of literature. In the absence of a consensus, the observational data reported are those which the *observer* thinks are most important, which is an important constraint. Plus, there is no guarantee that the above search, daunting in itself, would uncover the desired information. It is a little like having the maid tidy up after a crime before the investigators arrive.

4. Comments

All in all the general impression is only mildly satisfying. Emission from the vicinity of the polar caps seems to be offered no particular challenge from the data. The data do not seem to suggest that the magnetic inclinations are either preferentially small or large, something of a mystery given that alignment seems physically plausible. The complexity of the integrated pulse shapes and their division into numerous subpulses obviously suggests some sort of structuring, but the controlling agent is far from obvious. Auroral displays at Earth have complex but time-dependent structuring, whereas here the structuring is relatively stable and a characteristic "fingerprint" of each pulsar. Magnetic surface irregularities are a possibility, but again we presently have little idea of why they should even exist.

3.3 Other Suggestions

1. The Corotating Source

In the corotating source model (sometimes called the Smith model) the emission is attributed to a detached region in the magnetosphere that corotates with the star, typically rather close to the light-cylinder so that Doppler bunching would cause even an isotropically radiating source to appear pulsed. Geometrically the model is similar to that of Gold (1968); consequently many of the same problems must be faced: how is the region held in place and how is it powered? The original proposal by Smith (1969) has been followed up (Smith 1970, 1971a,b,c,d, 1973a,b, 1974, 1976) and has attracted considerable attention and elaboration (Manchester and Tademaru 1971; Zheleznyakov 1971; McCrea 1972; Zheleznyakov and Shaposhnikov 1972, 1975; Cocke et al. 1973; Cocke and Ferguson 1974; Ferguson 1973, 1976a,b, 1979; Ferguson et al. 1974; Zlobin and Udal'tsov 1975; Malov and Malofeev 1977; Malov 1979; Lyne and Smith 1979). This model enjoys a considerable following, but trails its mutually incompatible competitor, the hollow cone model. The latter has at least some similarity to the old standard model (except for the concept of multiple nested cones, an unexpected aspect in that theory). The corotating source model, however, has provided some impressive quantitative accounts for the variation of polarization and intensity in pulse profiles.

The angular momentum and energy budgets become of some interest in such models, because the energy loss from a rotator is given by

$$\dot{E} = \frac{1}{2} I \Omega \dot{\Omega}, \tag{1}$$

while the angular momentum loss is

$$\dot{L} = I \dot{\Omega}; \tag{2}$$

hence

$$\frac{\dot{L}}{\dot{E}} = 2/\Omega. \tag{3}$$

In contrast, photons can only carry angular momentum $\hbar$ but energy $\hbar\omega$, giving an angular momentum to energy ratio of $1/\omega$. For the magnetic torques carried by large-amplitude waves at the rotation frequency, the two frequencies are the same, and the balancing is automatic. Viewing the pulsar system as a black box, however, puts some restrictions on the possible source geometries. If, for example, we argue that the torque on the neutron star is produced predominantly by relativistic particle fluxes rather than by Poynting flux in large-amplitude waves, it immediately follows that the relativistic particles (electrons or photons) must have a virtual source essentially at the light-cylinder and beamed in the direction of motion.

There is no way that a gamma-ray source carrying the bulk of the energy output could be located near the surface of a pulsar: that would require an additional (nonrotational) energy source. Nothing would prevent a battery-powered source from emitting X-rays in any direction from any location, but if the source is a load on the rotating flywheel of the neutron star, the system must compensate elsewhere by having more angular momentum lost relative to the energy lost. On the other hand, the typical 10^{-5} of the total energy going into radio emission is hard to locate. This simple argument would seem to suggest that pulsars with a large fraction of their total output in energetic pulsed radiation (such as the Crab) would have to emit nearly at the light-cylinder.

a. *The Rotating Isotropic Source*

If a source of radiation in the magnetosphere of a pulsar is isotropic in a rotating coordinate system, a torque will be exerted on the system. Because to an external observer the photons will be blueshifted when the source approaches and be redshifted when the source recedes, the excess momentum will be $\gamma\beta$ of the rest system photon energy regardless of emission direction. However, the energy output will also be boosted by the transverse Doppler shift, γ. Thus we immediately have

$$\frac{\dot{L}}{\dot{E}} = \frac{\beta r}{c},$$

(4)

with r the emission distance from the rotation axis. Writing $\beta = \Omega r/c$ then gives the requirement that the emission region be near the light-cylinder distance if the energy loss and torques are to be consistent. If the photons were intrinsically beamed forward near the light-cylinder, the constraint on β would be relieved.

This result leads to a variant on Newton's water bucket experiment to determine inertial coordinate systems: One puts two light bulbs on the ends of long rods powered by a battery at the center. If the system is rotated, it slows down. Thus it seeks the inertial reference system. So would two charged spheres separated by a rod.

It remains unclear what constraints the ratio of angular momentum to energy loss might place on pulsars such as the Crab that emit several percent of their energy as very high-energy emission. Possibly these emissions are due to a different mechanism and indeed take place near the light-cylinder distance, but the above black-box picture has not been analyzed sufficiently, owing to uncertainty in the global current flow. If current flow is to a disk, for example, the disk angular momentum effectively "buffers" the system, and the energy/angular momentum balance is automatically taken care of in any self-consistent model. However, when one simply assumes that the emission is local near the surface and that all the radiation is effectively created there, the self-consistency is lost and the resultant system may well not make physical sense for an isolated neutron star.

2. Neutral Sheets

Natural structures that might furnish emission near the light-cylinder distance would be sheetlike discontinuities such as shocks (e.g., cresting of the large-amplitude waves) or neutral sheets (the structuring that might form about nodal sheets where the field amplitudes vanish). We will discuss physical models for this idea more specifically in sec. 7.4. The attractive features are (1) the relatively large surface areas for emission (the speed-of-light argument for the size pertains only to the thickness of the sheet), (2) the natural relativistic beaming by the portions of the sheet moving toward the observer, and (3) the natural preservation of the pulse because radiation from one sheet cannot overlap the neighboring sheets, assuming a large γ, until the sheets have undergone a huge expansion and thus presumably have ceased to radiate. The major difficulties seem to be the following: (1) The distinction between the north and south poles of an oblique rotator, which could be quite pronounced close to the object (figure 3.1), should become unimportant by the time one gets to the light-cylinder, suggesting naively at least that one should get pairs of nearly identical pulses rather than the single pulses so often observed. Observers have checked to be sure that the "single"-pulse pulsars are not actually producing a nearly identical pulse plus interpulse by averaging data over twice the period: the two average pulse shapes, one representing an average over the even-numbered pulses and the other being the average over the odd pulses, are found to be identical within statistical uncertainties. (2) The stable but complex multiple subpulse patterns (figure 1.3) seem more difficult to account for in a straightforward way than a single simple pulse would be. (3) The actual interpulse spacings are not always 180° from the main pulse, although again naive theory would argue that only the dipole magnetic field component should be dominant at the light-cylinder and hence one would expect rather precise alternation of poles even if the amplitude differences could be explained away. The highest-energy emissions (gamma-ray) from both the Crab and Vela pulsars do have almost equally spaced and nearly similar amplitudes for the main pulse and interpulse (see Maraschi and Treves 1974). The possibility that pulsed radiation is formed far from the pulsar may not yet be closed, although it is not presently in vogue.

3. Spectrum

Some phenomenology concentrates on accounting for the pulsar spectrum. The idea that coherent radiation comes from bunches is fairly simple to model because a bunch can be described with one or two parameters whereas a maser-active region would require more parameters and a more detailed model for the spatial structure. The rough spectrum that results from parameterizing the bunch with a single scale, $\lambda/2$, was illustrated in figure 2.22. The factor of 2 is to allow us to identify λ itself with the wavelength of the maximum (if one exists) in the spectrum. For wavelengths long compared to about $\lambda/2$, all particles in the bunch (N_R) radiate together and the incoherent single-particle curvature radiation is proportionately amplified. For wavelengths very much longer than $\lambda/2$, only the incoherent radiation is obtained. In between, there is diminished coherence where the "effective" size of the bunch that can radiate without destructive interference shrinks until only a single particle occupies that volume, on the average. Sturrock et al. (1975) discuss in more detail the expected spectrum for specific realistic bunch shapes. Cordes (1979a,b) discusses how one might infer the bunch shape from observation (see also Rickett 1975 and Cordes 1976). Similar spectra are presumably present in electron storage rings (Michel 1981), where the energetic electrons are stored in the form of small circulating bunches.

Such simplified spectra resemble typical pulsar emission in the sense of having a low-frequency cutoff in the radio, a steep decline in power with increasing radio frequency, and then a recovery (in a few cases such as the Crab pulsar) to a strong and presumably *incoherent* high-frequency source with again a steep decline (see, however, Sturrock et al. 1976 for arguments that the latter could also be coherent). The declining spectrum of the energetic ($\geq$ optical) radiation from the Crab pulsar is essentially a power law whereas an incoherent curvature radiation spectrum cuts off exponentially. The difference can be reconciled by assuming a spectrum of electron energies, because the low-energy part of the emission spectrum is largely independent of the electron energy spectrum and therefore a slower falloff can always be obtained. There are still problems in attributing the optical emission of the Crab pulsar to curvature radiation. Middleditch et al. (1983b) studied the infrared emission and found a rapid falloff in intensity ($\approx f^{2/3}$) at frequencies below the optical. Because curvature radiation itself would fall off more slowly at low frequencies ($\approx f^{1/3}$), they suggested this falloff as evidence of some additional mechanism such as synchrotron self-absorption. The same mechanism has been suggested for the turnover at low radio frequencies (Ochelkov and Usov 1984).

4. General Remarks

A number of other phenomenological proposals have been made, often to explain a single feature observed in only one or a few pulsars. Insofar as attempts to construct a basic theory go, the value of phenomenology decreases rapidly with the number of versions among which one is free to select. Once again, we admit a bias toward a first-principles effort in solving the pulsar mechanism. The number of known pulsars has passed 400. Yet neither the statistics of this rather extensive sample nor the properties of any specific pulsar seem yet adequate to reveal the underlying mechanism at work. The experience to date is therefore not too encouraging that pulsars will be more or less solved simply by examining the data in the context of general physical principles. One possible resolution of this impasse would be the formulation of an idealized, simple, self-consistent, physically complete model to compare with the available data. Another resolution, of course, would be the discovery of a pulsar so exceptional that its functioning would be obvious. It seemed as if the millisecond pulsars might serve that role, but apart from their short periods and necessarily weak magnetic fields, they seem similar to the more common pulsars. The eclipsing binary pulsar may yield important new insights.

4

Idealized Model: The Aligned Rotator

4.1 Introduction

We now turn to quantitative magnetospheric models, seeking to incorporate at least some of the dimensional analysis from chapter 1. The aligned rotator is a natural choice; the magnetic field is assumed to be a perfect dipole aligned with the axis of rotation as originally suggested by Goldreich and Julian (1969). Although a likely oversimplification, the aligned rotator is (a) a nontrivial physical system, (b) a model of intrinsic theoretical interest, and (c) a possible model for pulsars per se provided that the axisymmetry, which would otherwise preclude pulsations, could be broken without radically altering the model. If the aligned rotator were to radiate, then with (c) the goal of understanding the physics of the pulsar phenomenon would have largely been achieved. In any event, before we can analyze the more difficult but realistic problem of the oblique rotator, we must first understand the physics of the aligned case.

A large number of papers have been written assuming that an aligned rotator has the basic physical attribute of an actual pulsar, namely a source of directed radio-frequency emission. An aligned rotator would then be nonpulsed owing to the imposition of axial symmetry, but the pulsations could then be recovered simply by tilting the dipole moment (the implied existence of nonpulsing "pulsars" was suggested as a test of the model: Michel 1970b). The prototypical theoretical paper is likely to recycle, if the literature over the past 20 years is any guide, some or all of the following elements:

(1) The existence of a strong rotationally induced electric field is certainly correct if the star is conducting, is magnetized, and rotates.

(2) The charged particles are accelerated from the neutron star surface by this electric field. (The possibility of shielding by nonneutral plasmas enters here.)

(3) Gamma-rays are emitted by the charged particles following the curved magnetic field lines. It is hard to believe that this might not happen, but shielding of the electric fields could be very important.

(4) Extensive pair production by the gamma-rays in (3) results from conversion on the ambient magnetic field. As we have seen, gamma-rays from the

highest-energy pulsars such as the Crab and Vela support this assumption. However, the gamma-ray energy is such a sensitive function of the energy of the primary electron that a cascade might be difficult to maintain (sec. 2.3.6).

(5) The pairs provide current closure. No physically self-consistent model has yet been constructed to show how this might actually work.

(6) The pairs provide bunching. How such a mechanism might work remains poorly described.

(7) The pairs are essential (no pairs, no pulsar!). Given the strong fields around pulsars, it is quite plausible, perhaps even obligatory, that some pairs are created. The question is whether such production is essential to the pulsar phenomenon. The fact that an engine may make noise is largely irrelevant to the function; in the same way, pulsars could make pairs without pairs being essential. It is therefore a difficult theoretical choice to make. Observationally, the phenomenological death line (sec. 1.5.3), for example, might correspond to a threshold for pair production, but also could simply correspond to luminosity selection.

(8) Dynamical behavior is driven by loss of plasma beyond the light-cylinder. The large amount of unseen spin-down power suggests that there are indeed winds from pulsars, but an aligned rotator might or might not produce such a wind.

Roughly, the first, second, and last assumptions define the standard model for pulsar action, which now can be found in most new textbooks on astronomy: an oblique rotator with wiggly lines (radio emission) stemming from the magnetic poles. Perhaps such cartoons are broadly representative of pulsar action, but making such a model work has so far been illusive.

In what follows we will give first a straightforward analysis of the aligned rotator, and then discuss the former standard model.

4.2 Aligned Rotator Electrodynamics

The basic idea in the "standard" model is that the magnetosphere of a pulsar is entirely filled with plasma that is electrostatically pulled from the neutron star surface. Unfortunately for the model, one can show that complete filling need not take place, and as a consequence more physics must be introduced either to complete such filling or to permit pulsar action with only a partially filled magnetosphere, which would seemingly consist of an isolated *electrosphere* close to the neutron star (Michel 1985d). A basic unresolved physical issue is therefore whether complete filling of the pulsar magnetosphere with plasma is essential for pulsar action.

1. The Vacuum Solution

It is useful to start at the beginning and write down the solution for an aligned rotator without any magnetospheric plasma. We will take the magnetic moment to be a point dipole at the center of the star. It is then elementary electrostatics to calculate all of the potentials. A rotating conducting magnet induces an electric field. Some people are troubled by this concept because there are no $\partial \mathbf{B}/\partial t$ terms in a pure aligned rotator, so where does "induction" enter? The simplest answer is that the charged particles making up the rotator simply move in static fields in the inertial frame (Maxwell's equations as usually written are appropriate only to inertial frames, not rotating frames). If the particles are to rotate with the rest of the rotator, they must $\mathbf{E} \times \mathbf{B}$ drift at the corotational velocity of the solid body, and therefore the appropriate $\mathbf{E}$ field must either exist or be generated by charge separation if it does not. For the Earth, the spin axis (Ω) points north but the magnetic moment (M) points south (the *external* magnetic field points north, so the return flux inside the Earth is in the opposite direction). There are two extremes of how this flux is distributed: one possibility is to have a uniformly magnetized interior and another is to have a point dipole at the center with a dipole magnetic field everywhere inside or out. For the uniform interior magnetization model, a positive charge in the interior experiences a $\mathbf{V} \times \mathbf{B}$ Lorentz force toward the spin axis. Such a displacement of positive particles would create a polarization electric field $\mathbf{E} = -\mathbf{V} \times \mathbf{B}$ outward from the spin axis and negative surface charges in the equatorial zone. This electric field is a combination of that from a uniform charge distribution and a quadrupolar surface charge. Alternatively, with a point dipole at the center of the Earth, the internal magnetic field would be southward, requiring an *inward* electric field, which turns out to be the combined electric field of the net internal charge of the "point" dipole (it is easier to imagine a small finite sphere), its quadrupolar surface charge, and a volume quadrupolar charge separation either extending to infinity (beyond the surface) or terminating with a quadrupolar surface charge at the surface. Which is simply to say that we require an internal electrostatic potential (and an associated "space" charge in the interior)

$$\Phi = \Phi_0 a \sin^2 \theta / r, \tag{1}$$

where $\Phi_0 = \Omega a^2 B / 2$. Since all quantities are scaled by this voltage, we will normalize it to unity in the following expressions. The electric field and space-charge density are immediately given from the potential as listed in table 4.1. Matching the potential to vacuum monopole and quadrupole moments then gives the external fields. Although this solution is elementary, there are a number of noteworthy points to be made. First, the net space charge in the system is zero; both the surface charge and the internal volume charges are distributed as $1 - 3 \cos^2 \theta$, which integrates to zero over a spherical surface. Nevertheless, there is a net charge on the star because it has a monopole moment, and therefore this charge must be located at the magnetic dipole point source in this idealized model.

a. *Surface Charge*

Second, there is a surface charge. It is this surface charge (σ) that is pulled from the surface to form the magnetosphere. Note that $\mathbf{E} \cdot \mathbf{B}$ changes sign with σ, thus acting to pull electrons from the polar caps and positive particles from

TABLE 4.1—The Vacuum Solution (point dipole field)

Quantity	Expression	Surface Values at Equator[a]	Pole[b]
Inside star:			
Φ	$\sin^2 \theta / r$	$+1$	0
E_r	$\sin^2 \theta / r^2$	$+1$	0
E_θ	$-2 \sin \theta \cos \theta / r^2$	0	0
q/ϵ_0	$2(1 - 3 \cos^2 \theta)/r^3$	$+2$	-4
$\mathbf{E} \cdot \mathbf{B}$	0	0	0
Outside star:			
Φ	$\dfrac{2}{3r} + \dfrac{1}{3r^3}(1 - 3 \cos^2 \theta)$	$+1$	0
E_r	$\dfrac{2}{3r} + \dfrac{1}{r^4}(1 - 3 \cos^2 \theta)$	$+\dfrac{5}{3}$	$-\dfrac{4}{3}$
E_θ	$-2 \sin \theta \cos \theta / r^4$	0	0
q/ϵ_0	0	0	0
$\mathbf{E} \cdot \mathbf{B}$	0	0	0
Surface:			
$\sigma/\epsilon_0{}^c$	$2(1 - 3 \cos^2 \theta)/3$	$+\dfrac{2}{3}$	$-\dfrac{4}{3}$
$\mathbf{E} \cdot \mathbf{B}$ (average)	$2 \cos \theta (1 - 3 \cos^2 \theta)/3$	0	$-\dfrac{4}{3}$
Everywhere:			
B_r	$2 \cos \theta / r^3$	0	$+2$
B_θ	$\sin \theta / r^3$	$+1$	0

a) Here $f = 1$, $\theta = \pi/2$.
b) Here $f = 0$, $\theta = 0$.
c) The surface charge density σ is given from E_r (outside) $- E_r$ (inside).

the equatorial regions. As noted above, the Earth has $\mathbf{\Omega} \cdot \mathbf{M} < 0$, which is opposite that typically assumed for pulsars (because electrons radiating from the polar regions are assumed). A modest surface charge can be maintained on conductors under laboratory conditions since the work function is nonzero, but there is no way that a pulsar could maintain the large negative surface-charge concentrations that would be induced. As noted below in connection with the Ruderman-Sutherland model, there is a (questionable) possibility that the ion

TABLE 4.2—The Vacuum Solution (uniformly magnetized interior)

Quantity	Surface Values (expression)	Equator	Pole
B_r	$2 \cos \theta$	0	$+2$
B_θ	$-2 \sin \theta$	-2	0
Φ	$r^2 \sin^2 \theta$	$+1$	0
E_r	$-2r \sin^2 \theta$	-2	0
E_θ	$2r \sin \theta \cos \theta$	0	0
q/ϵ_0	-4	-4	-4
$\mathbf{E} \cdot \mathbf{B}$ (inside)	0	0	0
σ/ϵ_0 (surface)	$2 + \dfrac{7}{6}(1 - \cos^2 \theta)$	$-\dfrac{1}{3}$	$\dfrac{19}{6}$

work function is sufficient to maintain a positive surface charge, but for the moment we assume both species to be freely available.

Any solution of the standard model will require setting $\mathbf{E} \cdot \mathbf{B} = 0$ everywhere on the surface, because the assumption of free emission precludes binding them to form a surface charge. Note, in this respect, that image charges in a conductor are actually surface charges; thus, despite the fact that the neutron star is treated as a perfect conductor, the external magnetospheric charge distribution does not produce an additional "image" contribution: although the star is taken to be a conductor, $\mathbf{E}$ is not even normal to the surface, as one can see from table 4.1. The Goldreich and Julian (1969) solution simply corresponds to extending the interior (point dipole source) solution to infinity, with the space-charge density continuing smoothly through the surface. Rigid corotation of the magnetosphere is then an obvious consequence, because the corotional $\mathbf{E}$ and source $\mathbf{B}$ fields are simply extended to infinity. The only discontinuity at the stellar surface would then be in the density of neutral matter making up the neutron star.

b. The Central Charge

Let us now ask why there should be a huge charge associated with a point dipole (see, for example, Cohen et al. 1975). From E_r (outside) and Gauss's law we have a positive central charge

$$Q = 8\pi\epsilon_0 a \Phi_0/3, \tag{2}$$

which is of the order of 10^{12} coulombs (several kilograms of electrons!) for the Crab pulsar. This charge is actually distributed throughout the magnetic field source region, as we can see by replacing the dipole with a uniform interior magnetization as shown in table 4.2. We see that a uniform magnetic field in the interior actually becomes negatively charged while the surface becomes positively charged, by factors of, respectively, 2 and 3 times the net charge.

In a more realistic field model that did not have a discontinuity in B_θ at the surface, the net "central" charge would, of course, be distributed over a finite volume.

Of particular importance to theory is the fact that, unlike the laboratory situation, the potential and charge of a pulsar are not free parameters (in the zero work-function limit). Any attempt to alter the charge Q would produce a net surface charge, and any such charges would be lost into the magnetosphere. Thus, if the total system charge were to be more or less neutral, the magnetosphere would have to contain a negative charge excess of $-Q$.

2. Nonneutral Plasmas

The next logical step is to let the surface charge leave the neutron star surface and start to fill the magnetosphere. More properly, we might call this region the *electrosphere* because only finite volume will be filled with space charge, while much of the magnetosphere proper will remain empty. This somewhat startling property stems from the physics of nonneutral plasmas, which have come to become increasingly well known in the laboratory over the past few years, although still relatively unfamiliar to the astrophysical community. Consequently, some comments are appropriate. Briefly, the properties of totally nonneutral plasmas are the following:

(1) Nonneutral plasmas are entirely stable structures of finite extent trapped in combined electric and magnetic field configurations: they are confined in one axis by electrostatic forces and in the other two axes by $\mathbf{V} \times \mathbf{B}$ forces (e.g., by rotation).

(2) Nonneutral plasmas have slowly varying charge densities except at the surface, where the density drops almost discontinuously to zero (vacuum). The thickness of this transition is the Debye length. The magnetic field can cut right across such discontinuities. Such behavior is not possible for quasi-neutral plasmas (which, not being recognized earlier, misled workers to propose models lacking these essential features, i.e., the "standard" model itself).

(3) Nonneutral plasmas are *not* neutralized by attracting charges of the opposite sign (they *repel* charges of the opposite sign!). If a neutralizing charge is introduced into a nonneutral plasma, it will drift until it finds a surface with the vacuum and will then be accelerated away.

It might be useful to picture the nonneutral plasmas as a fluid in the sense of having a free surface, whereas the quasi-neutral plasmas behave more like gases in that they try to fill whatever the available volume might be.

Earnshaw's theorem tells us that one cannot have static configurations of electric charges in pure electrostatic fields. This theorem follows from $\nabla^2 \Phi = 0$. If we write out the numerical approximation of this differential equation in two dimensions (grid of spacing δ), we have

$$\Phi(x + \delta, y) + \Phi(x - \delta, y) + \Phi(x, y + \delta) + \Phi(x, y - \delta) - 4\Phi(x, y) = 0, \quad (3)$$

which simply states that the potential at any grid point is the average of those of the four surrounding points. The average cannot be a maximum or a minimum, so we immediately obtain the well-known result that an electrostatic field cannot trap a particle at a fixed point. Unfortunately, Earnshaw's "theorem" is one of the least robust in the sense that it fails if any of the explicit and implicit assumptions fail. Explicitly, there can be no extra forces. Implicitly, inertia is a force. Thus even a classical atom can trap electrons owing to their inertia. Or a charged particle can be trapped in an oscillating quadrupole electric field. Or a charged particle can be trapped in a static quadrupole electric field provided that there is also a magnetic field present. The last trap is called a magnetic Penning trap and has been a standard laboratory device for some time. Indeed, people have long known that a Penning trap could trap numerous charges of the same sign as well as just one. However, no one worked out what the resultant configuration would be, and the configuration was often casually described as a "cloud" of particles, implying a diffuse and indeterminate structure. Given a description that was completely misleading, the structure was not elucidated for years (until well after the discovery of pulsars!).

The simplest trapping geometry is in an axially symmetric electrostatic quadrupole with fields

$$E_x = \Phi_0 \frac{x}{a^2}, \tag{4x}$$

$$E_y = \Phi_0 \frac{y}{a^2}, \tag{4y}$$

$$E_z = -2\Phi_0 \frac{z}{a^2}, \tag{4z}$$

where a is the physical separation scale of electrodes having a potential drop of Φ_0 and a magnetic field

$$B_z = B_0. \tag{5}$$

The full motion of a charged particle in this trap is quite complicated, consisting of three components: an oscillation along the z-axis, the cyclotron motion around the magnetic field lines, and the $\mathbf{E} \times \mathbf{B}$ drift about the center of the trap in the x-y plane. Trying to imagine a large number of particles executing this motion while at the same time strongly scattering one another doubtless contributed to defaulting to the "cloud" picture. But the first and last of these three motions are essentially thermal in nature, so one can ask what happens at zero temperature, where only the $\mathbf{E} \times \mathbf{B}$ drift remains. The answer is that the particles form an ellipsoid of uniform density surrounded by vacuum! (The electrostatic force vector is a linear function of the distances along the three major axes in a uniform ellipsoid: Kellogg 1967, p. 195; MacMillan 1958;

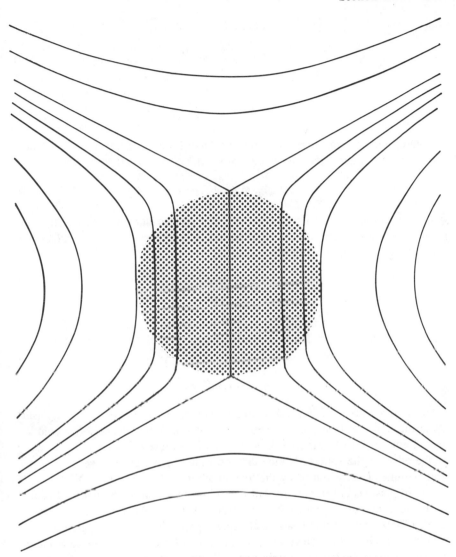

FIG. 4.1. Equipotential lines for a nonneutral sphere. Here the uniform charged sphere is centered in a quadrupole electric field. For the appropriate charge density, the equipotential lines are vertical within the sphere. If a vertical magnetic field is imposed, one satisfies $\mathbf{E} \cdot \mathbf{B} = 0$ and such a sphere of electrons (say) would rotate like a rigid body and be stably confined. From F. C. Michel, 1989, *Comments Ap.*, 13, 145 (figure 1). Reprinted by permission of Gordon and Breach Science Publishers Inc.

Durand 1964.) The simplest case is that of a *sphere*, as shown in figure 4.1. For a sphere of uniform density, the mutual repulsion of the particles is just given by the net charge interior to any radius; thus

$$E_r = Q(r)/4\pi\epsilon_0 r^2, \tag{6}$$

but

$$Q(r) = \frac{4\pi}{3} n_0 r^3,$$ (7)

so

$$E_r = \frac{n_0}{3\epsilon_0} r.$$ (8)

Because both the trap forces and the repulsive forces are linear with distance, one can set the net force everywhere along the z-axis equal to zero with

$$en_0 \equiv 6\epsilon_0 \Phi_0/a^2.$$ (9)

In other words, the trap parameters fix the charge density. The electrostatic force in the x-y plane is not balanced but is countered simply by rotation of the sphere with angular frequency

$$\Omega = \frac{3\Phi_0}{B_0 a^2}.$$ (10)

Note that the nonneutral plasma rigidly rotates in this trap, so there are no viscous forces at work. This latter relationship is much more transparent if we eliminate Φ_0 and write

$$n_0 = 2\epsilon_0 B_0 \Omega/e,$$ (11)

which is just equation 2.3.18, used to estimate the space charge around a pulsar! Note that the constancy of the density in equation 7 is essential. The solution would not work for a variable density. Therefore there *must* be a discontinuity at the edge of the sphere. Notice that the radius never entered because it is a free parameter, determined by the total number of particles in the trap.

These plasmas are important. For one thing, they can be trapped and cooled to the point of freezing into a Wigner crystal (Wigner 1934). Actually, Wuerker et al. (1959) showed that charged aluminum particles could be aggregated into "solids" much earlier. Nevertheless, for most of the intervening half-century physicists have tended to assume that a Wigner crystal simply represented a mathematical idealization that could not be realized in actuality. Earnshaw's "theorem" seems most likely a partial cause for this blind spot of such long duration.

The surface discontinuity cannot be infinitely sharp at zero temperature or with a finite number of particles: the structure simply freezes into a crystal (which cannot be exactly spherical). The usual estimate for the temperature at which such a transition takes place is when $\Gamma \approx 178$ (Ichimaru 1982), where

$$\Gamma \equiv \frac{e^2}{4\pi\epsilon_0 a_{\text{rms}} kT},$$ (12)

namely the ratio of the thermal energy to the mean electrostatic interaction energy. Here $a_{rms} \equiv n_0^{-1/3}$. At finite temperatures, the plasma will be "gaseous" (mainly in the sense of the particles not being strongly correlated with one another) and particles will be free to drift along the z-axis to the surface. There they will experience a restoring force because outside the sphere the electrostatic forces from the quadrupole field continue to grow but the self-repulsion now declines. Outside, the electric field gradient will return to normal from zero; thus the electric field parallel to **B** is zero at the surface and departs linearly thereafter, giving $E \approx \Phi_0 h / a^2$, where h is the height above the surface. For the distance h that a particle with energy kT can go, this electric field gives

$$h = \left(\frac{\epsilon_0 kT}{3 n_0 e^2} \right)^{1/2}, \tag{13}$$

and apart from a numerical factor, this is just the Debye length. The Debye length is roughly $7(T/n)^{1/2}$ cm, and for likely pulsar conditions near the surface ($T \approx 10^6$, $n \approx 10^{12}$) the Debye length is only 0.01 cm, negligible compared to the size of a neutron star.

The stability of nonneutral plasma configurations in general can be derived from conservation of the total (canonical) angular momentum (O'Neil 1980),

$$P_\theta = \sum (m\rho^2 \Omega + e A_\theta \rho), \tag{14}$$

where the first term is the particle contribution and the second is the particle-field interaction contribution. The sums are over particle indices (i.e., m becomes m_j, etc., if there are various masses, ρ becomes ρ_j to denote the axial position of the jth particle, etc.) which have been suppressed for clarity. Unless the system is rotating fast enough to be close to centrifugal instability (the so-called Brouillon limit), the first term is small compared to the second. In that limit, and for axial symmetry, we can write $A_\theta = B_0 \rho / 2$, giving

$$P_\theta = \frac{1}{2} \sum e B_0 \rho^2 = \text{constant} \times \sum \rho_j^2, \tag{15}$$

where the last step is possible only for a nonneutral plasma in which all the charges (e_j) have the same sign. Constancy of $\sum \rho^2$ is a strong constraint, because the escape of one particle to large distances requires all the rest to be confined even more tightly. External torques can, in practice, change P_θ and cause the configurations in an actual trap to evolve from prolate through spherical to oblate spheroidal rotators until the walls are finally encountered (Driscoll and Malmberg 1983). This evolution is generally attributed to field errors in the traps which resonantly excite the particles (Keinigs 1984). In the lab, however, one can apply systematic external torques to reverse this evolution. Note the difference from quasi-neutral plasmas, which would diffuse out of a trapping geometry (not the magnetic Penning trap) owing to collisions and instabilities; such stochastic processes cannot in general be reversed by "me-

TABLE 4.3—Comparison of Nonneutral and Quasi-neutral Plasmas

Nonneutral	Quasi-neutral
Only one sign charge	Essentially neutral except for waves and edge effects
Naturally stable trapping	Difficult to trap: instabilities and diffusion
Arbitrarily long trapping times	Short trapping times (except on astrophysical scales)
Occupy finite volumes	Act to fill all space, difficult to localize
Sharp interfaces with vacuum	No such stable interface exists
Density distribution is fixed	Density distribution is arbitrary
Freeze when Debye length small	Recombine when Debye length small

chanical" means. O'Neil and workers (Hjorth and O'Neil 1987; O'Neil and Hjorth 1985) have shown that thermal contact (between the single particle motions representing parallel versus perpendicular thermal energy) is suppressed in nonneutral plasmas.

Note that the plasma frequency also appears naturally in these traps, even though there is no neutral state to oscillate about: it is essentially the restoring frequency (within a numerical factor) with which a single particle would oscillate along the z-axis in the quadrupole electric field (the entire spherical configuration could oscillate with this frequency): $\omega_{well}^2 = e^2 n_0 / 3 m \epsilon_0$. Some contrasts between the two types of plasma are listed in table 4.3.

If these traps were unique to the laboratory and of no relevance to astrophysics, we would not be discussing them here. Consider the criterion for forming such a trap. If the trap is empty, we can calculate $\mathbf{E} \cdot \mathbf{B}$ to find that the locus for zero parallel force on the particle is just the x-y plane. The natural place to accumulate particles is wherever $\mathbf{E} \cdot \mathbf{B} = 0$. Such loci are surfaces in general and can trap only one sign of particle. As more particles are trapped, their self-repulsion becomes important and they are trapped in a finite volume. The simplest to understand cases are those with axial symmetry because the particles will be in motion, and this motion is simply bulk rotation in the above case. Some trapping configurations are listed in table 4.4.

TABLE 4.4—Some Trapping Geometries for Nonneutral Plasmas

Trap Geometry	Name	Plasma Configuration
B = const.; E = quadrupole	Penning trap	Rigidly rotating spheres/ellipsoids
B = const.; E = end electrodes	Malmberg trap	Rigidly rotating ellipsoids
B = 0; E = oscill. quadrupole	Paul trap	Frequency-dependent shapes
B = dipole; E = quadrupole	Rotating terrella (pulsar?)	Domes and disk
B = dipole; E = monopole	Charged terrella	Rigidly rotating disk
B = 0; E = monopole	Charged sphere (classical atom)	Kepler disk

How is it that quasi-neutral plasmas, rather than nonneutral ones, dominate our ideas of what a plasma consists of? Largely, because of experimental bias. If one wants to create a plasma, one can simply subject neutral gas to some ionization source, e.g., intense radio-frequency power. Having started with neutral gas, one naturally obtains a neutral plasma. What would be regarded as nonneutral plasmas are instead created as beams (Davidson 1974), as in the current that passes from the cathode to the anode in a vacuum diode (i.e., old-fashioned vacuum tube). Being intrinsically dynamic, such beams might well be regarded as transient structures and therefore not anything that could ever be in stable equilibrium. The transition from transient beam to stable configuration, however, is easily accomplished in the trap exploited by Malmberg and colleagues wherein a pulsed burst of a beam is trapped between the end electrodes of a long solenoid (figure 4.2). In astrophysical contexts, a similar progression can be traced. Given abundant ionization, one finds such quasi-neutral plasmas as those in the Earth's ionosphere. To find nonneutral plasmas, one must have both a natural trapping geometry ($E \cdot B = 0$ loci) and sufficiently weak ionization to fill such traps without shorting out the system. A rotating magnetized neutron star seems a reasonable candidate system, given the natural trapping regions for both signs of charge, the natural creation of nonneutral plasma as a beam drawn from the surface by the induced electric field, and a rotational energy supply that would be difficult to "short out." The outer planets of the solar system, such as Saturn and beyond, might arguably be nearer places astrophysically for finding nonneutral plasmas on large scales.

3. Space-Charge Shielding

As we will show, the vacuum aligned dipole intrinsically has two trapping surfaces, one for electrons (say) above the magnetic poles and one for positive particles in the equatorial plane. If particles are only available from the surface, they must fill these regions from where the magnetic field lines intersect the surface. The equatorial charges cannot fill much more than a snug torus around the star. The particles from the poles can extend to great heights, on the other hand. Historically, this understanding did not come so directly. An understanding of nonneutral plasmas, simple as they can be seen above to be, did not exist. The former standard model cannot work (Michel 1980) as originally thought.

It slowly became evident that completely filling the volume about the star, as shown in figure 4.3, with a stationary (corotating) charge distribution is not required, plausible as it seemed at first. As one can see in this figure, some field lines requiring positive space charge in the equatorial regions lead to negative space charge in the polar regions. Holloway (1973) argued that, if some of the positive equatorial charge were to be removed, it could not plausibly be replaced. There is no way to accelerate new positive charges from the polar regions without first driving all the negative charges to the surface. He concluded that the system would respond by splitting open along the $q = 0$ (*null*)

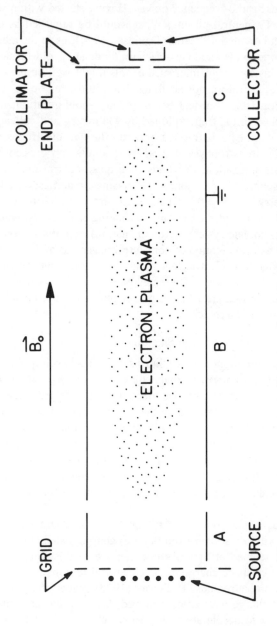

FIG. 4.2. Malmberg device. After deGrassie and Malmberg (1977). Electrons are injected from source at the left with cylindrical sections *A* and *B* grounded but *C* negatively charged. Electrons follow magnetic field lines to vicinity of electrode *C* and are electrostatically reflected. Before they can return to the source, electrode *A* is driven negative and the electrons are now trapped inside *B*. Electrostatic self-repulsion causes the electron distribution to rotate about the long axis. The system is extremely stable, and the electrons can easily rotate more than 10^{10} times without problem. From F. C. Michel, 1989, *Comments Ap.*, 13, 145 (figure 2). Reprinted by permission of Gordon and Breach Science Publishers Inc.

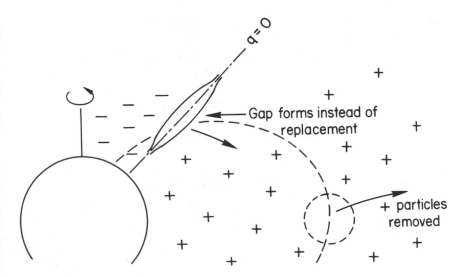

FIG. 4.3. Holloway's gap versus filled magnetosphere. Nonneutral plasma of one sign occupies the entire volume poleward of the line labeled $q = 0$, and plasma of the other sign occupies a volume near the equator. Note that large volumes of the equatorial plasma would have to have been accelerated from the surface *before* the poleward plasma was pulled from the star. The system could not possibly have emitted first the equatorial particles and then the poleward particles owing to the huge and inconsistent surface charge required between these steps. The total charge of each type is logarithmically divergent. If positive particles are removed from the equatorial regions of the Goldreigh-Julian model as illustrated, replacement from the surface seems impossible. Holloway proposed that the zero-charge surface splits to create a vacuum gap as shown. From F. C. Michel, 1982, *Rev. Mod. Phys.*, 54, 1 (figure 14); 1989, *Comments Ap.*, 13, 145 (figure 3), reprinted by permission of Gordon and Breach Science Publishers Inc.

surface in the space-charge distribution, and a *vacuum gap* would separate the two charge distributions as shown in figure 4.3. He did not, however, give any mathematical models, and the topology for such a perturbation has yet to be solved. My speculation is that the gap formed by removing some of the equatorial charge would not be locally confined but once started would extend to infinity.

Ruderman and Sutherland (1975) pointed out that for a suitable *charged* star rotating in a vacuum, an $\mathbf{E} \cdot \mathbf{B} = 0$ surface exists above the polar caps. They proposed then that the gap between this surface and the star could be vacuum, while beyond, one again had a filled magnetosphere solution. (This solution, however, still has the same pathologies as the gapless solution [Michel 1979b] and evidently doesn't exist anyway [Asséo et al. 1984].) Their emphasis was on the way in which the gap could modulate the particle acceleration above the polar caps. Unfortunately their model also contains a "gap" in relative particle density above the surface in the regions where the particles are accelerated but have not yet formed any pairs, a totally different issue having nothing in common with the physics of nonneutral plasmas.

In a later paper (Cheng et al. 1976) a Holloway-type gap is proposed as well, but again this type of gap is simply assumed and not modeled. Jackson (1976a,b) discusses a gap solution similar to that of Ruderman and Sutherland, but makes a quite different interpretation. He notes the $\mathbf{E} \cdot \mathbf{B} = 0$ surface as an accumulation point for charges, which he proposes to be only partially filled, which is, as we saw in sec. 2 above, a perfectly good solution. The complete solution, however, is gained when no further escape of particles from the surface is possible.

Pelizzari (1976) has examined the Störmer-like escape of particles from plasma-free rotators (i.e., all "gap") and was unable to eject both positive and negative charges at once. Salvati (1973) and Buckley (1976) considered the complementary possibility of regions devoid of magnetic fields(!).

It should be emphasized again that such gaps in nonneutral plasmas have a remarkable property. One has arranged charges and magnetic fields in such a way that as one passes through a region of space swarming with charged particles, following a magnetic field line along which the particles can move freely because it is an equipotential, one abruptly finds oneself in a vacuum. In other words, one has a true discontinuity in charge density (at zero temperature), going from a finite value to zero in an infinitesimal distance, as illustrated in figure 4.4. The remarkable fact that such discontinuities really exist is ambiguous in the above works. Jackson assumed that his trapped plasma had a density that declined exponentially toward zero. In Holloway's model one is free to assume the same thing, since no explicit model is given. He proposed that the zero-density interface is the site of gap formation, namely where the density has already feathered off to zero. As we now know, it is more characteristic for the density to go to zero abruptly. The Ruderman-Sutherland gap is really a transition from the acceleration zone to pair production (figure 2.6) and not a gap between nonneutral plasmas. Michel (1979b) reexamined the general question of how vacuum gaps might exist and developed a general formulation for constructing them in certain axisymmetric geometries, proposing that the field line potential distribution, which has been the source of so many difficulties as described above, might be modulated by the formation of appropriate gaps over the polar caps. Jackson (1976a,b) refers to the $\mathbf{E} \cdot \mathbf{B} = 0$ surface as an FFS (force-free surface). In figure 4.5 we illustrate the respective roles of the FFS's both as discontinuities and as accumulation regions. Such a system could easily be built in the laboratory and used to simulate an interesting and nontrivial N-body system (Michel and Freeman 1983, unfunded proposal). Here we consider a charged magnetized *nonrotating* sphere that emits some of its charge. We start with an FFS in the equatorial plane at which the particles accumulate. The resulting discontinuities between the particles and vacuum are the new FFS's. Thus the FFS appears in two distinct contexts: (1) a place in vacuum where charged particles of a certain sign can congregate, a property of the system in the absence of local space charge, and (2) a plasma/vacuum discontinuity. We will continue to use the term *discontinuity* to designate the latter. Accumulation of charges at an FFS (sense 1) splits it into a force-free *volume* bounded by

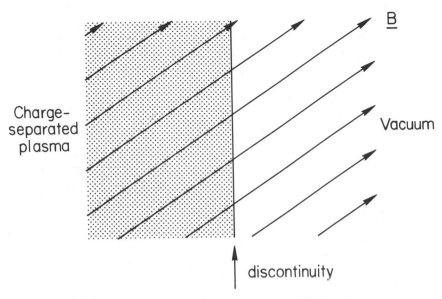

Charge-
separated
plasma

Vacuum

B

discontinuity

FIG. 4.4. Discontinuity in nonneutral plasma. The plasma density falls abruptly to zero to separate regions of finite space charge (but $\mathbf{E} \cdot \mathbf{B} = 0$) from regions of nonzero parallel field (but zero density). This is just a close-up of the surface of figure 4.1. The parallel field returns particles to the discontinuity. Such discontinuities can not be stable for a quasi-neutral plasma, because one or the other component would be accelerated away. At finite temperatures the discontinuity would be spread out over a thermal scale height which is also essentially the Debye length. From F. C. Michel, 1982, *Reb. Mod. Phys.*, 54, 1 (figure 15); 1989, *Comments Ap.*, 13, 145 (figure 4), reprinted by permission of Gordon and Breach Science Publishers Inc.

discontinuities (FFS's in sense 2). We cannot, of course, split a discontinuity (which is a surface with nonneutral plasma on one side and vacuum on the other) to end up with two more FFS's; adding particles to a discontinuity simply shifts its location by expanding the volume of trapped particles. In the charged rotating star model, however, the FFS can be of either type. Thus Jackson regarded the FFS as a *dome* over the polar cap which has accumulated (trapped) some charged particles. However, the charge will continue to accumulate from the neutron star surface until that surface itself becomes an FFS, which is just a roundabout way of saying that in general the filled volume extends outward from the surface, leaving no Ruderman and Sutherland "gap." To achieve yet further filling in order to eliminate the Holloway gap and give the filled magnetosphere solution illustrated in figure 4.3, we require a source of particles other than the surface of the star. Some possibilities, implicit in Rylov (1976) and explicit in Michel (1980), are pair production and the accretion of plasma from extrinsic sources. Thus the nature of the electrospheric structure around an aligned rotator consists of a dome of charge over the polar caps and an equatorial *torus* of opposite charge, the two enveloping the entire surface but not filling the magnetosphere (see figure 4.6). This result was finally confirmed directly by computer simulations (Krause-Polstorff and Michel 1985a,b). Trapping of

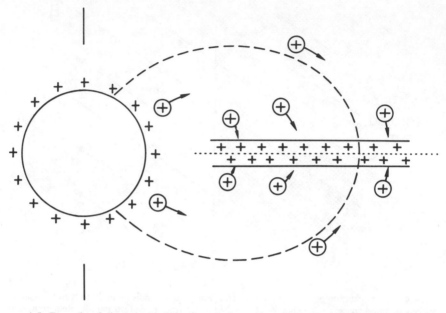

FIG. 4.5. Example of discontinuity formation. A charged nonrotating magnetized star loses some charge along field lines. These charges are repelled from the star to form a disk. The disk has finite thickness owing to self-repulsion of the charges. The disk is therefore bounded by two discontinuities (each a force-free surface) and is formed by splitting of the original vacuum force-free surface in the equatorial plane. Because $E \approx 1/r^2$ and $B \approx 1/r^3$, the disk rigidly rotates while the star remains essentially at rest. From F. C. Michel, 1982, *Rev. Mod. Phys.*, 54, 1 (figure 16); 1989, *Comments Ap.*, 13, 145 (figure 5), reprinted by permission of Gordon and Breach Science Publishers Inc.

the nonneutral plasma renders the issue of plasma loss at the light-cylinder largely irrelevant: there is now no plasma at the light-cylinder to lose. Such aligned rotators would be "dead" insofar as models for pulsars are concerned. Rylov earlier (1976, 1977) came to the same conclusions for the same reasons, and even calculated approximate shapes for the charged clouds. This early work seems to have been neglected because Rylov went on to postulate some unknown mechanism that would allow the equatorial particles to escape, which served only to make this model appear to be a peculiar version of the former standard model rather than what it was: a refutation of that model. There is no doubt, however, that Rylov appreciated the existence of stationary (inactive) aligned magnetospheres. Cheng et al. (1986a,b) also discuss discontinuities but attempt to maintain the quasi-neutral plasma view that these discontinuities must parallel the magnetic field lines. Wegmann (1987) argues that there is a mathematical inconsistency for this type of solution near the surface where the sign of the charge changes, but does not say what the correct solution might be.

The existence of a trapping region over the polar caps can be seen from the basic electrostatics of the aligned rotator. In table 4.1, the external potential

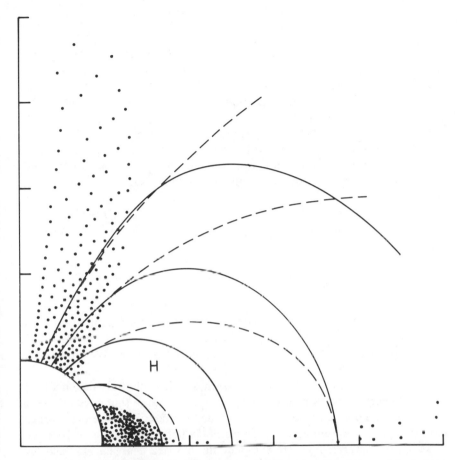

FIG. 4.6. Nonneutral plasma distribution about an aligned rotator. Computer model of plasma distribution about a rotating magnetized neutron star. Scale is linear with the unit sphere at the origin representing the neutron star (i.e., the source of the particles). Dashed lines are the dipole magnetic field lines. The dots are the "particles," actually rings of charge shown in cross-section about the rotation axis (vertical) and the equatorial plane (horizontal). The program has placed each charge at the potential minimum along that field line, and particles are taken from the surface until no surface charge is left on the star. All of the charges about the rotation axis have the same sign (negative, say), and all those about the equatorial plane have the opposite sign (positive, say). The solid lines are equipotential surfaces, which are seen to parallel the magnetic field lines in the charged volumes. This system is perfectly stable despite the huge void about the star. The nonzero parallel electric field in the vacuum regions acts to return negative particles to the polar domes and positive particles to the equatorial torus. A very wide, open Holloway gap (H) separates the electron dome from the positive disk. Pair production would reduce the gap between the two $E_{||} = 0$ surfaces, hence nullify the pair creation field (even assuming that a discharge) could be maintained without a source of primaries). From J. Krause-Polstorff, and F. C. Michel, 1985b, *M.N.R.A.S.*, 43p (figure 1); F. C. Michel, 1982, *Rev. Mod. Phys.*, 54, 1 (figure 16); 1989, *Comments Ap.*, 13, 145 (figure 6), reprinted by permission of Gordon and Breach Science Publishers Inc.

outside of the star is given by

$$\Phi = \frac{2}{3r} + \frac{(1 - 3\cos^2\theta)}{3r^3}. \tag{16}$$

This choice of potential corresponds to choosing the potential at the magnetic pole to be the same as that at infinity. The net charge on the surface of the star is, however, zero. Because any surface charge will eventually be lost, this value is a useful starting point. Of course this result corresponds to a large *dipolar* surface charge, and as charges are released from the surface, they see a potential well (given the constraint to follow magnetic field lines) along the polar axis, with a minimum at $r = \sqrt{3}$, where $r = 1$ is the stellar surface in these normalized units. This potential well can fill with electrons and corresponds to Jackson's dome, while the positive particles are trapped on the closed equatorial magnetic field lines.

It is useful to go back to a fully filled magnetospheric distribution and decompose the potential (which is simply $\sin^2\theta/r$ in our normalized units) into multipole moments. The result is, at $r = 1$,

$$\Phi = \frac{2}{3r} + \frac{(3 + 2r^5)(1 - 3\cos^2\theta)}{15r^3}, \tag{17}$$

where the first term is again the central charge, the second term (the "3") is the external quadrupole moment produced by the space charge inside $r = 1$, and the third term (the "$2r^5$") is the internal quadrupole moment produced by the space charge outside $r = 1$.

Equation 17 satisfies $\mathbf{E} \cdot \mathbf{B} = 0$ at $r = 1$; thus such a solution would not inject any more charges and would not require a surface charge (but does require an external quadrupole electric field). The r dependences are here only formal, however, serving to remind us of the multipole behavior of each contribution. If we go to larger radius, there is now more space charge inside and less outside, which increases the "3" in the second term and decreases the "2" in the third term. In fact each term decreases as $1/r$, and the sum is just $\sin^2\theta/r$. However, equation 17 provides the requirements for an alternative magnetospheric structure. The first two terms are boundary conditions fixed by the star; neither the central charge nor the internal space charge of the conducting star can be modified. The same external quadrupole component must also be present; otherwise $\mathbf{E} \cdot \mathbf{B} \neq 0$ inside the star. In a vacuum solution, this component is provided by the surface charge. If the plasma is pulled from the surface, the correct quadrupole component must instead be provided by the external (magnetospheric) charge distribution. For the vacuum (surface-charge) case, the potential in equation 16 corresponds to a point above the surface-charge layer, and in equation 17 to a point just below it (the same coefficient but an $r^{-3} \to r^{+2}$ change in r dependence). It follows therefore that every solution satisfying $\mathbf{E} \cdot \mathbf{B} = 0$ internally, owing to presumed high conductivity, must give the same contribution as the third term in equation 17 at the surface. If no surface charge can be maintained, then all the charges must be in the magnetosphere. Thus the requirement that $\mathbf{E} \cdot \mathbf{B} = 0$ inside the star is satisfied by *any magnetosphere whatsoever* that provides a contribution to the potential inside the star of

$$\Phi_{\text{magnetosphere}} = \frac{2r^2}{15}(1 - 3\cos^2\theta) \quad (r < a \equiv 1). \tag{18}$$

As a simple example, suppose that the magnetosphere consisted of two huge negative point charges floating over each polar cap. These charges $(-Q)$ see the potential of the star (the "3" term in equation 17)

$$\Phi_{\text{star}} = \frac{2}{3r} + \frac{1 - 3\cos^2\theta}{5r^3} \tag{19}$$

and one another; thus along the polar axis each charge sees an electric field

$$E = \frac{2}{3r^2} - \frac{6}{5r^4} - \frac{Q}{4r^2}. \tag{20}$$

Note that, as advertised, there is a natural trapping region above each polar cap at $r^2 = 9/5$ (for $Q = 0$; note that this value differs from the minimum at $r = \sqrt{3}$ when the external magnetospheric field, approximated by a distant quadrupole source as in eq. 16, is included). The full shape of this region (i.e., the force-free surface) is that of a sphere centered on the polar axis and crossing the axis at $r^2 = 9/5$ and $r = 0$ (the center of the star: Michel 1979b). (Note that eq. 16 overestimates the quadrupole potential of the star itself because here the surface-charge contribution is also included, and there is no such contribution once the particles are ejected to form the magnetosphere.) This "magnetosphere" is too crude to reproduce the potential (eq. 18) at the surface, except as the leading term in the multipole expansion of the field of two symmetrically located point charges. The latter condition is just

$$Q/r^2 = 1/15, \tag{21}$$

which when inserted into equation 20 with the condition $E = 0$ gives $r^2 = 1.889\ldots$, and $Q = 0.126\ldots$, hence a total system charge of $2/3 - 2Q = 0.415\ldots$. Thus we obtain a crude model for figure 4.6, collapsing each dome into a point charge and ignoring the equatorial torus. In this picture, then, the aligned rotator electrostatically traps negative particles above the polar caps and magnetically confines positive particles to an equatorial torus.

The precise structure and stability of such configurations has been shown directly by numerical modeling (Krause-Polstorff and Michel 1985a,b). Jackson (1979, 1980a,b) has shown several explicit closed magnetospheric configurations and proposed their possible relevance to the pulsar problem. These configurations do not satisfy the surface potential distribution $(\Phi = \sin^2\theta/r)$

given by a rigidly rotating neutron star, but they do correspond to differentially rotating stars. They nevertheless demonstrate that finite magnetospheres exist. Pilipp (1974) has shown that such solutions (with vacuum gaps) cannot link the magnetosphere to the star along magnetic field lines. Thus one requires regions of space charge that are not in rigid corotation with the star, such as shown in figure 4.5, where the disk must rotate while the star remains at rest. This result can be seen in a simple way for the aligned rotator if one neglects perturbations to the magnetic field. If one had rigid corotation everywhere, the connected regions would all have the same space-charge densities as for the totally filled magnetosphere solution. If there existed a solution with plasma of finite extent, it could be generated simply by deleting the appropriate volumes of charge. Because the problem is linear, we could then subtract the two solutions, which would now yield a supposedly $E = 0$ cavity surrounded by the remaining external quadrupolar charge distribution. But the quadrupole moment (and field) within such a cavity cannot be zero regardless of shape, contradicting the supposed existence of a null solution. Note in this respect that once the vacuum interface is specified on any finite surface element, it is specified everywhere (Kellogg 1967; Sunyach 1980). Thus, for example, for $\Phi = \sin^2 \theta$ surface potential and additionally $\mathbf{E} \cdot \mathbf{B} = 0$ at the surface, the potential is uniquely the vacuum solution (eq. 16). Hence, if one had a vacuum region above the surface at some point, the entire region would have to satisfy equation 16 and could not be truncated until another force-free surface (FFS) was encountered. Thus one can immediately discount the possibility of a vacuum gap right above the polar caps, because equation 16 has no other FFS along the polar axis (Asséo, Beaufils, and Pellat 1984). There is also a pathological solution with a conical FFS extending to infinity (Michel 1979b).

Mestel et al. (1976) have argued that, in the aligned case considered here, field lines may not cross the light-cylinder. The dome/torus solutions are certainly consistent with that conclusion, because there need be no light-cylinder. The conclusion of Holloway (1975) and Scharlemann et al. (1978 that steady unidirectional flow of completely charge separated plasma is impossible in the aligned rotator again points to the existence of static solutions. Pelizzari (1976) generalized the Störmer theory (see, for example, Rossi and Olbert 1970) to include accelerated motion in the strong electrostatic field outside a vacuum rotator. Normally, Störmer theory assumes an axisymmetric magnetic field and, because the azimuthal canonical momentum of a moving charged particle is conserved, one can assign allowed and forbidden regions for the particle without solving for its actual trajectory. Adding to this problem an axisymmetric electric field admits of the same sort of analysis. Pelizzari too found that positive charges could not escape but were trapped in a torus, whereas negative particles could be ejected to infinity at first but then become electrostatically trapped as the stellar charge grew owing to this loss. These results are shown in figure 4.7, where η is the dimensionless charge on the star in the same units as equation 16, which corresponds to the case $\eta = 2/3$. Note, in fact, that it is

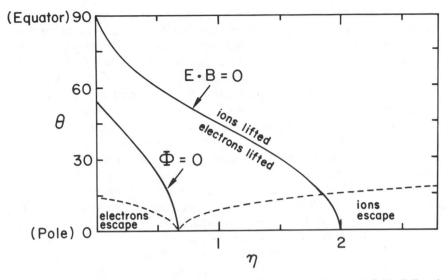

FIG. 4.7. Surface injection in Pelizzari theory, generalized to include electrostatic fields (Pelizzari 1976). The electrostatic field is that of equation 4.25 where the fixed charge of 2/3 has been replaced by a variable charge η. For $\eta = 0$, electrons are pulled from everywhere on the surface, but can only escape for $\theta < 15°$. For $\eta \geq 2$, corresponding to a highly positive stellar charge, one obtains a similar behavior for the ions. In no case can both species escape at the same time, and for η between 2/3 and 2 the ions cannot be injected even though they could escape (thus we cannot obtain an oscillatory solution about $\eta = 2/3$). From M. A. Pelizzari, 1976, Ph.D. thesis (figure 2). Reprinted by permission.

this value of the charge, taken to be a free parameter in Pelizzari's work, that corresponds to the nonescape condition for both electrons and "ions" (positive charges). If the star could have a smaller charge, electrons could escape and vice versa, but in no case could both be ejected at once (electrons escaping along the polar field lines while neighboring ions were accelerated to sufficient rigidity to escape the magnetic field lines).

a. Inertial Gaps

Holloway and Pryce (1981) have pointed out that, even in the Goldreich-Julian model itself, there must be gaps, owing to gravitational and centrifugal forces. These inertial forces are neglected relative to the electromagnetic forces, but still exist. Near the pulsar, gravity is dominant; pulsars are actually *slow* rotators compared to planets in the sense that typically the centrifugal force does not become important until a much larger number of stellar radii distant. Thus $\mathbf{E} \cdot \mathbf{B}$ cannot be exactly zero. For electrons to be trapped, there must be a parallel electric field component

$$\mathbf{E} \cdot \mathbf{B} = \frac{m_e}{e} \mathbf{g} \cdot \mathbf{B} \approx 10 \, \text{V/m} \tag{22}$$

supporting each electron against gravity. A positive charge introduced into this electron plasma would be accelerated to the surface by this same electric field. Thus, even for a filled magnetosphere, there must be a gap between the positive and negative space-charge regions; they cannot overlap, and joining them smoothly together would not allow the change in sign of E at the interface (unless a true charge sheet could be maintained). The global electric fields need to balance gravity and correspond to a slightly modified central charge, whereas to balance centrifugal force, an additional uniform charge density must be added. However, these perturbation terms have opposite signs for the electrons versus the ions, even in the corotating portions of the magnetosphere. Such gaps may not be too important, but they once again illustrate the discontinuous nature of charge-separated plasma.

At large distances, centrifugal forces will become dominant, and the likely regions of interest will be in the equatorial plane. Unless exactly neutralized, the major electric field will be that of the net charge on the rotator. This monopole field will result in rigid corotation to charges of the same sign trapped in the magnetic field (figure 4.5), and the centrifugal force from this rotation will have a weak component directed at the midplane. Consequently, *cold* particles of the opposite sign will sink to the midplane, whereas particles with enough velocity to drift out of the distribution will see the net charge of the aligned rotator and be accelerated to it. It is not known if this possibility of trapping both signs of charge beyond the light-cylinder could play an important role in pulsar dynamics. It seems a rather fragile effect.

b. *Pair Production Filling*

It has been casually suggested (unpublished) that pair production will act to fill in the vacuum regions; the negative dome actually attracts any nearby negative particles and vice versa, thus expanding the numbers of particles trapped and filling in the magnetosphere. (Pair production has already been overworked to form bunches and supposedly close currents.) Such a process would certainly work. However, the parallel component of the electric field vanishes as the two discontinuities approach one another (it is, after all, zero at the surfaces and therefore must be roughly of the form $\mathbf{E} \cdot \mathbf{B} \approx h(h_0 - h)$ between discontinuities a distance h_0 apart). The vacuum gaps cannot thereby be filled in by local pair production. Moreover, the system's electric field drops faster than r^{-2} with distance because the approximate closing of the gaps itself requires charge neutrality inside the radius at which the gaps are nearly closed (hence there is no r^{-2} dependence for the parallel component, which must go as a higher power). Thus the huge electric fields near the pulsar, which might plausibly support pair production discharges, quickly become enfeebled. (It would be worth simulating this possibility to be on the safe side, however.)

c. *Plasma from the Interstellar Medium*

For the aligned system to trap the electron dome, it is necessary that the total system charge be positive. Otherwise the electron at the most distant point of the dome would see no net monopole electrostatic field confining it to the system, and the quadrupole component always acts to eject it. In a mathematical sense, the solution for zero system charge probably requires a dome that extends to infinity along the z-axis. A number of workers (Mestel et al. 1985; Rylov 1984; Ruderman 1986) have suggested that the interstellar medium may act to discharge this net charge and thereby allow particles to escape to infinity, possibly reactivating this inactive solution. Others explicitly looked at this same possibility (Arons and Barnard 1983; Cheng 1985; Michel 1987a) and independently concluded that discharging against the outflow of Poynting flux (i.e., radiation pressure) was not possible unless the pulsar were almost perfectly aligned (within about 0.1°). For a typical pulsar with a nominal inclination, the interstellar medium is stood off at a distance roughly 100,000 times further away than the light-cylinder distance. Although the model is taken to be perfectly aligned, it is difficult to understand how the actual objects could be so perfectly aligned, especially when it is claimed that misalignment produces the pulsations.

d. *Historical Note*

It is interesting that we have, in a sense, come full circle. Birkeland in 1908 did a series of experiments involving bombarding an electrified terrella with cathode rays. In these experiments the electrons striking the terrella excited a phosphorescent coating. By accident, the vacuum of the apparatus was compromised, with the result that one could see the electron excitation of the background gas and therefore track the electrons. Not only did Birkeland simulate the aurora, but he also discovered that features reminiscent of the rings of Saturn would form in the magnetic equatorial plane. See especially figures 207, 223, 254 (showing Birkeland operating the discharge), 255, and 257. This locus is just the $\mathbf{E} \cdot \mathbf{B}$ trapping zone for such a configuration (figure 4.5). Unfortunately, he literally concluded that the Earth was charged positive and the Sun was charged negative so that cathode rays would flow from the Sun to the Earth to form the aurorae. The natural experiments to do in those days were with nonneutral plasmas. Now that we have become "sophisticated," great pains are taken to be sure that, in a simulation of the interaction of the solar wind with the Earth, the plasma is a neutral one.

4.3 The Former "Standard" Model

What was formerly taken to be the standard model basically consists of taking those simplifying assumptions that still promise to keep the problem interesting. These assumptions are (a) the magnetosphere is filled with a plasma such that $\mathbf{E} \cdot \mathbf{B} \equiv 0$ everywhere, (b) the particle motion consists of $\mathbf{E} \times \mathbf{B}$ drift across field lines plus free "sliding" along field lines, (c) stationary ($\partial/\partial t = 0$) plus axisymmetric ($\partial/\partial\phi = 0$) solutions exist, (d) particles are derived freely from the surface with possible contribution from pair production ignored, (e) the plasma is entirely charge separated (i.e., nonneutral), (f) gravity and thermal pressures are ignored compared to the electrodynamic forces.

Assumptions (d), (e), and (f) are corollary assumptions which follow naturally from the first three. The magnetic field source is taken to be a centered magnetic dipole moment aligned either parallel or antiparallel to the spin axis (assumption c). Assumption (a) requires some source of plasma, which is assumed to be free-field emission from the surface (zero work-function); pair production is typically ignored, since it is a supplementary plasma source expected to be important only for high spin rates and strong fields. The modest goal here would be to solve the basic physics of what happens if one rotates a spherical magnet that can freely emit plasma, not necessarily to model an active pulsar; then the critical role assigned to pair production in some models will be irrelevant insofar as this goal is concerned. As we have seen in the previous section, this goal has probably been achieved. Assumption (b) usually neglects gravity and centrifugal forces near the star, since these are all tiny compared with the electrostatic forces. It is sometimes forgotten, however, that such neglect only makes sense if the plasma is entirely charge separated, in which case a weak electric field component parallel to the magnetic field suffices to resist these inertial forces. However, for a two-component plasma (both + and − charges), one or the other component cannot be so supported. In this case one requires thermal support, and the temperatures would have to be large enough to resist the neutron star gravity ($\approx 10^9$ K for electrons, and m_{ion}/m_e times larger if the positive charge carriers are ions that must be thermally supported). Thermal support implies a potentially large radiation loss because an electron radiates its perpendicular energy almost instantaneously (eq. 2.5.5). Therefore the electron pressure tensor would become totally anisotropic, having only a component of pressure parallel to the magnetic field. Consequently the pressure would be nonzero at the pulsar surface, which cannot be much hotter than about 10^6 K, a temperature insufficient even to give the electrons a significant scale height, much less the ions they would have to support. Thus the corollary assumptions are made that for static regions of the magnetosphere, the plasma is charge separated and the particles have negligible thermal motion. Assumption (b) is not a good approximation for the Earth's magnetosphere, for example, because the particles have perpendicular energy, and, consequently, conservation of the moment invariant leads to mirroring, which traps particles in regions of weak field; the particles do not slide freely along field lines but are instead accelerated toward the weak field regions. Wang (1978) has proposed that anomalous resistivity might be invoked to retard the gravitational segregation discussed above. Endean (1972b) has questioned the drift approximation, asserting that $E > cB$ zones exist (but see Buckley 1977a and Burman 1977b). Goldreich and Julian (1969) were the first to pose the model in essentially these terms, and they suggested the general qualitative solution (still preferred by many today) shown in figure 4.8. The essential steps invoked are (1) the rigid rotation of the magnetosphere, (2) the fact that this rotational speed cannot exceed c, and (3) the plausible conclusion that therefore the magnetic field lines open beyond the light-cylinder distance and the plasma flows away

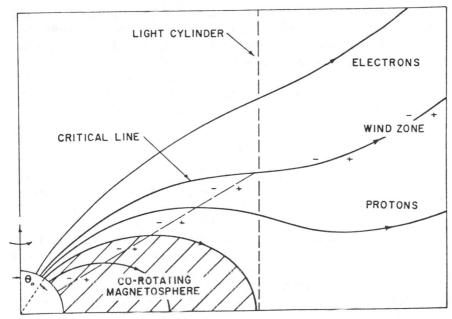

FIG. 4.8. Goldreich-Julian magnetosphere. Near the neutron star one finds only electrons above latitude 35° and only positive charges below. In the wind zone, positive particles supposedly flow away at low latitudes, but, as is easily seen, the field lines on which they are constrained to flow lead to the negative polar cap region. From P. Goldreich, and W. H. Julian, 1969, *Ap. J.*, 869 (figure 1).

from the system, requiring (4) reacceleration of the particles from the surface. In their paper, they repeatedly touch on the issue of whether their proposed model is "unique." The figure illustrates what was bothersome about the solution, namely that a straight line (the locus of $B_z = 0$, see eq. 2.3.18) starts out from the star and separates regions of positive space charge from negative regions. The awkward thing is that *open* field lines must thread from the star to infinity across this *null* line. In other words, the very field lines on which one hopes to find positive particles being injected to produce a net neutral wind (one cannot very well charge the star indefinitely) are rooted deep in polar cap regions having everywhere *negative* space charge. How can positive charges be pulled from the surface electrostatically without collapsing the entire negative space charge? And if it must collapse, what does a steady state solution mean? (See Gilinsky et al. 1970 for a point by point analysis of the Goldreich and Julian theory and also Goldreich et al. 1971.) As we have seen in sec. 4.2, one solution is that a nonneutral plasma is simply confined near the star.

1. The Current Closure Problem

In the next section, we will see that the former standard model was found to have many plausible physical properties. On closer examination, however, one property after another proved to be flawed. At this point, many still hold out hope that inclusion of pair production, oblique alignment, or some other consideration will relax an unsuspected unphysical constraint and permit one to assemble a fully self-consistent model. These deficiencies have not escaped the notice of the observers, and Sir Antony Hewish (1981) has dubbed this the *current closure problem*. However, some popular models (e.g., Ruderman and Sutherland 1975) are still cited frequently, although they have yet to show that they can satisfy the elementary requirement that the average net current from the pulsar be zero.

2. The "Pulsar Equation"

Here we abandon the vacuum solutions, suppose that the star is completely surrounded by a charge-separated plasma, and assume that this plasma is corotating in the equatorial zones near the star and is flowing outward along polar field lines. If one specifies the magnetic field lines by the enclosed magnetic flux (f) between any given field line and the axis of rotation, then one has in general a total current defined to be $\mu_0 A(f)$ flowing out of the pole within the surface of rotation bounded by the field line f. It follows immediately that the azimuthal field is given by

$$B_\phi = \Omega A(f)/\rho c, \tag{1}$$

where ρ is the axial distance. The electric current and enclosed magnetic flux between any two field line surfaces are conserved; hence it follows that (subscript m stands for *meridional* and indicates a vector in the θ-r plane)

$$\mathbf{J}_m = \frac{\mu_0}{c}\frac{dA}{df}\,\Omega\mathbf{B}_m. \tag{2}$$

In addition, we have the condition that the system be in electromagnetic force balance,

$$q\mathbf{E} + \mathbf{J} \times \mathbf{B} = 0, \tag{3}$$

or

$$qE_m + J_m B_\phi - B_m J_\phi = 0, \tag{4}$$

where

$$E_m = \Omega\rho B_m. \tag{5}$$

We can then factor out the (nonzero) factor B_m to obtain

$$\Omega\rho q + \Omega^2\mu_0 AA'/c - J_\phi = 0. \tag{6}$$

With our labeling of field lines according to included magnetic flux (f), one obtains (in cylindrical coordinates)

$$B_\rho = -f_z/\rho, \tag{7}$$

$$B_z = f_\rho/\rho, \tag{8}$$

$$\mu_0 J_\phi = (\nabla \times \mathbf{B})_\phi = \frac{1}{\rho}\left(f_{zz} + f_{\rho\rho} - \frac{1}{\rho}f_\rho\right), \tag{9}$$

and

$$q/\epsilon_0 = -\nabla^2\Phi = -\Omega\nabla^2 f = -\Omega\left(f_{zz} + f_{\rho\rho} + \frac{1}{\rho}f_\rho\right). \tag{10}$$

If one defines a scale length

$$a = c/\Omega, \tag{11}$$

one obtains (Michel 1973a)

$$f_{zz} + f_{\rho\rho} - \frac{1}{\rho}\left(\frac{a^2 + \rho^2}{a^2 - \rho^2}\right)f_\rho - \left(\frac{A'A}{a^2 - \rho^2}\right) = 0. \tag{12}$$

This equation is sometimes gratuitously referred to as the "pulsar equation." So far nothing has been specified about the function A. An obvious simplification would be for $A \sim f$, in which case the pulsar equation would be soluble by conventional eigenvalue techniques (see Scharlemann and Wagoner 1973, who independently derived this equation). Other independent derivations of equation 12 are given by Julian (1973) and by Cohen et al. (1973). The latter's derivation is more general, in that the plasma is taken to have both signs of charge carrier present (see also Cohen and Rosenblum 1972, 1973, Endean 1974, and Schmalz et al. 1979, 1980). Unfortunately, we see from equation 2 that $A(f)$ must have at least three zeros in the former standard model, two at each pole (no line current along the spin axis) and one on the equatorial plane (because $B_\phi = 0$ there by symmetry). Consequently equation 12 must be nonlinear. As a result, only a few special cases have been worked out, as discussed below.

3. Restricted Exact Solutions ($A = $ constant)

Because the space charge required for $\mathbf{E} \cdot \mathbf{B} = 0$ rigidly corotates, it in turn generates a current which modifies $\mathbf{B}$ from a pure dipole configuration. This effect was evaluated (Michel 1973b) and shown to lead to a cusp-like configuration at the light-cylinder which separates the closed field lines from the open ones. This solution corresponds to the choice $A = 0$ in equation 12, in which case one can solve for $f(z, \rho)$ by conventional separation of variables (Michel 1973b; see below). The resultant field line configuration is shown in figure 4.9.

Because the equations are now linear, one can solve for a monopole source

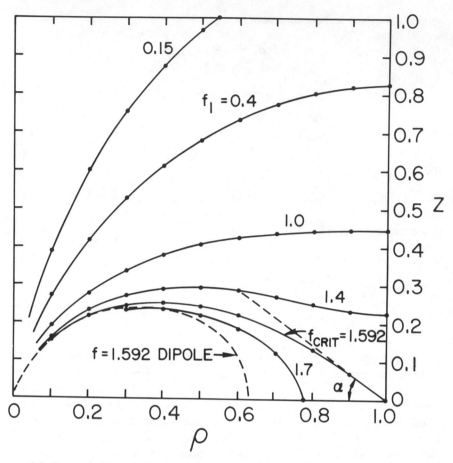

FIG. 4.9. Exact solution to the "pulsar equation" if outward flow is neglected. Here corotation of the space charge produces currents modifying the dipole magnetic field, which in turn modifies the space charge. Magnetic field lines are labeled such that the $f = 1$ flux line of the undistorted dipole would cross equator at light-cylinder distance ($\rho = 1$). Distortion causes an increased (factor of $f_{crit} \approx 1.592$) flux to cross light-cylinder. Note cusp angle $[\alpha = \sin^{-1}(1/\sqrt{3})]$ reminiscent of figure 2.3. From F. C. Michel, 1973b, *Ap. J.*, 180, 207 (figure 1b).

(using cylindrical coordinates as above),

$$f_0 = A + Bz/r, \tag{13}$$

where A and B are constants and simply differentiate the solutions to get a dipole source

$$f_0 = B\rho^2/r^3. \tag{14}$$

In addition to labeling the magnetic field lines, Δf also corresponds to the magnetic flux enclosed between the magnetic shells defined by rotating two field lines through 2π. One result can be obtained immediately, namely the cusp angle (α) formed when the dipole magnetic field lines are finally forced open by the rotating space charge. There one must have

$$2f_{\rho\rho} = -f_{zz} \quad \text{(at cusp);} \tag{15}$$

thus the angle $\alpha = \sin^{-1}(1/\sqrt{3})$. Although such opening is usually attributed to centrifugal force, inertial forces have been neglected by setting the Lorentz force equal to zero (eq. 3). It is the perturbation magnetic field of the rotating space charge, which finally becomes dominant at the light-cylinder distance, that produces this change in field topology.

a. *Separation of Variables*

Equation 12 is linear and separable, and thus we can write

$$f(z, \rho) = F(z)\psi(\rho), \tag{16}$$

where

$$F_{zz} = \lambda^2 F, \tag{17}$$

and

$$\psi_{\rho\rho} - \frac{1}{\rho}\left(\frac{1+\rho^2}{1+\rho^2}\right)\psi_\rho = -\lambda^2\psi. \tag{18}$$

Equation 17 has solutions $F \approx e^{\pm\lambda z}$, where the possible values of λ are determined by equation 18. Exactly the same method is illustrated in Panofsky and Phillips (1955) in solving for the potentials for a point charge in a cylinder, except there the eigenfunctions are just Bessel functions.

b. *The Light-Cylinder Eigenvalues*

Equation 18 is similar to the Mathieu equation in having five singularities at $\rho = 0$, ± 1, and $\pm\infty$. We can rewrite this equation as

$$(g\psi_\rho)_\rho + \lambda^2 g\psi = 0, \tag{19}$$

where

$$g = (1 - \rho^2)/\rho. \tag{20}$$

Accordingly, we can define a conjugate function ϕ as

$$\phi = g\psi_\rho/\lambda \tag{21}$$

and

$$\phi_\rho = -\lambda g\psi, \tag{22}$$

so

$$(g^{-1}\phi_\rho)_\rho + \lambda^2 g^{-1}\phi = 0, \tag{23}$$

and therefore ϕ satisfies the same equation as equation 18 except with the sign of the second term reversed. The g is the normalizing function, where

$$\int_0^1 \psi_n\psi_m g \, d\rho = \int_0^1 \phi_n\phi_m g^{-1} \, d\rho = \delta_{nm}h_n. \tag{24}$$

These equations can be solved numerically in the usual ways; the functions can be expanded in power series, the differential equations convert the coefficients of the power series into recursion relations, the recursion relations can be recast into continued fractions with which the eigenvalues can be evaluated, and the normalization factor h_n can be determined by numerical integration of the defining equation 24 (Michel 1973b).

c. Physical Results

Once this backup labor is performed (required simply because these are not tabulated functions like the Bessel functions), one recovers the solution

$$f_0(z, \rho) = \sum_{n=1}^{\infty} \phi_n(0)e^{-\lambda_n z}\psi_n(\rho)/h_n\lambda_n. \tag{25}$$

Differentiation with respect to z simply cancels out the λ_n in the denominator and gives the dipole solution:

$$f_1(z, \rho) = \sum_{n=1}^{\infty} \phi_n(0)e^{-\lambda_n z}\psi_n(\rho)/h_n. \tag{26}$$

In principle, one then plugs the computer-generated functions into equation 26 and plots out the results to generate figure 4.9. Interestingly, this straightforward approach fails miserably! One of the better kept secrets (or maintained ignorances) is that such formal eigenvalue expansions frequently fail to converge, usually just where one is interested in having an answer. For example, the *monopole* magnetic field line running along the x-axis evaluated at the light-cylinder is assigned the normalized value of unity, whereas the expansion gives

$$f_0(0, 1) \equiv 1 = \sum_{n=1}^{\infty} \phi_n(0)\psi_n(1)/h_n\lambda_n. \tag{27}$$

The terms in the sum are given in table 4.5. Clearly this is a divergent series! The effort in evaluating these terms to high precision appears wasted, given the nonconvergent series that result. If one departs from the $z = 0$ line, the exponential factors will eventually force the series to converge, but that will not

TABLE 4.5—Eigenvalues and Series Terms

n	λ_n	$\phi_n(0)\psi_n(1)/h_n\lambda_n$
1	3.219 517	2.060 988 627
2	6.336 203	−2.140 969 193
3	9.765 611	2.168 766 081
4	12.599 901	−2.182 555 357
5	15.736 585	2.190 719 318
6	18.874 629	−2.196 092 762
7	22.013 523	2.199 887 918
8	25.152 988	−2.202 706 425
9	28.292 855	2.204 880 122
10	31.433 018	−2.206 606 191

help in determining the value of the critical line in the dipole solution leading from the dipole to the cusp region, which is given by

$$f_{\text{crit}} \equiv f_1(0, 1) = \sum_{n=1}^{\infty} \phi_n(0)\psi_n(1)/h_n. \qquad (28)$$

Physically, this quantity determines exactly what field line from the polar cap first leads to the light-cylinder (previously we have estimated this location by assuming the magnetic fields to remain dipolar). But the dipole case will have each term multiplied by the increasingly larger values of λ_n and will diverge even more rapidly than the series in table 4.5.

Fortunately there is a very simple way of inducing the series to converge accurately. The method is based on the fact that the sum of the series up to any given number of terms (the *partial sum*) is an estimate of the value of the sum. If one can arrange these estimates so that they approach a fixed value even though the individual terms do not, one has at least a formal sum for the series. The latter can be accomplished by weighting the partial sums so that a weighted value is defined by

$$W_N \equiv \frac{1}{2}(S_N + S_{N-1}), \qquad (29)$$

where

$$s_N \equiv \sum_{n=1}^{N} a_n \qquad (30)$$

with a_n representing the terms in the series. The weighted sum W_N is just as good a limiting value for the series as S_N, and the oscillations of an alternating series will clearly be moderated. (Obviously this method—and summation in

TABLE 4.6—Convergence of Multiple Averages

$S_N{}^a$	$W_N(1)$	$W_N(2)$	$W_N(3)$	$W_N(4)$	$W_N(5)$	$W_N(6)$	$W_N(7)$	$W_N(8)$
1060988627								
	−9495970							
−1079980566		−2546748						
	4402475		−795796					
1088785515		955156		−271902				
	−2492164		251991		−98543			
−1093769842		−451173		74816		−37216		
	1589817		−102359		24111		−14459	
1096949476		246456		−26594		8299		−5702
	−1096905		49170		−7513		3055	
−1099143286		−148116		11568		−2189		1237
	800673		−26035		3136		−582	
1100744632		96046		−5295		1025		
	−608581		15445		−1087			
−1101961793		−65156		3121				
	478268		−9203					
1102918329		46751						
	−384767							
−1103687862								

a) Expected value of unity subtracted for clarity and remainder multiplied by 10^9 so that leading zeros need not be displayed; remaining series now sums to zero within the accuracy of the terms and number used.

general—only works for a divergent *alternating* series.) In fact, the Abel sum for the series $1 - 1 + 1 - 1 + \cdots$ of 1/2 is immediately obtained in this way. Even more can be squeezed out because in general the W_N will themselves oscillate about the limiting values, which in turn can be damped by again weighting successive terms to create a second generation of averaging by writing (we now write W_N as $W_N(1)$ to represent it as the first generation of averaging)

$$W_N(2) \equiv \frac{1}{2}\left[W_N(1) + W_{N-1}(1)\right] \tag{31}$$

and so on. The results of such averaging are shown in table 4.6. By inspection of the results, it is simple to identify the asymptotic limit to which the series is trending. In the table we have subtracted the expected value (unity) from the partial sums to illustrate how the deviations behave. The conventional mathematical statement regarding converging alternating series is that the partial sum is off by no more than the last neglected term. However, by averaging we convert a divergent set of partial sums to a convergent one, obtaining values every bit as valid an estimate for the series sum as the partial sums themselves. Because the successive averages also alternate, we know that the value of the series lies between successive overestimates and underestimates. Thus the first term of every successive average is an underestimate, the second an overestimate, etc. In this way we can determine with table 4.6 that the least overestimate is (in parts per billion) 1025, which is the last value after the sixth average; the smallest underestimate is −582 in the seventh average. Our best guess is that the value of the series is midway in between (+223), and we also know the maximum possible error (± 803), and therefore we have determined that $f_0(0, 1) = 1.0000002 \pm 8$. Clearly the tabulated values are entirely adequate and the correct value has indeed been reproduced.

This method is much more powerful than it might seem. Not only can we get precise results from an otherwise useless series, but we also get error estimates. Without error estimates, a formal sum for a divergent series is useless. In addition, if the numerically quoted values were not accurate to at least seven significant places, the regular pattern of alternating signs would break down, which would immediately tell us the actual numerical accuracy of the values. A few more terms would have told us if nine places were justified. Mathematicians have used such techniques largely to define *summability* for divergent series, but it is quite useful for physical problems as well. Often physicists will simply multiply in successive powers of a *convergence factor*, so that the series $1 - 1 + 1 - 1 + \cdots$ becomes $1 - x + x^2 - x^3 + \cdots$, the series expansion of $1/(1 + x)$ which can then be evaluated as $1/2$ in the limit $x \to 1$. Unfortunately this trick is of little use for *numerically* generated series. There is a puzzle with such a numerical convergence technique in that the successive numerical terms become successively irrelevant. Thus, if we had the next million terms of the series in table 4.5 calculated to an accuracy of 1 part in 10^7, they would be of no use whatsoever! Putting it another way, consider the obviously divergent series $1 - 2 + 3 - 4 + \cdots$, which nevertheless has the Abel sum $1/4$. We can establish this result with no more than the terms shown, so the rest of the series is similarly irrelevant. Which raises the question of how (or what implicit assumption) it is that only four terms can serve to constrain all the rest. Obviously if we looked at this series, we would guess that the next term was 5, but it has to be if the series can be summed by this technique (which is just one of a large class of summation methods).

There is a curious history connected with these solutions, beyond the fact that they seem to be reinvented on a regular basis (Mestel et al. 1979; Beskin et al. 1984). When I reported these results at a meeting on high-energy astrophysics at Caltech in October of 1972, a leading pulsar theorist stood up and announced that he had similar results "in his desk drawer," one of the more novel priority claims. Later, a leading pulsar theorist evidently claimed to everyone he could contact (except the author) that these results were suspect because there was a "current sheet in the equatorial plane." Of course any finite sum to equation 26 will leave an oscillating error in potentials that are equivalent to oscillating current errors in the equatorial plane. As we have seen, the sums do not necessarily even converge. In practice one either disregards the solutions where the convergence is obviously poor or (as was done) uses a weighting technique to accelerate convergence that effectively averages out these oscillations. If the solution actually corresponded to a "current sheet in the equatorial plane," the convergence technique would simply not work.

Mestel et al. (1979) have repeated this analysis with exactly the same results, although they claim their method (expanding the z-dependence as $\cos \lambda z$ rather than $e^{\pm \lambda z}$) to be superior (no "current sheet in the equatorial plane"). The amount of flux crossing the light-cylinder, compared to that for an undistorted dipole field, is given from the original expansion to be a factor

$$f_{\text{crit}} = 1.5918428 \pm 0.0000004, \tag{32}$$

which seems adequate precision for a "flawed" method (Mestel et al. 1979 obtain 1.592).

Hinata and Jackson (1974) have found pathological solutions for this same case corresponding to the presence of strong external magnetic fields surrounding the object. One can see that $A = $ constant also satisfies the same equations. This choice corresponds to a line current along the z-axis, hence simply the superposition of an azimuthal field, $B_\phi = $ constant$/\rho$. Scharlemann and Wagoner (1973) discuss the parallel solution for $A(f) = $ constant $\times f$, which again can be solved by separation of variables. As later noted by Michel (1975c), this choice for A requires a discontinuity in the equatorial plane joining two separate asymptotic solutions; thus one cannot simply "solve" equation 12 for this choice, because the solutions are not global but must be patch-wise matched across supplementary equatorial current sheets (which in this case would have to be real; see sec. 5.c below). Similar considerations probably follow for the nonlinear versions.

4. Exact Monopole Solution

A special solution to the "pulsar equation" (eq. 12) is the case for a monopole magnetic field. One has an exact solution (Michel 1973a), given by

$$V_m = c, \; V_\theta = 0,$$

$$\tan \xi = \Omega \rho / c, \tag{33}$$

$$B_\phi = \Omega f_0 \rho / r^2 c,$$

with the monopole field terms

$$f = f_0 z / r, \tag{34}$$

$$B_r = f_0 / r^2, \; B_\theta = 0.$$

Here ξ is the *garden hose* angle (imagine watering the lawn while spinning on one's heels), the angle between the magnetic field line and the radial direction,

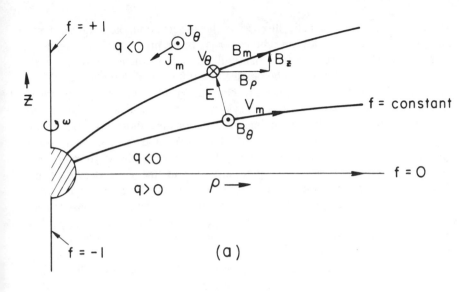

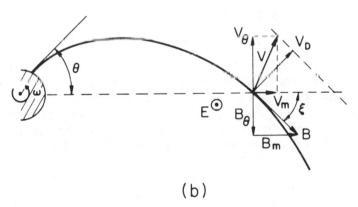

FIG. 4.10. Monopole field-line geometry. (a) Meridional projection. Symbols are standard except that q is the electric charge density and subscript m is the vector quantity in meridional projection. (b) Orthogonal projection onto plane normal to local E, namely the plane defined by the meridional vectors (all are parallel) and the θ direction. The plasma is obliged to have a specific V_m (and $\mathbf{V}_D = \mathbf{E} \times \mathbf{B}/B^2$ drift velocity) thereby resulting in the net velocity $\mathbf{V}$ constructed as shown. From F. C. Michel, 1973a, *Ap. J. (Letters)*, 180, L133 (figure 1).

as shown in figure 4.10. The field lines are wrapped backwards by the rotation whereas the plasma streams radially outward. These are precisely the properties expected for stellar-wind type solutions (see below), and they considerably reinforced confidence in this general approach. In terms of $A(f)$, this solution corresponds to the nonlinear choice

$$A(f) = (f_0^2 - f^2)/f_0, \tag{35}$$

where f_0 scales the magnetic flux. Thus the exact monopole solution corresponds to a quadratic (nonlinear) choice for A. In the previous choice, equation 12 was linear and hence the field multipolarity was mathematically irrelevant because the dipole solution can be obtained by differentiating the monopole solution, etc. If A is nonlinear, however, the solutions no longer can be expanded by superposition. There has been no systematic mathematical analysis of the pulsar equation. The only known analytic solutions are those given above for $A = 0$ or constant, the linear equation $A = -2f$, and the nonlinear choice appropriate for a monopole field (eq. 34). Because $A(f)$ physically represents the current flowing on field lines poleward of field line f, it is evident that $A(0) = 0$, and since the total current from the star must be zero, we also must have $A(f_c) = 0$, where f_c is the last open field line. Thus $A(f)$ must at least be quadratic to have two zeros and $A'A$ is therefore at least cubic in f. The monopole solution (eq. 35) has this basic property, for example, except that there is no f_c and the two zeros are therefore at the two polar caps (note that f is normalized differently for a monopole field [eq. 35] than for a dipole field [eq. 2.3.8], f being zero in the equatorial plane for the monopole rather than being zero on the polar axis as for the dipole).

5. Wind-Zone Solutions

a. *The Far-Zone Limit*

The monopole case is basically a distant wind-zone solution because it has no transition point from nonrelativistic to relativistic flow. More realistic wind solutions have been derived, first for the general case of a neutral plasma driven away from a pulsar by the rotating magnetic field (Michel 1969a,b) and then for the charge-separated case above (Michel 1974c). One assumes the meridional fields to be asymptotically radial, in which case equation 12 simplifies to

$$(1 - \mu^2)f'' - 2\mu(1 - \mu^2)f' + \frac{1}{2}\frac{dA^2}{df} = 0, \tag{36}$$

where $\mu = \cos\theta$, $f \to f(\mu)$ as $r \to \infty$, and $f' \equiv df/d\mu$, with $A = A(f)$ as before. A solution of this equation is seen, by direct solution, to be

$$f' = -A(1 - \mu^2), \tag{37}$$

from which one obtains once again the garden-hose relationship

$$\tan\xi = \Omega\rho/c, \tag{38}$$

giving asymptotically a perfect Archimedean spiral, as well as $V_\phi \to 0$, $V_m \to c$ and $q \to \epsilon_0 \Omega A(dA/df)/\rho^2$. When the equation is in this form, one need only choose a form for A to obtain a possible solution. Such solutions are not necessarily appropriate to a magnetic dipole central source at the origin.

b. *The Current and Charge-Density Paradox*

The asymptotic solution above was eventually noticed to have an intrinsic flaw (Michel 1975b). Indeed equation 37 has the simple physical interpretation

$$E_\theta / B_\phi = c,\tag{39}$$

which follows using the surface boundary condition $d\Phi = \Omega df$ (eq. 2.3.6). In other words, the particles are simply drifting outward in the $\mathbf{E} \times \mathbf{B}$ motion dictated by the dominant field components at large distance. However, J_ϕ (eq. 9) and q (eq. 10) do not, in general, vanish on the same field line, the asymptotic field lines now being designated by the value of $\mu = \cos \theta$ in place of f. For a charge-separated system, J_ϕ must vanish where q vanishes because these are conduction currents and therefore another important restriction is placed on the function A.

c. *Determination of $A(f)$*

It was subsequently shown (Michel 1975c) that in fact the linear choice

$$A(f) = -2f\tag{40}$$

avoided the above difficulty (or, more precisely, hid it). Here one has solutions reminiscent of the original monopole solution, but with a current sheet discontinuity in the equatorial plane. Although such a current sheet might well be considered artificial, it nevertheless resembles the qualitatively expected structure at large distances as shown in figure 2.3. The current sheet itself could be imagined to be the consequence of the mathematical idealizations, representing in fact a distributed volume current, possibly representing shock-heated plasma. Such a current sheet would map into the edges of the polar caps to form an *auroral* zone (see also Lovelace 1973). Although this field structure seems at least promising, problems even more severe remain. Moreover, it is not possible to connect this asymptotic solution to the corresponding near-zone solution (Scharlemann and Wagoner 1973).

6. Emission Regions

The assumption has long been that the emission is from regions near the star, leading to directed beaming along the local magnetic field lines near the polar caps. The alternative of emission elsewhere such as near the light-cylinder has been examined from time to time and has never been conclusively ruled out. The behavior of the critical plasma frequencies as a function of distance is shown in figure 4.11. For typical pulsar parameters, these two frequencies fall near the radio band as one reaches the light-cylinder distance (there is no a priori reason for this to happen, because the cyclotron frequency is entirely determined by B while the light-cylinder distance is determined entirely by Ω). However, for the Crab pulsar, these frequencies are still much larger than the

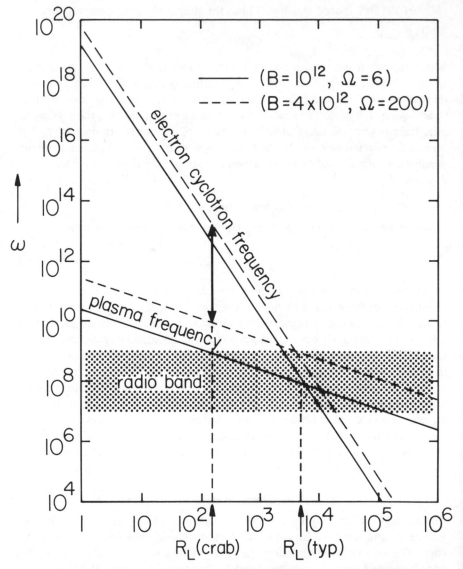

FIG. 4.11. Plasma and cyclotron frequencies versus radial distance, plotted for typical pulsar parameters (solid line) and Crab pulsar parameters (dashed line). Although these characteristic frequencies fall in the radio band (essentially 100 MHz to 10 GHz) at the light-cylinder for typical (1 s) pulsars, they are orders of magnitude higher for the Crab pulsar, giving no very clear support for light-cylinder emission of radio waves (the Crab pulsar is, however, an extremely inefficient radiator of radio waves).

radio. The Crab spectrum, moreover, is quite steep, with most of the intensity at the lowest observable frequencies.

7. The "Everted" Rotator

An instructive model, having only "academic" relevance to pulsars, is one in which the neutron star spins in an *external* uniform magnetic field rather than through an internal dipole magnetic field. Because the magnetic field model is effectively turned inside out, we will call this case "everted." The Goldreich-Julian type of solution is immediate because rotation in a uniform magnetic field produces an axial electric field requiring a uniform charge density. The cylindrical volume of magnetic field lines linking the rotator should be filled with this same rigidly rotating uniform charge. Because the electric field is everywhere axial, $\mathbf{E} \cdot \mathbf{B} \equiv 0$ and we therefore have a solution, an impossible solution! The solution is impossible, of course, because the system now has an infinite electrical charge.

The next guess one might make is that the system is neutralized by a sheath of opposite charge sitting just outside of the cylindrical volume. We can now imagine the time-dependent problem where the rotator was shielded instead by surface charges and surrounded by vacuum. Releasing the surface charges therefore would produce a *jet* of corotating uniform charge and the accompanying neutralizing sheath. One wonders what the velocity of this jet would be, but more interesting is the source of the sheath, which can only be located within a narrow neighborhood of the equatorial zone of the rotator. As the putative jet recedes, the electric field near the rotator should become very nearly axial once again, which means that the electric field required to pull neutralizing particles from the rotator vanishes. But the particles not only are required but must be injected at the jet velocity, so the electric field cannot become axial!

Alternatively one might look for a static solution, but it is hard to see how the most distant particles could end up with the electric fields exactly axial at their positions. Note also that the particle concentration in the sheath would have to be enormous (infinite in the usual approximation that the particles cannot cross magnetic field lines) and moreover the inner edge of the sheath would rotate with the rotator but the outer edge would not be in rotation, giving huge shears as well. We can estimate the maximum sheath thickness δ because the cyclotron radius of the injected particles would be $\delta = \gamma mc^2/eBc$ and the pole to equator electric potential drop would correspond to $\gamma mc^2 = e\Omega a^2 B$; hence $\delta \leq \Omega a^2/2c$ (about a meter for typical pulsar parameters). Much more work needs be done before we can confidently make statements about how such systems function. Collapsing the sphere to a disk yields a popular model for jets from a magnetized disk in an active galactic nucleus, with the same pathologies.

4.4 Problems with Standard Model

Scientific models can be divided into three types. In type "A," we have the so-called standard models wherein the model is so popular that inconsistencies with observation are discounted or alibied away. When such models hold sway, alternative ideas are unwelcome and suppressed to the degree possible. Examples are the quark model in high-energy physics and the radioactive decay model for supernovae (just because they are fashionable doesn't mean they are wrong). For type "B," there are several competing models struggling for recognition. In this case, the ability of any one model to explain the data is discounted or alibied away. Alternative ideas are still unwelcome, often on the basis that "there are already too many models around." Current examples are models for gamma-ray bursts and, with the demise of the former standard model, pulsars. Type "C" is called the "scientific method," in which there are ideally two contending models. On the basis of careful intercomparison of their strengths and weaknesses and ability to confront the data, one or the other is favored. New ideas are welcome. No examples come to mind.

Basically, the difficulties with the standard model arise from suspending disbelief that somehow the stationary space-charge distribution in the magnetosphere can be significantly different from the current flow charge density (figure 4.8). This problem was minimized, given that the model otherwise looked so self-consistent. As happens so often in astrophysics, it is so easy to be caught up in what the model is supposed to do that one is not sufficiently critical of details that do not work. If the aligned rotator had been assigned as an exercise to students of electricity and magnetism, they would not have concluded that it would generate a wind of relativistic particles. They would have been puzzled at first because the system would have proven very difficult to solve for if they did not understand the properties of the nonneutral plasma that would envelop the rotator (sec. 4.2). These properties were not understood at the time the model was first elucidated. The problem can be seen in figure 4.3: the space charge on the magnetic field lines changes sign, and therefore this space charge as a whole cannot flow anywhere without upsetting the force balance condition that it is there to satisfy. For about a decade, this fundamental failure of the model was discounted. People assumed that because the model was so attractive it must work, and therefore the challenge instead was to figure out how nature got around this seeming paradox. Also, the "cartoon approximation" did not help much because people wanted to show the transition to a wind zone on a scale similar to the neutron star size. The distance to the light-cylinder has the same proportionality to the size of the neutron star as the size of the Earth's orbit has to the distance to the nearest star! This unequal scaling tended to distract people from the absurdity of imagining close coupling between the inner regions near the neutron star and the regions beyond the light-cylinder. One might have wondered why one didn't just have approximately the corotation space-charge distribution in close and have something else happen at large distances. Even when Holloway showed that gaps could form in this magnetosphere, it did not immediately dawn on people that this could be in fact the "something else." Holloway himself introduced this effect in terms of an "improved" version of the standard model. Unfortunately, few were persuaded by the improvement, and the essential physics went unnoticed, only to be rediscovered independently by others. The following gives some explicit illustrations of how vexing it was to surmount the difficulties with the model, even at the time.

1. The Uncharged Field Line

We can rewrite equations 4.3.9 and 4.3.10 (Okamoto 1974; Pelizzari 1975) to give

$$\Omega \rho J_\phi / c^2 = -q - 2\Omega \epsilon_0 B_z, \tag{1}$$

which leads to an essential paradox if we adopt a picture such as figure 2.3 for the magnetic field configuration. Because the currents in the standard model only result from motion of the space charge, it follows that if $q = 0$ then $J_\phi = 0$, which from equation 1 gives the condition

$$B_z = 0 \ (\text{if } q = 0), \tag{2}$$

which in turn means that the uncharged field line must extend parallel to the equatorial plane. The dipolar field line that starts from the surface with $B_z = 0$ is buried deep within the corotation region, however; thus there would be no plausible way to detach that field line from the pulsar. For the monopole magnetic field case discussed above, there is no difficulty because there are no closed field lines and the field line in question is simply the one in the equatorial plane, hence automatically satisfying the constraint in equation 2. The same difficulty in the wind zone proves equally vexing near the star. Perhaps the simplest way out would be to have $\Omega = \Omega(f)$ and not constant, in which case additional terms would enter and possibly relieve the requirement that $B_z = 0$ along the null charge density line. However, this suggestion would mean that the field line potentials have a different value from that imposed at the surface and therefore $\mathbf{E} \cdot \mathbf{B} \neq 0$, violating a basic assumption of the model. Resistive effects within the star could strongly modify the surface potential. However, the current densities are actually not all that large for pulsars, but are comparable to those in electrical wires to home appliances, so the resistivity would have to be about 10^8 times higher than for normal metals for the resistive potential drops to become significant in altering Ω. Observationally, the consequent heat dissipation and blackbody radiation (soft X-rays) do not seem to be observed. For the Crab pulsar, the blackbody component of radiation is less than about 10^{-5} of the total output (Harnden et al. 1980). Thus the internal potential drops along field lines are probably much less than 10^{-3} of the accelerating potential and only a small correction to the standard model surface potential is possible. Another suggestion is that the plasma is not charge-separated (Okamoto 1975), in which case $J_\phi = 0$ and $q = 0$ need not occur on the same magnetic field line. It is difficult to see how such a quasi-neutral plasma can be pulled from the pulsar surface by an electrostatic field, but an accelerating field might alternate, first pulling negative particles, then positive, etc. (violating the

$\partial/\partial t = 0$ assumption). Pair production could admix some pairs into the primary beam and avoid complete charge separation; however, it is not obvious that the paradox is thereby resolved. The $q = 0$ magnetic field line at the surface is a site of zero particle emission, hence of no local pair production, so pairs would be unavailable just where they were most needed. Another problem is that the interstreaming between two charged species is known to be unstable and should be damped out over some time constant. If instead both components had the same velocity, the plasma would again behave as if charge separated. Salvati (1973) also examined this limit, concluding that either charge separation could not be complete or the magnetic field configuration would somehow not fill the entire space around the rotator. Similar considerations were raised by Buckley (1976, 1978) and Endean (1976). Jackson (1978b) discusses the effects of perturbations to the aligned magnetosphere, arguing that such solutions are unstable.

2. Monotonic Field Lines

A related problem is that a field line cannot curve over and approach the equatorial plane within the light-cylinder, as would be required for the wind field lines adjacent to the corotation region (figure 4.8). Such a field line would have $B_z = 0$ where it curves over and therefore

$$V_\phi = J_\phi/q = c^2/\Omega\rho > c \ (B_z = 0) \tag{3}$$

because the axial distance ρ at which the curve-over occurs is expected to be inside the light-cylinder. Thus the convection currents would have to exceed c even inside the light-cylinder.

3. Boundary Conditions at the Light-Cylinder

Scharlemann and Wagoner (1973) noted that the singular nature of the pulsar equation at $\rho = a$ meant that the B_z component was fixed there by the function $A(f)$. As a result, the equation is independently soluble inside and outside the light-cylinder. As pointed out by Ingraham (1973), this independence means that for an arbitrary current flow pattern as parameterized in $A(f)$ the field lines need not match up, violating $\nabla \cdot \mathbf{B} = 0$. They might still match up but with a discontinuity in slope, thereby requiring a current sheet. He suggested that $A(f)$ should therefore be determined by the condition that the field lines match without kinking. Numerical calculations by Pelizzari (1975) confirm the difficulty in matching field lines across the light-cylinder, as shown in figure 4.12. Pelizzari numerically solved the "pulsar equation," equation 4.3.12, for a number of trial functions, including ones of the form (see discussion in his thesis)

$$dA/df = -2(1 - f^n/f_{\text{crit}}^n), \tag{4}$$

where f_{crit} is defined in figure 4.9. Although the $n = 1$ case comes close to

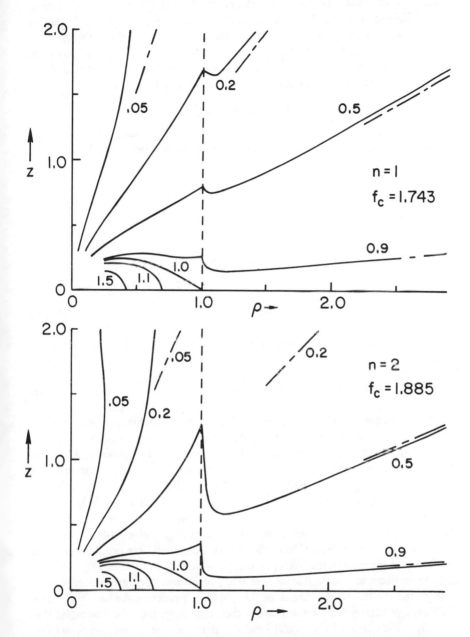

FIG. 4.12. Attempts to numerically solve "pulsar equation." for plausible functions of the form $dA/df \approx 1 - f^n/f_{\text{crit}}^n$, where f_c is the critical field line (see figure 4.9) leading to the cusp at the light-cylinder distance ($\rho = 1$). Note kinks and reversal of slope at the light-cylinder (vertical dashed line). From F. C. Michel, 1982, *Rev. Mod. Phys.*, 54, 1 (figure 12); M. A. Pelizzari, 1974, M.S. Thesis (figure 7), reprinted by permission.

appearing continuous, one sees that the physical behavior of the two solutions is quite different near the light-cylinder. One must also ask what would happen if an obstacle to the pulsar wind were present just outside the light-cylinder. Certainly the flow pattern inside the light-cylinder would be modified, and as the pulsar equation stands, this modification could result only by changing $A(f)$. There should therefore be continuous *families* of possible $A(f)$.

4. Wind-Zone Problems

It is reasonable to suppose that the pulsar wind flows radially away from the pulsar at large distances, in which case we have the asymptotic behaviors

$$V_m \to c, \tag{5a}$$

$$E, B_\phi, V_\phi \to 1/\rho, \tag{5b}$$

$$q, J_m, B_m \to 1/\rho^2, \tag{5c}$$

and

$$J_\phi \to 1/\rho^3. \tag{5d}$$

Consequently, the magnetohydrodynamic (MHD) assumption (zero net Lorentz force) reads

$$qE_m = -J_m B_\phi, \tag{6}$$

which is just equation 4.3.36. Morris (1975) first noted that this balance requires the asymptotic condition $V_m \to c$, which is impossible for particles with finite mass. Heuristically, this condition results because q is the source of E_m while J_m is the source of B_ϕ. Thus dimensionally we have $q^2 \approx J_m^2 \times$ constant, and the constant is simply $1/c^2$ because the reduction of the flow problem to one of pure electromagnetism automatically requires c to be the characteristic velocity. Although the plasma from a pulsar might be highly relativistic, $V_m < c$ and therefore the magnetic forces cannot quite cancel the electrostatic forces in equation 6 (see Buckley 1977a,b and ch. 9). An analogous problem arises in the theory of the solar wind, except there the wind is a quasi-neutral plasma, the qE_m term becomes negligible as a result, and instead pressure gradient terms counter the $J_m B_\phi$ forces. Because the pressure gradients decline faster than $1/\rho^3$ asymptotically, the assumption is then that there must be nonradial flow away from the equator to reduce B_ϕ and J_m. In the ultrarelativistic version of the standard (pulsar) model, this problem is concealed because the tiny difference between V_ϕ and c is ignored. For flows of real particles, the cancellation is not exact (Morris 1975) and consequently some nonradial flow must result. A more detailed analysis of this point is made in chapter 9.

5. The Transition Region

Several authors have suggested that the physics is incomplete unless a shock transition is included. Such a magnetohydrodynamic discontinuity is not included in the pulsar equation. A series of papers by Ardavan (1976a to e) has suggested that a shock wave exists at the light-cylinder. Aspects of this calculation have been challenged, however (Burman 1977a, 1980a,b). Most recently, it has been concluded that the discontinuities are not shocks (Ardavan 1982), suggesting instead a possible internal inconsistency in the underlying assumptions. A sizable literature exists pointing up problems encountered near the light-cylinder. Scharlemann (1974) thought in this connection that the complete charge separation might be an unrealistic assumption. Others have also been unable to avoid discontinuities at the light-cylinder (Buckley 1976; Henriksen and Norton 1975a) or related problems (Mestel 1973). Steady state flow equations are rather subtle to solve, however, because once one has imposed the steady state assumption one is no longer free to also arbitrarily choose the boundary conditions as well. Errors in the choice of boundary conditions generally result in mathematical peculiarities (infinities, etc.).

6. Curved Field Lines

Another elementary defect in the "pulsar equation" was noted by Scharlemann et al. (1978), namely that, if along a given field line one requires $J_m = q V_m$ where $V_m \approx c$ (at some distance above the surface), then in general V_m cannot be c elsewhere and still have q be the local space-charge density (q_0). In other words the idea that the plasma flows at roughly c everywhere is not consistent with the assumption of $\mathbf{E} \cdot \mathbf{B} \ll |E| \, |B|$ *everywhere*. Indeed, in the scenario of Mestel et al. (1979), it is assumed that $V_m \ll c$ well inside the light-cylinder with q almost exactly equal to the Goldreich-Julian value. The price then paid is that $|J_m|$ is small compared to $q_0 c$ and the torque on the star is much less than the observed torque (alternatively, the magnetic moment is $\approx c / V_m$ greater than the conventionally estimated value). On the other hand, if $V_m \approx c$, then *in general* q cannot be q_0 along a given field line at more than one point. In the special case of the magnetic monopole, both q and q_0 vary as r^{-2} and can be held in the fixed ratio $V_m / c \approx 1$. But if the field line curves, J_m / q_0 cannot stay in the fixed ratio $V_m / c \approx 1$ because q_0 has an additional dependence on the direction of a field line (the proportionality to B_z: eq. 2.3.18), and consequently $\mathbf{E} \cdot \mathbf{B} \neq 0$ must appear if the field line curves. This observation has led to the idea of *favorable* and *unfavorable* curvature, depending on whether $\mathbf{E} \cdot \mathbf{B}$ has the correct or incorrect sign to accelerate the local space charge. Scharlemann et al. (1978) concluded on this basis that the aligned rotator cannot have a steady, charge-separated flow solution (with $V_m \approx c$). They also pointed out that the favorably curved part of the polar flux tube of the *oblique* rotator could have consistent, ultrarelativistic, charge-separated flow out to distances of order the light-cylinder radius, and noted that at these

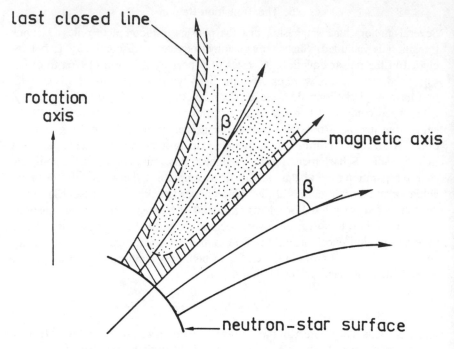

FIG. 4.13. Pair production discharge on favorably curved field lines. Since flow is assumed to be choked off owing to excess space-charge accumulation along downward-curved field lines, only the upward ones are assumed to be able to maintain particle injection. Dotted region represents a pair-dominated outward flowing plasma, while the hatched area is the acceleration region (see figure 2.5). After J. Arons, 1979, *Space Sci. Rev.*, 24, 437 (figure 1). Reprinted by permission of Kluwer Academic Publishers.

radii, the particle energy density can become comparable to the magnetic energy density, with current closure conceivably occurring through inertial forces. This drastic departure from the assumptions of the standard model has not received a consistent quantitative development as yet. A pulsar model (*with* pair creation) based on these considerations (Arons and Scharlemann 1979) is illustrated in figure 4.13. Here, no current flow and pair creation whatsoever are proposed along the unfavorably curved field lines, while pair production occurs at and above a well-defined surface along the favorably curved field lines.

These considerations can be seen rather simply for the *closed* dipolar field lines. Nothing prevents plasma from flowing from a source in one hemisphere to a sink in the other. Indeed, energetic photoelectrons from the Earth's sunlit ionosphere do flow to be deposited in the conjugate regions (other end of the same magnetic field lines). But the space-charge distribution is firmly fixed by the $\mathbf{E} \cdot \mathbf{B} = 0$ condition in the aligned rotator, whereas rotation of the Earth plays only a minor role in the overall electrodynamics of the Earth's magnetosphere. Thus a few test particles could be sent from one hemisphere of the aligned rotator to the other, but certainly not a current whose charge density would approach that of the required space charge. Certainly those charges could not themselves flow (especially because they have differing signs in differing regions!). The only way a significant amount of plasma could flow would be to have either a *neutral* plasma flow such that the net space-charge perturbation was essentially zero, or counterstreaming flows that again were bulk neutral (in other words, the neutralizing charge carriers could flow in either the same or opposite directions). In both of these cases, the flow will be *through* the stationary charge-separated plasma, which cannot itself move. Given that such relative streaming excites the two-stream instability, there would presumably be a large effective resistance to such flow on closed magnetic field lines. Whether or not one can mimic the stationary space charge by modulating the relative velocity of the flowing quasi-neutral plasma (as proposed for the case of the open field lines) has yet to be demonstrated. One problem is to keep $\nabla \times \mathbf{E} = 0$.

7. Numerical Modeling

Given these many difficulties in finding steady state solutions, why not just set up the full time-dependent equations on a computer and let the program run? Preliminary calculations of this sort have been done for the aligned rotator (Kuo-Petravic et al. 1974, 1975; Petravic 1976) and have been initiated for the oblique rotator (Smith et al. 1985). The former calculations numerically followed a linear spin-up of an aligned rotator from rest. The unexpected finding was that *the magnetic field lines remained closed beyond the light-cylinder.* Unfortunately, a number of compromises had to be made in this calculation; to avoid grid-size problems the calculations were begun not at the stellar surface but at $R_L/5$, and a rather large particle mass was adopted (order of 10^{16} eV) in order to track the particle trajectories. These two assumptions conspire to allow the particles to escape the magnetic field, owing partially to the injection scheme (Michel 1974a; see, however, Kuo-Petravic and Petrovic 1976) and partially to the fact that the gyroradii of these massive particles become comparable to the light-cylinder distance (Michel 1975a). The massive particles do not readily screen out the rotationally induced electric field and achieve high energies (sec. 2.3.6). Thus closure of the field lines beyond the light-cylinder would simply result from the free escape of very energetic massive particles, a feature which is not expected to simulate a pulsar magnetosphere accurately. Although there has been talk of performing calculations with more realistic particle masses, these calculations have not yet materialized, possibly owing to the expense and difficulty involved. An alternative approach would be to treat the charged particles as a fluid. In any case, their unexpected result, once again a closed magnetosphere, may also be suggestive of the static solutions of the previous section.

8. Laboratory Simulations

It would be impractical to rotate a spherical magnet (*terrella*) fast enough to induce an interesting electric field, but one could differentially charge the surroundings to produce the equivalent corotation electric field. If one attempts to scale down pulsar parameters to laboratory ones while still keeping sensible parallels (e.g., keeping the light-cylinder outside the terrella), one can show (Michel 1978b) that there is a fundamental minimum luminosity required of

$$L_0 = \frac{32\pi\epsilon_0 m^2 c^5}{e^2}.$$

For hydrogen ions this luminosity is 2.3×10^{24} ergs/s, and for electrons it is 7×10^{17} ergs/s. Note that these power requirements are entirely independent of design details (size of object, magnetic field strength, etc.). With existing technology, the electron luminosity might be obtained in a pulsed device, but one would then presumably need positrons as the neutralizing particles. Direct simulation seems unpromising, but the question should perhaps be reexamined.

9. Summary of Problems

The standard model seems beset by difficulties in all regions (near zone, transition region, far zone). Confusion still persists as to the simple classical behavior of a rotating magnet. This confusion is not diminished by the regular appearance of semiquantitative models which assume that the former standard model is "basically" correct. Such models are often appealing because they seem more realistic and relate more directly to the observational data. Consistency with observation is then turned around and the observations are claimed to support the model. Unfortunately, this approach is a false parallel to the practice in physics of choosing between two internally self-consistent but alternative theories on the basis of observation. Here the basic theory fails to be self-consistent, leaving no real choice at all.

10. Attempts to Salvage the "Standard" Model

A number of suggestions have been made criticizing the basic assumptions in the standard model while proposing modifications that could still be implemented within the overall picture.

a. *Two-Component Plasma*

A popular suggestion is that a two-component plasma is present (electrons plus ions or positrons to give a partially neutralized plasma). The apparent advantage of this suggestion is that electric current and charge density can, in principle, be decoupled so that one can be zero without obligating the other to vanish. The severe difficulty is that a two-component plasma is inconsistent with the model itself. One cannot pull both signs of a particle from the surface with a static electric field. It is conceivable that the accelerating field at the surface oscillates, first pulling electrons and then ions out to produce a quasi-neutral plasma. If it does oscillate, the $\partial/\partial t = 0$ assumption must be dropped. In other words, it would make more sense simply to seek oscillatory solutions than to impose a two-component plasma on a steady state system, since the latter two assumptions are mutually inconsistent. Moreover, many of the difficulties with the standard model seem to persist even if a two-component plasma is assumed. (The corotating regions, for example, would still be segregated by the inertial forces.)

b. *Pair Production*

An alternative hope was that pair production automatically provides for the neutralizing charges, because now the pairs can be produced even though only a single component of charge is accelerated from the surface (see, for example, Cheng et al. 1976 and sec. 8.2.3 for a survey of such models). However, as we saw in sec. 2.3.6.c, one cannot have a strong back-flow of oppositely charged particles from these pairs, and consequently one *cannot* significantly alter the current density that flows on any given flux tube over what would have flowed without pair production: typically just the corotation space charge times c. In fact, there is a general problem with any suggestion that strong currents flow, as in an auroral zone, because if the currents are charge-separated plasma, the excess space charge itself is grossly inconsistent with local $\mathbf{E} \cdot \mathbf{B} = 0$. On the other hand, if the plasma is quasi-neutral, it cannot be composed of back-flowing pairs for the reason noted above. The remaining alternative is that any concentrated currents flow from some source of plasma out to some other source of plasma (e.g., between conductors) such as the interstellar medium or a nearby disk. The interstellar medium, however, has difficulty getting past the effective radiation pressure in the pulsar wave zone (see sec. e, below).

This suggestion essentially begs the question since the standard model ignores pair production. After all, one can apparently choose parameters such that pair production cannot be important, which then leaves the same problems in explaining what happens. Indeed, one of the successes claimed for the pair production models is that cessation of this process could account for the lack of slow pulsars (Sturrock et al. 1976). But one should then be able to conclude from the inability of the aligned rotator to work as advertized (a classical mechanics problem) that pair production must exist in pulsars and at what rate (a quantum electrodynamical phenomenon). This eventuality would seemingly be profound, but still leaves open the question of how these systems function when "off." Alternatively, one might argue that the aligned rotator is a separate unsolved problem of no immediate relevance. In any event, no one has even shown that pair production solves the fundamental underlying current-closure problem.

c. *Closed Current Loops*

The standard model neglects radiation from the particles. If we approximate the plasma to be a continuous fluid, then a steady state system need not radiate. However, it has been suggested that the radiation plays an essential role (Mestel et al. 1976, 1979; Mestel and Wang 1979, 1982; Rylov 1982; Mestel et al. 1985;

Burman 1985; Fitzpatrick and Mestel 1988a,b). Here, the basic defect of the Gold (1968) model, namely that charged particles would drift across field lines as a result of radiation reaction (hence not be trapped as supposed), is taken to be an advantage. It is supposed that the particles come from the poles and move out to the light-cylinder, where they radiate and consequently move across field lines toward the equator. They then return along field lines to the star. Thus radiation reaction is argued to provide the electromotive force to drive a closed current loop out from the star, down along the light-cylinder, and back. This suggestion has been around for a while and has yet to be shown to be physically self-consistent. As noted before, this picture is uniquely at odds with the success in physics of treating radiation as a perturbation. Moreover, it relies on the drift approximation in magnetohydrodynamics, which in turn would require the particles to have infinite energy if $E = cB$, a feature completely at odds with the single-particle motions. A particle *will* gain energy when encountering such fields, but it is still ejected from the system and very little cross-field motion actually results, as discussed in sec. 9.2. Endean (1980) has pointed out that a nonaxisymmetric charge-separated magnetosphere must radiate as a result of rotation, and has proposed this effect as a pulsar model (Endean 1972a, 1981). Again, radiation reaction would require the magnetospheric charge distribution to circulate, and if loss were impossible, these currents would presumably be closed within the system (see also Holloway 1977 and Wang 1978). Recently Beskin et al. (1983) proposed a version of the Mestel work based simply on rigid corotation of the Goldreich-Julian space-charge model (Michel 1973b), with the assumption that the magnetosphere is filled, that wherever $E \geq cB$ the particles move perpendicularly across the magnetic field lines, and that all currents flow back to the star on the last closed magnetic field line. They also calculate a number of statistical properties for pulsars (Beskin et al. 1984), but since they simply assume that the radio luminosity is a fixed fraction (10^{-5}) of the total, these statistical results seem largely independent of their theoretical modeling. They also place particular importance on the "dimensionless" parameter $Q = 2P^{1.1}\dot{P}^{0.4}_{-15}$ (Beskin et al. 1986, eq. 6.10), which is not dimensionless (P is in seconds, while $\dot{P}_{-15}$, meaning period derivative scaled to units of 10^{-15}, is dimensionless). Nevertheless, all subsequent citations persist in referring to this quantity as "dimensionless."

d. *Formal Solutions*

Some recent formal advances have been made in which self-consistent charge-separated solutions are claimed to exist (Schmalz et al. 1979, 1980). At present these programs have not yet been fully carried through, so the nature of the solutions cannot yet be examined. They are more in the form of mathematical existence theorems.

e. *Particles from the Interstellar Medium*

If indeed the action of nonneutral plasmas is to shut down activity in the aligned model by trapping a dome and torus, the system will have a large net positive charge. A number of workers have proposed that the interstellar medium could discharge the star (Mestel et al. 1985; Rylov 1984; Ruderman 1986) thereby preventing such a large trapping potential to build up. In a related paper, Cheng (1985) has examined the possibility that interstellar dust grains penetrate the magnetosphere and provide a neutralizing source of ionization. This current source was proposed to explain the weak correlation between $P\dot{P}$ and pulsar space velocities (V) discussed in sec. 2.4.4; unfortunately, a naive application of the idea gives an *inverse* correlation because the *slowly* moving pulsars should accrete more dust grains; neglecting the motion of the grains, the gravitational sphere of influence extends out to $r \sim V^{-2}$, so the capture cross section declines as V^{-4}, more than compensating for the increased flux (V) of grains. It seems, however, that the interstellar medium (ISM) cannot provide significant neutralizing currents to activate a rotating magnetized neutron star as a pulsar, unless the magnetic moment is almost perfectly aligned with the spin axis (see Arons and Barnard 1983; Michel 1987a).

Let us assume that a rotating magnetized neutron star has indeed shut down owing to trapping of nonneutral plasma but is moving through the ISM. The problem of what happens when a magnetized object moves relative to an ambient plasma is a classical one in the magnetospheric physics of the Earth and other planets (see, e.g., Olson 1979). The simpler problem of what magnetospheric cavity forms about a dipole sitting at rest in an unmagnetized plasma has been solved to apparently high accuracy (sec. 10.3). Theory and observation concur that in general a magnetized cavity forms such that the magnetic pressure balances the plasma internal plus dynamic pressure at the interface (*magnetopause*), i.e.,

$$\frac{B^2}{2\mu_0} = P_0 + \rho v^2. \tag{7}$$

For pulsars the important term is probably the dynamic pressure, assuming the pulsar velocity to be of the order of 10^2 km/s. The plasma number density is conventionally taken to be about 0.03/cc (i.e., the electron concentration determined from dispersion of the pulsar pulses themselves). Near the pulsar the hard radiation from a hot neutron star could fully ionize the plasma to its neutral concentration, about 1.0/cc. These values are not too different from the solar wind density and velocity flowing past a planet like Jupiter which has a magnetic moment not too different from a pulsar (sec. 7.2). The characteristic magnetic field required to balance a 1.0/cc flow at 10^2 km/s is of the order of 5×10^{-5} gauss. This *standoff* field will be denoted B_{eff}.

In addition to the external pressure of the ISM, there are three characteristic electromagnetic stresses about a rotating magnetized object: the quasi-static magnetic pressure invoked in equation 7, the electrostatic tension, and the radiation pressure. The magnetic pressure is from the dipolar magnetic field component of the star, falling off as $1/r^3$, until one passes the light-cylinder distance, after which the field declines as $1/r$. For typical pulsar periods of about 1 s and a 10 km neutron star, the light-cylinder distance is about 5000 radii away.

The distance at which a nominal surface field of 10^{12} gauss would drop to B_{eff} can be expressed in units of $x_m R_L$ (m denotes magnetosphere), where

$$x_m^3 = (B_0/B_{\text{eff}})(a\Omega/c)^3 = S. \tag{8}$$

The right-hand side (S) has a value of about 1.6×10^5 for the nominal parameter values mentioned above ($B_0 = 10^{12}$; $B_{\text{eff}} = 5 \times 10^{-5}$; $a = 10^6$; $\Omega = 2\pi$ radians/s). Unless the pulsar moves very rapidly through a dense interstellar cloud, S is going to be a large number. If the object has a net charge (the issue addressed here), the electrostatic tension will terminate on surface charges at the magnetospheric interface. Indeed, if $E > cB$ at the interface, discharging currents could be pulled in across the magnetic field lines. In pressure terms, the condition $E > cB$ is equivalent to the electrostatic tension ($\epsilon_0 E^2/2$) exceeding the magnetic pressure ($B^2/2\mu_0$). The role of the electric field can therefore also be parameterized by the distance at which the electrostatic tension becomes comparable to the external pressure of the ISM. The characteristic electric field at the neutron star is the rotationally induced field ($E \approx a\Omega B_0$), and therefore any monopole component will be of the order

$$E = \zeta a^3 \Omega B_0/r^2, \tag{9}$$

where ζ is a number of order unity (or less) parameterizing the net charge of the system. Even the former standard model has a large positive central charge (sec. 4.2.1.b) with $\zeta = 1/3$; however, the infinitely extended space charge in that model also contributes to the radial (equatorial) electric field so that effectively $\zeta = 1$ on the equator. This electric field becomes comparable to cB_{eff} at $x_e R_L$, where

$$x_e^2 = \zeta S, \tag{10}$$

and again the system is scaled by the same large dimensionless factor S. Note that for $S > 1$ and $\zeta \approx 1$, $x_e > x_m$, which means that the magnetopause would be located where $E = cB$ (essentially the light-cylinder distance: R_L/ζ), and currents could be pulled in across the magnetosphere and then flow down along field lines to the neutron star. Such a process would be an appealing one to activate a pulsar by supplying plasma and electrical currents, but the role of radiation pressure must still be taken into account.

The radiation field produces an effective pressure driving charged particles away from the star. For simplicity one can decompose the magnetic dipole moments into a component parallel to the rotation axis and one orthogonal. A small inclination then corresponds essentially to the aligned rotator plus a small rotating dipole. Note that the radiation pressure is not zero along the rotation axis (and the magnetic field lines leading to "infinity"). The radiation field strength is scaled to the static field value at the light-cylinder, but now falls off as 1/r at larger distances; hence

$$B_{\text{wave}} \approx B_0 \sin \alpha a^3 \Omega^2 / c^2 r, \tag{11}$$

where α is the inclination of the magnetic moment from the rotation axis. Again comparing to the ISM, we obtain a characteristic distance $x_r R_L$, where

$$x_r = S \sin \alpha. \tag{12}$$

This distance is the largest of all, and therefore radiation pressure drives the ISM out to such large distances that the electrostatic tension is unimportant, unless of course the inclination angle is extremely small. Charged particles from the ISM cannot flow down the polar cap field lines against this pressure, and discharge of the system is prevented.

Could pulsars be nearly aligned? The inclination angle at which discharge could occur is given by the condition $x_r = x_e$; hence

$$\sin^2 \alpha \leq \zeta/S \leq 1/S, \tag{13}$$

and we therefore require magnetic inclination angles less than $S^{-1/2} \approx 0.0025$ radians. There are a number of observational and theoretical objections to such near-perfect alignment. A sizable misalignment is normally thought to be required to produce the sharp pulses observed if radiation is beamed out as either a cone or a fan beam, because the intrinsic beam widths are already of order $10°$. In any event, only a small fraction of pulsars would then be detectable for such tiny inclinations, making reconciling the pulsar ages with the supernova rate a serious problem (sec. 2.4). There would now have to be vast numbers of unobserved pulsars. From a theoretical point of view, electromagnetic alignment of the pulsars (see Good and Ng 1985 for a recent discussion) to this high degree seems unlikely because for faster spin rates the maximum inclination is smaller ($\sin \alpha \approx P^{3/2}$). Thus the alignment would have to take place first, and afterwards the discharge. The pulsar would then be observable. It is sometimes argued that there is a paucity of pulsars with periods less than 0.1 s, but that factor of 10 higher rotation rate translates (eq. 13) into $\alpha \approx 0.1$ milliradians. Parameters that seem forced for slow pulsars are seen to become even more implausible for the faster pulsars. The fastest pulsars present a somewhat different situation because it is possible that pair production processes could be active even at light-cylinder distances (Cheng et al. 1986a), at least for the Crab and Vela pulsars; otherwise it would seem astonishing that interstellar plasma could make its way upstream through the relativistic wind that the Crab pulsar is believed to emit (sec. 2.3) that lights up the entire nebula. Of course one is welcome to suppose that pulsars are somehow born with almost perfectly aligned magnetic dipole fields, but nevertheless they emit in a wide fan beam allowing them to be observed far off-axis. But such an assumption requires a large departure from the impressive body of phenomenology supporting emission from the vicinity of an inclined magnetic pole.

To estimate the effect of gravitational attraction on the plasma in the ISM, we integrate the hydrostatic equation

$$dP/dr = \rho d\phi/dr \tag{14}$$

using $P = K\rho^{5/3}$ to obtain the pressure as a function of distance, given the asymptotic pressure and density (P_∞, ρ_∞), to obtain

$$P = P_\infty(1 + r_g/r)^{5/2}, \tag{15}$$

where $r_g = 2\rho_\infty GM/P_\infty$ defines the gravitational sphere of influence outside of which gravity can be ignored ($B_{eff} = $ constant) and inside of which $B_{eff} \approx r^{-5/4}$. The ratio P_∞/ρ_∞ is the thermal velocity of the particles but can also represent that the space velocity of the pulsar is supersonic. In figure 4.14 $(P_\infty/\rho_\infty)^{1/2}$ is quoted as the characteristic velocity.

From these pressure balance arguments, it appears that radiation pressure prevents pulsars from being discharged unless they are nearly perfectly aligned.

4.5 Extensions of the Standard Model

Some researchers have attempted to leapfrog the issue of self-consistency. In these models the functioning of the aligned rotator model is taken as a given, and the analysis is built upon assumed properties of the standard model. The intention is basically to confront observation by describing how a pulsar might function, but on the basis of a physics not yet shown to be self-consistent. As a result, there is considerable confusion as to which model is "best." The aligned rotator models are typically directed at sorting out the self-consistency problem, while the extended models are aimed at explaining the data, but unfortunately in terms of a basic model not yet shown to work. The model originally proposed by Sturrock (1970, 1971a; see also Komesaroff 1970) and refined by Ruderman and Sutherland (1975) is one of the more important examples of the latter genre. An extensive discussion of this model can be found in the review by Sutherland (1979; see also Sturrock 1971b; Roberts and Sturrock 1972a,b; Ruderman 1976; Cheng and Ruderman 1977a,b,c, 1979).

1. Pair Production Cascade

In the corotating plasma of the standard model, there is always a local frame of reference in which locally $\mathbf{E} = 0$, at least within the light-cylinder. In the vacuum case, however, there are strong parallel electrostatic fields that cannot be removed by a change of reference frames ($\mathbf{E} \cdot \mathbf{B}$ being a relativistic invariant), which opens the possibility that a cascade breakdown could occur in these strong fields. The idea, first developed by Sturrock (1971a), is that electrons accelerated to high energies by the electric field would emit hard gamma-rays owing to their curvature radiation in following the dipolar magnetic field lines. These gamma-rays would in turn produce pairs in the intense magnetic field and thereby provide yet more electrons and positrons, and perhaps further steps of

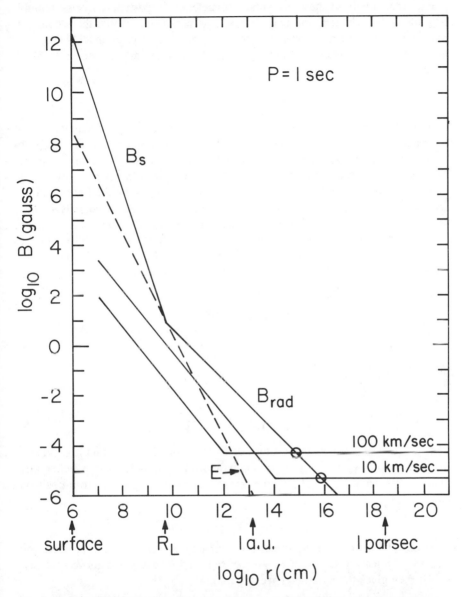

FIG. 4.14. Pulsar/ISM interaction parameters. Effective pressures plotted as magnetic field equivalents for a $P = 1$ s and $B = 10^{12}$ gauss pulsar. The quasi-static dipole field falls as $1/r^3$ out to $R_L = c/\Omega$, and thereafter the field falls as $1/r$. The monopolar electric field drops as $1/r^2$, of course (dashed line). A rather dense ($\approx 1.0/\text{cc}$) interstellar medium with a large characteristic velocity (directed or thermal) is balanced by the radiation pressure well outside of the light-cylinder at the circled locations. Gravitational concentration of the plasma plays no role (note the very compressed logarithmic scale). Sharp breaks indicate passage from one regime to another; these would, of course, be slightly rounded if calculated from the exact expressions, but again on such a compressed scale such rounding is not particularly important. From F. C. Michel, 1987a, *Ap. J.*, 312, 271 (figure 1).

pair production from these secondary particles. The particle energies rapidly degrade with each step, and only for extreme parameters would the secondary pairs result in tertiary pairs; hence the "cascade" terminates quickly. For example, a primary electron of energy γ_0 produces a gamma-ray (eq. 2.3.35) of energy

$$\gamma_p \approx 3\gamma_0^3\lambda_c/\rho_c, \tag{1}$$

in units of $m_e c^2$, where ρ_c is again the radius of curvature of the magnetic field lines and λ_c is the Compton wavelength. For a pure dipole magnetic field, the radius of curvature at the surface of a star of radius a is $\rho_c \approx (aL)^{1/2}$, where L is the maximum distance of the dipole field line from the origin. For pulsar models the maximum distance would typically be taken to be the light-cylinder radius, giving a curvature radius of about 10^7 cm for the Crab pulsar. The secondary gamma-ray can in turn produce pairs and, if above threshold, an electron-positron pair, each of which has a Lorentz factor $\gamma_1 \approx \gamma_p/2$. Thus the successive Lorentz factors are γ_0, $\gamma_1 \approx k\gamma_0^3$, $\gamma_2 \approx k(k\gamma_0^3)^3 = k^4\gamma_0^9$, and so on, where $k = 3\lambda_c/2\rho_0$, the coefficient in equation 1. We can rewrite this series in dimensionless form using

$$\gamma_k = \gamma_0\eta^{p(k)}, \tag{2}$$

where

$$\eta = k\gamma_0^2, \tag{3}$$

and

$$p(k) = (3^k - 1)/2. \tag{4}$$

Thus p takes on the successive values 0, 1, 4, 13, 40, 121, ..., and it is evident that, unless η is very nearly unity, the pair production quickly falls below any plausible threshold, $\gamma \geq 2$ being the threshold for creating an electron-positron pair.

a. *Radiation Reaction Limits*

The role of radiation reaction can be estimated by requiring the accelerating field to be at least of the order of $\gamma_0 m_e c^2/e\rho_c$ and repeating the above analysis, which then gives the parameter-free condition

$$\eta\gamma_0 < 9\pi\epsilon_0 mc^2\lambda_c/e^2 = 308, \tag{5}$$

the latter figure being just a numerical factor (9/4) times the reciprocal of the fine structure constant. Equation 5 is not really a firm limit since the accelerating field could be larger than estimated; however, very much larger values of $\eta\gamma_0$ clearly become suspect and demand very special geometries. In table 4.7 we have indicated those values of $\eta\gamma_0$ in excess of 10^4, which are suspect in that

TABLE 4.7—Energy Degradation in Pair Cascade

γ_0	$\gamma_1\,(\gamma_0)^a$	$\gamma_2\,(\gamma_1)$	γ_3
$\eta = 10^{-1}$	$(\eta = 10^{-3})^a$		
$(10^{14})^c$	$(10^{13})^c$	10^{10}	10
(10^{12})	(10^{11})	10^8	$[10^{-1}]^d$
(10^{10})	(10^9)	10^6	$[10^{-3}]$
(10^8)	(10^7)	10^4	$[10^{-5}]$
(10^6)	10^5	10^2	$[10^{-7}]$
$\eta = 10^{-2}$	$(\eta = 10^{-6})^a$		
$(10^{14})^c$	$(10^{12})^c$	10^6	$[10^{-12}]^d$
(10^{12})	10^{10}	10^4	$[10^{-14}]$
(10^{10})	10^8	10^2	b
(10^8)	10^6	$[1]^d$	b
10^6	10^4	$[10^{-2}]$	b
$\eta = 10^{-4}$			
$(10^{14})^c$	10^{10}	$[10^{-2}]^d$	b
(10^{12})	10^8	$[10^{-4}]$	b
(10^{10})	10^6	$[10^{-6}]$	b
10^8	10^4	$[10^{-8}]$	b
10^6	10^2	$[10^{-10}]$	b

a) Values corresponding to $\eta = 10^{-3}$ or 10^{-6} are obtained by reading this column as γ_0 instead of γ_1, the next column (labeled γ_2) as γ_1, etc.

b) Frequencies result which are below 10^{15} in our units (1.24×10^{30} Hz) and cannot propagate through interstellar space.

c) Values in parentheses correspond to $\eta\gamma_0$ products in excess of 10^4, which violate equation 5 below and are therefore unlikely from radiation reaction considerations.

d) Square brackets indicate termination of the cascade with photons of this energy which are too low to produce pairs of this energy.

they widely exceed the limit estimated in equation 5. Because the primary cascade gives photons of energy $\eta\gamma_0$, it is clear that a secondary cascade can be obtained only if the system operates near the radiation reaction limit. A tertiary cascade would occur only if $\gamma_0\eta^4$ exceeded unity, hence requiring $\eta \approx 10^{-1}$ but restricted to low values of $\gamma_0(\leq 10^5)$, an implausible condition on the magnetic field line curvature radius ($\rho_c \approx 10^{-1}$ cm).

A straightforward application to the Crab pulsar would (table 2.4), for $\gamma_0 = 10^5$, give $\eta \approx 10^{-6}$, assuming a rather small value $\rho_c = 10^6$ cm ($k \approx 10^{-10}$ cm/$\rho_c = 10^{-16}$; $\eta \equiv k\gamma_0^2 = 10^{-6}$). From table 4.7, we find $\eta = 10^{-6}$ heading the second group of rows of the second column and, reading down, we interpolate between γ_0 (see footnote a) of 10^6 and 10^4 and find that the photons produced have X-ray energies ($\gamma = 0.1$). Thus there would be no pair production. If instead we regard η as a free parameter, we require $\eta \approx 10^{-3}$ (again in the second column, but the first group of rows) to obtain Crab-like visible photons as the last conversion step. For example, primary electrons with $\gamma_0 = 10^6$ (interpolating again between the last two entries) would produce $\gamma_1 = 10^3$ photon and pair secondaries, the latter energetic enough to produce radiation with $\gamma_2 = 10^{-6} = 1$ eV photons but no accompanying pairs of course. Even with these rather extreme parameters (ρ_c is now $\approx 10^5$ cm), there would be only one stage of pair production, with those pairs then radiating visible light. If the injection energy could be boosted to $\gamma_0 = 10^{10}$, then $\eta \approx 10^{-3}$ (easily obtained at that energy: $\rho_c \approx 10^{13}$ cm $\gg R_L$) would produce a true cascade, first to $\gamma_1 = 10^6$ and then once again to $\gamma_2 = 10^{-6}$. If the absorption threshold were placed near γ_1, we would then obtain a spectrum extending from hard gamma-rays to the visible, as observed for the Crab. This possibility is therefore an attractive one, but at odds with the limitations imposed by radiation reaction (sec. 2.3.8).

The above analysis assumes the radiation to be curvature radiation; the presumably weak contribution from synchrotron radiation, owing to the finite pitch angles, has not been included. See Daugherty and Harding (1982, 1983), Rozental and Usov (1984), and Harding et al. (1989) for numerical simulations.

A regenerative mechanism for cascading was illustrated in figure 2.5 (Ruderman and Sutherland 1975), in which a downward-moving electron (remember, polar caps have the opposite sign to the Goldreich-Julian choice in the above model) formed at point (1) gains sufficient energy at point (2) to produce a gamma-ray that is absorbed at (3) to produce a pair, the electron almost immediately being absorbed by the surface and the positron reaccelerated in the opposite direction. This process must run just at threshold, with one particle (downward-moving electron or upward-moving positron) on the average being conserved in the gap. Exponentiation (one electron → two positrons → four electrons, etc.) would close the gap and vice versa. Note that this process also tends to be self-extinguishing since the cascade automatically seeks the least-curved field line, moving toward the poles in this figure. The space charge is generally zero in this model except for two sheaths at the top and bottom of the gap, and the current flow is modulated by keeping pair production near threshold in the gap rather than through space-charge limitation. Note also that the multiplicity (pairs produced for each oscillation of the "primary" particles in the gap) is just γ_0/γ_1 from energy conservation, hence of the order η^{-1}. A more realistic geometry for the gap is illustrated in figure 4.13. The cascade model in the Ruderman-Sutherland version is based on calculations showing that iron ions would be too tightly bound to escape from the crystalline lattice, owing to the expected strong magnetic fields.

2. Electric Current Production

A particularly attractive feature of the cascade model is the possibility that the cascades are localized within *sparks* rather than uniformly filling the entire polar cap. This model is thereby endowed with additional morphological detail which could well speak to the rich variety of behavior in observed pulsars. Cheng and Ruderman (1980) have expanded their model to include ion production from the polar caps, owing to the electron bombardment (see also discussion by Matese and Whitmire 1980). Arons (1981a,b, 1983a) has recently undertaken an analysis of how the cascade process operates. At the moment, however, pair production complicates the physics without greatly clarifying it. Current closure has not been demonstrated. Benford and Buschauer (1980) argue that bunching would grow too slowly in this model to account for coherent radio emission.

3. Massive Magnetospheres

One could add neutral plasma to the neutron star magnetosphere in addition to the space-charge plasma. By and large there has been little enthusiasm for such an approach since one cannot use the induced electrostatic field to lift a neutral component off the surface, and it takes about 100 MeV per nucleon to lift ions off a neutron star. (For example, if the 10^{41} electrons/s proposed by Shklovsky (1968) to be injected into the Crab Nebula were to come from a neutron star and were accompanied by a neutralizing component of nonrelativistic helium ions, then an energy comparable to the entire luminosity of the nebula would have to go simply into lifting those ions out of the gravitational potential well of the neutron star.) Nevertheless, there has been some motivation for studying the consequences of models having excess density (Roberts and Sturrock 1972a,b, 1973; Henriksen and Rayburn 1974; Pustil'nik 1977; Evangelidis 1977). The massive magnetosphere models are interesting in that the critical distance separating open field lines from closed field lines is moved inward from the light-cylinder. As a result, the stellar wind magnetic field is proportionately enhanced (i.e., a smaller magnetic field at the surface would suffice to give the observed slowing-down rate). Roberts and Sturrock (1973) also emphasize that the massive magnetospheres can give $n = 7/3$ instead of 3, which is in closer agreement with observation (Groth 1975a,b) of the Crab pulsar ($n \approx 2.5$). Also, the massive magnetosphere could be a source of plasma injection into the circumpulsar space (Scargle and Pacini 1971). In contrast, the total mass of a standard model magnetosphere would typically amount to only a few kilograms. The blackbody radiation rate would be greatly enhanced in such a magnetosphere, at least out to whatever the plasma frequency was at the effective photosphere.

The centrifugal opening of the magnetospheric magnetic field lines is an interesting nonrelativistic parallel to early pulsar models and has been studied, for example, by Pneumann and Kopp (1971), Shibata and Kaburaki (1985), and Shibata (1985). Kaburaki (1983) has examined the field distortion from a quasi-neutral magnetodisk.

4. Pulsar Extinction

The standard model does not embrace any specific radio emission mechanism, and consequently there is no distinction between rapid and slow pulsars, whereas observationally there is thought to be a paucity of slow pulsars. The statistics are not all that secure (see sec. 2.4), but a number of proposals have been made.

The perennial favorite is magnetic field decay (Ostriker and Gunn 1969b,e; Fujimura and Kennel 1980) although there are counterarguments (e.g., Holt and Ramaty 1970) and the expected problem with high conductivity (e.g., Ewart et al. 1975). The next supposition in terms of popularity is that pair production is essential for pulsar action, an idea dating back to Sturrock (1971a), with radio emission ceasing when the induced electric fields are too weak to produce pairs (Sturrock et al. 1976). Another idea appeals to the type of ions that might be available from the surface, which might at first liberate helium ions (Michel 1975d) until exhausted in carrying currents to infinity, after which only iron ions might be available for acceleration, leading to a transition in magnetospheric properties (Michel 1975a; Hill 1980). Later ideas involving disks (Michel and Dessler 1981a) led to another alternative because angular momentum transfer to disks eventually leads to rotation with the star and another critical spin frequency. Given that the decline of dipole spin-down power from pulsars is entirely adequate to account for the lack of observable slow pulsars, none of these mechanisms seems urgently required.

5. Neutral Magnetospheres

It is generally thought that particles in the form of a neutral plasma can be supported in the rotating magnetic field of a planet. The general idea is that the particles are held in place by the magnetic field, which inhibits motion across the field lines, and that other forces can then balance along the magnetic field lines. A classic example of this view has been developed by Alfvén et al. (1974), who point out that centrifugal forces balance gravitational forces in an aligned dipole magnetic field only if the particles have an azimuthal velocity such that

$$\Omega^2 \rho^2 = \frac{2}{3} \frac{GM}{r}. \tag{6}$$

This relationship is startlingly simple and attractive. It follows from simple resolution of the force components in a dipole magnetic field configuration, using the fact that the centrifugal force is directed axially outward (here ρ is the axial distance, with r the distance from the center of the planet) and the gravitational force is radial.

This relationship is also somewhat deceptive because Ω is no longer a constant along magnetic field lines but varies rapidly. Ferraro's law of isorotation is therefore violated. At what price? As Vasyliunas (1987) has pointed out, the price seems too high. To begin with, the electric field components across the magnetic field are immediately determined to have the form

$$E_r = -\sin(\theta)r^{-7/2}, \tag{7}$$

$$E_\theta = 2\cos(\theta)r^{-7/2}, \tag{8}$$

and it is straightforward to show that such an electric field is not curl-free.

To obtain a curl-free (steady state) electric field one must add a component of electric field *parallel* to the magnetic field. The latter is a field of the form

$$E_r = 4\cos\theta g(\theta)r^{-7/2}, \tag{9}$$

$$E_\theta = 2\sin\theta g(\theta)r^{-7/2}, \tag{10}$$

where

$$g(\theta) = (\cos\theta)^{1/4} \int d\,\theta(\cos\theta)^{-1/4}, \tag{11}$$

and the integral is indefinite (i.e., it includes an arbitrary constant of integration). The problem with a nonvanishing parallel component of electric field is that the Coulomb force is noninertial. Consequently the resulting *net* force on the plasma particles can only be balanced for one charge specie and not the other. To then hold the remaining charge specie in place, it is necessary to have pressure gradient forces. Such forces require thermal disequilibrium between the two species, because if they were at the same local temperature they would have the same pressure gradient and then the electric force imbalance could not be resolved for one without exacerbating the imbalance for the other. Consequently there are no zero-temperature equilibria for neutral particles trapped in a magnetosphere and no thermal equilibrium solution either.

If we ask about how the Van Allen radiation belts can be trapped, the answer is quite simple: they are not in thermal equilibrium. The belts are maintained by magnetic mirror forces, and such forces only act if particles have finite pitch angles to the magnetic field. Thermalization would always provide particles with zero or small pitch angle, and these would be precipitated to the planetary surface. It is *because* the Van Allen belts are rarefied that they can be maintained. Trapping large amounts of neutral plasma, however, is a contradiction in terms. It is similarly a contradiction in terms to try to fill a *volume* of space around a planet with randomly orbiting uncharged particles: collisions will collapse such a volume into a thin disk, with much of the material falling onto the planet.

Alfvén (1983) has used this "2/3 law" to interpret solar system systematics such as the distribution of gaps in Saturn's rings.

6. Dynamic Magnetospheres

An alternative approach to analyzing quasi-static rotating magnetospheres is the following: assume that all the perpendicular energy of the particles is immediately radiated away but that none of the particles lose their parallel energy. Next assume that the particles have negligible mass. The particles would therefore move at velocity c everywhere, which relieves us of having to know their energy along the field lines. Now let us try to fill the entire dipole magnetic field with particles accelerated from zero energy just above the surface at one hemisphere to just above the surface at the other hemisphere. Then we have the relatively simple problem of finding what space-charge density near the surface as a function of latitude field line (the density along a field line is now fixed if we assume that all the particles originate at the surface) will give a pure internal quadrupole moment ($\Phi = r^2 P_2(\cos\theta)$) inside the star. This condition is all that is required of the magnetosphere to give $\mathbf{E} \cdot \mathbf{B} = 0$ everywhere inside the star (sec. 4.2.3). Then one can ask where the trapping regions are, etc. At first it might be thought that such a program would find quite different solutions because the particles are in motion. But one must think about whether the solution would mean anything, because the electrostatic potential would now be determined everywhere, of course, and there is no way of requiring that the potential along a given field line be everywhere smaller (or larger) than the potential at the surface. We cannot very well let the particles have *negative* energy. Our guess then is that even if the particles are as "hot" as they can reasonably be, one will still get an electrosphere of finite extent, with the confining surfaces not being the discontinuities but rather being the surfaces within which the particles are energetically bound! However, this is just a guess. This author attempted such an analysis some years ago, and the mathematics seems quite tractable. The volume integrals can all be evaluated using recursion relations, so numerically the problem is fairly straightforward. Unfortunately the project got interrupted and given the later developments has not been taken up again.

5

A. Realistic Model: The Oblique Rotator

5.1 Introduction

1. Oblique Models

The aligned rotator problem may be regarded as a warming-up exercise for the problem of the oblique rotator (magnetic field axis at an appreciable angle to the spin axis). If one cannot solve the former, then how can one solve the latter with its added degree of complexity? Here the pioneering work is that of Deutsch (1955), who originally solved for the full **E** and **B** distribution in the vacuum case, suggesting that cosmic rays might be accelerated by the electric field component parallel to the magnetic field (see also Soper 1972). The oblique rotator, because its structure is manifestly time-dependent at the spin frequency, is a natural pulsar possibility. One idea is that the presumed polar cap emission from an aligned rotator can also be presumed for the oblique case, and it is this emission, sweeping past the observer, that produces the pulse. Of course, if the aligned case does not function to begin with, the above presumption fails. However, no one has come up with any other convincing reason for such a system to produce the sharp pulses actually observed. Some regard the radiation/wind from such a rotator as a beam that forms the sharp pulses when it interacts with surrounding material (sec. 7.4). Others invoke MHD shocks to "shape" the electromagnetic pulse, which would otherwise simply be a sine wave in the vacuum case (Michel 1971).

Some basic preparatory work has been accomplished. Ostriker and Gunn (1969a,d) treated the vacuum case and regarded particles emitted from the surface as a perturbation, showing that the particles would absorb the wave energy (see also Eastlund 1968, 1970, Cohen and Toton 1971, and Krivdik and Jukhimuk 1977). Mestel (1971) and Cohen and Rosenblum (1972) extended the Goldreich-Julian model (as a source of surface plasma) to the oblique case. Others have argued that these particles nevertheless strongly influence the electromagnetic field structure (Michel 1971). The near-zone equations have been formulated by Cohen et al. (1973), Parish (1974), Pfarr (1976), and Kaburaki (1978). Cohen et al. and Kaburaki give general-relativistic formulations. Considerable progress has been made toward developing a full dynamic model by Endean (1972b), Henriksen and Norton (1975b), Mestel et al. (1976), Burman and Mestel (1978), Jackson (1978a), Scharlemann et al. (1978), Avetisyan (1979), Kaburaki (1980, 1981), Davila et al. (1980), and Ray (1980).

At the moment, the overall status of the oblique rotator model seems to be in a curious sort of limbo. It is not clear whether the difficulties with the aligned rotator carry forward to the oblique case or whether those difficulties in fact result from the imposed geometrical constraint of alignment. In the extreme case of an exactly orthogonal rotator (figure 5.1) there are symmetric positive and negative space-charge regions above each magnetic pole and consequently current-conserving outward flow is conceptually much more plausible than in the aligned case. The lesson from the last decade or so of study of the aligned rotator, however, has been that first appearances can be misleading. The extra dimension introduced by obliquity might also simply make it more difficult to analyze and discover possible inconsistencies. It seems likely, nevertheless, that much more attention will be devoted to this model in the near future. Endean (1974) showed that the "pulsar equation" can be obtained as a special case of a more general oblique formulation. Beskin et al. (1983) also rederive the oblique, filled magnetosphere case. Kaburaki (1981, 1982, 1983) has examined the various inertial effects from the corotating *quasi-neutral* plasma inside the magnetosphere of an oblique rotator.

2. "Superoblique" Models

A simplified version of a completely oblique rotator would have a *line* dipole rotating about the line's axis. This would somehow constitute a two-dimensional pulsar model without cylindrical symmetry. Some preliminary evaluations have been made with such models (de Costa and Kahn 1982) although their usefulness has been questioned (Burman and Mestel 1979). At present, particle trapping, current closure, etc. have not been evaluated for such models.

5.2 Large-Amplitude Electromagnetic Wave Generation

It is a standard in textbooks to derive the electromagnetic wave generation from an oscillating dipole, be it electric or magnetic. However, this derivation is typically made from a formalism set up to do radiation problems in general, and frequently involves the introduction of ancillary concepts such as retarded potentials, which themselves require special study. For the simple dipole problem, such formalism is unnecessary, and we will give a direct derivation, which we will then build on to analyze the three-dimensional fields about rotating magnetized neutron stars. In particular, we need to evaluate the near fields because they are the ones that influence particles at or near the surface, before the particles have a chance to be ejected as wind particles, for example.

1. Oscillating Magnetic Dipole

The simplest case is a magnetic dipole in a fixed direction (z) that oscillates in amplitude sinusoidally. We will simply plug directly into Maxwell's equations

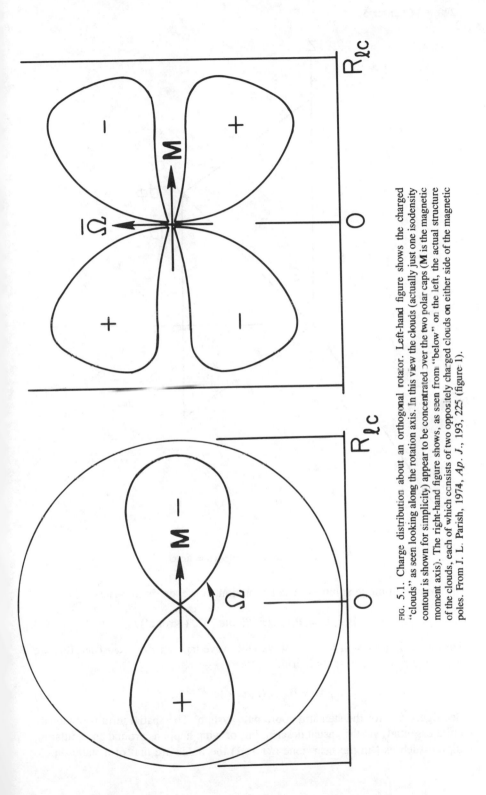

FIG. 5.1. Charge distribution about an orthogonal rotator. Left-hand figure shows the charged "clouds" as seen looking along the rotation axis. In this view the clouds (actually just one isodensity contour is shown for simplicity) appear to be concentrated over the two polar caps (M is the magnetic moment axis). The right-hand figure shows, as seen from "below" on the left, the actual structure of the clouds, each of which consists of two oppositely charged clouds on either side of the magnetic poles. From J. L. Parish, 1974, *Ap. J.*, 193, 225 (figure 1).

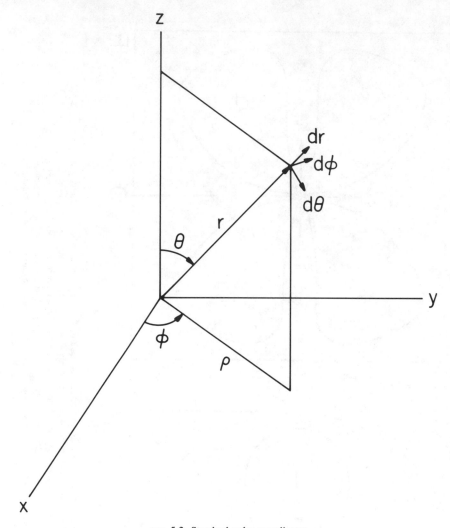

FIG. 5.2. Standard polar coordinates.

and find the unique solution. The temptation, however, is to write

$$\mathbf{B}(\mathbf{r}, t) = \mathbf{B}_{\text{dipole}}(r, \theta, \phi)e^{-i\omega t}, \text{ (wrong!)}, \qquad (1)$$

but the oscillation will launch a wave with wave number $ck = \omega$. Therefore we also require an extra spatial term,

$$\mathbf{B} = \mathbf{B}_{\text{dipole}}(r, \theta, \phi)e^{i(kr-\omega t)}. \qquad (2)$$

See figure 5.2 for the standard coordinate system. The spatial term is obviously quite essential, yet the casual descriptions of what happens around an oscillating dipole such as "in the near zone ($kr < 1$) the fields are that of a static dipole oscillating in sign" can entice one to neglect the fact that this quasi-static field is already nondipolar owing to the cos kr dependence. We need now to satisfy Maxwell's equations. One problem will be to satisfy $\nabla \cdot \mathbf{B} \equiv 0$, which is no longer true when the cos kr dependence is included. The next problem will be to include the induced electric field from $\nabla \times \mathbf{E} = -\partial \mathbf{B}/\partial t$. On the whole, $B \approx 1/r^3$ and differentiation with respect to time only pulls down a factor of $-i\omega$, so $E \approx 1/r^2$ to start with. However, the fields, viewed as a series in powers of $1/r$, have to stop at $1/r$ itself, because that corresponds to a constant flux of power away from the oscillating dipole. Thus we can write in general (with the seemingly wishful-but-fulfilled-hope that the θ-dependences remain global):

$$B_r = 2B_0 \cos \theta \left(\frac{1}{r^3} + \frac{a}{r^2} + \frac{0}{r} \right) e^{i(kr - \omega t)}, \tag{3}$$

and

$$B_\theta = B_0 \sin \theta \left(\frac{1}{r^3} + \frac{b}{r^2} + \frac{c}{r} \right) e^{i(kr - \omega t)}. \tag{4}$$

There is no $1/r$ dependence of B_r because such a term would violate $\nabla \cdot \mathbf{B} \equiv 0$. The electric field is entirely from induction and therefore is entirely azimuthal; hence, as noted above,

$$E_\phi = cB_0 \sin \theta \left(\frac{0}{r^3} + \frac{d}{r^2} + \frac{e}{r} \right) e^{i(kr - \omega t)}. \tag{5}$$

The various coefficients are entirely determined by the Maxwell equations, with

$$\nabla \cdot \mathbf{B} \equiv 0 = \frac{1}{r^2} \frac{\partial (r^2 B_r)}{\partial r} + \frac{1}{r \sin \theta} \frac{\partial (\sin \theta B_\theta)}{\partial \theta}, \tag{6}$$

requiring

$$(r^{-3}): b = -ik; \quad (r^{-2}): ika = -c, \tag{6a}$$

where the power of $1/r$ set to zero by these conditions is indicated in parentheses. The displacement current condition in vacuum is just

$$(\nabla \times \mathbf{B})_\phi \equiv \frac{1}{c^2} \frac{\partial E_\phi}{\partial t} = \frac{1}{r} \left(\frac{\partial (rB_\theta)}{\partial r} - \frac{\partial (B_r)}{\partial \theta} \right), \tag{7}$$

requiring (after replacing ω with ck)

$$(r^{-3}): 2a - b + ik = 0; \quad (r^{-2}): b = -d; \quad (r^{-1}): c = -e. \tag{7a}$$

Finally, the induction equation

$$\nabla \times \mathbf{E} \equiv -\frac{\partial \mathbf{B}}{\partial t} \tag{8}$$

has the two components,

$$\frac{1}{r \sin\theta} \frac{\partial (\sin\theta E_\phi)}{\partial \theta} = -\frac{\partial B_r}{\partial t}, \tag{9}$$

requiring

$$(r^{-3}): d = ik; \ (r^{-2}): e = ika \tag{9a}$$

and

$$-\frac{1}{r} \frac{\partial (rE_\phi)}{\partial r} = -\frac{\partial (B_\theta)}{\partial t}, \tag{10}$$

requiring

$$(r^{-3}): d = ik; \ (r^{-2}): d = -b; \ (r^{-1}): e = -c. \tag{10a}$$

Of these ten conditions, five are satisfied by $b = -ik; d = ik$, four are satisfied by $c = -e = -ika$, and the final one in equation 7a is satisfied by $a = -ik$, so altogether, our fields are explicitly

$$B_r = 2B_0 \cos\theta \left(\frac{1}{r^3} - \frac{ik}{r^2} \right) e^{i(kr-\omega t)}, \tag{11}$$

$$B_\theta = B_0 \sin\theta \left(\frac{1}{r^3} - \frac{ik}{r^2} - \frac{k^2}{r} \right) e^{i(kr-\omega t)}, \tag{12}$$

and

$$E_\phi = cB_0 \sin\theta \left(\frac{ik}{r^2} + \frac{k^2}{r} \right) e^{i(kr-\omega t)}. \tag{13}$$

2. Rotating Magnetic Dipole

Let us now consider the more interesting case of a magnetic dipole that is tilted and rotating. A rotating dipole can simply be considered to be the superposition of two orthogonal dipoles that are oscillating 90° out of phase in the plane of rotation, with a third steady component along the rotation axis. The third component can be added by superposition and will be ignored, leaving just the two out-of-phase oscillators. It is most convenient to rotate the dipole about the z-axis, which is inconvenient because in the above treatment we took the z-axis to be the dipole axis. So first let us tilt the static dipole into the x-y plane simply by letting $\sin\theta \to \cos\theta$; $\cos\theta \to -\sin\theta$ followed by a rotation through ϕ in that plane. This leaves the radial dependences unchanged but merely rearranges the angular dependences so that

$$B_r = 2\cos\theta \to 2\sin\theta, \tag{14a}$$

$$B_\theta = \sin\theta \rightarrow -\cos\theta \rightarrow -\cos\theta\cos\phi, \tag{14b}$$

and

$$B_\phi = 0 \rightarrow +\sin\phi, \tag{14c}$$

with

$$E_\phi = \sin\theta \rightarrow -\cos\theta \rightarrow -\cos\theta\cos\phi, \tag{14d}$$

$$E_\theta = 0 \rightarrow -\sin\phi. \tag{14e}$$

The angular coordinate is now set into rotation using $\phi \rightarrow \phi - \omega t$, and the replacements are simply $\cos\phi \rightarrow 1$; $\sin\phi \rightarrow i$, so the full rotating oblique dipole solution reads

$$B_r = 2B_0 \sin\theta \left(\frac{1}{r^3} - \frac{ik}{r^2}\right) e^{i(kr-\omega t+\phi)}, \tag{15a}$$

$$B_\theta = -B_0 \cos\theta \left(\frac{1}{r^3} - \frac{ik}{r^2} - \frac{k^2}{r}\right) e^{i(kr-\omega t+\phi)}, \tag{15b}$$

$$B_\phi = -iB_0 \left(\frac{1}{r^3} - \frac{ik}{r^2} - \frac{k^2}{r}\right) e^{i(kr-\omega t+\phi)}, \tag{15c}$$

$$E_\theta = +icB_0 \left(\frac{ik}{r^2} + \frac{k^2}{r}\right) e^{i(kr-\omega t+\phi)}, \tag{15d}$$

and

$$E_\phi = -cB_0 \cos\theta \left(\frac{ik}{r^2} + \frac{k^2}{r}\right) e^{i(kr-\omega t+\phi)}. \tag{15e}$$

To get the solution for an inclined dipole, all we need to add to this solution is a steady dipole component aligned along the z-axis, so for an inclination angle of ξ from the z-axis, we add a $\cos\xi$ of the static solution to $\sin\xi$ of the rotating solution. However, we must remember the all-important electric field induced in the neutron star by rotation of the conducting sphere through its own magnetic field. Thus we have for just the aligned terms

$$B_r = 2\cos\theta \frac{a^3}{r^3} B_0 \cos\xi, \tag{16a}$$

$$B_\theta = \sin\theta \frac{a^3}{r^3} B_0 \cos\xi, \tag{16b}$$

$$E_r = \omega(1 - 3\cos^2\theta)\frac{a^5}{r^4} B_0 \cos\xi, \tag{16c}$$

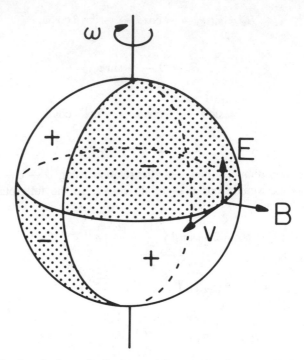

and

$$E_\theta = -\omega \sin\theta \cos\theta \frac{a^5}{r^4} B_0 \cos\xi. \tag{16d}$$

The absence of r^{-2} terms in E_r assumes that the total system charge is zero and that therefore there is a surface charge density canceling the central charge discussed in sec. 4.2.1.a.

3. Rotating Conducting Magnetized Sphere

The finite size of the conducting neutron star also modifies the vacuum near fields of the dipole radiation solution because the azimuthal electric field must vanish at the conducting surface. A canceling term of the form $E_\phi = \cos\theta \sin\phi$ is required which corresponds to the rotating induced quadrupole potential of the form (figure 5.3)

$$\Phi = -\frac{\sin\theta \cos\theta}{r^3}, \tag{17}$$

which gives the rotating electric fields

$$E_r = -6\frac{\sin\theta \cos\theta}{r^4} e^{i(kr-\omega t+\phi)}, \tag{18a}$$

$$E_\theta = 2\frac{\cos^2\theta - \sin^2\theta}{r^4}e^{i(kr - \omega t + \phi)}, \tag{18b}$$

and

$$E_\phi = i\frac{2\cos\theta}{r^4}e^{i(kr - \omega t + \phi)}. \tag{18c}$$

The origin of this term can be seen as follows. Suppose we surround the rotating magnetic dipole with a rotating conducting shell. In the shell, the particles move without being accelerated to either surface, and act as if they are performing an $\mathbf{E} = -\mathbf{V} \times \mathbf{B}$ drift. Because $V_\phi = \omega r \sin\theta$, we require the electric fields $E_r = \sin\theta \cos\theta r^{-2}$ and $E_\theta = 2\sin^2\theta r^{-2}$. These fields arise partly from induction in the shell (remember that Maxwell's equations apply to the *inertial* frame and are not valid in a corotating coordinate system) and partly from charge separation, which leads to the potential

$$\Phi = \frac{\sin\theta \cos\theta}{r} \quad \text{(in shell)}, \tag{19}$$

and at the surface of the shell the radial dependence becomes that of the rotating vacuum quadrupole, equation 17. Now, however, the rotating electric quadrupole itself radiates! The relevant electric and magnetic fields can be written in the form

$$E_r = f(kr)\sin\theta \cos\theta e^{i(-\omega t + \phi)}, \tag{20a}$$

$$E_\theta = g(kr)(\cos^2\theta - \sin^2\theta)e^{i(-\omega t + \phi)}, \tag{20b}$$

$$E_\phi = ig(kr)\cos\theta e^{i(-\omega t + \phi)}, \tag{20c}$$

$$B_r = 0, \tag{20d}$$

$$B_\theta = h(kr)\cos\theta e^{i(-\omega t + \phi)}, \tag{20e}$$

and

$$B_\phi = ih(kr)(\cos^2\theta - \sin^2\theta)e^{i(-\omega t + \phi)}, \tag{20f}$$

where the e^{ikr} factor is incorporated into f, g, and h. The common radial dependence and factors of i in equations 20b and 20c are imposed by $(\nabla \times \mathbf{E})_\phi = 0$ and in equations 20e and 20f by $\nabla \cdot \mathbf{B} = 0$. Then $\nabla \times \mathbf{E} = -\partial\mathbf{B}/\partial t$ gives two conditions,

$$kf = \frac{6h}{r} \tag{21a}$$

and

$$kg = \frac{1}{r}\frac{d(rh)}{dr},\tag{21b}$$

while the θ component of the curl of **E** equation gives

$$\frac{f}{r} - \frac{1}{r}\frac{d(rg)}{dr} = k^2 h.\tag{21c}$$

We know that f and $g \to r^{-4}$, so h starts as $\to r^{-3}$ and must stop at r^{-1}. Equation 21c can be written entirely in terms of h, so direct substitution as before quickly shows that

$$h(x) = \left(\frac{3}{x^3} - \frac{3i}{x^2} - \frac{1}{x}\right) e^{ix},\tag{22}$$

which, aside from a factor of i, is the second-order spherical Bessel function of the third kind, usually denoted h_2. In terms of this function, the fields become

$$E_r = \frac{6h}{r} \sin\theta \cos\theta e^{i(-\omega t + \phi)},\tag{23a}$$

$$E_\theta = \frac{1}{r}\frac{d(rh)}{dr}(\cos^2\theta - \sin^2\theta)e^{i(-\omega t + \phi)},\tag{23b}$$

$$E_\phi = i\frac{1}{r}\frac{d(rh)}{dr}\cos\theta e^{i(-\omega t + \phi)},\tag{23c}$$

$$B_\theta = h\cos\theta e^{i(-\omega t + \phi)},\tag{23d}$$

$$B_\phi = ih(\cos^2\theta - \sin^2\theta)e^{i(-\omega t + \phi)}.$$

These solutions were first provided by Deutsch (1955), who speculated that rapidly rotating magnetized stars might accelerate cosmic rays, an idea which resurfaces regularly. Thus we have essentially rederived these solutions. When properly normalized, the fields (**F**) can be written as

$$\mathbf{F} = \mathbf{F}(\text{aligned}) + \mathbf{F}(\text{dipole}) + \mathbf{F}(\text{quadrupole}),\tag{24}$$

where **F**(aligned) is given from equation 16, **F**(dipole) is given from equation 15 multiplied by

$$\frac{a}{k^2 h_1(ka)},$$

where

$$h_1 \equiv \left(\frac{1}{x^2} - \frac{i}{x} \right) e^{ix} \tag{25}$$

is the first-order spherical Bessel function within a factor of i, and F(quadrupole) is given from equation 23 multiplied by

$$\frac{a}{h(ka) + kah'(ka)}.$$

These normalizations are very nearly but not exactly just the obvious powers of a.

The above derivation "by hand" may seem inelegant. Clearly we could now go back and postulate an arbitrary multipole moment, write the differential equations in terms of the multipole constants, and show that the solutions correspond to second-order spherical Bessel functions of the third kind. On the other hand, we can here see where the individual terms come from, which can be helpful when trying to include such effects as plasma shielding.

5.3 Plasma Distribution about an Oblique Rotator

In this section we discuss the more realistic geometry of the oblique rotator, given that we have no a priori knowledge of the inclination angle (or even of whether pulsar magnetic fields are well described as a dipole!). The attractiveness of the completely oblique rotator is that the polar cap field lines lead directly to the wave zone. However, we will see that the ability of the system to accelerate particles from the surface is seriously diminished.

The status of the mildly oblique vacuum rotator seems also a critical issue. There are two logical possibilities: (1) that a slight misalignment leaves the system with complete trapping of nonneutral plasma just as in the aligned case, and (2) that a minor misalignment excites dynamical behavior. No particular physical argument has yet been advanced to support the second view, and the above results for even a completely oblique rotator do not seem encouraging.

1. Acceleration in a Vacuum Surrounding a Magnetized Rotator

As we see from the discussion leading up to equation 19, the electromagnetic fields inside the conducting star are essentially given by the corotational drift, dominated by $\mathbf{E} = -\mathbf{V} \times \mathbf{B}$. Note in this regard that $\mathbf{V} \neq \mathbf{E} \times \mathbf{B}/B^2$ as is sometimes supposed, because a component of $\mathbf{V}$ parallel to $\mathbf{B}$ is also required. In the steady state solutions, the fields that produce the latter vanish. The consequence, $\mathbf{V} = r\Omega \sin \theta \mathbf{e}_\phi$, is the resultant rigid rotation of the plasma with the star. If, as seems reasonable, nonneutral plasma is pulled from the surface to produce an electrosphere, the same conditions will exist within that electrosphere as well. It should be cautioned, however, that no explicit solutions yet exist for the electrosphere of an oblique rotator. Test particles outside the electrosphere will see a modified set of electric fields and consequently not necessarily drift with a rigid corotational rate.

In the corotating frame inside a rotator (i.e., moving with the $\mathbf{E} \times \mathbf{B}$ drift velocity), the particles see only a static magnetic field. Therefore the potential distribution is everywhere constant and can be set equal to zero. (As noted above, a parallel velocity component is typically required, which is not electromagnetically dictated in the steady state but arises, for example, from viscous drag between the charged particles and the neutral component of the rotator.) But we know that a quadrupolar charge separation is required to produce the corotational electric field. For all practical purposes the charge density is unchanged by transformation into a moving frame. So what has happened to the electric field produced by these charges? In other words, in the corotating frame we see a static magnetic field together with a nonzero space charge, but nevertheless the electric field is supposed to be zero! The resolution of this paradox is subtle and was given by Schiff (1939). Schiff notes a similar paradox posed in a question from Oppenheimer, who pointed out that two concentric spheres of equal and opposite charge have zero external electric and magnetic fields, but if one rotates these spheres together a nonzero external magnetic field will appear (the outer sphere creates a larger magnetic moment). If, however, one orbits the stationary spheres (to this observer the spheres again appear to rotate), no such magnetic field will be seen. So rotating the system gives different physical results than rotating the observer. Schiff shows that if one casts Maxwell's equations into a rotating frame, two additional terms appear, an apparent charge density and an apparent current density, both of which are independent of any actual charges moving about. Once one realizes that these terms exist in the noninertial corotating frame (in addition to the other obvious effects such as centrifugal force), the paradox vanishes. (On the nonuniqueness of "moving" magnetic field lines, see Vasyliunas 1972).

The analysis of more interesting problems is facilitated simply by knowing that the above system of "patching up" exists. We can now examine the question of how particles move in the vacuum external field of an oblique rotator. One could simply integrate (numerically) the particle trajectory in the time-varying Deutsch potentials seen in the inertial frame. But one would like a more intuitive approach. For one thing, the particles tend to move along magnetic field lines (as seen in the corotating frame), so one has already "integrated" one of the equations of motion simply with that observation. But in the vacuum case the particles are also accelerated along the magnetic field lines. Indeed, they tend to be pulled from the surface itself as is assumed in the standard model. In the oblique case the same thing happens, but now a question arises: can particles be accelerated off the surface at the foot of one field line and into the surface at the conjugate point? One's immediate intuition (correct) is that giving the particle a net energy gain in going from one point on the surface to the other must violate one of the conservation laws because the system would look the same after having accelerated the test particle. But tracking down how this arises can be confusing because in the inertial frame one has both induction and quasi-static electric fields to contend with, while in the corotating frame one has these "extra" Schiff terms. However, the argument is quite simple. We imagine that the space is not a vacuum but is filled with whatever space charge is required to give corotation. Specifically we require $\mathbf{E} = -\mathbf{V} \times \mathbf{B}$ with $\mathbf{V} = r\Omega \sin \theta \mathbf{e}_\phi$, and the charge density is simply the divergence of this electric field. For a rotating aligned dipole magnetic field the electric fields correspond to quadrupolar fields. For a 90° oblique rotator, the potential required is of the form $\sin \theta \cos \theta / r^2$ whereas the vacuum quadrupole has the same angular dependence but $1/r^3$ radial dependence. At this point we need to note an important point: *The electrostatic potential in the corotating space charge is zero.* Thus the correct potential in the vacuum system is a vacuum quadrupole minus the corresponding pseudoquadrupole that would have been generated by the corotating plasma (all the remaining induction terms are independent of the existence of this plasma). In other words, one sees a potential

$$\Phi \approx \sin \theta \cos \theta \left(\frac{1}{r^3} \quad \frac{1}{r} \right). \tag{1}$$

By definition this potential is zero at the surface ($r = 1$) so that particles from the surface are indeed accelerated to form an electrosphere if none were there to start with, but lose all their energy once they approach the surface again, regardless of whether they exactly follow a given magnetic field line.

The bottom line is that Maxwell's equations apply to an inertial frame, and the useful concept of "moving" field lines is, nevertheless, gratuitous. Maxwell's $\mathbf{E}$ and $\mathbf{B}$ fields refer to a fixed inertial coordinate frame.

Another way of obtaining the above result is simply to calculate $\mathbf{E} \cdot \mathbf{B}$ for the dipole radiation fields given from equation 5.2.15. A direct calculation reveals that all terms cancel except the $1/r^3$ magnetic dipole terms and the $1/r^2$ induction electric field terms, which read (after taking real parts)

$$B_r = 2 \sin \theta \frac{1}{r^3} \cos (kr - \omega t + \phi), \tag{2a}$$

$$B_\theta = -\cos \theta \frac{1}{r^3} \cos (kr - \omega t + \phi), \tag{2b}$$

$$B_\phi = \frac{1}{r^3} \sin (kr - \omega t + \phi), \tag{2c}$$

$$E_\theta = \frac{1}{r^2} \cos (kr - \omega t + \phi), \tag{2d}$$

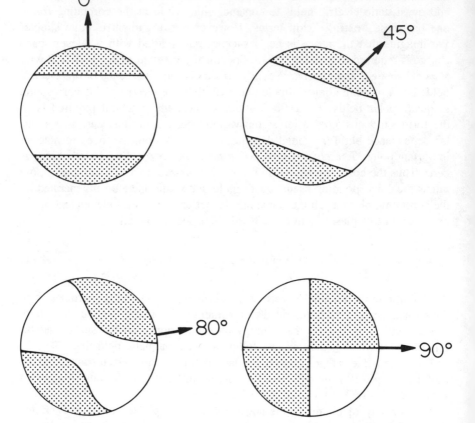

FIG. 5.4. The surface-charge distribution versus inclination, as viewed from side, for rotators of varying obliquity.

and

$$E_\phi = \cos\theta \frac{1}{r^2} \sin(kr - \omega t + \phi), \tag{2e}$$

which give

$$\mathbf{E} \cdot \mathbf{B} = \frac{\cos\theta}{r^5} \text{ (dipole only)}, \tag{3}$$

which gives the parallel electric field seen by a test particle outside any electrosphere. This accelerating field is analogous to the same field for the aligned case, which acts to create the electrosphere. Thus in figure 5.3 the negative surface charge is pulled from the (upper) surface to create an attached negative "cloud" while a positive cloud is pulled from the other (upper) surface. These two clouds then repel one another to stabilize the configuration. In fact it is this repulsion that is the induced quadrupole moment on the star, which in turn modifies equation 3 to

$$\mathbf{E} \cdot \mathbf{B} = \cos\theta \left(\frac{1}{r^5} - \frac{1}{r^3} \right). \tag{4}$$

Unlike the aligned rotator components, *there is no direct force at the surface of the star to accelerate particles*. Thus the particle energy increases not as $\Phi_0(h/a)$ but as $\Phi_0(h/a)^2$, where h is the height above the surface. For a typical pulsar with $\Phi_0 \approx 10^{12}$ V, electrons would not become relativistic until $h \approx 10$ cm, which means that there would be a significant gravitational potential barrier to surmount. These trapping regions above the surface should still eventually fill with particles, particularly those thermionically emitted from the surface.

It would be interesting to calculate the stability of this system, because it is possible that particles could be accelerated out along the polar magnetic field lines, which here are in the equatorial plane. Furthermore, neutralizing particles would come from the same polar cap. The weakness of the accelerating potential at the surface seems, however, a discouraging aspect.

Insofar as the trapping of an electrosphere is concerned, the above forces simply act to tilt the dome and torus distributions. As the tilting progresses, the two domes presumably become two of the charge clouds that characterize the completely oblique case, and the torus splits to form the other two. This fully three-dimensional problem remains to be simulated. It is hard to illustrate these cloud structures anyway, but figure 5.4 shows the surface boundaries between different signs of surface charge, which is just

$$(3\cos^2\theta - 1)\cos\xi + 3\sin\theta\cos\theta\cos\phi\sin\xi = 0. \tag{5}$$

Note that even for inclinations of 80°, the entire polar cap will be shielded with plasma of one sign, which has turned out to be a basic problem with the standard model.

6

The Disk Model

6.1 Introduction

Some years back, my long-time colleague Alex Dessler was working on *Voyager* spacecraft data from Jupiter which revealed that there was an entire torus of plasma about Jupiter in the orbit of the inner Galilean satellite Io which seemed to be stimulating magnetospheric activity. Previously, attention had been centered on interaction with Io itself (Goldreich and Lynden-Bell 1969) as the sole agent of activity. Dessler wondered out loud if something parallel might be taking place with pulsars, namely if there might be something else in the pulsar system besides just a rotating magnetized neutron star. At this time I was struggling with the problem of neutron stars cloaking themselves with nonneutral plasma and not being active, but I resisted Dessler's suggestion, arguing that workers in the field would simply not consider a more complicated model. After a year's sabbatical in France thanks to the Guggenheim foundation and the stimulation of working with René Pellat, however, I became quite enthusiastic. Dessler's "something else" could hardly be anything but a disk of matter; its presence would not be directly revealed in pulsar timing data; if there at all, it could be an arguably ubiquitous phenomenon (necessary to avoid even more problems with the pulsar birthrates); and having a conducting disk in Kepler orbit would almost certainly produce electrodynamic activity, given that there would now be large mismatches in electrostatic potential between the disk and the star. The question was whether such a system might look like a pulsar, and in my enthusiasm, it occurred to me that the community would collectively kick itself for failing even to consider such an obvious possibility. My original assessment turned out to be the correct one.

Given that it has proven so difficult to theoretically model activity from an isolated rotating magnetized neutron star, one might naturally begin to wonder if indeed such a model is sufficient or even correct. Trimble (1983), for example, has pointed out that numerous "curious stars" have turned out to be binaries or more complex objects and not single objects. Yet for some curious sociological reason the aligned rotator continues to attract devoted theoretical effort, despite the serious self-consistency problems and the fact that this model

302

has been shown to admit of stationary solutions of no interest insofar as pulsar activity is concerned. The same solutions resolve the previous self-consistency problems. This insistence on pursuing a dubious model is doubly curious because it is clear that something must break the axisymmetry to give pulses. The obvious place to which to devote effort then would be the oblique model, which is certainly richer in possibilities but has hardly been analyzed to the degree that one could confidently exclude it. Still, there are difficulties with this model too. The trapping of nonneutral plasma is a general effect, not confined to aligned geometries. If one takes an aligned rotator shielded with nonneutral plasma and makes it slightly oblique, the plasma on open field lines is still bound to the system by the net system charge, while the plasma on closed field lines is trapped as always, whether quasi-neutral or nonneutral. Only at inclinations approaching 90° can this trapping geometry break down. However, a requirement that all pulsars be nearly orthogonal rotators would in itself introduce a number of observational and theoretical problems. Observationally, it would be hard to account for pulsars with large duty cycles and one would expect the polarization swings of pulsars with interpulses almost always to have the same sign as the main pulse (there does seem a bias in this direction). Theoretically, one lacks a mechanism to produce such orthogonal misalignment. There is not a scrap of evidence that orthogonal magnetization is the natural result of any dynamo process, and if neutrons stars are formed with a variety of magnetic inclinations, most would then be inactive unless something caused them to become orthogonal. (Our understanding of dynamo processes is, unfortunately, minimal.) In the presence of pulsar activity, with currents flowing to and from the neutron star, misalignment torques are at least arguable, but in the absence of such torques (i.e., before the system evolved to become a pulsar), the only torque known to come into play is that from the magnetic dipole radiation, which is also known to cause alignment, not misalignment (except in very special cases such as a triaxial neutron star which do not lead to orthogonal misalignment anyway). So one is in a logical trap unless oblique rotators function at almost all inclination angles. Although the trapping argument seems to cast doubt on that possibility, there is still the possibility that the additional terms in the Deutsch fields might drive particles from the system. It would be very valuable to explicitly check out this possibility. Informal estimates, however, are not too encouraging. For one thing, it is difficult to drive particles near the star directly across any static aligned component of an inclined dipole magnetic field, which is larger than either the induced or radiation magnetic field components by factors of $c/a\Omega$ and $(c/a\Omega)^2$, respectively (about 5000 and 25,000,000 for a 1 s pulsar), as deduced in sec. 5.2. The problem is that, in the usual paradigm, the particles must start at the neutron star surface, where the fields are well described by a quasi-static magnetic field in the corotating frame. But a careful analysis remains to be done.

Given this situation, it seems prudent to examine other possibilities, such as the possibility that the system is more complicated than thought. This idea is an old one, harking back to ideas of creating the pulsed emission through interaction of the pulsar with the surrounding medium, for example. Clearly at some distance there is an interaction with the interstellar medium; otherwise pulsars would have swept the galaxy clear of ionized gas and one wouldn't even have a dispersion measure to deal with. This complexity (extraneous material at a distance) is a given. Which leads one to question how empty the near environment of a pulsar might be. The casual expectation is that it should be a very hard vacuum at least out to the light-cylinder and probably far beyond. After all, a supernova is such an energetic event that everything must have been blown away when the neutron star was formed. If literally true, this would argue against the existence of the neutron star. On closer examination, it becomes quite ambiguous what happens in the near vicinity of a forming neutron star and what evolutionary processes follow the formation event. If we simply postulate that some matter is left behind, there is little doubt about what form it must take: an orbiting Kepler disk. Disks are, of course, commonplace elsewhere in astrophysics. The binary pulsating X-ray sources are presumably just such systems, a neutron star surrounded by a disk. The major difference is that the latter systems also include a companion star to maintain the disk while the neutron star in turn accretes the disk. Indeed, this model is so standard that one gets the impression that the natural relationship between a neutron star and a disk is for the neutron star to be in the process of actively consuming the disk. But it is far from clear that an *isolated* neutron star/disk system will end up with the disk consumed. In fact, the interaction of the neutron star with disk material beyond the corotation distance will cause the disk to be driven away from the neutron star. And the existence of two conducting elements in a small system linked by magnetic field lines is a natural dynamic system carrying currents. Indeed, one problem with such a system is that it might be *too* dynamic!

It seems worthwhile to analyze the physics of neutron star/disk systems on general principles, not just to model pulsars, because they are standard models for the binary pulsating X-ray sources (ch. 10).

6.2 The Electrodynamics

It has been argued (Michel and Dessler 1981a) that the radio pulsars and the X-ray pulsars differ mainly in that the latter are surrounded by an accretion disk while the former are surrounded by a disk of matter of some different origin (see also Pineault 1986). One specific possibility is that the matter is left over from the formation event. In such a model energy extracted from the rotation of the neutron star by interaction with the disk produces pulsar luminosity. An alternative mechanism for forming a disk would be angular momentum conservation in the collapse phase of the neutron star, some matter being shed to carry off excess angular momentum. Another mechanism would be trapping of material falling back from the supernova rather than formation in place. Yet another would be matter left over from disintegration of a companion, as in the models for type I supernovae. Parallels between the radio and X-ray pulsars were long ago suggested by Tucker (1969).

As we have seen, in the former "standard" model for pulsars, the rotationally induced electric field of a rotating, magnetized neutron star pulls plasma off the surface and ejects it beyond the light-cylinder to form a relativistic stellar wind. Pulsar emission is then associated with the acceleration of new particles to maintain this wind, assuming that rotation self-excites the pulsar emission. However, there is serious doubt that the standard model even functions this way, as discussed in chapter 4. The charge-separated magnetosphere (*electrosphere*) may have finite extent and not reach out to the light-cylinder at all. If it doesn't, the standard model does not obtain for an aligned rotator. If the oblique rotator shuts down in a similar manner (which has yet to be firmly demonstrated), even the Ostriker and Gunn (1969a,d) picture of particle acceleration by giant dipole waves fails, since there are no particles in the wave zone to be accelerated. The standard model would then predict no more than the emission of unobservably low-frequency waves, as was originally supposed (Pacini 1967).

Astrophysical radio sources, solar system particle acceleration, and planetary radio sources seem all to be associated with local relative motions of plasma that drive current loops and excite electromagnetic radiation. (In contrast, the former standard model sought to do without such relative motion.) In solar system magnetospheric physics, the idea that particle acceleration involves plasmas in relative motion is an old one (e.g., Dessler and Juday 1965). Also, the thought that Jupiter may utilize some physical mechanisms relevant to pulsar emissions has occurred to several workers in the field and has been pursued with varying degrees of enthusiasm (e.g., Burbidge and Strittmatter 1968; Douglas-Hamilton 1968; Dowden 1968; Hill and Michel 1975; Kennel and Coroniti 1975; Braude and Bruk 1980; Michel 1979a). The early attempts to understand the decametric radio emissions from Jupiter were largely focused on the innermost Galilean satellite Io owing to the correlations discovered by Bigg (1964). It was not realized until after the *Voyager* flyby of Jupiter in 1979 that the physics of the Jovian magnetosphere is dominated by the presence of a torus of plasma surrounding Jupiter at the orbital distance of Io, as discussed in Vasyliunas and Dessler (1981) and references contained therein. It now seems that Io acts only as a perturbation on a system that is active regardless of the position of Io. Note that Saturn has a famous and well-defined disk, and is an aligned rotator in that its magnetic axis is almost exactly aligned with its spin axis (Acuna et al. 1980; Smith et al. 1980). Yet Saturn has nevertheless been shown to exhibit spin-periodic pulsations of its radio emissions similar to those exhibited by Jupiter (Kaiser et al. 1980; Warwick et al. 1981). Although there are significant differences between the Jupiter and Saturn systems, the phenomenon seems a general one. It is perhaps ironic that an early pulsar model was indeed an Io analog (Burbidge and Strittmatter 1968) but failed because the only object that could tolerate the tidal forces of a neutron star would be another neutron star (Douglas-Hamilton 1968). The effect of such tides would be to spread the matter into a disk, in itself a possible object for energizing the system. But everyone at the time thought that a discrete satellite was needed, presumably the reason the idea was dropped.

A disk model does not necessarily require a new set of emission mechanisms;

it merely needs to rescale the parameters somewhat and fix the geometry more narrowly. No specific radiation mechanism need be endorsed. The possibility is that, even if the spin, magnetic, and disk axes are all parallel, pulsar action might obtain, in parallel with the pulsed emissions from Jupiter and Saturn.

1. The Magnetic Field

In any disk model, the interesting range of distance scales is between the neutron star radius and the light-cylinder distance, which has a geometric average of about 7×10^7 cm for typical pulsars. Even a relatively trivial amount of matter (we will use a value of $10^{-5} M_\odot$ as a benchmark) uniformly filling a sphere of such a size would have a mean density of almost 10^5 g/cc. It is therefore clear that any significant amount of matter will be degenerate and in the form of a disk, and therefore a conducting disk. The important service that a disk could provide is to duct electrical current out from near the neutron star (where the currents are constrained to flow along field lines) to relatively large distances where the magnetic field has declined to the point that currents can flow readily across the field lines.

a. *Simplest Model*

In essence, one would have two coupled unipolar generators: the neutron star and the disk, as shown in figure 6.1, the disk acting here as a load. As we have seen, corotation of plasma with the stellar surface gives a potential that, if projected along dipolar magnetic field lines to the equator, would give

$$\Phi_{NS} = \Omega B_0 a^3 / 2r + \Phi_{0S}, \tag{1}$$

where Φ_{0S} is the neutron star potential relative to zero at infinity. The disk potential itself, assuming Keplerian rotation

$$\Omega^2 = GM/r^3, \tag{2}$$

is

$$\Phi_D = (GM)^{1/2} B_0 a^3 / 5r^{5/2} + \Phi_{0D}, \tag{3}$$

where Φ_{0D} is the disk potential relative to zero at infinity. It is clear that there can be only one magnetic field line along which the two potentials agree. If sufficient plasma exists in such a system to couple the neutron star and the disk electrically, closed currents must flow. With a disk, the natural scale length is no longer the light-cylinder distance but is now at or near the *corotation distance* at which a particle in Keplerian orbit rotates with the same period as the star:

$$R_c \equiv \left(\frac{GM}{\Omega^2} \right)^{1/3}. \tag{4}$$

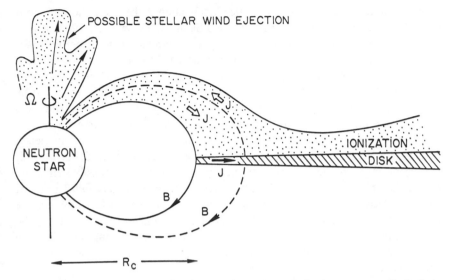

POSSIBLE STELLAR WIND EJECTION

Ω

NEUTRON
STAR

IONIZATION
DISK

B

B

J

J

R_c

FIG. 6.1. Sketch of disk/neutron star interaction. A current loop couples the two systems dissipatively, thereby heating the disk, to maintain electrical coupling, and the polar caps, to provide a stellar wind (at least in the case of more energetic pulsars such as the Crab). A Pedersen current shown in the disk exerts a torque on the disk and star, while an axial component into the disk segment shown resists centrifugal forces. The arc drawn between the disk and the star is presumed to be the source of the coherent radiation. Such a simple model would also be a source of considerable thermal radiation, which would rule it out observationally. From F. C. Michel and A. J. Dessler, 1981a, *Ap. J.*, 251, 654 (figure 1); F. C. Michel, 1982, *Rev. Mod. Phys.*, 54, 1 (figure 22a);

No electromagnetic forces are required to enforce corotation at this distance. If the standard assumption that the pulsation period is locked to that of the neutron star is adopted, the interaction with the disk is assumed to take place just outside the corotation distance. Combining equation 4 and the definition of the light-cylinder distance gives the dimensionless parameter

$$\zeta \equiv \frac{R_L}{R_c}. \tag{5}$$

Standard model parameters can be rescaled to give disk parameters. For the pulsar in the Crab Nebula (PSR 0531 + 21), $\zeta = 9$; for a 1 s pulsar, $\zeta = 30$; and for one of the slower pulsars known (PSR 0525 + 21), $\zeta = 45$. The most economical assumption is that a disk would extend inward to the corotation distance, although it is possible that the disk could be driven further away by the magnetic torques. The interaction of the more slowly orbiting material in Kepler orbit with the rotating neutron star will cause the neutron star to slow down. But a disk cannot spin up in return; rather, the torques act to push it to larger distances where it orbits even more slowly. Suppose that the interaction region on the disk is a localized region (spot) projected along field lines from the star to the disk, this spot rotating (on the average) with the neutron star regardless of the local disk velocity. Such localization of the interaction region is observed in radio emission from Jupiter and Saturn (Dessler et al. 1981; Kaiser et al. 1980). Such spots, or, more generally, longitudinal inhomogeneities, have a theoretical justification in terms of magnetic field variations on the neutron star surface, which maps outward into the magnetosphere (Dessler 1980a,b; Dessler et al. 1981; Hill et al. 1981), and the degree of localization could be nonlinearly enhanced (e.g., concentrated locally, similar to the sparks suggested by Ruderman and Sutherland 1975). Pulsed emissions are produced even by such a highly symmetric rotator as Saturn (Kaiser et al. 1980), so the phenomenon exists independent of whether a generally accepted theoretical model is at hand. A disk not only modifies several parameters important to the possible emission mechanisms but also may introduce a richer set of physical possibilities: (a) if plasma is derived from the disk as well as from the pulsar surface, the current flow can be neutral; (b) moving the characteristic distance inward from the light-cylinder rescales many of the pulsar parameters, decreasing the minimum required magnetic field strength, for example; and (c) the disk itself becomes a potential site for particle radiation.

One can estimate the maximum torque on the disk; the maximum shear exerted by the magnetic field cannot be greater than about $B^2/2\mu_0$. Integrating this maximum torque over the entire disk (from about R_c to much larger distances) and multiplying by the rotation rate then gives the maximum power output from the neutron star,

$$L \approx \frac{\pi B^2 a^6 \Omega}{3\mu_0 R_c^3}.$$ (6)

This luminosity scales as $L \approx \Omega^3$ instead of $L \approx \Omega^4$ as implied by equation 2.3.13. The disk torque would be comparable to the wind torque, given the same B, for $R_L = R_c$ (rotation at breakup).

b. *Modified Model*

Although the model described above is simple and dynamic, the bulk of the energy would be transferred from the pulsar to the disk and it would be quite difficult not to heat the disk in the process. Given that the slowing-down luminosities are much larger than the radio luminosities (order $10^5 \times$) and correspond to power outputs of order 10^{30} ergs/s or greater, it would be difficult not to have substantial luminosities from such disks in the X-ray or optical. Seward and Wang (1988) give many nondetections of pulsars at X-ray luminosities 10^{-3} below the spin-down luminosity. One way out of this difficulty is to suppose that the magnetic field lines are largely excluded from the disk, as shown in figure 6.2.

Roberts and Sturrock (1973) proposed that matter from the neutron star could accumulate at the corotation (or *force balance*) distance. They then assumed that the magnetic field was dipolar out to this distance and radial (monopolar)

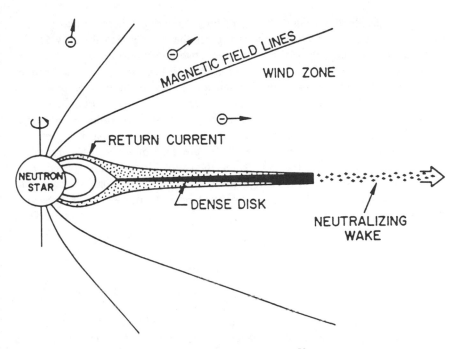

FIG. 6.2. Modified interaction if magnetic field lines are ejected. If finite conductivity is insufficient to limit the current flow, $\mathbf{J} \times \mathbf{B}$ forces should eject the magnetic field lines. The steady state solution might then be as shown, where only a restricted magnetic flux crossed the disk. Because $\Delta \Phi = \Omega \Delta f$, the flux reduction also reduces the available electromotive force. This would return us to the geometry of figure 2.3, with the disk acting the role of a *neutral sheet* in the equatorial plane. The field structure is identical to that proposed by Roberts and Sturrock (1973), which gives a deceleration parameter $n = 7/3$. From F. C. Michel, 1983b, *Ap. J.*, 266, 188 (figure 1); 1982, *Rev. Mod. Phys.*, 54, 1 (figure 22b).

beyond. That the field became monopolar was attributed to the eventual insta-bility of plasma collected near the corotation distance and then ejected, carrying the field lines along with it. They did not propose that a persistent physical disk would maintain such an open configuration against magnetic reconnection. A transition at the corotation distance from dipolar to monopolar field lines has two effects: (1) it gives a value $n = 7/3$ for the braking index, and (2) it re-duces the surface magnetic field required to account for the magnitude of the slowing-down rate. Oddly, the first point was made in Roberts and Sturrock (1973), but not the second. The magnetic field at the light-cylinder determines the slowing-down rate if the predominant torque in the system is the Poynting flux. For a vacuum rotator, the surface field is reduced by a factor a/R_L at the light-cylinder. For the disk case, the reduction is instead a factor of a/R_c to the inner edge of the disk, followed by a factor of R_c/R_L for the monopo-lar portion out to the light-cylinder, a net factor of $a^3/(R_c R_L^2)$. The surface magnetic field required for a given slowing-down rate is reduced by the factor ς. The total energy output is essentially proportional to $B_L^2 R_L^2$ (the magnetic energy density flowing at the velocity of light through a sphere a light-cylinder radius in diameter), which immediately gives the usual proportionality Ω^4 for the vacuum case and $\Omega^{10/3}$ for the disk case (two powers of R_L are replaced by R_c in the latter case). Thus, for rotational energy loss, n is 3 for the vacuum case and 7/3 for the disk case. Inconsistencies in a space-charge separated wind from a pulsar (Michel 1975a) could be circumvented if the field lines were separated by a neutral sheet (Michel 1975b), with the remaining wind structure being uniquely determined. A material disk would serve such a purpose. The topology of magnetic field lines in the presence of a disk is a first-order uncertainty not only here but also in the theory of binary X-ray sources (ch. 10). Kundt and Robnik (1980) have given analytic expressions for two- and three-dimensional magnetic dipoles confined by a diamagnetic disk (see also Aly 1980 and Lipunov 1978).

2. Disk Electrodynamics

a. *Gap Solutions*

It would be ideal to have self-consistent nonneutral plasma solutions for the disk/rotator system, but these remain to be calculated even for the axisymmetric system. A simple calculation can be made (Michel 1979a) for the space charge of a dipole magnetic field with variable Ω (i.e., Kepler shear) on the flux lines f. Because Kepler shear gives $\Omega \approx r^{-3/2}$ while $f \approx r^{-1}$, we have $\Omega \approx f^{3/2}$. Assuming plasma everywhere and therefore equipotential magnetic field lines, the charge density is

$$
\begin{aligned}
q &= -\nabla^2 \Phi(f) \\
&= -\Phi_f \nabla^2 f - \Phi_{ff} \nabla f \cdot \nabla f \\
&= \frac{\Omega}{r^3} \left(2(1 - 3\cos^2 \theta) - \frac{f\Omega_f}{\Omega}(1 + 3\cos^2 \theta) \right).
\end{aligned}
\tag{7}
$$

Thus $f\Omega_f/\Omega = 3/2$, and a disk threaded by dipole field lines would have the same sign of charge surrounding it as would be found in the equatorial plane for the vacuum solutions.

b. *Energetics*

Even given only a rudimentary physical analysis of how currents might flow and what the field structure might be, one can still make some estimates. The power delivered to the disk by magnetic torques turns out to be a small fraction of the spin-down energy output. Disks would not then be readily detectable owing, say, to thermal X-ray radiation. For simplicity, we will assume axial symmetry, although presumably pulsed emission arises from current flow to preferential regions of the neutron star (e.g., regions controlled by the surface magnetic field structure: Michel and Dessler 1981a).

An axisymmetric system might oscillate in time, thereby introducing secondary periodicities (e.g., marching subpulses), but the high stability of the average period suggests that the pulse window must be physically tied to the neutron star. The polar cap area can be estimated assuming that the field line to the inner edge of the disk is approximately dipolar; hence the polar cap area is

$$A_{PC} = C_1 \pi a^3 / R_c = C_1 \pi a^2 \sin^2 \theta, \tag{8}$$

where C_1 is a correction for the distortion of the magnetic field (we will assume $C_1 = f_{crit} \approx 1.6$: eq. 4.3.31) and θ is the colatitude of the auroral zone bounding the polar cap. The space-charge density just above the polar cap is estimated as usual,

$$q_{PC} = 2\epsilon_0 \Omega_0 B_0. \tag{9}$$

The total current into (or out of) the polar caps therefore has the magnitude, counting both polar caps as before,

$$I_0 = \frac{4C_1 \pi a^3 \epsilon_0 \Omega_0 B_0 c}{R_c}, \tag{10}$$

and in addition the magnetic flux into the wind zone from each pole is just

$$2\pi f_0 = B_0 A_{PC} = \frac{I_0}{4\epsilon_0 \Omega_0 c}. \tag{11}$$

In our notation, $B_z = (\partial f / \partial \rho)/\rho$, etc., in which case f, which labels the field lines, has units of flux per radian of longitude and hence is 2π less than the actual physical flux. From equation 11, we estimate the stellar wind torque on the star:

$$T_W = \frac{4\pi \Omega_0 f_0^2}{\mu_0} = \frac{I_0^2}{16\pi\epsilon_0 c \Omega_0}, \tag{12}$$

which is presumed to be the dominant energy-loss term for the rotating neutron star. The current out of the polar caps would return through the disk (figure 6.2), therefore requiring a current density

$$J = \frac{I_0}{4\pi R H}, \tag{13}$$

where R is the radial distance and

$$H = H_c \left(\frac{R}{R_c} \right)^{3/2} \tag{14}$$

is the disk height above the midplane (i.e., the half-thickness), with H_c this height at the corotation distance. In sec. 6.3.3 we will estimate H_c to be about

2×10^6 cm. The electric field is therefore

$$E = \frac{J}{\sigma}, \tag{15}$$

and the total potential drop is

$$\Delta \Phi_0 = \int_{R_c}^{\infty} E \, dR = \frac{I_0}{6\pi\sigma H_c}, \tag{16}$$

where the upper limit is placed at infinity because E falls so rapidly with distance (assuming that the conductivity σ is more or less constant). These relations ignore components of the return current that leave the disk at $R > R_c$ and flow to the neutron star along field lines nearly paralleling the disk rather than through the disk itself. Thus the potential drop is overestimated. The power dissipated in the disk is then

$$L_D = I_0 \Delta \Phi_0 = \frac{I_0^2}{6\pi\sigma H_c}. \tag{17}$$

Comparing equation 17 with equation 12, we see that the fraction of energy going into the disk is simply

$$\frac{L_D}{L_W} = \frac{2\epsilon_0 c}{3\sigma H_c}. \tag{18}$$

Assuming a conductivity of the order of that for normal metals, $\sigma \approx 10^8$ mho/m (Flowers and Itoh 1976 give substantially *higher* conductivities), and a nominal disk thickness of, say, a kilometer, only about 10^{-13} of the total power would be dissipated in the disk. For the Crab pulsar at a total power output of 10^{38} ergs/s, that would imply only 10^{25} ergs/s from the disk. The quantity Δf_0 gives the width, d, of the auroral zone for an aligned dipole field, because that flux is also given by

$$2\pi\Delta f_0 = B_0 2\pi R_{PC} d, \tag{19}$$

and hence the return current density in these regions is just $f_0/\Delta f_0$ times greater than the current density over the polar caps, or

$$J_{az} = 2\epsilon_0 \Omega_0 B_0 c \left(\frac{f_0}{\Delta f_0} \right). \tag{20}$$

The implied values for d are too small to be physically plausible. If we used this value of Δf_0, we would find

$$\frac{f_0}{\Delta f_0} = \frac{4\epsilon_0 c}{3\sigma H_c}, \tag{21}$$

or again about 10^{-13}. The standard space-charge density already corresponds to about 10^4 A/cm^2, and boosting it to 10^{17} A/cm^2 gives a current density that would almost certainly be highly resistive if there were any counterstreaming or wave-particle scattering. As noted above, such large currents are physically impossible unless they consist of neutralized flow, so counterstreaming would be *required*.

3. The Auroral Zone

Here we examine the physics of the zone wherein the current out of the polar caps and back through the disk finally returns to the neutron star. Given that the disk should be highly conducting, only a narrow range of magnetic flux would provide all the voltage drop necessary, assuming that the field lines are approximately equipotential near the star. That assumption, as we have seen above, would require that the auroral current sheet be extraordinarily thin at the stellar surface ($\approx 10^{-6}$ cm) and carry huge current densities ($\approx 10^{17}$ A/cm^2, or about 1 ampere per square Ångstrom!). Collisional resistivity alone should significantly modify the behavior above the auroral zone. Consequently a large resistive drop should appear over the auroral zones, in which case it is not the neutron star but rather the disk that determines the potentials on those field lines. The auroral zones must therefore expand to widths of the order of centimeters as more magnetic flux links the disk to the star (regardless of how conducting the disk might be), and a significant amount of energy must be dissipated in the auroral zones themselves (comparable to the observed pulsar radio luminosities, as we will see). The total slowing-down power, which presumably drives a relativistic stellar wind, would be considerably larger.

a. *Electron Energies*

With a discharge between two conductors, space-charge limitations are not the governing factor for the current flow and particle energies. Let us parameterize the scattering properties of the intense return current with a fixed scattering cross section σ_A. For the sake of illustration, take $\sigma_A \approx 10^{-16}$ cm^2 (a more or less minimal scattering of electrons on, say, partially stripped iron atoms). Wave-particle interactions could generate quite different effective cross sections. It turns out that, *regardless of the value chosen for the cross section*, the electron energies are automatically regulated to be of the order of the electron rest-mass energy. This result is as follows: the electron conductivity is given by

$$\sigma = \frac{\epsilon_0 \omega_p^2}{\nu}, \tag{22}$$

where ν is the collision frequency; hence $\nu \approx c/\lambda$, where we take the electron velocity to be essentially c, and λ is the mean free path. However, the electric field in this region is given by

$$E = \frac{J}{\sigma}, \tag{23}$$

and the energy gained between collisions is

$$\epsilon_{\text{ave}} = eE\lambda, \tag{24}$$

giving altogether $(\omega_p^2 = e^2 n / \epsilon_0 m)$

$$\epsilon_{\text{ave}} = mc^2. \tag{25}$$

In other words, the energy is independent of λ provided $\lambda \ll$ the system scale, and $\Delta\Phi \gg mc^2$. These conditions are not necessarily obtained in the laboratory because long mean free paths imply high conductivities, hence low electric fields, and vice versa. This argument has nothing to do with pair production, despite mc^2 appearing as the characteristic energy. Pair production presumably would take place and degrade a current of energetic electrons into a current of oppositely moving electrons and positrons. Let us simply assume that surface ions (Fe, say, or He if available: Michel 1975d) are liberated by bombardment and form a neutralizing component from the neutron star surface (see Cheng and Ruderman 1980) or from the disk, depending on the sense of current flow. It seems surprising that the electron energy might be so restricted, but as discussed in sec. 8.2.6, relatively small Lorentz factors may be required if circular polarization arises from curvature radiation.

b. *Auroral Zone Scales*

The width of the auroral zone in such a model seems insensitive to the pulsar parameters, being about 5 cm for our assumed parameters. Consider the controlling conductivity to be on the *Birkeland current* in the auroral zone (Michel and Dessler 1981a) owing to the constriction of the field lines (increasing the current density) near the star and consequent high collision rates. The current density in the auroral zone is given by

$$J = \frac{I_0}{2\pi a d \sin\theta}, \tag{26}$$

where d is the width of the current sheet, taken to encircle at least partially the magnetic pole. The effective conductivity is then

$$\sigma = \frac{e^2}{mc\sigma_A}, \tag{27}$$

and the potential drop is

$$\Delta\Phi = \frac{J}{\sigma}\frac{a}{2}, \tag{28}$$

where the (unimportant) factor of 1/2 comes from integrating along field lines assuming an r^{-3} decline in field strength. The potential drop is also given from

$$\Delta\Phi_{\text{az}} = \Omega_0 \Delta f_{\text{az}}, \tag{29}$$

where

$$\Delta f_{az} = B_0 a d \sin \theta \cos \theta \tag{30}$$

(again, a factor of 2π converts Δf into the total magnetic flux trapped between field lines f and $f + \Delta f$). Equation 29 gives the potential across the auroral zone assuming that the disk acts essentially as a shunt, ignoring the tiny potential drop across the disk (eq. 16). The $\cos \theta$ factor in equation 30 can also be set equal to unity, in which case these two relationships for $\Delta \Phi_0$ can be solved for d, giving

$$d^2 = C_1 mc^2 \epsilon_0 a \sigma_A / 2e^2. \tag{31}$$

Except for the geometrical coefficients, this length is the cross section times the neutron star radius divided by the classical electron radius. For nominal values of C_1, a, and σ_A,

$$d = 4.74 \, \text{cm.} \tag{32}$$

The auroral power zone dissipation is then

$$L_{az} = I_0 \Delta \Phi_{az}, \tag{33}$$

or about 10^{34} ergs/s for the Crab and 10^{28} ergs/s for a typical pulsar. The latter value is consistent with observed levels of radio emission from pulsars (for the Crab pulsar, such power output would be consistent with that in giant pulses but not that in typical pulses). Such estimates are only illustrative, however, owing to the rather arbitrary choice for σ_A.

c. Scaling Laws

One can use the observed P and $\dot{P}$ to eliminate most of the unknown parameters. The wind torque, if assumed here to be dominant, is given from equation 12 to be proportional to I_0^2, as one would expect on general principles. The auroral zone luminosity is instead

$$L_{az} = \frac{I_0^2}{4\pi \sigma d \sin \theta}. \tag{34}$$

The only scale variable that enters is $\sin \theta$, which scales as $(a/R_c)^{1/2}$ or $\Omega^{-1/3}$. Thus the auroral zone luminosity scales as $\dot{P}/P^{8/3}$ instead of $\dot{P}/P^3$ for the total luminosity. Note that there are two scalings; P and $\dot{P}$ determine I_0 and thereby L_{az} for each of a collection of pulsars at a given epoch. The total current I_0 out of any given pulsar declines (eq. 10), and consequently the wind torque gives a slowing-down torque such that $\dot{P}P^{1/3} \approx$ constant instead of the light-cylinder prediction of $\dot{P}P \approx$ constant.

d. *Accelerating Potentials: Energetic Particles*

The values of $\Delta\Phi$ here are quite large for pulsars such as the Crab ($\approx 10^{12}$ eV) but decline to $\approx 10^{9}$ eV for typical pulsars (Michel 1981, figure 4). Again these seem generally plausible values, and they may even scale the gamma-ray emission. The flux of particles that could emit gamma-rays (if energetic enough) is plausibly not too sensitive to details of the model, probably some fraction of I_0. The likely sensitivity is on particle energy loss, which scales as energy to the 4th power if scaled by curvature radiation. Accordingly, particles accelerated in the fringing fields of the auroral zones would have energies of the order of $e\Delta\Phi_{az}$. A phenomenological relationship such as

$$L_\gamma \approx K\Delta\Phi_{az}^4 \qquad (35)$$

with $K = 1.3 \times 10^{-10}$ gives $L_\gamma \approx 10\%$ of the total spin-down power output for the Crab and 0.5% for Vela, roughly as observed (L_γ is taken to be the integrated high-energy spectrum). Another likely candidate for energetic emissions given this extrapolation would be PSR 1930+22, with an expected gamma-ray luminosity of almost 10^{33} ergs/s (see figure 6.4). Unfortunately, this pulsar's dispersion measure indicates that it is about 16 times further away than the Vela pulsar. Ögelman et al. (1976) set an upper limit of around 7×10^{35} ergs/s for its gamma-ray luminosity. For the remaining pulsars, the rapidly declining values of $\Delta\Phi_{az}$ would rule out observation of other closer pulsars.

If neutron stars are formed at maximum rotational velocity with an attached disk, the potential differences between the neutron star and disk would be of the order of ΩBa^2, where $\Omega \approx 6000$ radians/s, $B \approx 10^{12}$ gauss, and $a \approx 15$ km; hence particle energies would be of order 1.3×10^{20} eV. Because the magnetic field drops rapidly with distance, particles with energy $\approx 1/4$ of this value (Michel 1981) cannot be retained in the magnetosphere and they escape. (Jupiter, for example, is a source of electrons with energies ≈ 30 MeV, comparable to its 170 MeV pole-to-equator potential drop: L'Heureux and Meyer 1976; Chenette et al. 1974.) Such systems would be sources of extremely energetic particles.

e. *Blackbody Radiation*

Direct particle heating by bombardment of the disk can be estimated to be the particle flux times the average energy of the particles; hence

$$L_{bombardment} = I_0 mc^2, \qquad (36)$$

which would be significant only for rapid pulsars such as the Crab (see table 6.1). An important question is how much mechanical work is done on the disk. Equation 17 gives the resistive electrical dissipation in the disk, but there is also mechanical $\mathbf{J} \times \mathbf{B}$ work done on the disk, some of which acts to lift the disk out of the local gravitational potential of the neutron star. The magnetic

TABLE 6.1—Nominal Disk Model Parameters

Parameter	Crab Pulsar	Typical
Spin-down power	5×10^{38} ergs/s	7×10^{32} ergs/s
Radio power	2×10^{34} ergs/s	8×10^{28} ergs/s
Bombardment power	1.3×10^{28} ergs/s	4×10^{26} ergs/s
Disk/pulsar mass	1.0×10^{-5}	1.0×10^{-5}
Neutron star radius (a)	1.0×10^6 cm	1.0×10^6 cm
Polar cap radius	3×10^5 cm	1.0×10^5 cm
Auroral zone width (d)	5 cm	5 cm
Mean free path (AZ)	0.3 cm	94 cm
Corotation distance (R_c)	1.7×10^7 cm	1.7×10^8 cm
Inner disk thickness	2×10^6 cm	6×10^7 cm
Midplane disk density	1.1×10^6 g/cc	1.1×10^6 g/cc
Ambient particle concentration	6×10^{11}/cc	6×10^9/cc
Discharge particle concentration	3×10^{16}/cc	1.1×10^{14}/cc
Surface magnetic field (B_0)	3×10^{11} gauss	0.9×10^{11} gauss
Corotation magnetic field	5×10^7 gauss	2×10^4 gauss
Disk magnetic field (normal)	9×10^2 gauss	1.0 gauss
Surface gravity at disk	8×10^{10} cm/s^2	3×9^{10} cm/s^2
Total current (I_0)	3×10^{15} amp	3×10^{12} amp
Ejectability ratio (e)	1.3×10^6	3×10^{12}

flux through the disk given from equation 30 (not eq. 19, which was estimated ignoring the large drops expected in the auroral zone) is then of the order of

$$B_z = \frac{\Delta f_{az}}{R_c^2}, \tag{37}$$

and the torque on the disk is of the order of

$$T_D = I_0 B_z R_c^2, \tag{38}$$

given a power transfer of about

$$L_D = \Omega_0 T_D = L_{az}. \tag{39}$$

Thus the power output would be dominated by that in the auroral zone, which in turn is presumed here to be comparable to that in radio emission. These power levels seem consistent with those inferred by Knight (1982) for the Crab pulsar.

TABLE 6.2—Pulsar Disk Bolometric Magnitudes (apparent)

Pulsar	S_{400} (mJy)	m_{bol} (max)	T_{surf} (K)	Obs.
0833−45	5000	22.75	10,500	23.7 (blue)
0329+54	1400	24.13	13,200	...
1749−28	1300	24.22	8700	...
0950+08	900	24.61	3200	...
0531+21	800	24.74	14,000	16. (pulsed)
1641−45	375	25.56	16.300	...

f. Disk Luminosity

A disk heated at a rate comparable to the radio-frequency luminosity provides an estimate of the bolometric luminosity of the disk directly from the radio brightness of the pulsar. A mean flux (S_{400}) of 1 mJy at 400 MHz translates into an apparent bolometric luminosity of +32.00 (Allen 1963). The brightest pulsar, Vela (PSR 0833-45), with an S_{400} of 5000 mJy (Manchester and Taylor 1981), would therefore have an apparent magnitude of 22.75. If the pulsar beam is pencil-like, one overestimates the power output by a factor of about 5 (1.75 magnitudes), which would further reduce the brightness to 24.5 magnitudes. The candidate object (Lasker 1976) is 23.7 magnitudes, in between these two estimates. Estimated luminosities for the brighter pulsars are summarized in table 6.2. In the case of the Crab, this emission would be overwhelmed by nonthermal emission from the pulsar. If viscous heating were the dominant energy source, we would not expect the object to be seen as a radio pulsar.

g. Radiation Mechanisms

Given disk scalings, attention is naturally drawn to the intense currents that should flow in the auroral zones. The energy output is about right if a reasonable fraction of the available energy is transformed into radio emission. The physical scale of the auroral sheets is reasonably consistent with the bunch sizes required for coherent emission. The maximum coherence factor would be roughly of the order

$$N_{coh} = n_{az}d^3, \tag{40}$$

or about 3×10^{18} for the Crab and 1.5×10^{16} for a typical pulsar, assuming density fluctuations comparable to the mean density. Typically one requires $N_{coh} \approx 10^{11}$ (sec. 8.2.6); thus rather minor fluctuations would be adequate to provide the observed levels of coherent radio emission. The radiation from such a current sheet would naturally be free to propagate with little dispersion because it would be in the thin relativistic wind flowing from the polar cap. A sheet-like current could provide a geometry automatically ensuring a dense source (to provide coherence) while at the same time the signals, once produced, could propagate in the neighboring thin wind plasma without being reabsorbed (figure 2.23). The radiation could even be refracted forward by this relativistic wind (Barnard and Arons 1982).

6.3 The Disk

1. Origin

Theorists are rather confident that disks surround neutron stars in pulsating X-ray sources. The source of material is generally taken to be Roche-lobe overflow from a companion star. Such disks are usually called *accretion disks*, although accretion is now gratuitously used to modify *disk* in general.

In the case of an isolated pulsar, there would seem to be no such source of matter, hence no disk. However, if pulsars are born in supernova events, there are a number of alternatives.

a. *Disks from Disrupted Companions (type I supernovae)*

In type I supernovae, often modeled as a contact binary of two white dwarfs, one of the two stars is pushed over the Chandrasekhar limit and the other incinerated. But do we know that *all* of the companion is ejected? If not, we already would have matter left in orbit around the neutron star, which, being well within the Roche limit and insufficient to be self-gravitating, would spread out into a disk. Although disruption of companions has been suggested to explain the isolated millisecond pulsars (assuming a spin-up scenario: Ruderman and Shaham 1985), it would be very difficult in such a scenario to explain how the resultant disk of matter was expelled.

b. *Disks from Angular Momentum Conservation*

For type II supernovae, usually modeled to be the end point in the evolution of a massive star with or without companions, numerical modeling is much more detailed but is usually done assuming spherical symmetry. That simplifying assumption cannot be strictly correct or pulsars would not be rotating. Indeed, the amount of angular momentum contained in the presupernova core is difficult to assess. The core could easily rotate too fast to form a neutron star, in which case matter would be spun off during the collapse phase. The obvious fate of the matter would be to form a disk. Durisen and Tohline (1984) and Tohline and Williams (1988) give calculations showing how rapidly rotating condensing stars shed excess angular momentum by forming detached disks orbiting a central object. Even cores that are not rotating so fast will provide matter with high enough specific angular momentum to stall at distances outside the neutron star.

c. *Disks from Late Infall Following a Type II Supernova*

As a simple model, consider a star following the formation of a neutron star in its core. Energy deposited in the star will start it expanding with a velocity proportional to the distance from the center. As long as the density declines with distance, a sufficiently large velocity gradient leads to ejection of all of the mass beyond some value M_0 to infinity. Inside, matter would move out at first and then fall back within a characteristic time T_{fb}, where

$$\left(\frac{T_0}{T_{fb}}\right)^{2/3} = \frac{V_0^2 - V^2}{V_0^2};$$ (1)

here V_0 is the free-fall collapse time to form M_0 for matter starting from rest. For velocity directly proportional to distance (a so-called *Hubble flow*, which should be well approximated about a week after the event), the velocity profile can be converted into a distance profile and written as $(r_0^2 - r^2)/r_0^2$, which in turn can be written as a mass profile taking as a first approximation a constant-density star, in which case we finally get (expanding the expressions about their zero values)

$$\left(\frac{T_0}{T_{fb}}\right)^{2/3} = \frac{2}{3}K\frac{M_0 - M}{M_0},$$ (2)

where K corrects for density variation in the star and other possible velocity profiles, but is nevertheless of order unity. Again we obtain $M \to M_0$ within a time of order T_0, even with the approximations (the factor of 2/3 is somewhat gratuitous). Differentiating gives

$$\dot{M} = K\left(\frac{M_0}{T_0}\right)\left(\frac{T_0}{T_{fb}}\right)^{5/3}.$$ (3)

For these estimates, $T_0 \approx 1$ s and $M_0 \approx 1.4\ M_\odot\ 3 \times 10^{33}$ g. The neutron star reaches very nearly its asymptotic mass within a few seconds, but nevertheless there can be a substantial mass waiting to fall in. From equation 2 we see that even after a year there is $1.5 \times 10^{-5}\ M_\odot$ left gravitationally bound to the neutron star. When this mass falls in, it is structurally unimportant to the neutron star. (For just such reasons, it is unlikely that the neutron star formed when SN 1987A produced the neutrino pulse could be followed 4 hours later by collapse to a black hole.) The rate of infall is important because the dynamic pressure of the infall perturbs the radiation pressure from the neutron star assuming it is rotating (Ω) and magnetized (B). Balancing the gravitational potential energy gained by the infall against the pulsar luminosity gives

$$\dot{M}\frac{GM_0}{R_L} = L.$$ (4)

With nominal numbers, a 10 ms pulsar with a Crab-like magnetic field of 4×10^{12} gauss would give $L = 6 \times 10^{40}$ ergs/s and $R_L = 1.5 \times 10^7$ cm, hence from equation 4 a critical infall rate of $\dot{M} = 1.4 \times 10^{22}$ g/s, which from equation 3 corresponds to $(K = 1)\ T_{fb} = 6.3 \times 10^6\ T_0 \approx 0.14$ years. From equation 2 we see that the mass waiting to fall in at this time is $M_0 - M \approx 6.6 \times 10^{-5}$ $M_\odot$. With these masses and scales, the cross-sectional density of the disk would be of the order of 10^{12} g/cm^2, as discussed in more detail in Michel (1988e). The estimated infall masses for a variety of possible initial conditions are shown in figure 6.3, with the shaded area a conservative estimate of the likely initial values. For pulsars with very weak fields, the rather long fallback times raise doubts as to the validity of the assumptions. How fallback might work in a binary system is a more complicated question, and might point to why only one of the two presumed neutron stars in the PSR 1913 + 16 system acts as a pulsar.

2. Evolution of the Disk

Once formed, a disk is not easy to get rid of. Many workers take their clue for disk evolution from the behavior of the pulsating X-ray sources, which must accrete matter rapidly from their disks to be observable. Must neutron stars in general eat up any surrounding disk? The disk in the X-ray source case is not isolated, but is continuously perturbed by the cascade of in-falling matter from the companion plus tidal and electromagnetic perturbations from both stars, which can easily mimic a huge effective viscosity (Michel 1984a, sec. 10.4). These perturbations are absent in an isolated neutron star (or even one in the tightest known binary system), so rather than consuming the disk, the neutron star acts to try to eject it from the system. There is no particular basis for insisting on generalizing the *apparent* large viscosity in complex accreting systems to the case of an isolated disk quiescently orbiting a single neutron star. Because the specific angular momentum of objects in Kepler orbit scales as $r^{1/2}$, a neutron star can never eject its disk by transferring angular momentum to it; it is a bottomless sink for angular momentum. The Earth cannot eject the Moon despite the action of tidal forces that cause the Moon to recede.

The disk is estimated above to have a mass of order 10^{-5} $M_\odot$ and a density comparable to that of the presupernova core ($\approx 10^6$ g/cc) owing to the intense tidal forces near the neutron star; hence the disk is electron-degenerate and highly conducting. If a disk were essential for pulsar action, then such interaction probably would not blow away the disk, despite what one might casually assume. The disk would simply move out to a position where the interaction was sufficiently reduced. Also, it is somewhat difficult to blow away a 10^6 g/cc disk with radiation pressure, regardless of where it is. For a young pulsar such as the Crab, the inner edge of the disk would now be at about 1.7×10^7 cm ($\approx 17\,a$), about 10 times closer than the conventional light-cylinder distance (1.5×10^8 cm) and well within the Roche limit (1.7×10^{11} cm) for normal matter. A neutron star would couple magnetically to the disk plasma and act to force the disk plasma into corotation, whereas neutral matter in the disk is in Keplerian motion. Such differential motion would lead to heating of the disk and maintenance of electrical contact between the neutron star and disk. Because such interaction would be well within the light-cylinder, radiation per se would carry away rather little angular momentum. The system should evolve

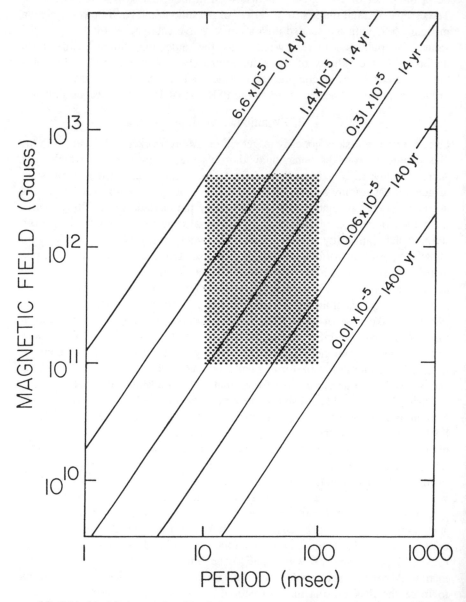

FIG. 6.3. Critical in-fall times as function of neutron star magnetic field and spin period. After the indicated times, in-fall should be forming a disk with an eventual mass as shown. At earlier epochs the in-fall is presumed to be accreted onto the neutron star simply by following dipolar magnetic field lines to the surface. From Michel (1988c). Reprinted by permission from *Nature*, 333, 644. Copyright © 1988 Macmillan Magazines Ltd.

at essentially constant angular momentum, with the neutron star slowing down while simultaneously driving the disk away to greater distances.

As noted above, Burbidge and Strittmatter (1968) recognized the parallelism between Jupiter's emissions and those from a pulsar. At that time it was thought that Io alone excited the emission, and it was immediately noted (Douglas-Hamilton 1968) that a satellite with an orbital period of the order of pulsar periods would be inside the Roche limit and hence be broken up. Such a system would speed up, not slow down, as energy was radiated. Then came the failure to detect any objects whatsoever in orbit about a pulsar. Finally, it was theorized that newly formed pulsars are so active as to account for the supernova outburst itself (Ostriker and Gunn 1971; Bodenheimer and Ostriker 1974). This theoretical activity seemed to provide an internally consistent picture wherein the pulsar simply blew its surroundings clear of matter at birth. The only firm observational fact, however, is that pulsars do not seem to have planets. Disks were not ruled out. The Roch-limit argument is irrelevant to disks, and the idea that newly born pulsars are so fiercely bright as to be capable of ejecting any nearby matter is simply that: an idea yet to be supported by observation. To the contrary, the main puzzle has been that most young supernovae have no detectable pulsars in them, not that they contain exceptionally energetic pulsars. Much the same situation obtains for the theory of X-ray burst sources (Lewin and Joss 1981), which invokes not only an accretion disk about a neutron star but also a low-mass companion star, neither of which is readily detectable. Because the bursters are not seen as pulsators, one loses the fine timing information that would make a binary system evident.

3. Disk Structure

Basically, the overall mass of the disk broadly determines the disk structure, within fairly narrow uncertain ties. The exact mass turns out to make little difference within a wide range of likely choices. One cannot rule out highly structured disks, and such uncertainties lead to very model-dependent estimates for conductivities and viscosities.

a. *Kepler Disks*

The simplest idealization for a disk is one in Keplerian orbit. In this case, the effective gravitational potential (so-called geopotential) is

$$\phi_{\text{eff}} = GM \left(\frac{1}{r} - \frac{1}{R} \right) \approx GM \frac{h^2}{2R^3} + \text{neglected terms}, \tag{5}$$

where $r^2 = R^2 + h^2$, h being the height above the midplane and the r^{-1} term being the potential of the neutron star (we ignore the gravitational attraction of the disk itself, assuming the disk to be very much less massive than the primary). The R^{-1} term (R = axial distance) represents the equivalent potential corresponding to the centrifugal force, which balances in the plane. Such a disk, filled with an adiabatic fluid, satisfies

$$\phi_{\text{eff}} = \phi(\rho),\tag{6}$$

where ρ is the particle density. Thus the density is constant (a maximum) in the midplane and falls to zero at some equipotential surface; hence from equation 5 we estimate a disk thickness

$$H \approx R^{3/2}.\tag{7}$$

Integrating the hydrostatic equation,

$$\nabla P = -\rho \nabla \phi_{\text{eff}},\tag{8}$$

we obtain, for $P = \kappa \rho^{5/3}$,

$$\frac{P_0}{P} = \frac{2}{3} \phi_{\text{eff}} = \frac{GMH_s^2}{3R^3},\tag{9}$$

where P_0 and ρ_0 are the midplane values while H_s is the surface height above the midplane. For nonrelativistic degenerate electrons, $\kappa = 1.3 \times 10^{12}$ dynes/cm^2 at $\rho = 1$ g/cm^{-3} in processed matter (He, C, O, $\cdots$,Fe), so $\rho = H^3$, and therefore the integrated surface density

$$\sigma = \left(\frac{3\pi}{8}\right) \rho H s \approx H^4.\tag{10}$$

The density and therefore the mass of the disk are consequently sensitive functions of the disk thickness. Conversely, the disk thickness would only vary by a factor of 10 in going from $M_D \approx 10^{-9}$ to $M_D \approx 10^{-5}$ (in units of the neutron star mass). As we have seen, the disk thickness would not seem to play a very essential role in the electrodynamics anyway. The important issues are, rather, the mass and evolutionary dynamics of the disk. The total mass of the disk is given by

$$M_D = \int \sigma 2\pi R \, dR = \frac{2\pi\sigma_C \int R^{3/2} dR}{R_C^{1/2}},\tag{11}$$

where σ_C is the value at the inner edge of the disk (σ falls to zero, of course, at the physical edge, but to minimize fine detail we approximate the disk by simply truncating it at each edge; σ_C would be evaluated at something like the knee in a plot of σ versus R). The integral in equation 11 is heavily weighted at large distances and therefore insensitive to the inner limit of integration at R_C if the disk is reasonably extended. The outer edge of the disk might be placed at some distance comparable to that of the presupernova core radius (R_{snc}). Presumably matter beyond that distance will have been largely ejected in the formation event, if that is the limiting factor. We cannot, moreover, make the disk arbitrarily thick; otherwise H, which grows with increasing R, would exceed R itself. Thus we have a physical limitation on the extent of the disk of roughly

$$R < R_C^3 / H_C^2. \tag{12}$$

For the Fermi temperature to exceed, say, 200 eV one needs an effective potential of $(200 \text{ eV})/(2 \times 10^9 \text{ eV}) \approx 10^{-7}$ assuming two nucleons per unit charge, and since $m \equiv GM/c^2 \approx 1$ km, one can estimate $H_s^2 m / 2R_C^3 \approx 10^{-7}$. For the Crab pulsar, $R_C \approx 1.7 \times 10^7$ cm, and adopting $a = 10^6$ cm gives $H_C \approx 7 \times 10^4$ cm near the inner edge of the disk. Alternatively, we can parameterize the system in terms of the thickest (and most massive) disk that could plausibly be put into the system, namely $R_{max} = R_{snc}$ or a maximum thickness, H_m, at the inner edge, of

$$H_{max}^2 = R_C^3 / R_{snc}. \tag{13}$$

Again, for the Crab pulsar, taking $R_{snc} \approx 10^8$ cm gives $H_{max} \approx 7 \times 10^6$ cm, over 100 times that necessary for degeneracy. H_C scales as $R_C^{3/2}$, so in general one needs disks roughly 1% the maximum thickness to have extensive degeneracy, hence more massive than $\approx 10^{-11} M_\odot$ (see below). Evaluating the mass in equation 13 gives (note from eqs. 9 and 12 that $\rho \approx R_{max}^{-3/2}$)

$$M_D = \frac{2}{7} (GMR_{snc}/3K)^{3/2}, \tag{14}$$

which gives a maximum mass of about $10^{-3} M_\odot$ taking $R_{snc} \approx 10^8$ cm. The point here is not to propose such a massive thick disk, but rather to establish a baseline by which the term "thin" can be quantified. A dimensionless thickness h can be introduced where

$$H \equiv h H_{max} \tag{15}$$

is the physical thickness of the disk, and the resultant disk mass is therefore, from equation 10,

$$M_D = h^4 M_{max}. \tag{16}$$

Thus, for a disk of mass $\approx 10^{-5}$ (Michel and Dessler 1981a), we find $h \approx 0.3$, consistent with $h > 0.1$ for degeneracy. (Nondegenerate disks might well serve to excite the pulsar action, but it looks as if the most likely physical state would be degeneracy.) Note that the outer edge of the disk refers to the principal mass distribution; a fairly light component could extend far beyond R_{snc} owing to the action of viscous and electrodynamic interactions (e.g., an analog of Saturn's E ring).

b. *Nulling and Turnoff*

In the present version of the disk model, the disk plays more the role of a spectator in that most of the angular momentum is carried away from the neutron star by a stellar wind rather than by transfer to the disk by means of electrodynamic torques. In the latter case, the disk would have had to retreat to fairly large distances before it could come into corotation with the neutron star. If instead stellar winds slow the neutron star, the disk will eventually come into corotation even if it does not retreat at all from the pulsar. For example, if we place the outer edge at $R_{\text{snc}} \approx 3 \times 10^8$ cm (slightly expanded, owing to the nonzero torques applied), the orbital period is still only 2 s. The angular momentum of a $10^{-5} \, M_\odot$ disk of such dimensions is of the order of that for a 1 s neutron star; thus only for strong coupling to an initially fast Crab-like pulsar would the disk be greatly expanded beyond R_{snc}. Such a configuration would be paradoxical because the inner edge, which we argue should be at the corotation distance, would then overlap the outer edge. It seems likely that such an eventuality could cause the pulsar activity to turn off; as the disk narrows into a ring, it ceases to be effective in shorting out the field lines. Moreover, at least part of the disk will now have to fall inside the corotation distance, which means that charged particles will be lost down the field lines to the neutron star surface until the disk is largely consumed. The possibility of such a phase opens the question of significant gamma- and X-ray luminosities from extinct or near-extinct pulsars (sec. 11.2). There seems, however, little observational requirement for such turn-off (sec. 2.4).

c. *Ejection of a Disk*

The critical issue concerning disks at this level of their development is whether they can persist, or exist to begin with. The major considerations are therefore the direct dynamic forces (could a pulsar wind blow away a disk?) and viscous forces (will the disk dissipate itself?). The first issue can be confronted by assuming, at the formation event, the disk to be fully threaded by the field lines from the neutron star. The early relativistic wind (or outward flow from the formation event) could then pull back magnetic field lines until they made an angle of about 45° to the disk surface. The outward surface stress would then be roughly $B^2/2\mu_0$ on the disk, balanced by the gravitational attraction on the disk by the star, or σg, where σ is the mass per unit surface area of the disk. Because the magnetic field declines at least as fast as r^{-2}, as does g, it is evident that the maximum effects will be at the inner edge of the disk ($\approx R_c$). We can therefore directly calculate an *ejectability factor*

$$e = B^2/2\mu_0 \sigma g \tag{17}$$

from the disk model. As can be seen in table 6.1 these values are always small compared to unity. A direct dynamic ejection of the disk by an energetic newborn pulsar seems not to be expected. The large values of e close to the pulsar have another significance, however, since they provide a mechanism for ejection of field lines from the disk. The flux tubes should be unstable to interchange (if their emergent magnetic field lines are swept back by a stellar wind), with the strongly magnetized surface elements of the disk exchanged outward and replaced by weakly magnetized elements until the disk magnetization is reduced to some equilibrium value, such as estimated above. Most pulsars are simply not energetic enough to blow away a disk of, say, mass 10^{-5}. Such a disk could absorb all of the neutron star angular momentum simply in expanding. If all of the pulsar spin-down energy went into picking up disk particles and depositing them at infinity with zero velocity, about 10^{46} ergs would be needed to lift the entire disk out of the gravitational potential of the neutron star. But that is, in fact, all the available energy that a 1 s pulsar has! At the existing spin-down power outputs ($\approx 10^{32}$ ergs/s), an absolute minimum of 3 million years would be required to dissipate the disk, even if energetically possible. The actual time scale should be significantly greater because one can load only so many particles onto a flow without stagnating it. The added particles must therefore be accelerated to energies not too different from those of the unperturbed wind particles, which in turn are probably not too different from those implied by figure 6.4, hence implying a reduction by another factor of $\approx 10^3$ for typical pulsars (a 3×10^9 year dissipation time scale) and $\approx 10^6$ for the Crab at its *present* power output.

d. *Viscosity*

We turn next to the role of viscous forces, a somewhat controversial issue, owing to the existence of X-ray sources and modeling in which they are fed by accretion disks. The complex and uncertain dynamics of such systems are frequently lumped into *effective* viscosities, which have to be quite large compared to ordinary molecular values. The effective viscosity, to be important, must be on the order of

$$v_{\text{eff}} = R^2/T, \tag{18}$$

where R is the characteristic size of the system (e.g., the disk radius) and T is the time within which the matter in the disk must be processed (i.e., replaced by overflow from a companion star). For example, $T = 1$ year and $R \approx 1$ $R_\odot$ gives $v_{\text{eff}} \approx 2 \times 10^{14}$ cm^2/s, which is comparable to the viscosity needed to describe the slow plastic *creep* of the Earth's mantle! By definition, such numbers would require an extremely rapid evolution of a pulsar disk. Because the system lacks a stellar companion relentlessly trying to fill the neutron star's surroundings with plasma, it is not evident that the appropriate viscosity will be significantly different from the rather small values expected from kinetic theory. For pulsars, the critical viscosity is that corresponding to $R \approx R_c$ and $T \approx P/\dot{P}$, hence critical viscosities of $\approx 3 \times 10^3$ cm^2/s for the Crab pulsar to four orders of magnitude larger for the longer-period pulsars having long apparent lifetimes.

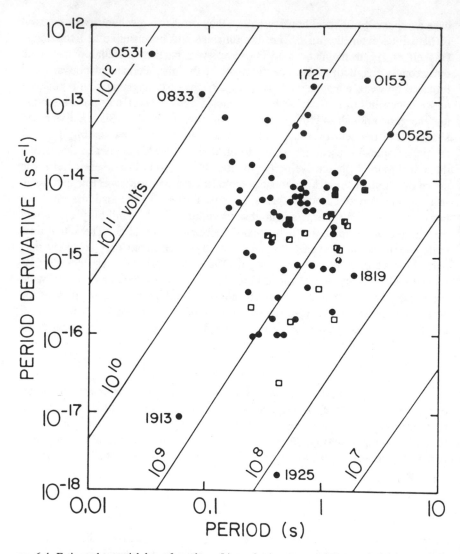

FIG. 6.4. Estimated potential drops for pulsars. Lines of constant auroral zone potential drop (volts). Note the rapid decline for (presumably) older pulsars. PSR 1930 + 22 would have the next highest drop from those of the Crab and Vela pulsars. From F. C. Michel, 1983b, *Ap. J.*, 266, 188 (figure 4).

These are huge viscosities for known fluids. The expected kinematic viscosity at the densities proposed is, in fact, of the order of 10 or less. No theoretical estimate puts it as high as even 10^2 (see Durisen 1973, figure 2; Flowers and Itoh 1976, eq. 107; Nandkumar and Pethick 1984; Itoh et al. 1987). The dynamical viscosities are often quite large (10^7 poise), but this increase is offset by the concomitant large densities (10^6 g/cc). Thus, isolated disk evolution may be largely independent of microscopic viscous forces. Macroscopic sources of viscosity (turbulence, convection, or entrained magnetic fields: see Paczyński 1976, Vila 1978, and Coroniti 1981, respectively) might play important roles at certain epochs, but lacking any predictive theory at the moment, one can do little more than remain alert for possible future developments. As discussed in sec. 10.4, there is increasing interest in the possibility that shocks propagating in the disks produce the required effective viscosity. In any case there is no reason to suppose that phenomenology developed for one class of object (accreting X-ray sources), which remains unsupported by theory, need be applicable to the quiescent disks that might encircle neutron stars. The probable consequence of having intermediate values of the effective viscosity need not be dramatic from a kinematic point of view. Indeed, for viscous forces to be important they have to be greater than the electromagnetic forces that act to drive the disk away from the pulsar. The most likely consequence of viscous torques exceeding electromagnetic torques would be ablation of material from the inner edge of the disk and the deposit of that material on the neutron star. For active pulsars, it seems most reasonable to assume that there is little or no accretion because the electrodynamic torques dominate and maintain the inner edge of the disk at or near the corotation distance; the outer edge of the disk then retreats under the combined effects of these two torques. Presumably, one has instead an X-ray source when accretion cannot be stopped in this manner. A tiny amount of accretion would serve a useful role in maintenance of electrical contact between the inner edge of the disk and the neutron star. Thus viscosity per se need not cause the disk to be consumed by the neutron star during the pulsar phase; most of the disk mass and angular momentum is at large distances, near the outer edge of the disk, and the disk simply expands away from the neutron star while the inner edge is maintained near the corotation distance. The disk cannot be removed by such a process; it simply becomes thinner. As we have seen, the electrodynamics is insensitive to the disk thickness as long as the disk can be taken to be a good conductor.

4. Disks in Close Binary Systems

Would a disk have observable effects in a binary system, particularly a relatively close system such as PSR 1913 + 16 where even general-relativistic effects are detectable? A disk about a star would give the system a sizable quadrupole moment, which is well known to imitate general-relativistic precession of the line of apsides. Misner et al. (1973, p. 1112) give the quadrupole correction, which corresponds in General Relativity to

$$\text{precession rate} = 1 + \frac{Ja^2}{ma_0(1 - e^2)}, \tag{19}$$

where unity corresponds to the standard general-relativistic precessional rate, J is the dimensionless quadrupole moment, a is the stellar size, m is the stellar mass (in gravitational length units, $m \equiv GM_{star}/c^2$), and a_0 is the orbital semi-major axis. The disk mass estimate is about 10^{-5} of the neutron star, the neutron star has $m = 2$ km, the disk radius is guessed to be about 100 neutron star radii, and the neutron star radius is usually taken to be about $5m$, so altogether J is roughly 3 (i.e., comparable to the neutron star rotating at breakup), potentially a very important correction. But the remaining factors are of the order of $4m/a_0$, which for PSR $1913 + 16$ is only about 10^{-5} (i.e., comparable to the first-order gravitational corrections). Consequently, the effect of even such an extended object as a disk is effectively a *second-order* correction in general relativity, whereas the precession itself and all the other famous tests of general relativity are *first-order* relativistic corrections. It is only because the Sun is so large that a possible solar oblateness would contribute significantly to the precession of the orbit of Mercury. There is consequently no worry that the disk will interfere with the use of PSR $1913 + 16$ for testing general relativity. Unfortunately there is also correspondingly little chance of detecting a disk on the basis of its quadrupole moment, unless very close neutron star binaries are discovered.

6.4 Consequences of Disks

From an observational point of view, a disk system (figure 6.2) provides a potentially rich variety of morphological variation, which one can compare with the similarly rich variety of pulsar properties. Michel and Dessler (1981a) discuss several such points; for example, that alignment of spin and magnetic axes need not affect pulsar activity, that ion confinement in the pulsar surfaces need not hinder current flow, that long nulling periods could be attributed to long time scales associated with the disk, that the magnetic field may be markedly reduced from previous estimates (typically to as low as 10^9 gauss, as inferred from the drifters), that the braking index would be characterized by a limit near $n = 7/3$ (see sec. 6.2.b) rather than a fixed value of 3, and that some *residuals* in the timing data would naturally arise in a disk model, owing to small variations in the location of an active spot. Finally, the lack of a one-to-one association between pulsars and supernovae can even be understood in terms of plasma obscuration, and further suggests an explanation for the giant pulses from the Crab pulsar and for its comparative dimness as a radio source (although the latter could be incorporated into nondisk models as well). Of particular interest would be detection of weak X-ray emission (presumably the bulk energy output is mainly in the form of a stellar wind; otherwise the slowing-down rates would imply luminosities, if all in electromagnetic radiation, not seen by existing UV or X-ray satellites).

1. Alignment

Contemporary X-ray pulsar theories, like radio pulsar theories, assume that pulsation comes about as a result of beaming. Thus, for the beam to sweep periodically past the observer, one assumes that the spin and magnetic field axes are at some significant angle to one another. It has long been known that this misalignment produces a torque on a pulsar that acts to align its magnetic axis with its spin axis (Michel and Goldwire 1970; Davis and Goldstein 1970), yet it is generally supposed that such an aligned rotator would not produce pulsed radiation, owing to its symmetry. The aligned rotator has generally been regarded as a simplified mathematical model for analysis, and, even should such a model generate radiation, this radiation would be seen as pulsed only if some misalignment were then introduced. Why then do pulsars not turn themselves off, or at least not show some evidence of alignment (Manchester and Taylor 1977)? Goldreich (1970) has suggested, alternatively, that centrifugal distortions are locked into the star and slowly relax, owing to creep, thereby giving the star sufficient triaxiality that even if it were aligned, it would not remain so and in some cases would even align orthogonally. Whether or not the neutron stars actually align also needs to be reexamined because a disk could induce precessional torques, and in the same way such torques can lead to equilibrium misalignments. The planets Jupiter and Saturn act much like pulsars. They each have a plasma disk (Jupiter's disk is the Io plasma torus, and Saturn's plasma disk is well outside its famous ring system), they rotate, and they have a rather dipolar magnetic field. The magnetic moment and spin axis of Saturn in particular are almost perfectly aligned. It is observed that the satellite Io simply modulates the intensity of the pulsed decametric emission from Jupiter and that pulsed emission would exist given only the Io torus (i.e., a disk). In the cases of both Jupiter and Saturn, emission of electromagnetic radiation (and, in the case of Jupiter, emission of relativistic electrons into interplanetary space) is modulated with a first harmonic dependence on planetary rotation period (a 10 hour modulation) rather than a second harmonic dependence (a 5 hour modulation), as would be expected if dipole tilt were the controlling physical factor. Thus it is possible that a disk model does not require misalignment to give pulsed electromagnetic emission (see also Vasyliunas and Dessler 1981 and references contained therein). In any case, the observational fact (for Jupiter and Saturn) is that gross misalignment is not necessarily a prerequisite for pulsed emission. Alignment of pulsars need not be tantamount to pulsar extinction. Clearly, alignment could be yet another argument for eliminating long-period pulsars (if needed), but there seems to be no particular observational support for this idea (Manchester and Taylor 1977).

2. Ion Confinement

Ruderman and Sutherland (1975) have pointed out that the strong magnetic fields proposed for pulsars change the lattice structure of the neutron star crust in such a way as to greatly increase the ion work-function, especially if the surface is composed of iron as is often assumed. As a consequence, it is much more difficult to remove positive particles by the field-emission process usually invoked. Indeed, in the Ruderman and Sutherland model it is supposed that only electrons are pulled from typical pulsars, with pair production providing positrons to neutralize the stellar wind. (Otherwise the overall system charge would grow to the point that no more electrons could be emitted.) Later work has obtained progressively lower ion work-functions (sec. 1.7.3.a), but even if emission of ions were entirely suppressed from the neutron star surface, a disk would be a source of charged particles of both signs. Instead, the emission would be from normal (albeit degenerate) matter in a relatively low magnetic field.

3. Nulling

Roughly 30% of the 300 pulsars detected so far are observed to *null*, that is, to fall abruptly below detectable luminosity for tens and even hundreds of pulse periods before abruptly reappearing. The abruptness of the transition from pulsing to the nulled state is easy to understand since, in the standard model, the particles are relativistic and so are all of the characteristic velocities. Thus the whole system can readily start up or turn off within one pulse. But how does it remain off so long? Apparently there is a second, relatively long time constant yet to be identified. The rapid start-up and turn-off time constants should not be greatly changed by the existence of a disk according to the above argument. However, the nulling time may well be quite different because the disk surface is largely ordinary matter under ordinary conditions, and the interaction region could cover relatively large areas. Thus plausible physical effects (depletion, saturation, convection, etc.) can have relatively long time constants without appreciably changing the turn-on/turn-off time constants. The latter may even be shortened if a disk edge is ζ times closer than the light-cylinder.

4. The Magnetic Field Strength

In the standard model one attributes all of the energy output to magnetohydrodynamic torques; and hence one deduces the magnetic field given in equation 1.3.8. For one of the slowest known pulsars (PSR 0525+21), this estimate gives an extremely large magnetic field of 12.4×10^{12} gauss, approaching the critical magnetic field ($B_{crit} = 44 \times 10^{12}$). Do pulsars really have fields that intense? Alternatively, if we rescale according to sec. 6.2.b, we will have fields of 0.046 and 0.030×10^{12} gauss for PSR 0531+21 and PSR 0525+21, respectively.

5. Drifting Subpulses

The interaction region is not necessarily exactly at the corotation distance, so the disk material is in motion through that region. Moreover, because the region must cover a finite area, there will also be a velocity shear across it. As a result of this velocity discrepancy, there should be one or more secondary periodicities introduced into the system. It is qualitatively clear that long secondary periodicities are obtained when the interaction distance is almost exactly at the corotation distance. Such periodicities are observed as drifting subpulses. The observed drift rate of subpulses is extremely slow, amounting to only a few degrees per pulse; hence $\Delta V/V \approx 10^{-2}$ or, equivalently, $\Delta r/r \approx 10^{-2}$.

Pulsars that do not show drifting could then simply be pulsars with somewhat larger values of $\Delta V/V$. As a practical matter, rapid drifters would not be seen as drifters owing to aliasing if the drift rate per period much exceeded the pulse window. Taking the eight observed drifters listed by Manchester and Taylor (1977) and dividing their magnetic fields by the appropriate factor $2\zeta^{3/2}$ gives a geometric average of 2.4×10^9 gauss at the surface of the neutron star.

6. Braking Index

The braking index, when it can be measured, seems to be consistently smaller than the standard scaling, which gives $n = 3$. It is difficult to modify this value because the magnetosphere is so thin in the standard scalings that plasma effects simply cannot be important insofar as n is concerned. As pointed out by Roberts and Sturrock (1973), a disk model gives instead a braking index of $n = 7/3$, provided that the inner edge exactly tracks the corotation distance.

7. Residuals

The braking index is often difficult to measure because there are significant *residuals* that represent a departure in the actual slowdown from any smooth theoretical slowdown that has a time constant comparable to the pulsar age. Also, there may be occasional glitches. However, the interaction region with a disk, as outlined above in conjunction with the drifters, would quite reasonably not be rigidly fixed on the disk but could move around a bit, which would vary the torque. Plasma density/conductivity changes too would vary the torque slightly about some average value, possibly leading to such residuals. Because the distribution of matter in the disk must vary (it could even be composed of numerous rings, for example), the variability represented by the residuals could vary from pulsar to pulsar, as observed.

8. The Supernova Connection

It has long been a puzzle why, if pulsars are indeed formed in supernova explosions as generally assumed, more pulsars are not found in, or very near, recent supernova remnants. In the standard model it is difficult to see why the pulsar should not be seen soon after the nebula becomes transparent to radio waves (about 20 years: Michel et al. 1987). A newly formed disk, however, could be either too close or too far from the neutron star-too close in the sense that a strong interaction with the neutron star floods the system with relatively low energy plasma, snuffing out any coherent radio emission. Such an object would probably be a weak version of the X-ray pulsars found in binary systems. Such strong interaction would eventually deplete the nearby disk by ablation, bringing such a phase to an end. In the case of a disk that is too distant, hence too weak an interaction, one must wait for internal viscosity to expand the disk into the near proximity of the neutron star. In either case there would be an evolutionary time constant involved that would have to pass before the neutron star could be seen as a pulsar. Observations of the Crab pulsar, which is famous for being associated with a known supernova, are not inconsistent with this picture. The Crab pulsar is, on the average, a feeble radio source compared to its total energy output, yet it is characterized by occasional *giant pulses*. Both phenomena could be attributed to blanketing of the emission, with fluctuations in blanketing that would permit a strong local source to break through from time to time. Blanketing also seems qualitatively consistent with the long-period variations in radio luminosity (the optical is quite steady) observed by Rankin et al. (1974) and their observation that all radio components, the main pulse, precursor, and interpulse, are modulated together with zero phase lag. In addition, the Crab is presently unique in exhibiting variations in dispersion measure (Rankin and Counselman 1973), again consistent with the presence of anomalous amounts of plasma. Linscott and Backer (1982) report observations implying that the giant pulses from the Crab pulsar appear in phase delayed pairs, as if a single pulse were reflected off a conducting surface and the observer saw both the original pulse and the reflected one. The Crab pulsar is frequently interpreted as being an almost orthogonal rotator with the spin axis near the plane of the sky, so this would be favorable disk geometry, assuming of course the interpretation is correct. The supernova association problem is sometimes resolved by supposing a large fraction of pulsars ($\approx 80\%$) to be invisible owing to beaming effects (Manchester and Taylor 1977). As noted in sec. 2.4, this assumption creates birthrate problems, and pulsars such as PSR 0950 + 08 radiate detectably throughout $360°$ of phase (Hankins and Cordes 1980); thus a putatively bright, young pulsar should be detectable to the observer at essentially all orientations, even assuming a strong beaming modulation. Unless one happens to be viewing along the spin axis, which could only account for a tiny fraction of the cases, the radio emission should still be spin modulated, although possibly giving a broader pulse shape.

9. Theory

The upshot of these considerations is that a number of observational problems seem less puzzling in a disk model. As noted in detail in chapter 4, the cloaking by nonneutral plasma should shut down the standard model and something will be required to replace it or significantly modify it in any event.

7

Alternative Models

7.1 Introduction

In view of the problems with the basic physics of the rotating magnetized neutron star model, it is probably worthwhile to review briefly various other suggestions that have been made. In any event, given the tendency for history to repeat itself, it is worthwhile to list some of the ideas already considered and, to varying degrees, discarded.

7.2 Jupiter as a Pulsar

The similarity between the pulsar phenomenon and the Jovian decametric emissions was noted almost immediately after the discovery of pulsars (Burbidge and Strittmatter 1968; Warwick 1969). At the time, it was thought that Io was entirely responsible for exciting the Jovian emissions, owing to the unexpected correlation between that satellite's orbital position and radio emissions (Bigg 1964), which was theoretically attributed entirely to the satellite (Goldreich and Lynden-Bell 1969).

Actually, it was known all along that there was a *non-Io* component to the radiation (see the reviews by Gehrels 1976; Carr and Gulkis 1969), now thought to be due to a torus or disk of plasma more or less filling Io's orbital path, but much less startling than the Io component. Consequently the preoccupation with Io may have been misleading (see Carr et al. 1983, figure 7.17) given that the non-Io component is quite pulsar-like. In any event the analogy itself was quickly dispatched by the observation that such a body would have to be within the Roche limit (Douglas-Hamilton 1968), assuming that the orbital period was the pulse period. The Jupiter parallel has surfaced repeatedly (Mertz 1974; Kennel and Coroniti 1975; Michel 1979a; Braude and Bruk 1980), although the later efforts tend to try to fit Jupiter into the standard model. Indeed, Kennel and Coroniti (1975) thought that a convection type of model might be promising in view of what they termed the "grave difficulties" with the standard model, while Michel (1979a) suggested that some essential physical element might be missing, given that other energetic radio sources seem to contain magnetized plasmas in relative motion (see Sedrakyan 1970a,b, who suggested differential rotation as an energy source, also Akasofu 1978 and Lu 1976). Orbiting (binary) neutron stars would not suffer the Roche-limit difficulty and were proposed (Saslaw et al. 1968; McIlraith 1968; Aldridge 1968) and dismissed (Pacini and Salpeter 1968). Again, orbiting systems should speed up, not slow down, as energy is radiated.

The source of the pulsar-like emissions from Jupiter remains something of a mystery, but seems phenomenologically related to an *active sector* in the Jovian magnetosphere, which in turn seems related to the mapping of magnetic field inhomogeneities at the surface into the magnetosphere, a possibility discussed in Michel and Dessler (1981a). A neutron star containing a pure, centered (often tilted) dipole is almost universally assumed in pulsar theory. Yet it is presumed that in the stellar collapse that forms the neutron star, significant nondipolar magnetic fields must be compressed and frozen into the neutron star (Elsner and Lamb 1976). That is, a realistic star should include quadrupole, octapole, and higher-order magnetic moments, even though we expect the dipole term to dominate at large distances. These surface field deviations from a pure dipole are often referred to as magnetic anomalies.

1. Magnetic Anomalies

In figure 7.1, we show two magnetic flux tubes A and B having equal cross-sectional areas at the distance of the disk. If we follow these flux tubes back to the surface of Jupiter, we see that one foot of flux tube B, which enters a region of anomalously weak field, has a larger cross-sectional area than the feet of the other flux tubes. Thus, in this specific example, the area of the northern side of the disk defined by flux tube B is magnetically connected to a larger area of Jovian surface than at any other longitude of the planet. The larger area of the northern foot of flux tube B allows a maximum charged-particle exposure of a disk-like *ring current* within this restricted longitude range. If the current is limited by the resistivity at the foot of the flux tube, the height-integrated conductivity could be a maximum at the northern foot of flux tube B, where the magnetic field is weakest. In addition, bombardment of the surface by particles from the ring current would be enhanced since magnetic mirroring and other hydrodynamic effects tend to choke the flow into rapidly converging field lines (i.e., there is less flux to a high field region than to a low field region, given the same source region). Consequently, the maximum current flow links the disk to the weakest field region on the surface.

2. Currents to a Disk Aligned along the Magnetic Field (Birkeland currents)

In addition to the direct currents that flow in response to the electromotive force created by differential rotation of the star and disk, there is a second, orthogonal current system created by outward radial motion of the plasma, i.e., the possibility that a corotating magnetospheric convection pattern is established, such as that observed for Jupiter (Dessler et al. 1981). The greater contact between flux tube B and the surface leads to a greater plasma content within this

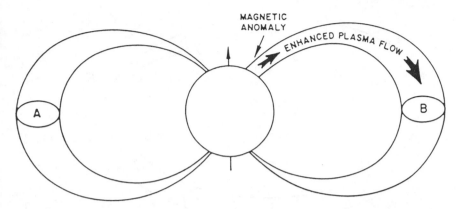

FIG. 7.1. Magnetic anomaly model. Weak-field regions on the surface increase the surface area exposed to a unit area of ring current (or disk): consequently current flow is least restricted and maximum ionization is produced by surface bombardment. From F. C. Michel and A. J. Dessler, 1981a, *Ap. J.*, 251, 654 (figure 2).

flux tube as compared to flux tubes at other longitudes. Because the equatorial portion of flux tube B is outside the corotation distance, plasma that is swept into corotation is stressed outward by centrifugal force. This outward force is balanced by the interaction between a longitudinal current (J_ϕ) component and the magnetic field so that

$$\mathbf{J}_a \times \mathbf{B} = \sigma (\mathbf{\Omega} \times \mathbf{r}) \times \mathbf{\Omega}, \tag{1}$$

where $J_a = (1 + \frac{3}{2}s_r^2)\sigma\Omega^2 rB$, σ is the plasma mass per unit length along the ring, and s_r is the ratio of cyclotron speed to corotation speed (Dessler 1980a). Because σ generally varies with longitude, the axial current J_a varies with longitude. In order that $\nabla \cdot \mathbf{J}_a = 0$, a corotating Birkeland current J_B (a current aligned along the magnetic field) flows such that the current density is related to the longitudinal gradient in plasma density:

$$J_B = \left(1 + \frac{3}{2}s_r^2\right)\left(\frac{d\sigma}{d\phi}\right)\frac{\Omega^2}{B}. \tag{2}$$

These Birkeland currents flow on either side of the magnetic anomaly, and they are also possible sites of radio emission and sources of enhanced coupling between the disk and the magnetic anomaly (see figure 7.2). The Birkeland currents close through the surface/ionosphere where the finite conductivity leads to an impressed electric field that causes the plasma in flux tube B to $\mathbf{E} \times \mathbf{B}$ drift outward. An electric field pattern ensues that causes plasma on the opposite side (of Jupiter or a neutron star) to drift inward, thus establishing the magnetospheric convection pattern. Because the origin of the asymmetrical mass loading that causes the convection is the magnetic anomaly, the convection pattern rotates with the system. The power extracted by this mechanism from the

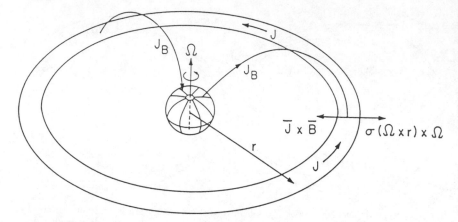

FIG. 7.2. Birkeland current system. The current system resulting from centrifugal stress on a spot or incomplete segment (*ring current*) of disk plasma. The current J is required to keep the plasma from being immediately expelled. The currents close as Birkeland (*field-aligned*) currents to the star. From F. C. Michel and A. J. Dessler, 1981a, *Ap. J.*, 251, 654 (figure 3).

kinetic energy of rotation of the central object can be shown to be (Dessler 1980b; Eviatar and Siscoe 1980)

$$P = \frac{1}{2} \dot{M} \Omega^2 (R_w^2 - R_c^2), \tag{3}$$

where $\dot{M}$ is the rate of mass outflow from the ring and R_w is the effective limit of corotation beyond which the plasma simply flows away as a wind. There are therefore two sources of power output: (1) the magnetic coupling between the disk and the magnetic source, which takes place even if axial symmetry obtains, and (2) the centrifugal ejection of asymmetrically distributed plasma, leading to equation 3. If we assume, for the sake of illustration, that the two are roughly comparable, we can estimate the mass-loss rate from the disk. The simplest estimate for R_w is just R_L, in which case, assuming $R_c \ll R_w$, the power output is simply $\dot{M}c^2$. For a pulsar model, it is the outflowing stellar wind that becomes relativistic; the disk itself is essentially in Keplerian orbit at R_L. This dimensional argument was used in Michel and Tucker (1969). As noted in the disk model, even for the Crab pulsar the loss rate would correspond to only $10^{-9} M_\odot$/year, requiring a minimum disk mass of the order of $10^{-6} M_\odot$ in order that the disk last for the typical lifetime of a pulsar.

7.3 Oscillating Objects

1. Oscillating White Dwarfs

It is ironic that there is fairly little literature on the oscillating white dwarf model, considering that it was the most heavily favored one to begin with. Many influential astrophysicists at the time were loath to adopt an object, the neutron star, that might well not exist. But white dwarfs, although known to exist, simply could not oscillate or rotate rapidly enough. Heroic assumptions were required to get a period even approaching 1 s (which is about average for a pulsar). Higher harmonics were suggested as a mechanism to get shorter periods (Ostriker and Tassoul 1968), but as it became clear that the pulsars were such excellent clocks, the idea of a pure high-harmonic oscillation came to seem forced, and any remaining pockets of resistance were more or less swept away with the discovery of a 33 ms pulsar in the Crab Nebula (see Henry 1968; Israel 1968; Durney et al. 1968; Faulkner and Gribbin 1968; van Horn 1968; Cocke and Cohen 1968; Kundu and Chitre 1968; Simon and Sastri 1971; and Black 1969). This program seemed doomed even early on (Skilling 1968; Bland 1968). It was also suggested that the oscillations might be confined to the white dwarf atmosphere (Black 1969; see also sec. 7.6.)

2. Oscillating Neutron Stars

An oscillating neutron star suffers from the opposite problem of a white dwarf: it should oscillate too fast (≈ 1 ms at small amplitudes) and would become unbound at large amplitudes (Thorne and Ipser 1968). Such models have been widely discussed (Hoyle and Narlikar 1968; Israel 1968; Baglin and Hayvaerts 1969; Harrison 1970; Stothers 1969; Papoyan et al. 1973). Heintzmann and Nitsch (1972) have estimated the damping times for such oscillations to be of the order of months to years. Finzi and Wolf (1968) proposed even before the discovery of pulsars that the Crab Nebula could be excited by a vibrating neutron star. It is certainly physically plausible that neutron stars oscillate, particularly if they are born in a collapse, but such oscillations may be damped out rather rapidly. In any event observations have not yet been made that point strongly to such oscillations, with the possible exception of the half-millisecond pulsations briefly observed in SN 1987A.

3. Rotation Plus Oscillation

Chiu and Canuto (1969a,b) proposed a pulsar model, basically the rotating magnetized neutron star, in which oscillation of the star would excite maser action in the near magnetosphere (see sec. 8.3.1; there is an extensive literature); see also Kumar 1969, Yukhimuk 1971, Vladimirskii 1969, and Van Horn 1980). At present, there is not much enthusiasm for a scenario in which the energy source is in oscillations with rotation providing the timing.

7.4 Sheet Discontinuities

A number of theories appeal to radiation from sheet-like structures such as standing shock waves and magnetic neutral sheets.

1. Shock Waves and Neutral Sheets

Michel and Tucker (1969) proposed that the pulsar generates a magnetically driven supersonic wind which eventually must shock and become subsonic in

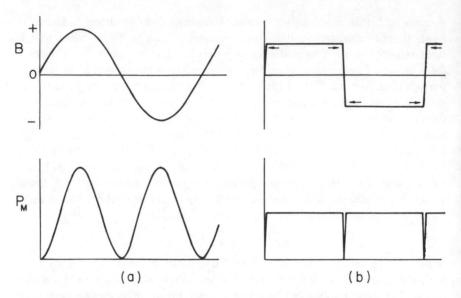

FIG. 7.3. Steepening to shock transition. The wave is shown in the plasma rest frame wherein the average electric field over each half cycle is zero. (a) the sinusoidal case wherein the total pressure is unbalanced if the plasma is cold. The large magnetic field pressure can only be balanced by dynamic pressure, namely the acceleration of plasma toward the zero-field (neutral sheet) regions. (b) The result of the acceleration process. Plasma streams into the neutral sheets and a step-function structure would be in static equilibrium were some magnetic energy dissipated to heat the plasma at the neutral sheet. In actuality the approach to equilibrium would be complicated by propagation of shock waves into the uniform field regions, heating of the plasma at other than the neutral sheet, etc., but (b) illustrates a physically consistent simplification, unlike (a). Such a transition would presumably take place near the light-cylinder, where the electromagnetic fields first begin to become wave-like (a). From F. C. Michel, 1971, *Comments Ap. Space Phys.*, 3, 80 (figure 1). Reprinted by permission of Gordon and Breach Science Publishers Inc.

response to external pressure, requiring a standing shock wave. Neutral sheets convected outward were proposed to be sites of coherent particle excitation upon crossing the standing shock. This model would have to operate at large distances from the pulsar and at quite low particle densities, making sufficient coherence more difficult to attain. The presumed transition zone between a corotating plasma and a stellar wind is an alternative region that has attracted considerable attention because such a zone could well contain a shock wave (see figure 7.3) and, in a charge-separated flow, such a shock could produce coherent motions of charged particles over a scale of the order of the shock thickness. For the Crab pulsar, for example, this circumstance is not implausible given that the magnetic field at the light-cylinder is $\approx 10^6$ gauss while the electron energies are $\approx 10^{11}$ eV (inferred either from space-charge limited flow [Michel 1974c] or from the nebular synchrotron radiation [Shklovsky 1968]), which gives a cyclotron radius of ≈ 3 m. The number of particles in a cubic meter at the light-cylinder of the Crab pulsar should be $\approx 10^{14}$, giving a limiting effective temperature of $T_{\text{eff}} = NE/k \approx 10^{29}$ K, which, at a surface area scaled by the light-cylinder, would be about adequate to account for the brightness temperature of the Crab pulsar (Michel 1971). Note that the brightness temperature limitation is considerably relaxed because the radiating elements are expanded from small bunches over the polar caps to a broad sheet (see also Bertotti et al. 1969a,b, Kardashev 1970, Endean and Allen 1970, Ferrari and Trussoni 1973, and Stewart 1977). Fujimura and Kennel (1979) have given numerical solutions for such shock transitions.

2. Current Sheets

A closely related model was suggested by Lerche (1970a,b,c, 1971), who modeled the neutron star's electromagnetic field propagating in a vacuum, separated from external plasma by a current sheet which oscillated as the neutron star rotated, producing coherent emission (see also Dokuchaev et al. 1976). Tademaru (1971), on the other hand, proposed that outwardly expanding charge sheets were the emission source (see also Grewing and Heintzmann 1971).

Unless the pulsar is embedded in a dense cloud, the equilibrium position at which the finite pressure of the interstellar medium balances that of the ultra-low-frequency radiation of the pulsar is of the order of 0.01 pc. The difficulty of putting the source of the radio-frequency emissions at such large distances is that variations in the ISM would vary this position and lead to fluctuations in the apparent arrival times. Some small *restless behavior* is indeed observed in pulsar pulse arrival times, but of very small amplitude. How exactly the wave would interact with the ISM is an interesting question. Normally one would assume that the wave would simply be reflected because the frequency is well below the plasma frequency of the ISM (a few kHz). However, the frequency is *so* low that a fundamental assumption of plasma wave propagation is violated, namely that the perturbation to the particle properties is just that, a perturbation. In fact, the electrostatic field is maintained at these low frequencies so long that the electrons would be accelerated to relativistic energies, even at 0.01 pc. We examine these issues quantitatively in chapter 9. It is likely that the consequence is absorption of the wave rather than reflection, but this issue has not been carefully addressed.

3. Double Layers

The role of electrostatic double layers has not been paid much attention until recently (Williams et al. 1986). Such layers could play a role in the accretion-type sources because the in-falling plasma must be decelerated to rest, and given that the cross sections for electron and ion scattering are significantly different, large electric fields will be generated to maintain near charge-neutrality. The possibility that particles are accelerated in pulsars by such layers has been mentioned in Joshi et al. (1984) and Tajima and Dawson (1979). Double layers have not been invoked for generating radio-frequency pulsar emission but are noted here for completeness.

7.5 Unconventional Models

1. Volcanos

Dyson (1969b) suggested that matter ejected from the interior of a neutron star gives an inhomogeneous plasma distribution about the star. This interesting proposal has largely been neglected although volcanos might be commonplace and could serve to concentrate current flows into localized regions of the neutron star magnetosphere. Even a minor upwelling might be sufficient to produce a surface feature and an alternative to magnetic field inhomogeneities. In the case of the volcanic satellite Io in the Jovian moon system, Gold (1979) has suggested that the volcanic plumes interact with the Jovian magnetosphere to concentrate electric discharges (see also Peratt and Dessler 1988).

2. Starspots or Flares

Early ideas on pulsar emission attempted to draw parallels with solar activity, which also produces considerable coherent radio emission. (See Fujimoto and Murai 1972, 1973, Apparao and Chitre (1970), Ostriker 1968, Glencross 1972, and Smoluchowski 1972; the first paper also proposes ejection of matter from the neutron star.)

3. Other Proposals

Somewhat related proposals were made by Kovalev (1979, 1980). In his models coherent radiation escapes through cracks in the pulsar crust. Kaplan and Eidman (1969) suggested that parts of the surface might move with relativistic velocities (see also Kumar 1969, Tsygan 1977, Karpman et al. 1975, and Lu 1976).

7.6 White Dwarf Hypothesis

The possibility that pulsars might be neutron stars was mentioned in the discovery paper. However, the more influential astrophysicists at that time were understandably reluctant to seize on a hypothetical object before exhausting the known stellar objects that might be pulsar candidates. Thus early models proposed that pulsars were white dwarfs (e.g., Ostriker 1968). It became impossible, however, to reconcile the observed pulsar periodicities with such an object.

A "white" dwarf is, of course, simply a star in which electron degeneracy pressure supports the matter (mainly the ions) against their self-gravitation (see, for example, Chandrasekhar 1984). Although white dwarfs are hot enough to be luminous and visible, their internal temperatures are usually small compared to the Fermi temperature throughout most of the star. Consequently the structure can be calculated to a good approximation simply by ignoring the temperature altogether. The mass, radius, or central density determines uniquely the other two, so one basically has a one-parameter family of stars. A burnt-out Sun would become a dwarf star about the size of the Earth. But for only a slightly more heavy dwarf, about 1.4 $M_\odot$, the radius would shrink to zero. Unlike planets, where accreting more matter leads to a larger planet, white dwarfs have an *inverse mass-radius relationship*, wherein the larger the mass the smaller the white dwarf and vice versa (until one gets to such small masses, around 10 times Jupiter's, that the matter is nondegenerate). White dwarfs in contact binary systems are therefore unstable to mass transfer because the donor actually expands, which acts to accelerate mass transfer to the shrinking recipient.

What happens is that for more massive stars the degeneracy pressure must be higher to hold up the outer layers; hence the Fermi temperature (as well as the density) must be higher to provide that pressure. The electrons begin to become relativistic if required to support a large enough stellar mass, which in turn makes them too compressible to resist gravitation. It is not difficult to prove that a self-gravitating fluid satisfying the adiabatic law

$$P = \kappa\rho^\gamma, \tag{1}$$

with

$$\gamma = 4/3 \tag{2}$$

(i.e. relativistic particles or photons), is neutrally bound: a radially inward push causes it to shrink to a point, and an outward push disperses it to infinity. However, when the electrons were nonrelativistic, the star had a finite *binding energy* and there is no way for the star to restore that deficit when the electrons become relativistic; thus at a certain point the star dynamically collapses to a point where the *nucleons* become degenerate—the neutron star—and the same physics of degenerate pressure support is replayed one more time (and possibly the last; see sec. 7, below). The limiting masses of the two objects are essentially determined by the heaviest mass particle (m):

$$M_{max} \approx 1/m^2, \tag{3}$$

and therefore, even before correcting for nuclear forces, etc., the maximum masses of white dwarfs and neutron stars are necessarily similar, despite the different source of degeneracy pressure for support (in both cases, it is essentially the nucleons that must be supported against gravity). Note that the existence of very massive particles beyond the nucleons would not, in itself, provide a more massive, stable stellar object. The neutron star is consequently regarded as being a final end point for stars. A more massive object is presumed to collapse to a black hole (see next section).

Note that neutron stars, viewed as end points of stellar evolution, are actually *endothermic* insofar as the chemical potential of the nucleons is concerned. In other words, the highly publicized role played by nuclear "burning" ultimately serves not so much as an energy source as it does a delaying action in staving off as long as possible the conversion of hydrogen into neutrons. The net energy output therefore all comes from gravitational binding, not nuclear binding. Since white dwarfs are so easily modeled, there is little doubt about what their minimum vibrational periods might be, and the minimum rotation period is of the same order of magnitude. This period is easy to estimate, since the characteristic propagation velocity of a compressional wave is of the order

$$V^2 \approx P/\rho \approx kT/m_i, \tag{4}$$

where the white dwarf pressure is from nonrelativistic electrons while the mass density is from the nuclei; thus, if we take the maximum Fermi temperature that the electrons could have and yet not be fully relativistic,

$$kT_F \leq m_e c^2, \tag{5}$$

we find for the propagation velocity

$$V \leq 7 \times 10^6 \, \text{m/s}. \tag{6}$$

Ordinary white dwarfs have radii the size of the Earth ($\approx 7 \times 10^6$ m) and therefore a vibrational period of about 1 s. In fact, Melzer and Thorne (1966) concluded that the minimum period was more like 8 s. This work, having been done before the discovery of pulsars, was not influenced by any pressure to find shorter periods. Smaller (more massive) white dwarfs could oscillate more rapidly but are not stable against collapse for the reasons described above and for another reason: as the electrons become more energetic (not necessarily relativistic), inverse beta decay can lead to capture on the nuclei composing the star, which drops the pressure and causes the star to contract, further elevate the electron energy, and lead to a runaway collapse. Much effort was devoted to refining (and reducing) the lower limit of vibrational period so that the shortest-period pulsar then known (PSR 0950+08 with a period of 0.253 s) could also be explained. The discoveries of the Vela (PSR 0832–45) and Crab (PSR 0531+21) pulsars with periods of 0.089 and 0.033 s, respectively, dashed these hopes. These objects, if white dwarfs, would have to be 10 to 30 times smaller than normal dwarfs and the electrons would unquestionably be highly relativistic (i.e., the Crab and Vela would not be stable configurations but would be collapsing to become neutron stars).

7.7 Black Holes

Black holes are not strictly alternative models for pulsars, but they are alternative objects which supernovae might produce instead of neutron stars or to which neutron stars might be converted in the process of accretion, as is presumably happening in the binary pulsating X-ray objects (ch. 11). On a much larger scale, massive black holes have been postulated to exist at the centers of active galactic nuclei (AGN). Within the usual astronomical uncertainties, one can estimate the total power output from an AGN and the total mass of the galaxy to conclude that the conversion efficiency of mass into energy must approach 10% or be faced with a number of unpleasant alternatives because the activity could not have been going on as long as inferred (e.g., as deduced from the extent of associated phenomena: radio halos or jets).

As with so many astrophysical models, we have again a case where there was one possibility under consideration for a long time with no very viable alternative, with the result that it came to be adopted and any alternatives, even if discovered later, became "speculative." Of course, one reason for the lack of alternatives could also be that the physics was incomplete (unfortunately this observation has come to be something of a "motherhood" statement, made to show that authors are open-minded whether or not that is the case). A good example is the powering of stars. The only significant energy source originally known was gravitational contraction. Indeed, if gravitational contraction had been more nearly sufficient, it would certainly have been *the* accepted explanation. If gravitational energy could have accounted for, say, 10% of the Sun's age, people would probably have simply insisted on revising that age and on insisting that other estimates conform to it. This would have caused the geologists and others immense headaches because then their time scales would have been hard to reconcile. Fortunately, gravitational theory is so precise that it could not be fudged to look like it worked. People therefore became persuaded, correctly, that there were other, subatomic, processes providing the heat. The story with black holes has the same elements, but the resolution is not yet so clear. Again, the gravitational picture is the preferred one. The realization that a star has a finite upper mass if held up by electron degeneracy has its parallel for nucleon degeneracy, suggesting logically that there is a finite upper mass for neutron stars. In fact, one can show (Møller 1952) that even a star made of formally *incompressible* matter (which violates causality because pushing on one side of a piece of incompressible matter would have to instantaneously move the other side) has a finite upper mass. (This latter mass is *not* the mass at which the gravitational radius equals the physical radius.) Furthermore, there is the well-known Schwarzschild solution for gravitating objects in which the matter is entirely uncharacterized. So on the one hand one has a physical crisis and on the other hand one has a candidate solution.

Nevertheless, there was for a long time a reluctance to accept the collapse of physical stars to a singularity of the gravitation field equations, namely the Schwarzschild solution, which actually corresponds to a *vacuum* solution without mass: the major theoretical "advance" has been to rename these solutions *black holes*. Purists differentiate between "true" singularities (Schwarzschild) and the space-time about a star collapsed to a black hole, because the latter is "evolving": the e-folding time for stellar collapse is of the order of 1 ms, so a black hole formed in the early galaxy differs from a true singularity by about 1 part in $10^{-100,000,000,000,000,000,000}$, a rather fine distinction.

There are issues of whether the general-relativistic equations are correct to all limits. Quantum electrodynamics, for example, gives paradoxical behavior for particles with too large a charge and too small a radius. Such particles seem not to exist in nature but for reasons not necessarily involved with electrodynamics per se. Thus there seems to be some larger unity of nature that avoids infinities in locally observable physics (which is hardly surprising even if fun to imagine). The discovery that black holes should radiate (Hawking 1974), with the consequence that small black holes would vanish in a burst of radiation, actually exacerbates the concern over whether such objects could actually exist. Consider, for example, the conservation laws for elementary particles. Electrons and nucleons are separately conserved by all presently known forces, including gravity. Yet, if one can create even in principle a black hole and evaporate it, one has obviously violated these conservation laws with the very forces that *preserve* them! We would then have a global inconsistency, because one could either let a mini-black hole evaporate or add a hydrogen atom and then let it evaporate. Thus the existence of black holes means that a hydrogen atom has a quantum mechanical chance (of perhaps small but nevertheless finite probability) of itself evaporating into photons. It then follows for relativistic quantum mechanics that an isolated proton can decay into photons and a positron. But the argument for the inevitability of black holes is based on the assumption that the nucleons locked up in a star are conserved!

Some popular theories involving strong interactions predict proton decay, and therefore no problem exists because a lump of uncharged matter will turn into pure energy (photons) if simply left long enough. That possibility implies two very interesting things: (1) if one "turned off" the interaction leading to proton decay, gravity would also have to "turn off" (otherwise the paradox would return and proton decay would be incidental) and therefore the theory would have to automatically give general relativity to be logically complete (which does not yet seem to be the case although string theories may provide such a connection); and (2) if a lump of neutral matter could convert into pure energy, stars would not necessarily collapse to black holes to begin with!

We will illustrate this situation with two assumptions consistent with established physics: (1) that the path to formation of a black hole involves the compression of ordinary matter to high density and pressure, and (2) that a burning process beyond ordinary nuclear processes (which stop at iron) exists involving *strange particles* or perhaps accelerated proton decay. The first assumption reflects the properties found in stellar evolution calculations of black hole formation, and the second has already had considerable discussion in the high-energy physics community in terms of "quark" matter. Together the assumptions suggest that neutron stars might explode rather than collapse to black holes. It should be evident that such a possibility suggests radically new scenarios for activity in galactic nuclei, for gamma-ray burst sources, etc.

1. Another Stage of "Nuclear" Burning?

The proposal that a neutron star over the Oppenheimer-Volkoff limit collapses to a black hole is based on assumed physics. Stars stave off gravitational collapse by nuclear burning. Nuclear burning terminates at the iron-group elements. A neutron star is worse than burnt out; it has actually used gravitational energy to endothermically produce the neutron-rich matter in its interior. Thus there should be no further source of energy. In a sense, we are in much the same position Eddington and others were in before nuclear burning itself was elucidated. At that time, no burning process whatsoever was known that would liberate significant energy other than gravitational contraction (with the unpleasant consequence that the Sun could only be about a million years old, equally unsatisfactory to geologists and the clergy by about three orders of magnitude—but in opposite directions). So they just guessed that one existed. The specific guesses were not too bad (see Eddington 1926), but the burning process virtually had to exist even if the mechanisms could only vaguely be perceived. Eddington seems to have preferred the conversion of hydrogen into helium as a power source but couldn't fathom how four protons and two electrons could be crowded together to form an alpha particle, particularly at the comparatively low central temperatures of stars. The competing idea was electron-proton annihilation. A modern parallel is the proton-decay issue, albeit not insofar as powering stars goes.

Given that Eddington was opposed to the idea of white dwarfs being able to collapse to a black hole, he would almost certainly have felt vindicated by the present thinking that neutron stars are instead the result if anything is left at all. He would doubtless have opposed in turn the idea of neutron stars collapsing to black holes, even if again unable to articulate a precise physical mechanism that might work.

With the neutron star, we once again seem out of *known* burning processes (leaving aside speculations on proton decay). Or are we? The idea that neutrons and protons or combinations thereof are the ultimate stable structures of matter is, in fact, not at all obvious. We know that the neutron and proton are simply the least massive members of an octet of baryons (companions to the Λ, the three Σ's, and the two Ξ's) as shown in figure 7.4. In addition there is a family of 10 *resonances* starting with what used to be called the *3,3 resonance* in nucleons (now the Δ) and culminating with the Ω^- particle, which is not really a resonance at all but a distinct, weakly decaying particle (to elementary particle physicists, such particles are considered to be "stable": see the compilations of elementary particle properties that appear episodically in *Reviews of Modern Physics*). Even richer but less clearly resolved bumps and resonances occur at yet higher energies. Because this complex structure is found at substantially higher energies, it has tended to be disregarded insofar as "ordinary" nuclear physics goes (as if the neutron itself were irrelevant because it is energetically "uphill" from the electron and proton). Physics at high density is anything but "ordinary." The high Fermi energy of a degenerate nucleon renders the Λ and other strange particles stable above a few times nuclear density (Ambartsumyan and Saakyan 1960; Bethe and Johnson 1974), and we should wonder what stabilities might exist for a host of possible new

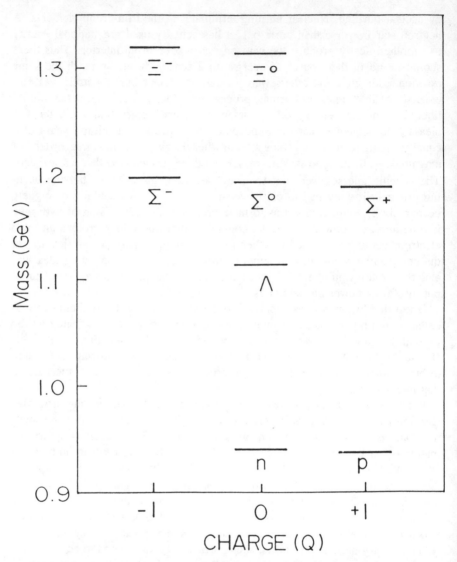

FIG. 7.4. The baryon octet. Conventionally, these approximately degenerate multiples are assigned quantum number S (*strangeness*) with integer values of 0, -1, and -2, respectively, with increasing mass. From F. C. Michel, 1988b, *Ap. J. (Letters)*, 327, L81 (figure 1).

complexes that seem quite inaccessible in the laboratory. Indeed, it has been speculated from the quark model that ordinary nuclear matter could be compressed into a degenerate quark plasma and that this *quark matter* might even be stable relative to nuclear matter (Witten 1984). Alcock et al. (1986) have even entertained the idea of "burning" neutron star matter to quark matter. But if quark matter is indeed stable, so might be complexes of finite size that have high symmetry properties (Chin and Kerman 1979). Strong limits to the existence of such quark "nuggets" have been given by Madsen (1988). A parallel in conventional nuclear physics is the spin zero, isospin zero alpha particle. Burning to such complexes could, in principle, release enough energy to disintegrate a collapsing neutron star rather than form a black hole (Michel 1988a). Again, we are in somewhat the same position as workers at Eddington's time: we have to guess, and our first guesses may be poor ones. Still, two obvious candidate complexes exist, both patterned after the high stability of the alpha particle. Because the proton and neutron are fermions, it is possible to assemble two pairs of them together so that all four are in the lowest (tightest-bound) s-wave state, despite the Pauli exclusion principle. As a result, the alpha particle has exceptionally tight binding. If we back up and note that there are actually eight distinct baryons, the natural generalization of the alpha particle would be what we will term an octet particle constructed of two each (spin up + spin down) of all eight baryons. The mass of such a particle is poorly known. A considerable "cost" of 200 MeV/nucleon goes into just the higher-rest-mass energies of these baryons compared to the nucleons. On the other hand, with an octet particle one would have 16 strongly interacting particles all sitting in the lowest s-wave states in their mutual overlapping potential walls. The available phenomenological models for nucleon forces simply do not extrapolate that far! The binding of such a complex could well reduce its mass below that of 16 ordinary nucleons, in which case the transition would be exothermic, as estimated in Michel (1988b) and shown in figure 7.5. Conversely, one might complain that nucleons are just three quarks in a bag (the MIT bag model: DeGrand et al. 1975) and the octet particle should dissolve into one large bag full of quarks. The Pauli exclusion rules are slightly different for quarks. In the bag model, the natural analog of the alpha particle would be a particle (*qualpha*?) made up of all three colors of the spin-up plus spin-down pairs of the up, down, and strange quarks (in baryonic terms, a hypernucleus of an alpha particle plus an Ω^- spin pair, for example). Indeed, stable dibaryon states of the nucleon and Ω^- have been predicted (Goldman et al. 1987). The octet and qualpha particles would have interesting properties. They would both have spin zero and zero net electromagnetic charges (and zero magnetic moment). As a result, they would be invisible massive particles. At this point, they represent objects about which little can be said with confidence. The fact that little can be said with confidence, however, undermines the assumption that neutron stars have no remaining source of energy (and therefore *must* be pushed over into black holes if they become too massive).

2. Neutron Star Detonation

If burning to form new complexes (octets, qualphas, or whatever) can take place, fusing or accreting neutron stars may simply explode rather than become seed black holes. A very natural scenario follows from these considerations. Transition to quark matter (and stable complexes) probably takes place

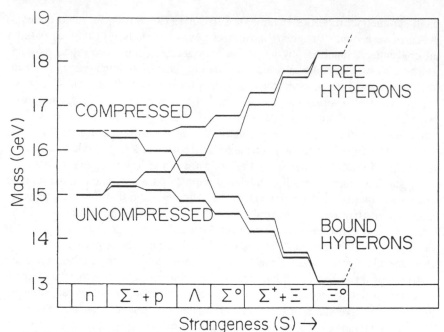

FIG. 7.5. Possible octet-particle energy level diagram. Here unsaturated (proportional to N^3) forces are assumed. The "uncompressed" curve leading up to "free hyperons" simply shows the rest-mass energy cost of replacing the particles indicated. The order is roughly that imposed by the high electron Fermi energy which enforces near bulk-neutrality. For combined replacements such as $\Sigma^- + p$, the two steps represent replacement by baryons with the mean mass. The "uncompressed" to "bound hyperons" shows how the N^3 binding would produce an activation threshold which would render ordinary matter stable. The "compressed" to "bound hyperons" shows how a high nuclear Fermi energy (here ≈ 130 MeV) can remove the activation threshold and permit direct cascading by $\Delta S = 1$ decays to the bound state (here shown bound by $\approx 15\%$). From F. C. Michel, 1988b, *Ap. J. (Letters)*, 327, L81 (figure 2).

at 10-20 times nuclear (Chapline and Nauenberg 1977; Baym and Chin 1976; Freeman and McLerran 1977). These densities are almost reached at the centers of stable neutron stars (according to the calculations, and depending sensitively on the equation of state as discussed in sec. 1.7) and would quickly be reached for a neutron star pushed over the Oppenheimer-Volkoff limit and beginning to collapse. Conversion into stable complexes would take place within typical strangeness-changing weak interaction decay times ($\approx 10^{-10}$ s) whereas the collapse would take place on the free-fall time scale ($\approx 10^{-4}$ s). Consequently the central regions of the neutron star could be burnt almost instantaneously, which would compress (and cause the burning of) the neighboring regions. A detonation wave would therefore propagate outward and consume much of the star (Michel 1988b). This scenario effectively recycles a popular type I supernova model (Sutherland and Wheeler 1984) wherein a white dwarf is pushed over the Chandrasekhar limit but burns its remaining fuel so fast that it explodes rather than collapsing to a neutron star. The resultant explosion might even look like a gamma-ray burst (Paczyński 1986; Goodman 1986; Michel 1988c), in which case black holes might never form under the expected circumstances.

3. Neutron Star Fusion in Clusters

Let us make the *Ansatz* that formation of a massive black hole involves the compression of matter to extreme densities and pressures. As Møller (1952) has shown, this is essentially what happens if one constructs stars of mathematically incompressible matter: stable configurations exist only up to a critical mass, at which point the central pressure becomes infinite. For any physically sensible equations of state satisfying causality, both density and pressure must diverge. Many people are indifferent to this technical point, being content with the natural expectation that if enough mass is concentrated in a small enough volume, one must get a black hole. Others even believe black holes can form from low-density matter. A large amount of work has been done recently on the collapse of star clusters (neutron stars) to massive black holes (Kochanek et al. 1987), so one is invited to imagine that a more or less uniform density cloud of stars simply shrinks until it is enveloped by the event horizon. Although seductive, this assumption fails to answer the simple question, "Where did the horizon come from?" The event horizon is, after all, a topological feature and one does not simply make a transition from one topology to another (a sphere to a torus, say) without intermediate states such as pinching the poles of a rubber sphere together until they touch, coalesce, and then open to form the torus. So the most plausible expectation is that the event horizon starts somewhere and grows to its final size. But the only way one can create a smaller black hole is with larger densities, because the mean density goes as r^{-2} and therefore the initial horizon must be initiated at large density. Thereafter the mean density can decline as more matter is entrapped.

If this line of thinking is correct, neutron stars are the path by which black holes of any size can be formed, so our attention naturally turns to where that matter finally achieves arbitrarily large density: the center of a neutron star that has been pushed over the Oppenheimer-Volkoff (1939) limit. (There is a certain "hall-of-mirrors" aspect here because on the microphysical level the transition to infinite density involves somehow forcing particles "inside" one another, because no known particle has a size or Compton wavelength remotely as small as its intrinsic gravitational size.) We will ignore this paradoxical detour, as is customary, although it is now commonly assumed that interesting things might happen at densities corresponding to the *Planck mass*. The Planck mass is where the Compton wavelength of a particle (h/mc) would be comparable to its gravitational radius (Gm/c^2); hence a mass

$$m_{\text{Planck}} = \left(\frac{hc}{G}\right)^{1/2} = 5.4 \times 10^{-5}\,\text{g}, \tag{1}$$

corresponding to a length 4.0×10^{-33} cm, neither of which are particularly intriguing values. After decades of wondering what such numbers might correspond to, elementary particle physicists have now confidently declared that *grand unification* of elementary forces takes place at the corresponding Planck energy 3×10^{28} eV, an idea that should be reasonably secure from direct observational challenge for the foreseeable future.

Neutron stars may be the unique pathway to forming black holes. One can never pack discrete objects together closely enough to get higher than average densities: before that they collide and coalesce. A natural nontrivial consequence is that putative formation of a massive black hole in a cluster of normal stars would be preceded by formation of individual black holes rather than direct collapse to a black hole. But there is no known star that can with certainty *directly* collapse to a black hole. Instead supernovae explode leaving a neutron star remnant if anything. Thus rapid nuclear burning and stellar evolution must produce prodigious numbers of neutron stars in the cores of massive clusters that replace the normal stars, and subsequently these fuse (or accrete) until they become black holes. Quinlan and Shapiro (1989) have shown that neutron star explosions prevent a central black hole from forming.

Thus neutron star explosions would provide a power source in active galactic nuclei to replace that from a massive black hole, because the gamma-rays would be absorbed in the optically thick surroundings, thereby providing a power source at the 10% of Mc^2 level (if neutron stars are to be disrupted), as required. Similar suggestions concerning highly bound baryon systems were made by Bodmer (1971) in terms of what he called *collapsed nuclei* (see also Terazawa 1981, 1989). There too the suggestion was made that quasar powering might be supplied in neutron star collapse prior to forming a black hole.

4. Related Issues

The bottom line is that astrophysics, high-energy particle physics, and general-relativistic theory (not just big bang cosmology) are converging on some very fundamental and interrelated issues. It may be that our present-day world view will continue into the future, or it may be that it will evaporate as quickly as the turn-of-the-century world view of physics as an activity mainly involved in determining the physical constants to ever higher precision. It is interesting that Lattimer and Schramm (1976) have pointed out that a possible resolution for the origin of the so-called r-process elements could be obtained simply by decompressing neutron star matter (see also Meyer 1989). Neutron star detonations would be such a source. Indeed, a very similar suggestion has been put forward by Eichler et al. (1989). Also, it is interesting that big bang cosmology is hard-pressed to begin a universe with pure matter and end up with ordinary matter, and not a matter/antimatter mix. Somehow pure energy ends up as ordinary matter. The obvious converse is that ordinary matter must under some circumstance become pure energy. That circumstance might be high densities, just as in the case of most big bang scenarios. Thus the existence of black holes, while hardly ruled out, is not necessarily a given. Whether particles such as the octet necessarily exist is irrelevant. At this point one would have to be fantastically lucky to predict how nature would resolve this issue, given the plethora of possibilities (and given that we have nothing like a secure overall picture of high-energy physics).

8

Radio Emission Models

8.1 Introduction

Although much of the emphasis in this book has been on the global electro-dynamics of pulsar models, historically that aspect got rather little attention. The former standard model was assumed to be essentially correct, and workers concentrated instead on what took place in the likely emission regions. For the most part, theoretical attention was (and is) centered on current flow regions near the surface of the neutron star. It seems secure that conduction currents of some sort must flow into and out of the neutron star and such currents will be more concentrated near the star, regardless of where they eventually flow. How these flows are distributed is an essential morphological question, but from the standpoint of explaining the radio-frequency emissions, it is mainly important that they *exist*. As we saw in chapter 1, we can make plausible estimates of what the current strengths are.

Unfortunately, the apparent generality collapses when one tries to go further. What sort of particles, for example, carry the current? The current could be simply a charge-separated flow (e.g., all electrons), it could be augmented with positron/electron pairs, or it could contain counterstreaming charges of the opposite sign (ions or positrons). Which of these options seems the more reasonable depends sensitively on the global model, so the idea that one can solve one element of the system without solving the rest is questionable. Assumptions about the current are implicitly assumptions about the global system and not at all model-independent. Granted, if one had such a complete theory of radiation from plasmas that the entire range of possibilities could be handled at once, then one could look at the radiation properties of pulsars and see that only one unique system would suffice. Unfortunately, astrophysics is not currently done this way and such an all-encompassing theory does not exist in any event. If one simply passes a large current through a plasma in the laboratory, one indeed gets coherent radio-frequency emission, but the theory of such experiments is only slowly being worked out. The difficulty is that the plasma behavior is driven nonlinearly and only recently have workers begun to understand how to handle nonlinear evolution. It is rare for people to take the trouble to analyze a model and then discard it (they might declare someone else's model no good, of course). So astrophysical models almost always "work" because there are enough unknown parameters involved that can be chosen conveniently, and in good faith, to give a rough fit to observation. All of the above current models can, within the existing state of theory, be made arguably to work. There must be at least a hundred pulsar theory papers announcing in the abstract that the proffered theory is in "good agreement with observation." In this chapter, we review some models of and suggestions pertaining to the radiation processes themselves.

8.2 Radiation Models

1. Curvature Radiation from Flux Tubes near a Star

Here we examine the possibility that the radio-frequency emission results from charged-particle bunches moving in circular arcs without asking what causes the bunching per se. It will also be attractive to assume that these arcs are curved magnetic field lines near the surface which are inhomogeneously illuminated (*flux tubes*). This picture is more or less the Ruderman and Sutherland (1975) idea of sparks. Curvature radiation by bunches has long been a tempting mechanism to theorists (Komesaroff 1970; Sturrock 1971a; Tademaru 1971; Buschauer and Benford 1976, 1980; Buschauer and Benford 1977a,b; Michel 1978a), and has been reexamined along the lines developed here in Michel (1987d).

a. *Basic Considerations*

We have the following constraints: (1) The width of the pulse determines the minimum possible Lorentz factor, which is also sometimes argued to be *the* Lorentz factor. (2) One can calculate the radiation *rate* for a particle of this Lorentz factor moving in a circular arc of some radius ρ_c (e.g., following a segment of the magnetic field line). (3) Assuming the particles are lost, they will only be seen once and the time to radiate is limited to be of the order of ρ_c/c, so multiplying the rate by this time gives the energy given up per particle. (4) This energy-loss rate can be boosted by the maximum possible coherence factor, which is the number of particles assumed to be moving together in a bunch, where a bunch is the maximum space-charge density in a cube of radius 1 wavelength λ at the maximum in the spectrum (about 1 m, say). (5) Given this maximal amount of synchrotron radiation that can be squeezed out of one particle, we then multiply by the maximum particle flux, which is the space-charge density times the polar cap area times c. The only free parameter is now ρ_c, and if ρ_c is estimated to be of the order of the neutron star radius, we get a luminosity (see below) of about 5×10^{26} ergs/s for a typical pulsar, which is about adequate on the average. However, assumption (1) above can be relaxed to give more favorable estimates (see sec. f below). Assumption (5) can be relaxed if the current can be neutralized (but seemingly not by pair production, as discussed in sec. 2.3.6).

b. *Other Options*

If we use the radius of curvature for dipole magnetic field lines, $\rho_c \approx (aR_L)^{1/2}$, we lose almost two more orders of magnitude for a typical pulsar, which makes the calculation seem pointless. Alternatively, we can boost this estimate by reducing ρ_c, but even going to $a = 1$ m is barely adequate and certainly a hard-to-accept limit; one can hardly go to curvatures smaller than the bunch size and maintain coherence, and such highly curved field lines would require intensely magnetized spots of size ρ_c, which could presumably not last long in the fairly resistive normal matter crust.

In principle, we can increase the particle flux by including electron/positron pairs, but now we are faced with the problem of separating the two; coherence results if the bunch has a large net charge, not because the bunch contains a large number of particles per se. Such separation must take place within ρ_c to make any difference, and the separation must at least exceed the wavelength λ of the coherent radiation. Thus the velocity difference for relativistic particles must amount to $\delta\beta \approx \lambda/\rho_c$, which requires that one of the Lorentz factors be no larger than $\gamma \approx (2\rho_c/\lambda)^{1/2}$ or about 10^2. The energy output per particle can exceed that carried by the particle as long as the particles are being accelerated while they radiate.

The Ruderman and Sutherland assumption that pair production plays an essential role forced them to assume an accelerating potential much larger than what turned out to be possible owing to space-charge limitations (sec. 2.3.6) and in fact owing even more to "poisoning" from the putative pair production itself (Arons and Scharlemann 1979). Worse, the underlying global model is the "standard" model, which seems to fail owing to trapping of the space charge (sec. 4.2). A large value of the Lorentz factor was inferred ($\gamma \approx 800$) for the radio-emitting particles (plus a set of *primary* particles with huge Lorentz factors) along with a tight curvature for the field lines ($\rho_c \approx 10^4$ m) in order to get sufficient radiation within what they assumed would be a narrow height directly above the surface.

c. *Circular Polarization*

For the most part, circular polarization has been neglected in the theoretical work, which has instead been preoccupied with finding a model that did *anything*, much less produce coherent radiation. Even in discussions by observers (Manchester and Taylor 1977; Smith 1977), circular polarization is briefly noted as an odd phenomenon having some curious systematics. To quote from Manchester and Taylor (1977), "The detailed behavior [of circular polarization] is complex, but at times a sense reversal appears to occur close to the center of the subpulse" (p. 55). This signature is *not* frequency dependent, as might be expected if it arose from propagation or plasma emission processes. In the Ruderman and Sutherland model the amount of circular polarization would be nil because circular polarization of one sign from particles moving on one field

TABLE 8.1—Some Pulsars with Simple Circular Polarization Signatures

Pulsar	P	$\dot{P}$	log Efficiency	B
0329 + 54	0.7145	2.04959	3.32	12.09
1451 − 68	0.2634	0.9878	5.25	11.21
1508 + 55	0.7397	5.0327	5.56	12.29
1857 − 26	0.6122	0.16	3.22	11.50
1933 + 16	0.3587	6.00354	4.42	12.17
2020 + 28	0.3434	1.89549	5.33	11.91

line would be canceled by radiation of the other sign on neighboring field lines. Worse, such a model cannot directly give the orthogonal mode changes observed in some pulsars (abrupt changes in the position angle of the linearly polarized component by about 90°) because the curvature radiation produces geometrically fixed polarization which would vary smoothly as the star rotated. Given these difficulties in directly accounting for pulsars polarization properties, the exact nature of the emission mechanism has tended to be relegated to be some comparatively obscure plasma emission or propagation effect (e.g., mode coupling between ordinary and extraordinary modes: Manchester and Taylor 1977, pp. 225-26). Rankin's (1983a) compilation of pulsar polarization patterns shows a number of pulsars clearly displaying such a signature, such as PSR 1451–68, PSR 1857–26, PSR 1907+02, PSR 1933+16, and PSR 2002+31. Some pulsars in which circular polarization has been explicitly discussed are listed in table 8.1. Pulsar 2020+28 shows right to left at orthogonal jump (data from Rankin 1983a; PSR 1508+55 from Manchester and Taylor 1977). None of the other properties of these pulsars seem particularly remarkable, the periods being clustered around fairly typical values.

One can obtain significant circular polarization, however, simply by adopting smaller values of γ. If one picks a number like 3° for the half-width of the subpulses (the integrated profile is generally much broader), one obtains a Lorentz factor of only 9. Because the radiation rate scales as γ^4, the incoherent radiation rate per particle is rather low, ultimately requiring implausibly small radii of curvature (ρ_c) for the magnetic field lines in order to increase the rate to useful values, even after boosting the values with the highest possible coherence factors. Moreover, the incoherent spectrum now extends up only to frequencies of order $\gamma^3 c/\rho_c$, which is an even more demanding constraint on the curvature.

It is important to note that this polarization signature is just what would be expected from curvature radiation. Solokov and Ternov (1968) give the circular polarization intensity of synchrotron radiation viewed at an angle to the plane of motion as

$$V = 64\xi/\pi\sqrt{3}(1 + \xi^2)^3 \qquad (1)$$

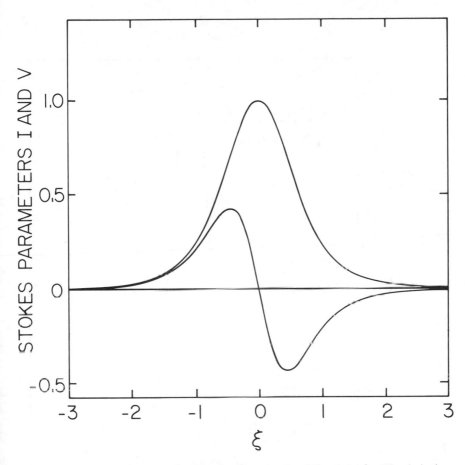

FIG. 8.1. Synchrotron pulse shape and polarization. Intensity of total (I) and circular (V) polarization as function of normalized viewing angle (see text).

while the total intensity is

$$I = (7 + 12\xi^2)/(1 + \xi^2)^{7/2}. \tag{2}$$

These equations then predict a maximum circular polarization intensity at a percentage polarization of 61.3% for $\xi = 1/\sqrt{5}$, where $\sin \phi = \xi/\gamma$ is the sine of the viewing angle. (The percentage of circular polarization rises to 100% at large angles, but the relative intensity is insignificant there.) This pattern is plotted in figure 8.1. For comparison, these polarizations are shown as measured for PSR 2002+31 in figure 8.2.

Rotation of the star will rotate the plane (in which radiation from a curved flux tube lies) past the observer. Viewing this plane edge-on, the observer sees linearly polarized emission. Viewing from above the plane, the observer sees

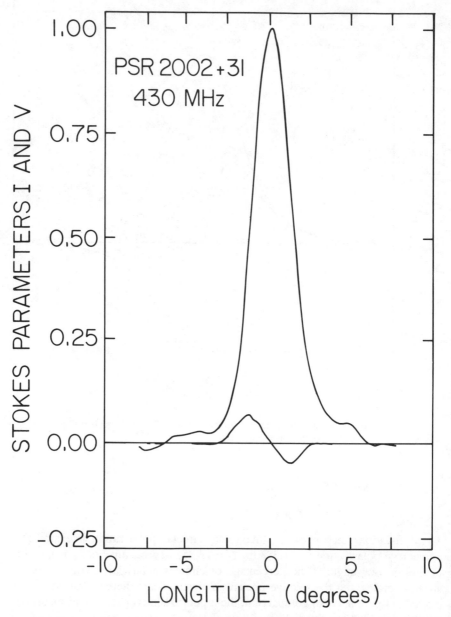

FIG. 8.2. Linear and circular polarization of PSR 2002 + 31. Note the similarity to figure 8.1, although not too many such cases are known. After J. M. Rankin, 1983a, *Ap. J.*, 274, 344 (figure 6).

elliptically polarized radiation of one-handedness and the opposite handedness when viewing from below.

The much more celebrated *position angle swing* of the linearly polarized component would follow exactly as in the hollow cone model because one preferentially sees radiation from field lines having $\phi \approx 0$, and these are constantly changing as the star rotates. Some swing would be seen even if only a single field line were illuminated because the plane of the field line twists as well as rotates past the observer if the spin axis is not orthogonal to the line of sight. If the spin axis were orthogonal to the line of sight *and* the radiation were entirely from a single field line lying in the meridional plane, there would be no significant change in position angle; there is negligible change in position angle associated with changes in ϕ alone.

d. *Fan Beams*

If the radiation is due to curvature of the magnetic field lines, then geometrically the length along which radiation can take place is comparable to the radius of curvature (ρ_c). The natural expectation is emission in a fan beam spread out over about a radian in the plane of particle motion, but having a natural width of only about $1/\gamma$ to each side of the plane. Such a picture is very difficult to incorporate into the hollow cone model because such emission from the field lines surrounding the polar cap would *fill* the cone and produce a beam of about a steradian, inconsistent with observation; to avoid such filling, it has been necessary to implicitly assume emission from only a narrow range of height. Such phenomenology is a problem for theory because a short emission region both reduces the energy output proportionately and makes it even more difficult to identify an adequate bunching mechanism because bunching must then take place so quickly. Narayan and Vivekanand (1983b) argued from the statistics on position-angle swings that not just the emitted beam must be fan-shaped but the emission region at the surface must itself be fan-shaped. The simpler interpretation is that the fan beams are those expected from discharges along isolated flux tubes. Fan beams would help solve another problem, namely that of reconciling the pulsar birthrate with that of its likely progenitor, the supernova (sec. 1.5.10). With a fan beam we can reasonably suppose that we observe most of the pulsars rather than suppose that a large majority radiate in directions we cannot observe.

e. *Satisfying Basic Constraints*

The basic criteria that any model must meet are *not* generally considered to be circular polarization or beam shape but (1) sufficient radio power output, (2) reasonable spectrum, and (3) narrow, but not extremely narrow, pulses. Naively satisfying criterion (3) leads to failure in (1) and (2). A simple alternative is to ignore (3) for a moment and instead satisfy (1) and (2). Constraint (2) turns out to be the most stringent because the resultant radio spectrum must extend at least up to what is observed. For curvature radiation, the cutoff frequency is entirely determined by ρ_c and γ. Assuming that near the star higher magnetic multipoles dominate and consequently $\rho_c \approx a$, where a is the neutron star radius (see also Barnard and Arons 1982), and choosing an upper cutoff frequency $f_{max} \approx 3\gamma^3 c / 4\pi\rho_c$ large enough not to seriously conflict with observation, namely about 10^{10} Hz, imposes Lorentz factors of the order of 100, intermediate in magnitude between what would be required ($\gamma \geq 9$) to satisfy (3) and the Ruderman and Sutherland estimate ($\gamma \approx 800$). Extending the cutoff to 10^{11} Hz would increase γ by only a factor of 2.

Returning to the pulse-width question, there are a number of straightforward options. A finite angular extent for the flux tube would broaden the pulse, for example. Such broadening leads, however, to depolarization, with a tenfold broadening giving roughly a tenfold reduction in circular polarization from the theoretical level of around 61% to about 6%. Circular polarizations of 20% or so are not uncommon, so that some *nondepolarizing* broadening mechanism would also be required. One such mechanism would simply be a tilt in the long axis of the fan beam so that the observer did not pass through the beam along the narrow axis but along an intermediate axis. Also, the beams stay in the observer's line of sight longer when the observer happens to be at higher latitudes, which also can give an apparent broadening. A companion mechanism is broadening of the original beam by propagation effects, because a mechanism that simply spreads the ray trajectories out more widely need not cause depolarization. Finally, the natural beam width is wider at frequencies below the cutoff (a potentially observable parameter). It is therefore possible that the pulse-width constraint can be resolved separately, in which case the basic spectral and power constraints for coherent radio emission (e.g., Michel 1978a) might be met.

f. The Radio Luminosity

Given that the typical pulsar radiates about 10^{-5} of its spin-down energy in the radio, one can calculate what coherence factor is required and reinterpret that factor in terms of the density fluctuations $\delta n/n$ required in bunches whose maximum factor would be the space-charge concentration $n = 2\epsilon_0 \Omega B_0 / e$ in a volume of a cube with side $(\lambda/2)$ proportional to the wavelength at the maximum in the spectrum (we take λ to be of the order of a meter). The radio power is simply the above factor times the radiation rate for incoherent curvature radiation

$$\dot{W} = e^2 \beta^4 \gamma^4 c / 6\pi\epsilon_0 \rho_c^2 \tag{3}$$

for a single particle moving at velocity βc in a circle of radius ρ_c, with $\gamma = (1 - \beta^2)^{-1/2}$ as usual. The total output per particle is this rate for a time of order ρ_c/c. Putting in numbers we get

$$\delta W = (10^{-9}\,\text{eV})\gamma^4 / \rho_c \,(\text{m}) \tag{4}$$

as the incoherent synchrotron energy radiated per particle, where $\rho_c(\text{m})$ is the radius of curvature in meters. Synchrotron radiation is confined to a cone angle $\delta\theta = \cos^{-1}\beta \approx 1/\gamma$, which for 5° (half-width) corresponds to a Lorentz factor of about 12. The coherence factor for a 1 s pulsar corresponds to an electron concentration of about 6×10^{10}/cc in a 1 m cube, or 6×10^{16}. We then obtain

$$\delta W = (1.2 \times 10^{12}\,\mathrm{eV})/\rho_c\ (\mathrm{m}) \tag{5}$$

for the coherent synchrotron energy loss per particle. Given the loss rate per particle, we can estimate the total output current to be the space-charge concentration times c times the two polar cap areas; the radius of the latter is given for a dipole magnetic field by a^3/R_F, where R_F is the fiducial distance (traditionally assumed to be the light-cylinder distance $R_L = c/\Omega$). For a 1 s pulsar, the loss rate is 2.5×10^{30} electrons/s, to give a total luminosity of $3 \times 10^{42}/\rho_c\ (\mathrm{m})$ eV/s, or $5 \times 10^{30}/\rho_c\ (\mathrm{m})$ ergs/s. The magnetic field strength is estimated from the period and period derivative in the usual way (sec. 2.3.3):

$$B_0 = \left(\frac{3Ic^3 P\dot{P}}{8\pi^2 a^6}\right)^{1/2} \tag{6}$$

with $I \approx 10^{45}$ g cm^2 the estimated neutron star moment of inertia.

Assembling the above factors gives a radio luminosity of

$$L_r = \frac{\delta n}{n}\frac{1}{6}\frac{\epsilon_0 \Omega^3 B_0^2 a^3 \lambda^3 \gamma^4}{\rho_c}. \tag{7}$$

In principle then one can work backwards from the observable parameters the period, period derivative, radio luminosity (estimated from S_{400} and the dispersion measure), wavelength of spectral maximum ($\approx\lambda$), spectral high-frequency cutoff ($f_{\max}$), and pulse width structure/polarization to calculate the magnetic field, the accelerating potential drop, the particle Lorentz factor (γ), the curvature of the magnetic field lines (ρ_c), and the degree of bunching in the beam ($\delta n/n$). Here we substitute guesses for ρ_c, λ, and $f_{\max}$ and we adopt a nominal power output or efficiency. If complex fields near the surface are invoked, one might expect a shorter radius of curvature comparable to the size of the star; hence $\rho_c = 10^4$ m. We have two constraints on ρ_c: (1) the synchrotron spectrum must span the observed range, which is nontrivial for small values of γ, and (2) *any* energy-loss mechanism giving some $\dot{W}$ operating over a distance ρ_c must give the observed radio emission, which puts an independent restriction on ρ_c. The estimate $\rho_c \approx a$ gives a radio luminosity of 5×10^{26} ergs/s (assuming complete bunching but the above small value of γ), which is still a bit low but approaching the right order of magnitude. Allowing for higher values of γ then gives a more favorable fit. The inferred parameters for a pulsar with $P = 1$ s, $\dot{P} = 10^{-15}$, and $L_{\mathrm{radio}}/L_{\mathrm{total}} = 10^{-5}$ are given in table 8.2. It is encouraging that the density contrast $\delta n/n$ can be relatively small, because that gives some room for a spectrum of bunch sizes.

TABLE 8.2—**Typical Pulsar Parameters**

Input parameters:	
P (s)	1.0
$\dot{P}$	1.0×10^{-15}
L_{radio}/L_{total}	1.0×10^{-5}
Spectral cutoff f_{max} (Hz)	1.0×10^{10}
Curvature radius ρ_c (m)	1.0×10^{4}
Spectral maximum λ (m)	1.0
Conventionally derived parameters:	
L_{total} (ergs/s)	3.9×10^{31}
L_{radio} (ergs/s)	3.9×10^{26}
Polar magnetic field B_0 (gauss)	1.0×10^{12}
Electron concentration n (/cc)	7.0×10^{10}
Maximum coherence factor $(n\lambda^3)$	7.0×10^{16}
Light-cylinder distance R_L (m)	4.8×10^{7}
Polar cap radius ρ_0 (m)	1.4×10^{2}
Corotation radius R_C (m)	1.7×10^{6}
Particle-loss rate (/s)	2.8×10^{30}
Parameters derived here:	
Lorentz factor γ	111.
Density contrast $\delta n/n$	6.7×10^{-4}
Minimum beam radius ρ_f (m)	4.5
Incoherent radiation rate (eV/s)	0.46
Coherent radiation rate (eV/s)	2.7×10^{12}
Energy emitted/particle (eV)	8.9×10^{7}
Accelerating potential drop (V)	1.46×10^{8}

One might have supposed that the particles' Lorentz factors are restricted to small values by the coherent radiation reaction, but here the accelerating potential drop is comparable to $\gamma mc^2/e$ (obviously the variation in γ along the field line should be taken into account, but the parameters will still be of the same order). In such a model *no* highly relativistic particles are produced locally. Heating of the surface by any downward precipitating particles (e.g., from pair production) would be essentially at the same rate as the radio emission rate, which in turn corresponds to a small continuum luminosity ($\approx 10^{-6}\ L_\odot$). Such luminosity is an important constraint given that pulsars would be visible optically or in the X-ray if their luminosities were comparable to their spin-down luminosities, but would generally not be visible if their luminosities were comparable to their radio luminosities.

g. *Flux Tube Scales*

There are theoretical constraints on what these flux tubes might be like. The electric current out of a pulsar scales as the space-charge density times the polar cap area times the velocity of light. The immediate consequence is that a flux tube carrying this much current must be as large as the polar cap itself, unless the current density is augmented by the counterstreaming of negative and positive charges. One cannot simply adopt a significantly larger space-charge density in the flux tubes without causing the electrodynamics to be dominated by the flux tube instead of the star. Counterstreaming is itself an attractive source of free energy to drive particle bunching. Viewed instead as a discharge between two conductors, the beam could plausibly be partially neutralized with charges of one sign from the neutron star and the other from a disk (sec. 6.2.3). Neutralization is essential if the beam size is to be less than the polar cap size, which is necessary in turn if several beams are to be installed in that general area to account for complex pulse shapes. The alternative of creating electron/positron pairs to partially neutralize the beam becomes rather problematical not only because it does not seem to work but also because if the Lorentz factors are actually low, no pairs can be produced (sec. 2.3.7), unless, as Ruderman and Sutherland suggest, there is a very energetic primary component to generate the pairs as secondaries with such Lorentz factors. The issue of beam size becomes particularly acute for a fast pulsar such as the Crab, which should have a relatively large polar cap yet has narrow, sharp features in the pulse profile.

The minimum beam size can be estimated by noting that the perturbation magnetic field (δB) at the surface of a flux tube should not exceed the ambient field; otherwise it would not be a current discharge guided by the neutron star magnetic field but would dominate the magnetic field structure, a companion argument to why the space charge cannot much exceed the ambient values. With the standard estimate of the electric current out of the polar caps, one finds that $\delta B/B \approx 1$ at the light-cylinder distance, which is equivalent to the usual expectation that one moves from quasi-rigid corotation to an outflowing wind at such distances (hence from a poloidal to toroidal magnetic field geometry). Closer, the surface magnetic field increases inversely with the flux tube radius (r_f) while the ambient magnetic field increases as $1/r_f^2$, so $\delta B/B$ is proportional to r_f (i.e., the perturbation magnetic field becomes unimportant near the star, where the flux tube radius is of order of the polar cap radius, ρ_0). If we take a different flux tube size, ρ_f, the perturbation field scales inversely as the ratio, so putting these factors together, the perturbation field would be unity at a fiducial distance where

$$\rho_f \approx \rho_0 (R_F/R_L), \tag{8}$$

which simply states that unneutralized flux tubes cannot exist if the fiducial distance is out at the light-cylinder. This provides yet another way of understanding why nonneutral plasmas would shut down pulsar action (sec. 4.2). If we take the corotation distance, $R_C = (GM_{NS}/\Omega^3)^{1/2}$, as the fiducial distance, as in the disk model, the flux tubes could be as small as 3.5% of the polar cap radius, or about 5 m if circular in cross section. It seems more likely that they are arranged in sheets similar to the auroral precipitation seen on Earth. The particle concentration could then be 800 times larger. The flux tubes could, of course, be larger than this lower limit. The value of γ might then be related to flux tube dimensions.

It is possible that coherent radio emission is curvature radiation from bunches in which only one or a few geometrically fixed flux tubes are radiating into a fan beam. These field lines would have to be basically the main (dipolar) magnetic field lines from the polar caps except near the star where local field inhomogeneities are required to bend and twist the flux tubes idiosyncratically. In such a model, the polarization signature for a single beam is the same as that for synchrotron radiation from a single particle in a circular orbit: circular polarization that changes sign as a distant observer line of sight passes through the plane of the orbit. To get the energetics to make sense, a narrow elementary beam must be broadened by the finite physical dimensions of the flux tube (at the expense of depolarizing the circular component down to about 6% for a nominal factor of 10 broadening) and possibly by some nondepolarizing propagation mechanism such as having a tilted fan beam (the locally distorted field line need not lie in the meridional plane). The emission would (Michel 1987d) then be in the form of a fan beam. Possibly overlap of the beams from distinct flux tubes accounts for the sudden orthogonal mode changes or reversals of circular polarization sometimes observed.

If the subpulses are indeed elementary structures (i.e., discharges on narrow curved flux tubes near the stellar surface), spectra of individual discharges could perhaps be isolated and might be more diagnostic of the coherent radiation mechanism than the overall spectrum, which might consist of numerous overlapping sources. Some work along these lines has already been done by Krishnamohan and Downs (1983), albeit only at one frequency (i.e, without spectral information). They found that the integrated pulse shapes of the Vela pulsar (PSR 0833–45) varied systematically with individual pulse intensity, and they thereby deconvolved the otherwise seemingly simple mean pulse profile into four components (assumed Gaussian) which they found to radiate essentially independently of one another. Taylor et al. (1975) studied the modulation index of pulsars, a measure of the pulse-to-pulse variability, and found that this index varied with pulsar phase. Krishnamohan and Downs argued that this variation of the modulation index suggested that a multiplicity of poorly resolved components was probably common for pulsars. They also found that Gaussian fits were poor at the tails of the profiles, in the sense that the Gaussians cut off too fast, suggesting indirectly that the synchrotron profile (eq. 2) might give a better fit. They also found that one of the four components required a linear polarization roughly orthogonal to the rest (here required to give depolarization rather than mode changes). To avoid having some components fall on top of one another, the above authors suggested that they were dispersed in altitude as much as 500 km, but fan beams could produce the apparent overlap of separated sources. The theoretical difficulty in locating the emission regions at such large distances from the star is that the particle concentrations (hence coherence factors) are 10^5 times smaller and the field line curvatures necessarily 50 times larger, factors which are not compensated for by the shorter period and larger-than-average magnetic field of this pulsar, and which are out of the question for typical pulsars.

2. Pair-Plasma/Beam Instability near the Surface

The prototype here, the Ruderman-Sutherland model (1975), does not differ too much geometrically from the above except for the assumption that the radiating medium is a pair plasma which has been created by very energetic particles accelerated from the pulsar. The pair plasma is itself taken to be essentially two interpenetrating beams of electrons and positrons in their local rest system (both being convected away from the neutron star). Except for the slower growth rate imposed by relativistic motion away from the star, similar features to the above model can be incorporated, such as concentration of the flows into sheets or filaments (flux tubes) and boosting of the particle concentration above the space-charge separation minimum. The problem with the model as originally elucidated is its failure to identify where these currents might be flowing to. But the concept of a pair-plasma/beam instability could certainly be exported to other model geometries given that the input assumption is simply a very relativistic beam moving on curved field lines, a rather general concept. As noted earlier, it is difficult to get a sufficiently energetic beam to produce a copious pair plasma for the slower pulsars.

A careful self-consistent simulation of pair production has yet to be made, so it is difficult to include here much analysis of this important possible model. Typically the pair plasma is simply modeled by assuming one Lorentz factor for the electrons and one for the positrons. Judicious choices of density, density fluctuations (bunching), and Lorentz factors apparently provide sufficient power output to account for very active pulsars such as the Crab.

3. Discharges in Gaps

Ruderman has pioneered a long study of the possible discharges that might occur in gaps (e.g., sec. 4.2) in pulsar magnetospheres, exemplified by Cheng et al. (1986a). These models are directed at explaining activity in the very energetic pulsars (Crab and Vela), whose spectra extend to quite high energies. Whether the physics generalizes to pulsars as a whole is not presently clear.

8.3 Bunching Mechanisms

1. Klystron-Type Mechanisms

The simplest model involving ad hoc particle bunches is the model proposed by Gold (1968). What, however, would produce the bunches in a self-consistent model? In the standard vacuum model, there is simply a direct acceleration and ejection of the particles from the neutron star, so there is no beam instability to be excited (e.g., by the two-stream instability). Goldreich and Keeley (1971) argue that particles moving in an arc (actually in a circle, for simplicity of analysis, but here simulating curved dipolar magnetic field lines) are unstable to clumping owing to the radiation reaction of one upon the other. Buschauer and Benford (1978) and Asséo et al. (1983) have carefully reconfirmed the analysis of this effect. The growth rate S is given from (Goldreich and Keeley 1971, eq. 27):

$$\frac{S}{\omega_0} \approx \left(\frac{\Gamma\left(\frac{2}{3}\right) r_c N_0}{2\,(3)^{1/3}\, \rho_c \gamma^3} \right)^{1/2} n^{2/3}(1 + i\sqrt{3}), \tag{1}$$

where $r_c = 2.82 \times 10^{-13}$ cm is the classical electron radius, N_0 the particle density in the beam in particles/radian, ρ_c the radius of curvature, γ the Lorentz factor, n the circular harmonic corresponding to the bunch size, and ω_0 the circulation frequency (for a ring). Let us concentrate on the dimensionless ratio $r_c N_0/\rho_c$. For polar cap ejection, $N_0 = \dot{N}/c2\pi\rho_c$, where $\dot{N}$ is the particle flux out of the pulsar, and $\rho_c \approx (aR_L)^{1/2}$ for the curvature of dipole field lines. For the Crab pulsar, $\dot{N} \approx 1.1 \times 10^{34}$ and $\rho_c \approx 1.2 \times 10^7$ cm; thus $r_c N_0/\rho_c \approx 1.1 \times 10^{-4}$. The only factor that can increase the growth rate is n, the harmonic that is excited, which must be of the order of $2\pi\rho_c/\lambda$ (where λ is the bunch size, which in turn must be of the order of a meter to give the observed spectra); hence $n^{2/3} \approx 8.3 \times 10^3$, which helps considerably. Using $\Gamma(2/3) = 1.35411\cdots$ gives then a growth rate of order $60\gamma^{-3/2}$.

The clear difficulty is that the particles can scarcely be relativistic! Furthermore, for slower pulsars, the expectation is that ρ_c increases and $\dot{N}$ decreases. If one uses the much larger particle flux rates of 10^{40} particles/s, the growth rate obviously improves dramatically, but this would have to be an electron/positron beam, which would require a complete reanalysis of this bunching mechanism.

Larroche and Pellat (1988a) have reanalyzed this geometry and concluded that rapid growth is only obtained for plasmas with a sharp spatial gradient (again, as one might expect for currents flowing as sheets). There is some debate on this point, however (Beskin et al. 1988; Larroche and Pellat 1988b). These more realistic estimates give, for optimal gradients of the order of 100 cm, large growth rates of order 3×10^4/s using Ruderman-Sutherland parameters. In fact, the situation may be more favorable in laboratory storage rings, where S/ω_0 is still small but storage times are longer, giving $ST \gg 1$ and hence more potential of observing such bunching (Michel 1981). Alternatively, it has been proposed that radiation reaction at the pulsar surface immediately causes bunching by nonlinearly suppressing particle emission from the surface (Michel 1978a), which may effectively be equivalent to the above mechanism in that it would rely on the same physical effect, radiation reaction of one part of the beam on the other. These are all probably variations on the klystron mechanism in some sense (see also Rylov 1976 and Sturrock 1971a). The radiation characteristics expected from such bunches are discussed by Saggion (1975), Eidman (1971), Cox (1979), Epstein (1973), Sturrock et al. (1975), Buschauer and Benford (1976, 1977a,b), Shklovsky (1970a), Pacini and Rees (1970), Tademaru (1973), Michel (1978a) (see figure 2.22 above). Rylov (1978) discusses a model with pair production (see below) wherein the return beam is excited by passing through a trapped space-charge region.

2. Turbulence

The distinction between bunching of particles and plasma "turbulence" (the latter is surely one of the all time favorite buzz words in astrophysics) is often one of style. However, the latter term makes sense if the radiation is indeed from a spectrum of inhomogeneities instead of from a few discrete ("monochromatic") bunches. The following authors specifically refer to radiation from *turbulence* excited in the plasma: Layzer (1968), Coppi and Ferrari (1970a,b, 1971), Ichimaru (1970), Coppi (1972), Buckee et al. (1974), Khakimova et al. (1976), and Hinata (1976a,b,c, 1977a,b, 1978, 1979).

3. Pair Production Models

In the models where positron-electron plasma is invoked, one has counterstreaming of the two equal-mass particles, hence some version of the two-stream instability, hence bunching (Cheng and Ruderman 1977a; Hardee and Rose 1976, 1978; Arons and Smith 1979; Mikhailovskii 1980). In general, such beam-plasma instabilities have received considerable attention (Elsässer and Kirk 1976; Elsässer 1976; Hinata 1976a,b; Melrose and Stoneham 1977; Buschauer 1977; Rylov 1978; Lominadze et al. 1979a,b; Lominadze and Mikhailovskii 1979; Hardee and Morrison 1979; and Asséo et al. 1980b). Benford (1975) has argued that the beam is unstable to filamentation. Other discussions based on the assumption of pair production are given by Parker and Tiomno (1972a,b), Hinata (1973), Al'ber et al. (1975), Daugherty and Lerche (1975, 1976), Jones (1977b, 1978, 1979, 1980b,c), Kirk and ter Haar (1978), Lominadze et al. (1979a,b), Sturrock and Baker (1979), and Kundt and Krotscheck (1980).

4. Two-Stream Instability

A standard way of bunching particles is to invoke the two-steam instability wherein the free energy in counterflowing particles is transferred into electrostatic oscillations. In the standard model, with all the particles flying away from the star, the streaming is outwardly directed, and therefore the relative streaming is in the rest system of this outward flow. Consequently there is significant difficulty in getting the instability to act fast enough for bunching still to occur near the star, where the density is still large. Usov (1988) discusses these difficulties and the nonstationary phenomena that can circumvent them.

Alternatively, if there is a discharge between the star and a disk, for example, the growth rate becomes somewhat irrelevant: the fluctuations simply grow to such an amplitude that radiation reaction removes energy as rapidly as the instability takes energy out of the directed motion in the discharge. Of course the amplitude cannot be arbitrarily large. Normally it is thought that the resultant oscillations at the plasma frequency cannot radiate because the plasma itself is essentially opaque to waves at the plasma frequency. This is true for homogeneous plasmas, but not inhomogeneous ones, because oscillations within a skin depth (order c/ω_p) couple to the external vacuum and are seen as oscillating dipoles on the surface. For thin sheets or filaments of plasma, then, the radiation can escape to the sides. For discharges into something like an auroral zone on the pulsar, these are just the structures to be expected, as illustrated in figure 2.23. Melrose (1978) has considered pulsar radiation induced by oscillating parallel electric fields.

5. Other Mechanisms

Ruderman (1981) has explored the role of instabilities between counterstreaming Fe ions and backstreaming electrons. These electrons could come either from a pair production region or from stripping of incompletely ionized Fe primaries (see Cheng and Ruderman 1980). Arons and Smith (1979) propose a shearing instability which would operate even in a completely charge-separated magnetosphere.

8.4 Maser Mechanisms

1. Chiu and Canuto

One of the first mechanisms proposed was a true maser mechanism through population inversion. Here the states in question are the Landau orbitals. The pumping to produce this population inversion was assumed to be driven by an oscillation of the neutron star, with rotation providing the timing mechanism (Chiu and Canuto 1969a,b; Chiu and Occhionero 1969; Chiu et al. 1969; Chiu and Canuto 1970; Chiu 1971; Chiu and Canuto 1971; Canuto 1971). The high brightness temperatures asserted were criticized at the time (Roberts and Fahlman 1969; Simon and Strange 1969), although at least one attempt has been made to resuscitate the model (Virtamo and Jauho 1973, 1975; see also Casperson 1977 and Mertz 1974).

2. The Russian (Georgian) School

Except for the efforts of Chiu and Canuto, research in the United States has tended to assume that coherence is due to particle bunching. The Russians, however, have proven to be quite interested in the maser mechanisms. Broadly speaking, the idea is that electrons in a strong magnetic field quickly radiate away their transverse energy. This leaves behind a very anisotropic distribution function which exhibits certain instabilities. See Ginzburg et al. (1969a,b), Takakura (1969), Ginzburg and Zheleznyakov (1970a,b), Coppi and Ferrari

(1970a,b), Eastlund (1970), Tsytovich and Kaplan (1972), Zheleznyakov and Suvorov (1972), Kaplan and Tsytovich (1973a,b), Sazonov (1973), Suvorov and Chugunov (1973), Ginzburg and Zheleznyakov (1975; this paper is mainly a review paper), Machabeli and Usov (1979). See also Blandford (1975) for a critical, but not necessarily unfavorable, assessment of maser processes. (See Kawamura and Suzuki 1977 for a rather different maser process.) Kaplan et al. (1970) have even examined the possibility that maser action in the Compton downscattering of energetic photons (by energetic electrons) into radio photons might work. However, Zheleznyakov and Shaposhnikov (1979) note that if the energy density of the radiating particles is less than that of the magnetic field (the usual assumption, consistent with dimensional estimates), then coherent curvature radiation is ineffective for maser action (see also Blandford and Scharlemann 1976).

3. Stimulated Linear Acceleration Radiation

Cocke (1973) found a maser-like action in the electron acceleration region, assuming simply a uniform acceleration of electrons to ultrarelativistic velocities while following curved magnetic field lines. Melrose (1978) extended this work, finding it a promising emission mechanism, whereas Kroll and McMullin (1979) argued that the amplified emission is almost completely suppressed by propagation effects. However, Arons (private communication) has pointed out that Cocke considered a stream of electrons accelerated by a uniform electric field (E) from zero energy at zero height above the surface and thereby attaining energy eEh at height h. Then Cocke used an Einstein A and B formalism to calculate an emission rate, with the implicit assumption that the resultant particle population was inverted in energy-space. For this to work, it is necessary that particles have a range of energies *at the same place*, which is not the case. In this model, the stream has a δ function distribution in momentum and is not inverted, in the sense required by the Einstein A and B formalism. From a physical point of view, there is no free energy for collective emission: the beam propagates through a vacuum (in effect), not with respect to any other medium, so there are no instabilities.

4. Cherenkov Emission

Cherenkov emission was touched on in sec. 2.5.4. See Charugin (1975) and Kolbenstvedt (1977).

9

Winds and Jets from Pulsars

9.1 Introduction

As we have already seen, the quasi-neutral plasmas (usually created by ionizing neutral gas particles) are quite different from nonneutral plasmas (created by injecting particles of only one sign into a trapping field geometry). It comes as no surprise that the winds of these two different plasmas have significantly different properties.

1. Nebulae about Pulsars

It is widely presumed that a relativistic wind streams away from pulsars because the spin-down power of pulsars vastly exceeds that of their radio-frequency emission (roughly a factor of 10^5). How this wind is generated, how it propagates, and how it interacts with the nebula are all topics of continuing uncertainty, largely owing to a paucity of observable consequences of these phenomena. Most of the analytical work has been devoted to propagation, with relatively limited dimensional analysis applied to generation and termination. An early model (Rees and Gunn 1974) applied to the Crab Nebula simply assumed that the particles were shot out radially, carrying with them a *frozen-in magnetic field* until the flow shocked at some distance from the outer shell of the nebula, taken to be a distance of about 0.1 of the nebular size to match a supposed hole in the optical continuum emission claimed to be of about that size (the emission is actually complex and differs dramatically with polarization). Later work found that the flow should be so relativistic that it cannot shock in the usual sense and that it does not significantly slow down. All the power would be in the Poynting flux, and one has in effect a laser beam with a few particles swept along with it. To fix this problem, Kundt and Krotscheck (1980) and Kennel and Coroniti (1984a) proposed that the outflow was much more heavily loaded with plasma. Then the flow would significantly slow at the shock and be stagnated against the outer shell. It is quite possible that pair production could produce this excess of particles. Michel (1984b) reexamined the space-charge limited wind solutions and found (Michel 1985a) solutions in which the wind does not shock so much as it diverts into collimated outflows

370

which can be highly penetrating and extend as jets to large distances. A jet-like structure has been seen in the Crab Nebula (Fesen and Gull 1983), but it does not seem to be aligned with the pulsar (see also Benford 1984). Jets were long ago theorized for active galaxies (Lovelace 1976). Observation suggests that the development of such nebulae is more complicated than usually modeled. For example Braun et al. (1987) conclude that the structure of Cas A apparently consists of decelerated ejecta that are punctured by condensed fragments of slower-moving material that is now catching up with this outer shell. Any pulsar wind would then flow in a very complex environment.

As noted above, there is substantial direct observational evidence that the Crab pulsar excites the nebulosity, contributing both energetic particles and magnetic fields. The emission of such plasma, a generalized wind analogous to the solar wind but much more powerful, has been a subject of investigation for a long time. One of the early pulsar theories appealed to radio emission excited when such a wind is shocked (Michel and Tucker 1969). This model stimulated the extension of the existing nonrelativistic pressure-dominated solar wind solutions (Dicke 1964; Weber and Davis 1967; Modisette 1967) into the relativistic magnetically driven regime (Michel 1969a,b [see Goldreich and Julian 1970 for an elaboration]; Henriksen and Rayburn 1971; Belcher and MacGregor 1976; Kennel and Pellat 1976; Okamoto 1978; Nakamura 1980; Nerney 1980; and Kennel et al. 1983). It is a bit complicated to sketch through the wind derivation although the results are fairly easily understood. Basically, the magnetic field pattern rotates rigidly until the corotational velocity approaches the transverse Alfvén velocity (i.e., a radially propagating shear wave). This condition provides one critical point (Dicke 1964). A second critical point is where the radial flow velocity equals the compressional Alfvén velocity; for pressure-free flow (negligible temperature), the second critical point is asymptotic to infinity (Goldreich and Julian 1970). A central role in these calculations is played by the dimensionless parameter

$$\sigma = e\Omega Ba^2/mc^2 \text{ (monopole)}, \tag{1}$$

which is much larger than unity for relativistic flows, and vice versa. Physically, σ is what the Lorentz factor of the particles would be if, asymptotically, all the Poynting flux went into the particles. However, the MHD flow equations imply that the particles get a Lorentz factor of only $\sigma^{1/3}$. The total luminosity scales as σ^2 and the particle flux as σ; thus most of the energy output is in the Poynting flux in the relativistic limit. Another significance of σ is that it is the pole-to-equator electrostatic potential drop in units of the electron rest mass. Thus an electron emitted from the pole (here we are thinking in terms of an aligned rotator) would have to reach the equator to obtain a Lorentz factor equal to σ, which is understandably unlikely. Indeed, for applications to the dipole magnetic field case, the potential drop is more plausibly restricted to that over the polar caps (not the full pole-to-equator drop), and a more realistic estimate is therefore given by replacing a^2 with the polar cap area; hence

$$\sigma = e\Omega B(a^3\Omega/c)^{1/2}/mc^2 \text{ (dipole).} \tag{2}$$

One can also regard σ as an observationally inferred parameter, irrespective of theoretical considerations. In the calculations described above, the outstreaming plasma was taken to be neutral rather than charge-separated, which leads to some differences ($\gamma \approx \sigma^{1/2}$ in the latter case: Michel 1974b), although the physical idea is simply to treat the plasma as a conducting fluid entraining magnetic field lines (see also Henriksen and Rayburn 1972).

Kennel and Coroniti (1984a,b) and Kennel et al. (1983, eq. 3.5) use a different definition for σ, dividing the σ as defined above by $\gamma_{\text{injection}}$. In effect, they assume that the particles are already injected with their full energy and not particularly energized in the wind zone. For the Crab pulsar, the particle energy is then taken to be about 10^{12} eV versus a polar cap potential drop of about 10^{16} V, which would give a σ_{KFO} of about 10^4; however, they also adopt a number flux larger by 10^4 (presumable from pair production), giving σ_{KFO} about unity. But now the electron rest-mass energy is important because energy is required to produce these pairs, so in fact they take

$$1 + \sigma_{\text{KFO}} = \frac{\text{Poynting Flux}}{\text{Particle Energy Flux}}, \tag{3}$$

where σ_{KFO} can now be a small number. In applying this approach to the Crab Nebula, they seek solutions with a small σ_{KFO} so that the flow is nonrelativistic, which requires reducing the Poynting flux in the wind and reducing B at the light-cylinder to much less than usually assumed. Instead, they magnetize the nebula by accumulating this weak field over the age of the nebula, 900 or so years. There are some potential problems, although this view of a hot nonrelativistic wind would solve the important problem of how the nebula can stagnate the very relativistic flow conventionally estimated. The large B usually assumed at the light-cylinder comes from the spin-down torque. With energetic particles injected from the pulsar onto weak fields, the Alfvén-Dicke radius would be very close to the star, and it is not clear that the torque would not come out too small, in which case such solutions would correspond to discharging a battery at the pulsar rather than getting the energy out of rotation. Indeed, therein lies a problem with the "hot" wind solutions: where does the energy come from to heat the wind in the first place?

This torque problem (if indeed it is a problem) could be resolved by saying that the Crab pulsar is a nearly oblique rotator (for which there is observational evidence in that the polarization swing of the main pulse is in the same sense as that of the interpulse), and that it is the instability of the large-amplitude electromagnetic wave (which would be emitted in a vacuum) beyond the light-cylinder that consumes the magnetic wave field and instead accelerates the particles, with only the small component of magnetic field that is aligned with the rotation axis convected out into the nebula. This would allow a large magnetic field (giving the torque) at the light-cylinder but a small magnetic field at distances large compared to the light-cylinder.

Such a process would have to energize the particles rather cleanly (if they were thermalized in the 10^6 gauss fields near the light-cylinder, they would radiate with time scales of less than 1 ms). In any case, such a model is interestingly different from the simple MHD wind theories. In particular, there would then be only virtual critical points because the conversion of magnetic field energy into particle energy certainly has not been included in the MHD theories, and therefore one does not have an adiabatic flow through critical points near the light-cylinder but a quite different process. It is not clear that the resulting flow has to satisfy any critical point constraints at all!

9.2 Steady State Magnetohydrodynamic Solutions

1. The MHD Model

The only way to reduce the MHD equations to one-dimensional flow is to replace the magnetic field with a monopole field, simulating the case for field lines that have been pulled out almost radially by flow away from a more realistic magnetic field topology (e.g., dipolar). In this approximation and assuming radial flow, Maxwell's equations become

$$E_r = 0, \tag{1}$$

$$E_\theta = K(\theta), \tag{2}$$

$$B_\theta = 0, \tag{3}$$

and for steady state conditions ($\frac{\partial}{\partial t} = 0$)

$$E_\phi = 0, \tag{4}$$

and azimuthal symmetry ($\frac{\partial}{\partial \phi} = 0$)

$$B_r r^2 = \text{constant} = F. \tag{5}$$

a. *The Faraday Condition*

Finally, we have the so-called Faraday condition, which is given simply by setting the θ component of the Lorentz force equal to zero (the MHD approximation); hence

$$r(V_r B_\phi - V_\phi B_r) = rK(\theta) \tag{6}$$

$$= -\Omega r^2 B_r = -\Omega F. \tag{7}$$

The latter step assumes that one can follow the radial field lines right down to the surface. In that case conditions near the surface determine $K(\theta)$, which in turn determines conditions at large distances. (Note from eq. 7 that the monopole approximation actually requires K to be a constant.) Unfortunately, the former step is not obvious because the field lines are unlikely still to be radial near the surface. Equally important, the approximation itself is inconsistent. As $r \to \infty$, $E_\theta \sim 1/r$, $B_\phi \sim 1/r$, $V_r \sim$ constant, $V_\phi \sim 1/r$, and $B_r \sim 1/r^2$. Thus the $V_\phi B_r$ term in equation 6 is asymptotically negligible compared to the other terms, yet we have already neglected another term, the inertial term $d(P_\theta)/dt$, where P_θ is the particle moment in the θ direction. In other words, we are invited to keep one small term but to neglect another small term! However, this assumption is made in practically every MHD flow model published to date. In the asymptotic charge-separated plasma case discussed below, we lose any information on V_ϕ and B_r, and only by solving rather uncertain inner boundary conditions can we get the necessary information to flesh out the θ dependencies. At this point we regard equation 6 to be a safe assumption in the asymptotic limit where the $V_\phi B_r$ is unimportant, but are concerned about the extension to equation 7 and using the equation to determine the above product.

The remaining equations necessary to carry through the analysis are the laws for conservation of mass

$$\rho V_r r^2 = \text{constant} = f, \tag{8}$$

where f is the mass-loss rate per steradian, for conservation of momentum,

$$(\gamma \rho V_r + E_\theta B_\phi / \mu_0) r^2 = \text{constant} = \mu f, \tag{9}$$

where μ is the total energy per particle, and for conservation of angular momentum,

$$(\gamma \rho V_r V_\phi - B_r B_\theta / \mu_0) r^3 = \text{constant} = \delta f, \tag{10}$$

where δ is the angular momentum per particle. These relationships are sometimes written with $\gamma \rho_0$ instead of ρ everywhere, in which case ρ_0 is the proper density. The dimensionless parameter σ is defined in these terms by

$$\sigma = \frac{\Omega^2 F^2}{\mu_0 f c^2}. \tag{11}$$

These equations then give a set of algebraic relationships which typically consist of physical quantities being given from expressions of the form N/D, in which case the vanishing of D requires N to vanish at the same place (the critical points). From this the full solutions can be obtained with one further condition, namely that there is a critical point at infinity because these equations have zero pressure. The critical points correspond to points at which the flow velocities exceed local characteristic velocities. Thus the flow starts out slowly and below all characteristic velocities and, in a simple thermal expansion away from a star, accelerates through the speed of sound at the critical point. Other solutions would be possible, but they are time dependent and presumably settle down to stationary conditions when the correct behavior occurs at the critical points by propagating any differences as inward and outward propagating waves. In the cold rotating solution the flow is driven instead by *magnetic slinging* of the particles (e.g., Michel 1969b; Henriksen and Rayburn 1971) so the particles accelerate out to the point at which the transverse Alfvén wave speed equals the flow speed. This critical point determines the angular momentum flux out of the star. Because there is no pressure other than magnetic, the flow asymptotically approaches the compressional Alfvén velocity (there is no difference between the expressions for these two velocities in the zero pressure limit, but the first is fixed by B_r and the second by B_ϕ; see figure 9.1). This point was missed by Michel (1969b); lacking the one remaining condition necessary to fully determine the flow, it was assumed that the system acts to minimize the net torque. Goldreich and Julian (1970) then pointed out the missing critical point at infinity and established that the minimum torque hypothesis was equivalent to requiring passage through this critical point. However, the latter is important because it is sufficient to establish the energy condition alone (whereas the discussion is necessarily quite involved and indirect in the minimum torque discussion).

b. *Derived Quantities*

First let us obtain the asymptotic Lorentz factor. Let us note that in the frame of reference moving with the flow, the magnetic field is $B_0^2 = \frac{1}{\gamma^2}B_\phi^2 + B_r^2$, but in this limit $B_r/B_\phi \to 0$ so the second term vanishes at infinity. The proper Alfvén velocity in units of c is just

$$u_a^2 = \frac{B_0^2}{\mu_0 \rho_0 c^2}. \tag{12}$$

Unfortunately, B_ϕ is indeterminate unless we invoke the Faraday condition, which at large distances reads (eq. 7, neglecting the asymptotically unimportant $V_\phi B_r$ term)

$$B_\phi = -\frac{\Omega r B_r}{V_r} = -\frac{\Omega F}{r V_r}, \tag{13}$$

which on substitution into equation 12 becomes, writing $u \equiv \gamma V_v/c$ for the proper radial flow velocity,

$$u_a^2 = \frac{\sigma}{u}. \tag{14}$$

Therefore equating u and u_a at this critical point gives the asymptotic flow proper velocity $u = \sigma^{1/3}$. This result follows entirely from equations 5, 7, and 8, and requires the Faraday condition.

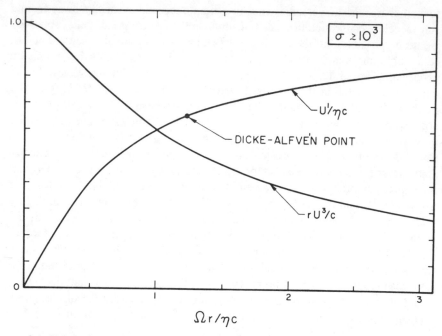

FIG. 9.1. Universal curve for magnetic sling acceleration. To extreme relativistic velocities. The radial proper velocity is plotted in units of its asymptotic value ($U^1 \mapsto \eta_c$, where $\eta^3 = \sigma$), while the distance scale is in units of η_c/Ω (c/Ω = corotation limit). The critical point is located at the corotation limit and is therefore lost into the origin in the limit $\eta \to \infty$. Note that the azimuthal proper velocity is plotted in units of c (not η_c as for the radial velocity), and consequently the rise from zero to about 1 in the vicinity of the critical point is not shown in this limit. The Dicke-Alfvén point remains fixed in this figure, located as shown. The second solution, which crosses at this point, is essentially a vertical line and has been omitted. From F. C. Michel, 1969b, *Ap. J.*, 158, 727 (figure 1); 1982, *Rev. Mod. Phys.*, 54, 1 (figure 23).

Another physical quantity of interest is the torque on the rotating star. The result is disarmingly simple and intuitively appealing: the torque exerted is simply that corresponding to rigid corotation of the mass flow as it moves outward on the radial magnetic field lines to the Alfvén critical point (Dicke 1964). However, this prescription is only interesting in the nonrelativistic limit where the particles carry significant angular momentum. In the relativistic limit the critical point is essentially moved out to the light-cylinder, but all the angular momentum is carried by the Poynting flux and is dominated by that term in equation 10. The torque per steradian is therefore

$$\frac{dT}{do} = \frac{F^2 \Omega}{\mu_0 c} \tag{15}$$

based on the estimate that the azimuthal magnetic field scales as

$$B_\phi = -\frac{\Omega F}{cr}, \tag{16}$$

with the radial velocity approximated by c. All of these estimates, however, rely on the Faraday condition.

c. *Restrictions*

Although a monopole magnetic field is an unphysical assumption globally, it is usually argued that it provides a simple idealization that should be close to the true behavior. But this argument is not obvious. For example, if the magnetic field is dipolar, B_r is zero by symmetry in the equatorial plane. However, if we accordingly set F to zero, all our estimates vanish! A more realistic modeling would, however, involve the neglected field components E_r and B_θ even with azimuthal symmetry, and there seems to be no additional constant of the motion(s) available to treat the two-dimensional extension that a realistic model would require.

2. Test Particle Acceleration

Here we will treat just the acceleration of a single particle in rectilinear crossed fields. We will compare that response to the above estimates as a useful first step to a self-consistent nonneutral wind and also for the case of acceleration in large-amplitude waves.

a. *Rectilinear Case*

We will take the electric field (E_0) to be in the y-direction, and the magnetic field (B_0) to be in the z-direction; the particle drift is then in the x-direction. The equations of motion can be integrated without recourse to any assumption of nonrelativistic velocities, and are

$$F_x = mc\frac{d}{dt}(\gamma\beta_x) = ev_y B_z \equiv ecB_0\beta_y \tag{17}$$

and

$$F_y = mc\frac{d}{dt}(\gamma\beta_y) = eE_0 - ev_x B_z \equiv eE_0(1 - \beta_x), \tag{18}$$

where we have assumed the limiting case $E_0 = cB_0$ to simplify the second equation. The characteristic cyclotron frequency for a nonrelativistic particle is just $\omega_c \equiv eB_0/m$, and our equations now become

$$\frac{d}{dt}(\gamma\beta_x) = \beta_y\omega_c \tag{19}$$

and

$$\frac{d}{dt}(\gamma\beta_y) = (1 - \beta_x)\omega_c, \tag{20}$$

which we will now solve exactly. Note that the time scale could be rescaled to eliminate ω_c but we will leave this explicit scale in place. Because the energy comes only from moving across equipotentials, we have

$$\dot{W} = \dot{\gamma}mc^2 = E_0 v_y = \beta_y c E_0 \tag{21}$$

or

$$\dot{\gamma} = \beta_y \omega_c, \tag{22}$$

which can be integrated immediately to give (for the initial condition $\beta_x = \beta_y = 0$)

$$\gamma = 1 + \frac{y\omega_c}{c}. \tag{23}$$

Substituting into equation 17 then gives exact derivatives on either side, hence the integral of the motion

$$\beta_x = 1 - \frac{1}{\gamma}. \tag{24}$$

The other velocity component is given from

$$\frac{1}{\gamma^2} \equiv 1 - \beta_x^2 - \beta_y^2, \tag{25}$$

which, after eliminating β_x with equation 24, gives

$$\beta_y^2 = \frac{2}{\gamma - 1}. \tag{26}$$

We can eliminate γ with equation 23 and thereby calculate the trajectory from

$$\frac{dy}{dx} = \frac{\beta_y}{\beta_x} = \left(\frac{2c}{y\omega_c}\right)^{1/2}, \tag{27}$$

to obtain

$$x = \frac{y^{3/2}}{\lambda^{1/2}}, \tag{28}$$

where $\lambda \equiv 9c/2\omega_c \approx 1.6 \times 10^2 B$ (gauss) cm. The energy gain is simply

$$\gamma = 1 + \frac{9}{2}\left(\frac{x}{\lambda}\right)^{2/3}. \tag{29}$$

Note that although the particle is carried along with the Poynting flux as one would expect, the energy gain becomes increasingly less efficient with distance. The assumption that the input energy was zero is not essential to obtaining these exact relationships.

b. *Radial Flow Case*

We now consider radial flow. The major change is that the divergence of the equipotentials and the tendency for flow in any direction to become asymptotically radial combine to essentially defeat energization of the particles to Lorentz factors as high as σ, unlike the rectilinear case, where, given a sufficiently long run in the system, particles will eventually cross all of the equipotentials. The radial and meridional Lorentz force equations become, asymptotically, with the fields E_θ and B_ϕ,

$$\frac{d(\gamma\beta_r)}{dt} = \lambda\beta_\theta \tag{30}$$

and

$$\frac{d(\gamma\beta_\theta)}{dt} = \lambda\left(\frac{1}{\beta_0} - \beta_r\right), \tag{31}$$

where β_r is the radial velocity in units of c, the quantity $\beta_0 \equiv E_\theta/cB_\phi$ (we will argue that $\beta_0 \to 1$ and is essentially the asymptotic velocity in the flow direction), and

$$\beta_\theta \equiv \frac{rd\theta}{cdt}, \tag{32}$$

while

$$\lambda = \frac{eB_\phi}{m}. \tag{33}$$

One integral of the motion is the particle energy; thus

$$d\gamma = r\lambda\beta_0 d\theta, \tag{34}$$

with $r\lambda$ a function only of θ, so equation 30 can now be integrated over dt using

$$\dot{\gamma} = \beta_0\lambda\beta_\theta \tag{35}$$

to give the second integral of the motion (starting the particle from rest with $\beta_r = 0$, $\gamma = 1$)

$$\beta_r = \beta_0\left(1 - \frac{1}{\gamma}\right). \tag{36}$$

We can now calculate the particle energy by solving for β_θ using the definition for γ

$$\gamma^2 = (1 - \beta_r^2 - \beta_\theta^2)^{-1} \tag{37}$$

to write

$$\beta_\theta^2 = (1 - \beta_0^2) + \frac{2\beta_0^2}{\gamma} - \frac{1 + \beta_0^2}{\gamma^2}. \tag{38}$$

If $\beta_0 \neq 1$, it might appear that the dominant term would be the first term, which would indeed be the case in the laboratory where the electromagnetic fields can be varied arbitrarily. For a wind, β_0 is the asymptotic flow velocity, with the result that the fields are not arbitrary but must be self-consistently adjusted to give this ratio. This term is then of the order of $1/\gamma_\infty^2$ and negligible in the ultrarelativistic limit. The second term therefore gives $\beta_\theta \approx \gamma^{-1/2}$, which scales the bulk of the acceleration process. Remembering that $r\lambda$ is essentially a constant ($\approx \sigma$) and that for relativistic motion we can approximate $dt = dr/c$ lets us write equation 35 as

$$(r\lambda)\frac{dr}{r} \approx \frac{d\gamma}{\beta_\theta} \approx \gamma^{1/2}d\gamma, \tag{39}$$

so $\gamma \approx (\sigma \ln \frac{r}{r_0})^{2/3}$.

It is easy to get a wrong result here. For example, if we assumed that β_r and β_θ approach constants, which is essentially the drift approximation where we set the acceleration vector parallel to the particle trajectory (assuming a fixed asymptotic velocity direction), we would get $d\gamma/dt = \lambda$ and a linear acceleration with distance. What happens in fact is that β_θ quickly rises as the particle accelerates more or less directly across the magnetic field lines; then when $\gamma \approx 2$ it reaches a maximum value and thereafter declines toward a value of the order of γ_∞^{-2}, as shown in figure 9.2.

Let us recap here. The MHD approximation, which is suspect here, gives wind particle energies of order $\sigma^{1/3}$, whereas the charge-separated wind picture gives wind particle energies of order $\sigma^{2/3}$. In addition, the space-charge limited flow approximation requires particles to have energies of the order of $\sigma^{1/2}$. There are at least three different particle energization steps, which complicates the discussion: (1) the particle energy at injection into the wind zone (the above $\sigma^{1/2}$ is only a generic estimate), (2) the wind zone acceleration by $\sigma^{2/3}$, and finally (3) the acceleration of the particles when the wind interacts with the surroundings. If the radial wind picture held for all distances, interstellar space would be charged to huge potential differences as these winds extended out from the pulsars, which is ultimately preposterous because the ambient particles would absorb all of the Poynting flux at some point.

Although the calculation of test particle acceleration is straightforward, a number of estimates have been given in the literature. Goldreich and Julian (1969) quote $\gamma \approx \sigma^{1/3}$, but do not give any details of the calculation. Ostriker and Gunn (1969a, eq. 58) give $\gamma \approx \sigma^{2/3}$ but with a radial dependence of $(1 - \frac{r_0}{r})^{2/3}$ rather than logarithmic, as they give elsewhere (Gunn and Ostriker 1969, eq. 10). Buckley (1977a,b) suggests, on the basis of a sketchy argument, that

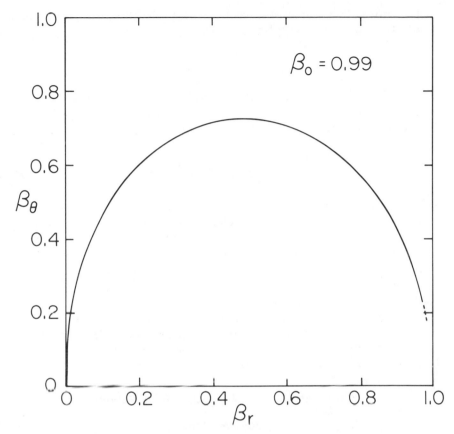

Fɪɢ. 9.2. Single-particle trajectory in velocity space. Numerical integration shows that trajectory is cycloidal.

$\gamma \approx r$ and therefore $\gamma \to \sigma$. As noted above, the MHD picture gives $\gamma \approx \sigma^{1/3}$. Asséo et al. (1975) give general formulae (their eqs. 33 and 34) from which our results can be recovered in the long-wavelength limit. Previously Kennel et al. (1973) obtained $\gamma \approx \sigma$.

It seems entirely implausible that test particles introduced into an MHD flow gain more energy than the particles in the flow itself, so something is wrong with one or the other of the above two limiting analyses. The natural suspicion is the MHD approximation, which is at the ragged edge of applicability and requires the questionable simplification of assuming a monopole magnetic field. Simulations with a more realistic model would cast light on whether that is indeed the case.

3. Phase Locking in Waves

We can use the above relationships to see if, when the fields are given by a large-amplitude electromagnetic wave, the particles will drift significantly in phase. Here we examine just the rectilinear case. The phase change relative to a wave propagating at c is just given from

$$\Delta L = \int_0^y (1 - \beta_x)\, dy, \tag{40}$$

which is just the distance the particle slips behind while going a distance y across equipotentials. But $1 - \beta_x = 1/\gamma = 1/(1 + y\omega_c/c)$, so the integral immediately gives (remember this is the rectilinear case, not radial expansion)

$$\Delta L = \frac{9}{2}\lambda \ln \left[1 + \frac{9}{2} \left(\frac{x}{\lambda} \right)^{2/3} \right]. \tag{41}$$

Given the Fermi criterion that logarithms are rarely larger than 5 to 10, even a wave from a millisecond pulsar would be very much longer than the above slip distance, so phase locking is essentially assured (Ostriker and Gunn 1969d; they also calculate the radiation produced when the particles are first accelerated). Even for laboratory scales, phase locking is feasible and has been proposed as a way of accelerating energetic particles in laboratory laser beams (e.g., Katsouleas and Dawson 1983).

In the above analyses, the particles gain essentially the same energy when injected into the wind *regardless* of distance from the pulsar. Clearly that cannot be true. The modulating effect of the external medium on the acceleration will be addressed, to a limited extent, in sec. 9.3.

9.3 Steady State Charge-Separated Solutions

1. Nonneutral Models

In the MHD picture the particle charge densities and currents are automatically taken into account under the assumption that sufficient plasma is available to provide whatever is required. If the source of the plasma is acceleration of charged particles from the stellar surface, however, the plasma will be nonneutral and then the connections between mass density, charge density, and current flow are rigorously fixed. This is the same problem as that with the "pulsar equation," which started by assuming that the particles moved with essentially c, but then led to a paradox with particle acceleration or deceleration depending on whether the magnetic field lines had favorable or unfavorable curvature. The simple resolution was just that solutions existed in which the particle velocities were *zero*, which corresponded to trapping of the nonneutral plasma. *Initially* particles would be propelled into a vacuum about the rotator with huge accelerations, but that does not ensure that acceleration will continue once the magnetosphere is loaded (and even this example is artificial, because the neutron star would never be surrounded by a vacuum to begin with). The difference between MHD and nonneutral pictures is not whether one or the other approximation is best but whether a totally different physics is involved, because the

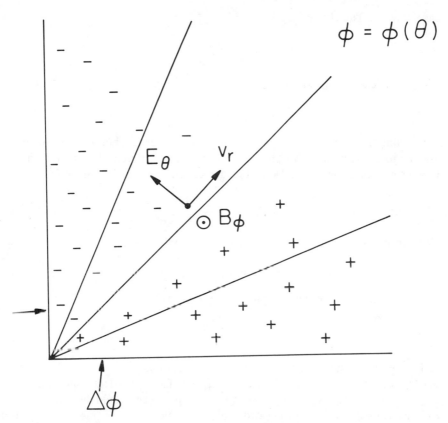

$$\phi = \phi(\theta)$$

FIG. 9.3. Self-consistent relativistic flow. Particles move radially outward on equipotential surfaces, executing $\mathbf{E} \times \mathbf{B}$ drift in the fields E_θ and B_ϕ which result self-consistently from the space charge and associated convection currents.

MHD picture assumes quasi-neutral plasma while the nonneutral plasma is a completely charge-separated plasma.

For axially symmetric configurations (e.g., aligned rotators), the electrostatic potential distribution in the radially outflowing wind illustrated in figure 9.3 would be

$$\Phi = \Phi(\theta), \tag{1}$$

in which case

$$E_\theta = -\frac{\Phi'}{r}, \tag{2}$$

where the prime indicates differentiation with respect to the argument (here θ) as usual. The net potential drops are simply given from equation 1 or 2 as the case may be. The charge density is

$$\rho \equiv en = \frac{\epsilon_0 (\Phi' \sin \theta)'}{r \sin \theta}, \tag{3}$$

and the convection current is

$$J_r = \rho V_r. \tag{4}$$

Our assumption of fully charge-separated plasma is central, because the currents are due to convection of this plasma (as opposed to differential particle drift as in a quasi-neutral plasma). The resultant magnetic field is then given from

$$\mu_0 J_r = \frac{(B_\phi \sin \theta)'}{r \sin \theta}. \tag{5}$$

The particle drift in these crossed fields supposedly satisfies

$$E_\theta = V_r B_\phi \, \text{(MHD)}. \tag{6}$$

However, the above Maxwell equations give instead

$$V_r = \frac{c^2 B_\phi}{E_\theta} \tag{7}$$

and therefore would require $V_r \equiv c$. This inconsistency comes from imposing MHD assumptions even for $E \approx cB$, when in fact the MHD approximation requires $E \ll cB$. As already discussed, the particles should be accelerated to large Lorentz factors (of the order of $\sigma^{2/3}$). Nevertheless, most of the energy flux remains in the Poynting flux of the fields, although the particles gain substantial amounts of energy themselves.

The power output from the system is given by

$$L = \int_0^\pi \Phi J_r r^2 \sin \, d\theta, \tag{8}$$

and dimensionally we can solve for σ from

$$\sigma^2 \approx L \frac{e^2}{m^2 c^5 \epsilon_0} = \frac{L}{0.693 \times 10^{16} \, \text{ergs/s}}, \tag{9}$$

which for the special case of pulsars (so far nothing has been said about what it is that is rotating) can be written

$$\sigma \approx 2.4 \times 10^{15} \left(\frac{\dot{P}}{P^3} \right)^{1/2}, \tag{10}$$

where P is the period in s and $\dot{P}$ is the dimensionless slowing-down rate, with the spin-down luminosity estimated assuming a neutron star moment of inertia of 10^{45} g cm^2.

An alternative way of calculating the power output from MHD calculations is to calculate the torque on the rotator, which in turn depends on terms (B_r, V_ϕ, etc.) that fall off more rapidly than r^{-2} and which are therefore neglected asymptotically. To treat the problem near the rotator, however, requires specification of how it is that the wind is formed, and because an aligned rotator (which would be the ideal object to satisfy this symmetry) traps plasma, one would have to go to the more complicated disk model at a minimum. To date, such calculations have not been performed. The neutral sheet discontinuity model (Michel 1975b) corresponds to a choice of potential of $\Phi = \Phi_0$ $\tan^2 (\theta/2)$ ($0 < \theta < \pi/2$), where the neutral sheet (disk) limits the validity of this equation to only the upper hemisphere. This solution is conservative in the sense that it satisfies the constraint that each field line have the same potential in the wind as it has on the stellar surface. In this solution the neutral sheet was a mathematical one. No one has figured out how to produce such a neutralization current spontaneously from an isolated pulsar. Recently Fitzpatrick and Mestel (1988a,b) rediscovered that it is useful to suppose that some sort of neutral sheet transports currents. They argue that this is a spontaneously formed plasma sheet and not the physical disk of material postulated by Michel and Dessler (1981a).

a. *Particle Pickup and Pair Production*

A further complication involves the interaction of the outgoing wind and the surroundings. The plasma wind should flow past any neutrals without perturbation. However, the neutrals may be in a steady state between ionization and recombination, in which case the ions and electrons are incorporated into the flow, as well as the neutrals as they become ionized. Incorporation of such particles into the flow effectively increases the particle flux. Because the particles gain $\sigma^{2/3}$ of the energy in the nonneutral limit, increasing the particle flux by a factor of $\sigma^{1/3}$ would put most of the energy in particles, so the character of the flow would have to change in that case. Similarly, a production of $\sigma^{1/3}$ positron/electron pairs per primary electron at the source would similarly change the flow. This crossover from the space-charge wind to (presumably) an MHD wind has not been carefully examined. With the above estimates, one needs about 10^4 pairs per primary from the Crab pulsar and about 100 per primary in typical pulsars to so load the wind.

2. Termination of the Wind

Although the angular dependence of the electrostatic potential in the wind may depend on the detailed nature of the source, the overall energetics seem rather insensitive to such details. On the other hand, interaction with a distant plasma may be sensitive to this dependence, especially for any quantitative description. Accordingly, we will choose a mathematically tractable assumption, namely that the potential distribution is quadrupolar, just as it would be if the surface potential of an aligned rotator were radially projected outward. The motivation, however, is mathematical simplicity. Thus

$$\Phi = \Phi_0 P_2 (\cos \theta), \tag{11}$$

where P_n is the standard Legendre polynomials of order n. Suppose at some distance b there is located a conducting grid that allows the particles to pass but modifies the fields; then we need to set Φ to a constant at that distance. The "grid" of course is just an attempt to simulate the interaction with external plasma, with the conductivity of the grid representing the effect of wholesale pickup of charged particles by the wind flow. The obvious modification to the radial potential is to add a vacuum quadrupole that sets $\Phi = 0$ (or a constant) at $r = b$; hence

$$\Phi = \Phi_0 P_2 \left(1 - \frac{r^2}{b^2}\right). \tag{12}$$

As simple as this modification is, it yields some interesting consequences (figure 9.4). To first order, the fields are (within the constant factor Φ_0)

$$E_r = -\Phi_0 \frac{r}{b^2}(1 - 3 \cos^2 \theta), \tag{13}$$

$$E_\theta = \frac{3\Phi_0 \cos \theta \sin \theta}{b} \left(1 - \frac{r^2}{b^2}\right), \tag{14}$$

and the magnetic field, neglecting to first order any effects on either field from nonradial motion, is

$$B_\phi = \frac{3\Phi_0}{rc} \cos \theta \sin \theta. \tag{15}$$

Here B_ϕ is ignored because it drops off at least as fast as r^{-2}. Owing to this modification of the fields, we no longer necessarily have $E = cB$ and in fact defining

$$K = E^2 - c^2 B^2 \tag{16}$$

allows us to distinguish zones of pure $\mathbf{E} \times \mathbf{B}$ drift ($K < 0$) from acceleration regions ($K > 0$). The boundary between the two will of course be at $K = 0$, which is just the locus of

$$\frac{r^2}{b^2} = 18 \cos^2 \theta \frac{1 - \cos^2 \theta}{1 + 3 \cos^2 \theta}. \tag{17}$$

These two regions are shown in figure 9.5. The fields used to calculate equation 17 are only approximate, but qualitatively this separation into two regions of

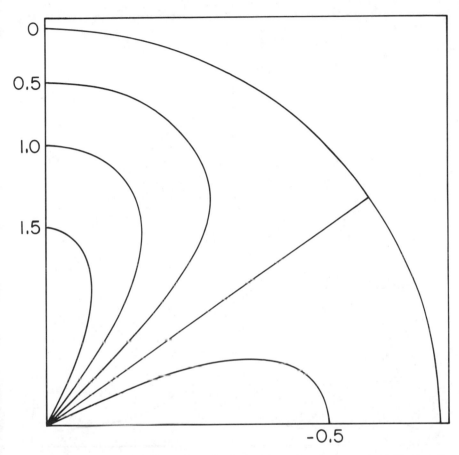

FIG. 9.4. Closure of equipotential surfaces. Equipotentials as given from equation 9.12. From F. C. Michel, 1985a, *Ap. J.*, 288, 138 (figure 1).

positive and negative K must take place. The exact topology is less important than the existence per se of these regions. It is easy to see that the $K > 0$ regions lie along the poles and the equator because B_ϕ must vanish there (no line currents and by symmetry, respectively), whereas the electric field cannot vanish owing to the closure of the equipotential surfaces as illustrated in the figure.

a. *The Synchrotron Region (K < 0)*

In the synchrotron region the drift approximation should be valid because the particles have an average drift velocity $E/B < c$. But the particles themselves were presumably injected with relativistic energies (e.g., Michel 1974c; Fawley et al. 1977), and if their drift velocity is less than c they must then be executing cycloidal drift orbits. Consequently they must also be emitting synchrotron

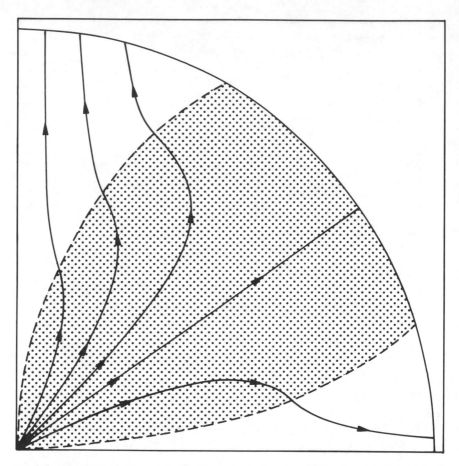

FIG. 9.5. Flow pattern in the nebula. Shaded region is the subrelativistic $K < 0$ drift region, and unshaded region (within the circular shell boundary) is the relativistic $K > 0$ acceleration region where particles gain energy logarithmically. Sign of the space charge changes across the straight line which leads to a stagnation point on the boundary. Dashed line is the locus of equation 17, solid lines with direction arrows are the flow lines. From F. C. Michel, 1985a, *Ap. J.*, 288, 138 (figure 2).

radiation. We therefore find an alternative resolution to the paradox stated by Kennel and Coroniti (1984a,b), namely that the usual Poynting flux-dominated relativistic wind is difficult to stagnate. This difficulty led them to propose a much denser wind. Our analysis could, however, be extended to the dense wind case. Note that a hole in the synchrotron emission around the pulsar in the Crab Nebula (Scargle 1969), attributed to the preshock wind in the MHD picture, follows here simply because the particles have little cyclotron motion even within the $K < 0$ regions until they are at fairly large distances within the nebula. As long as K is nearly zero, the particle trajectories have only a small sinusoidal component, which would cause them to radiate only weakly.

Aschenbach and Brinkmann (1975) believe the central X-ray regions of the Crab Nebula to resemble a doughnut tilted to the line of sight, not inconsistent with such a forced response of the flow to intervening matter.

It is interesting that the polarization of the pulsar in the Crab Nebula in the optical has the same position angles (159°) as the continuum radiation in the optical and in the radio (Kristian et al. 1970; Wright and Forester 1980). This observation is consistent with curvature radiation at the pulsar from the lowest-lying magnetic field lines, which would give a polarization electric vector paralleling the magnetic spin axis. In the distant wind zone, the magnetic field lines would be wound up and perpendicular to the spin axis, but now energetic particles would spiral about these magnetic field lines and the emission would be perpendicular to these asymptotic magnetic field lines, hence again parallel to the spin axis of the pulsar. Thus the polarization in the nebula would be expected to parallel the polarization at the pulsar, which seems to be the case.

b. *The Acceleration Region (K > 0): Jets*

From figure 9.5 we see that the $K < 0$ region envelops the origin (pulsar) and therefore virtually all the particles pass through the synchrotron region before being injected into the wind. In so doing, they must follow equipotential surfaces, which in turn duct them through the synchrotron region and out into the acceleration regions ($K > 0$) (see also Goldreich and Julian 1969). Once in the acceleration region, particles follow asymptotic trajectories so that only a component of the electric force counters the magnetic force; hence roughly (for the exact relationship, see sec. 9.2)

$$E \sin \eta \approx cB, \tag{18}$$

where η is the angle between the electric and velocity vectors. The particles that exit along the polar axis are concentrated in three stages. First, the closure of the equipotential surfaces systematically ducts the particles toward either the poles or the equator. Second, the acceleration zone further concentrates the particles. Third, the particles emerge from the shell directed to a focus at an axial distance $3b$, where again b is the radius of the shell. But stages 2 and 3 are to some degree artifacts of the perfectly spherical geometry assumed for the shell. Benford (1984) has discussed on rather general grounds the way in which a jet would interact with the nebular and interstellar medium. Accretion-driven jets have been discussed by Kaburaki and Itoh (1987), jets driven by electromagnetic waves by Baker et al. (1988), and jets from magnetized disks by Kundt (1982).

c. *Observational and Theoretical Consequences*

The expected beam density from the Crab pulsar would be about 10^{-3} particles/cm^2/s and correspondingly smaller for the other pulsars. Such a beam might escape attention. This picture seems inconsistent with the odd jet-like feature observed by Gull and Fesen (1982) because the axis of that feature does not intersect the pulsar but is significantly offset. Benford (1984), however, suggests that a more sophisticated treatment of the nebular interaction could remove this discrepancy. There seems to be a paradoxical difference between threshold attempts to model the Crab Nebula on MHD grounds and the single-particle picture presented here. It is also interesting to note that a toroidal synchrotron region has already been inferred from the data by Aschenbach and Brinkmann (1975) and Brinkmann et al. (1985). We would expect that an MHD and single-particle treatment of one and the same problem would give substantially consistent results. It may be that the single-particle picture reflects the appropriate relaxation of the usual spherical symmetry assumed in most MHD treatments, in which case it is not really a question of one or the other being the correct one. We note that the magnetic field does not accumulate in the nebula in this picture. Moreover, the escape of cosmic rays would be direct and not diffusive through tangled magnetic field lines (although interaction with filaments might modify that conclusion). Circular polarization has been suggested for the Crab Nebula at about the 1% level (Rees 1971a) but was not seen by Landstreet and Angel (1971), who give about $+0.3 \pm 0.4\%$.

The physical consequences of this flow diversion seem twofold. First, for relativistic particles to pass through the drift region, they must have an average velocity less than c, and therefore they must execute cycloidal paths, in which case they will radiate synchrotron radiation (otherwise they do not because their orbits are almost rectilinear). Second, they are diverted into cylindrical and disk flows instead of radial flows. This change represents an interesting response to the confining medium, which for radial flow would eventually succeed in smothering the flow. But if the flow becomes cylindrical, it can propagate to arbitrary distances, reminiscent of astrophysical jet phenomena (Michel 1987c), as shown in figure 9.6.

Manchester and Durdin (1983) have concluded from a study of the morphology of supernova remnants that there appears to be localized excitation of the nebula as if the central pulsar were emitting two jets that excited the emission rather than an isotropic wind.

9.4 Waves in the Wind

In a previous section we saw that test particles introduced into a large-amplitude wave simply experience phase locking and ride with the wave, slowly gaining energy but not enough to share significantly in the wave energy budget in the space-charge limit. We also saw that if a large enough neutralizing component was added (order of $\sigma^{1/3}$ ion/electron or electron/positron pairs per initial space charge), indeed the particles should start to share significantly in the wave energy. What this implies is that the plasma starts to back-react significantly on the wave. Analyses of this case typically show failure of phase locking coupled with severe instability of the wave. What exactly this means insofar as the winds expected from pulsars are concerned remains unclear. For the bulk

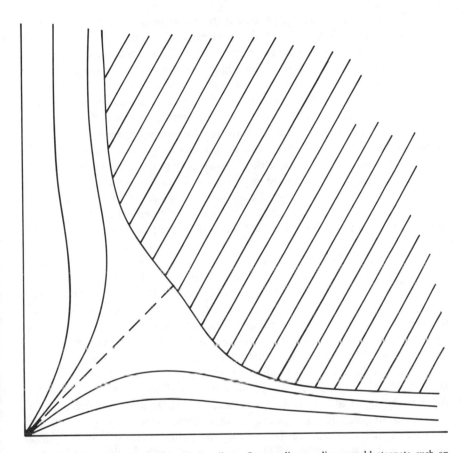

FIG. 9.6. Self-consistent ducting by surroundings. Surrounding medium would stagnate such an outflow were it not for the natural tendency to reorganize the outflow into jets and a disk outflow. This reorganization prevents the decline of dynamic pressure with increasing distance and allows the jet to push through the confining medium. It is possible that the disk flow could be choked off, in which case the entire flow would be in the form of a net-neutral jet. From F. C. Michel, 1987 c, *Ap. J.*, 321, 714 (figure 3).

of the pulsars, the wind (assuming it in fact exists) is invisible and therefore it seems plausible that the charge-separated limit may be close to the correct description. In any case there is no external evidence that the wave energy is transferred to the particles, which one would expect would produce radiation in the process. Insofar as the Crab pulsar is concerned, the assumption persists that the particle output is much above the space-charge limit. Conversion of wave energy into radiation in this case could be evidenced by the emission of a large fraction of the output in the pulsed high-energy (optical to gamma-ray) radiations seen, so in broad terms the idea seems possible. Kundt (1981b) has suggested that just the gamma-rays may be created in such a fashion, with the X-rays and optical coming from near the pulsar. Arons has emphasized the implausibility of γ emission too near the pulsar owing to the opacity of the magnetic field there (sec. 2.7). On the other hand, dissipation of the wave would eliminate one source of the nebular magnetization. This problem can be resolved by attributing the nebular field to the aligned component of the magnetic field, which could be frozen into the wind but would not be dissipated. The remaining problem is the one that has haunted wind-zone emission theories since Michel and Tucker (1969), namely how does one get interpulses that differ from main pulses and other structural details? Naively, one would attribute the interpulse to the second node in the wave (sec. 3.3.2), but one would expect that to be indistinguishable (unless perhaps the plasma supply to those field lines differed).

For an oblique rotator, the stellar wind should in fact be a large-amplitude wave (Ostriker and Gunn 1969a,d), as well as a wind. The important case of vacuum solutions was covered in sec. 5.2. Here we comment on the plasma effects and point out that the waves themselves do not appear to be stable. Indeed, the original Ostriker and Gunn (1969a) proposal was that energy was systematically transferred from the wave to the particle (see Fischer and Straumann 1972; Usov 1975; Drake et al. 1976; Buckley 1977a,b). In their proposal, however, the particles were treated as a perturbation (i.e. the particles were simply accelerated from rest by the radiation pressure), raising the question of whether relativistic waves might not be formed (once the particles became sufficiently energetic). This question has been studied by Max and Perkins (1971, 1972) and Max (1973), who concluded that relativistic waves would indeed propagate in an electron-ion plasma (see also Ferrari 1972, Ferrari and Trussoni 1971, 1974, 1975a,b, Onishchenko 1979, Stenflo 1980, Luheshi and Stewart 1979, Ashour-Abdalla et al. 1981, Leboeuf et al. 1982). Kennel and Pellat (1976) modeled a simple relativistic wave in an electron-positron plasma, but concluded that radiation reaction by the particles would damp the wave in about 10 wavelengths (Asséo et al. 1978). Moreover, it was later discovered that the two-stream instability of the counterflowing electrons and positrons should destroy the wave within one oscillation (Asséo et al. 1980a). Yu et al. (1984) find that circularly polarized waves are more stable.

There are basically two possible types of wave; either waves that propagate *through* a plasma with phase velocity greater than c or waves that convect the plasma along with them (phase velocity less than c: Kulsrud 1972). In the first case, the particles are obliged not only to become relativistic but also to reverse transverse flow direction as the wave cycles pass them, hence the radiation reaction loss and, in the case of an electron-positron plasma, the strong two-stream instability. For phase velocity less than c, one has really a wind in disguise, since one can now transform into the flow frame of reference, and the "wave" can actually be any combination of plasma and magnetic field such that the total pressure is constant. A simple example is to have all the particles in sheets separating compartments of alternating magnetic field (Michel 1971)-a set of marching neutral sheets. It can be shown easily, however, that this system cannot propagate far either. To separate two compartments of field strength B

and $-B$, a surface current density

$$j = \frac{2B}{\mu_0} \equiv \int J\,dx \tag{1}$$

must flow (dx is the transverse distance through the neutral sheet). But the required surface density of relativistic particles is then

$$\sigma \geq \frac{j}{ec}. \tag{2}$$

For radial expansion away from the pulsar, the surface density σ declines as r^{-2} whereas the azimuthal magnetic field entrained between two sheets declines only as r^{-1}. Thus σ and j cannot be in fixed ratio (once more a current/charge density problem!). Presumably the magnetic fields would simply reconnect and transfer magnetic energy into particle energy. Indeed, Coroniti (1990) refers to such a wind as being *magnetically striped* and makes a phenomenological estimate of its evolution with distance. In his analysis, the neutral sheet thickness grows until the magnetic fields between them are entirely annihilated, giving a transition from a high σ wind to a low σ, MHD wind. If an aligned component were also convected out (not discussed), the alternating sheets would contain different net magnetic fluxes, and therefore the annihilation would be incomplete leaving behind this wound-up aligned component. For moderate inclination angles, this process would result in a rough equipartition of electromagnetic and particle energy. The time evolution is important here, given the transverse Doppler slowing of the magnetic merging rates, and this interesting issue deserves further study.

There does not seem to be a consensus as to what exactly a pulsar emits, yet there is apparently a fairly strong magnetic field to be explained in the Crab Nebula. If this field cannot be transported from the pulsar in the form of waves, it must move in the form of a wind. The charge-separated wind, however, also has theoretical difficulties. A quasi-neutral wind seems indicated, but it is not easy to see how the standard model would produce such a wind.

9.5 Plasma Pickup by Pulsar Electromagnetic Waves

Clearly, plasma pickup must take place from a companion star in the case of PSR 1957 + 20 or from filamentary condensations in the case of the Crab Nebula. We have already taken a look at this process. One approach is to examine the conservation laws for uniform flow where plasma is picked up and equilibrated in some *interaction region* (black box). We do not need to know the details if the conservation laws are sufficiently constraining (that is why we can solve for the jump conditions across a shock wave without solving for the microphysics in the shock transition itself). Thus we add mass at some rate f (in g/cm^2 s, say), which gives us for the mass flow after pickup (see figure 9.7):

$$\rho V = f. \tag{1}$$

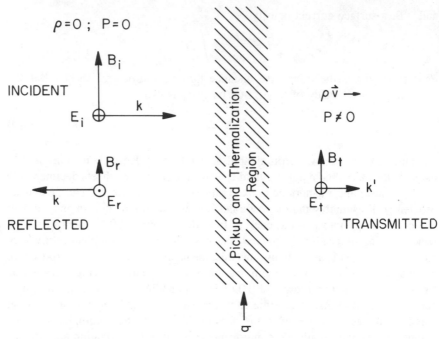

$\rho = 0$; $P = 0$

FIG. 9.7. Pickup of plasma by an electromagnetic wave. Incident wave is partially reflected and partially transmitted, with pickup taking place within a region small compared to the wavelength (here the wavelength is taken to be infinite). The transmitted wave then satisfies the condition $\mathbf{E} = -\mathbf{V} \times \mathbf{B}$, and physically the pressure cannot be zero because pickup works by first accelerating the particles in the direction of $\mathbf{E}$ rather than in the $\mathbf{k}$ direction.

Although one might think that adding particles to a vacuum electromagnetic wave would just sweep them away, one would actually need a reflected wave to satisfy the conservation laws. For large-amplitude waves we can, to first order, take the waves to have zero frequency. Thus to balance momentum we have

$$\frac{1}{2\mu_0}(B_i^2 + B_r^2) = \frac{1}{2\mu_0}\left[(B_i - B_r)\frac{c}{V}\right]^2 + (\rho V^2 + P) \tag{2}$$

and for energy flux balance

$$\frac{1}{\mu_0}(B_i^2 - B_r^2) = \frac{1}{\mu_0}[(B_i - B_r)]^2\frac{c}{V} + \left(\frac{1}{2}\rho V^2 + u + P\right)V, \tag{3}$$

where B_i is incident wave magnetic field, B_r is that of the reflected wave, and V is the plasma velocity after pickup. These terms are standard except perhaps for the way we have incorporated the momentum of the magnetic field onto the plasma, which can be traced back to continuity of the electric field tangential to the pickup plane. Here u is the energy density of the plasma, which is just $P/(\gamma - 1)$ for a perfect gas, with $\gamma = 5/3$ for a monatomic gas. For a given

B_i and f, we have two equations and *three* unknowns: B_r, V, and P. Because particle pickup initially puts particles into cycloidal orbits, we know that $P/\rho V^2$ is of order unity, but to do better would seemingly require detailed knowledge of the microphysics. Interestingly, numerical evaluation of the solutions shows that only for

$$\rho V^2 + P = \frac{1}{2}\rho V^2 + u + P, \tag{4}$$

corresponding to the fixed ratio

$$\frac{P}{\rho V^2} = \frac{3(\gamma - 1)}{2(2 - \gamma)}, \tag{5}$$

does one obtain solutions valid for all values of f. Otherwise we find divergent (unphysical) results at small f if $P/\rho V^2$ is to one side of the above ratio and at large f on the other side. It is physically implausible that a large-amplitude wave cannot pick up a few particles (small f), and for larger f we are in effect just reflecting the wave from a dense plasma (and letting that plasma recoil slightly before creating new plasma in the space evacuated). So we expect solutions for all f and in fact have a perfectly well determined one. Thus we can calculate how hot the plasma becomes when picked up, etc. These equations can also be extended (approximately) for finite frequencies provided that the wavelength is longer than the interaction region. With the above condition, we find that

$$\left(\frac{B_r}{B_i}\right)^2 = \frac{c - V}{c + V} \tag{6}$$

and we have the valuable result that the reflected wave is determined by the plasma velocity.

These equations do not exactly hold, of course, in the extreme relativistic limit and must be appropriately reformulated to handle the case of test particle pickup. In that limit, the particle term in the momentum equation is replaced by

$$\rho V^2 + P \rightarrow \gamma^2 V^2 (e + P) + P \tag{7}$$

and in the energy equation by

$$\left(\frac{1}{2}\rho V^2 + u + P\right) V \rightarrow \gamma^2 (e + P) V, \tag{8}$$

where e includes the rest-mass energy density of the particles, $e \equiv mnc^2 + u$, and the mass flux is given by

$$\gamma mnV = f. \tag{9}$$

Note that in the relativistic limit, $\gamma \gg 1$ and again the particle terms must approach one another. Now, however, it is ambiguous what part of the energy term dominates. With the cold pickup assumption ($P \leq mnc^2$), we get $\gamma^2 \approx 1/f \approx \sigma$, and therefore $\gamma \approx \sigma^{1/2}$, which again indicates that the particles do not directly equipartition with the wave.

9.6 Cosmic Ray Acceleration in Shocks

The possibility that astrophysical shock waves play an important role in accelerating cosmic rays has received considerable attention (see, for example, Cassé and Paul 1980), and has a long history (Darwin 1949; Parker 1958; Colgate and Johnson 1960; Hoyle 1960; Shatzman 1963; Fisk 1971; Axford et al. 1977; Krimsky 1977). See Blandford and Eichler (1987) for an extensive review. Cosmic rays are examined from the standpoint of how a local particle population with a very high energy power-law tail can be created, not necessarily with the intention of accounting for the cosmic ray spectrum observed at Earth; as noted in sec. 2.7.4 there seem to be significant difficulties in achieving the second objective. These shocks could be shocks from a supernova explosion, a termination wind shock from plasma flow out of a pulsar, an accretion shock about a pulsating binary X-ray source, etc. The spectrum of the accelerated particles can be calculated from general macroscopic principles (Michel 1983a), with results that are in accord with the microphysical calculations of Bell (1978) and Blandford and Ostriker (1978), but perhaps are more readily understood without a specialized background. Broadly speaking, the theoretical description of particle acceleration by a shock wave is as follows: (1) some seed cosmic rays either preexist or are produced by the shock itself, (2) the shock wave causes them to be compressed, thereby energizing them, (3) some cosmic rays diffuse back across the shock wave to be reaccelerated, owing to some *nondissipative* scattering process in the interstellar medium, and (4) a few cosmic rays end up being reaccelerated many times. Particle acceleration by a shock wave has the important property of creating a power-law spectrum, which can enormously increase the maximum energy of these particles. In contrast, the usual types of spectra (i.e., Boltzmann) that result from statistical processes have *exponentially* decreasing probabilities of producing very energetic particles. The essential condition leading to the power-law spectrum is that the cosmic rays are relativistic and do not thermalize with either themselves, the interstellar medium, or the scattering centers at an important rate. Thermalization with the scattering centers-generally taken to be magnetic field irregularities or waves-would lead to classic second-order Fermi acceleration (ignored here to underscore the shock acceleration process which is a first-order Fermi acceleration). It is very important to this mechanism that the particles not significantly scatter one another; otherwise the acceleration fails and simply heats the particles, which is what happens in ordinary gas-dynamic shocks. In other words we need particles that are essentially invisible to one another and that effectively scatter on very massive entities (e.g., magnetic field inhomogeneities) so that their energy remains essentially intact after each scattering.

1. The Power-Law Spectrum

The fact of a power-law spectrum (number of particles within a unit energy spread dN/dE proportional to $E^{-\mu}$) determines its spectral index, and for a number of plausible assumptions one obtains a *spectral index* $\mu \leq 2$. For spectral index larger than 2 the total energy of the particles is concentrated at the *low*-energy end of the spectrum, while for spectral index smaller than 2 it is concentrated at the *high* end, in which case the particles could become an arbitrarily deep sink for the energy in the system. In contrast, if the particles were simply adiabatically compressed by a shock, the particle energy would only be boosted by the relatively minor factor of $R^{1/3}$, where R is the compression ratio across the shock. For ordinary gas-dynamic shocks, $R \leq 4$, although larger compression ratios are possible across radiative shocks. Considering the startling difference in physical consequences, it is helpful to be assured that a power-law spectrum indeed results. Only in the limit of relativistic particles will it be assured that a power law results, and accordingly we restrict this discussion to that limit.

a. *Random Walk across a Moving Surface*

Let us reduce the shock acceleration model to its essentials and thereby confirm acceleration to a power law. Imagine a single particle diffusing about in a scattering medium and that some mathematical surface passes by at a uniform velocity. What is the probability that the surface is ultimately crossed once, twice, etc.? The above question expresses the essential statistics of cosmic ray acceleration by shocks, even though crossing the surface need have no physical effect. A velocity change is, of course, the physical agent for compressing and accelerating the cosmic rays; however, once this energization is assigned (at each shock crossing) the remaining complications introduced by the velocity change are not essential to the power-law spectrum, as we will see. First a minor point: the number of crossings will always be an odd number, since a diffusing particle will always be overtaken *eventually* by a steadily advancing surface and this overtaking side of the surface is where one finds particles that have crossed an even number of times. For example, if the particle is overtaken by the surface and subsequently scatters back across the surface, it has crossed twice and is once more being overtaken. It now follows that the probabilities (P) for crossing $2n + 1$ times and $2n + 3$ times ($n = $ integer) satisfy

$$\frac{P_{2n+3}}{P_{2n+1}} = \text{constant}, \tag{1}$$

which then gives us the exponential distribution of the mth crossing:

$$P_m \approx e^{-m/M}. \tag{2}$$

Because the energy of a relativistic particle is boosted by a fixed fraction (i.e.,

Doppler-shifted) each time the particle is returned, it too exponentiates, although not necessarily by the same factor:

$$E_m \approx e^{-\beta m/M}. \tag{3}$$

These two relations then immediately give a power-law dependence for the number of particles within a certain energy range versus their energy (Bell 1978). Equation 1 simply restates the basic assumption in a random walk process: *each succeeding step is independent of the nature of previous steps.* If the particle manages to reencounter the surface, it has returned to the same initial condition, and the fact that it has previously crossed the surface is irrelevant insofar as the probability of a future encounter is concerned. Thus equation 1 follows directly. One can now see that the velocity difference between the two sides of the surface in the case of a shock, which we ignored at the outset, must be irrelevant to the expectation of an exponential distribution. Of course the constant in equation 1 will usually depend on this velocity difference, but it is still independent of n (or m), which is all that is required. This argument is essentially the one given by Bell (1978). In this simple mathematical model, there is no dependence on n. For application to astrophysical problems, a dependence on n may arise because the particle velocity and mean free path depend on the energy (hence on n). These are important questions in general, but if we restrict ourselves to relativistic particles, the velocity change becomes unimportant. Moreover, variation in mean free path will be seen (in the next section) to drop out. Thus, for relativistic particles, acceleration to a power law seems assured.

a. *Exponential Spectrum from Diffusion Equation*

Here we provide an alternative way of obtaining the exponential distribution of crossing probabilities given by equation 1. Imagine a steady flux of particles moving past a stationary surface. What is the proportion of incident particles versus particles that have crossed the surface one or more times? Balancing convection against diffusion gives

$$u\frac{dn}{dx} = D\frac{d^2n}{dx^2}, \tag{4}$$

where the $\partial n/\partial t$ part of the convection term vanishes for steady flow in the shock frame. Here $n = n(x)$ is the particle density, u the bulk flow velocity, and D the diffusion coefficient. The general solution of equation 4 is

$$n(x) = A + Be^{kx}, \tag{5}$$

where $k = u/D$. This profile is shown in figure 9.8a, plotted such that n_0 (density of particles that have not yet crossed the surface) and n_c (particles that have crossed one or more times) satisfy

$$n_0 + n_c = \text{constant}, \tag{6}$$

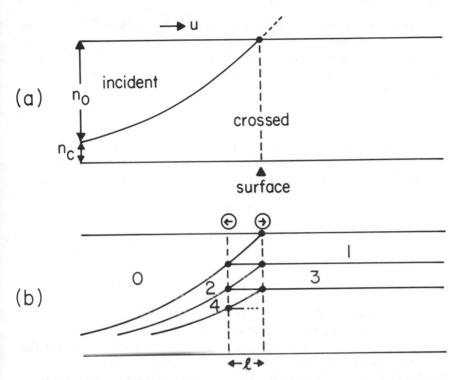

FIG. 9.8. Distribution of accelerated particles. (a) Incident particles flowing toward a hypothetical surface versus particles that have crossed that surface. The latter diffuse upstream against the flow, replacing incident particles that diffuse downstream (hence moving faster than the average flow rate). The total number of particles is constant (here the surface does nothing to them), and the exponential curve proportions the vertical height according to the relative densities. (b) Surface split by 1 mean free path to represent the fact that a particle must diffuse this distance before it can recross the surface. Thus the surface denoted ($\rightarrow$) converts all incident (n_0) particles into once-crossed particles (n_1). Crossing surface ( ) to the right no longer counts as a crossing. These diffuse upstream to be converted into twice-crossed particles (n_2) at surface ($\leftarrow$), which leaves only those within the band labeled "1." These, however, flow downstream to become (n_3), diffuse upstream to become (n_4), etc., yielding an interleaved distribution of crossing experiences as shown. From F. C. Michel, 1983a, *Ap. J.*, 247, 664 (figure 1).

which must follow since the surface need not physically do anything to the particles (in this simple model). The surface itself must be located at the point where the particles have all crossed at least once. It might seem surprising that the density of incident particles (n_0) vanishes at the surface, because there would then seem to be no incident particles left to become *crossed* particles (n_c)! This miniparadox comes from the fact that the diffusion equation blurs the full nature of the particle motion by lumping the (relativistic) particle velocity c and the mean free path λ into the diffusion coefficient $D = c\lambda$. The incident source is really the *finite* number of particles located one mean free path upstream of the surface that are just about to cross the surface within a time $\approx\lambda/c$. These elementary comments bring us to a seemingly straightforward question: "How many times on the average do the particles cross the surface?" This question is unanswerable from equation 4.

We know that some particles cross only once while others cross twice, thrice, etc. Thus we should be able to divide up n_c according to the number of crossings, which downstream would read

$$n_c = n_1 + n_3 + n_5 + \cdots. \tag{7}$$

But the ratio n_1/n_3, for example, is a dimensionless number, and we cannot construct a dimensionless number if we are limited to the parameters u and D of equation 4. Indeed, the diffusion equation "solves" this dilemma by requiring

$$n_1 = 0,$$

$$n_3 = 0,$$

$$n_5 = 0, \text{ etc.!!}$$

As we saw, n_0 vanishes at the surface since the surface *destroys* uncrossed particles by converting them into *crossed* particles. The density therefore vanishes at a sink. But the surface also destroys once-crossed particles by converting them into twice-crossed particles, etc. Thus the density n_1 is zero at the surface because the surface is a sink, and n_1 is zero everywhere else because the surface is the *only* source. The same follows for all other crossing numbers, except n_0, which is the upstream source. The sum in equation 7 thus consists of an *infinite* number of *infinitesimal* terms that add up to a finite result. We can resolve this nonphysical result simply by noting that a particle that has just crossed the surface must go one mean free path (on the average) before it can be scattered back across the surface. Thus we can simply imagine the surface to be split into two surfaces, one mean free path apart. One surface only counts if crossed from left to right, and vice versa. This gives the interleaved distribution shown in figure 9.8b. We now see that, from equation 5,

$$n_1 = n_0 - n_0 e^{-kl} \approx n_0 kl,$$

$$n_3 = n_0 e^{-kl} - n_0 e^{-2kl} \approx n_0 kl e^{-kl}, \tag{8}$$

and in general

$$n_{2m+1} = n_0 e^{-mkl} - n_0 e^{-(m+1)kl} \approx n_0 kl e^{-mkl},$$

and once more we find that the distribution is exponential. The distribution must be exponential, because the pattern for the n_3 and succeeding profiles must match the n_1 and succeeding profiles if multiplied by n_1/n_3, which is just the reciprocal of the recrossing probability, equation 1. Note that the dimensionless ratio $kl \approx u/c$ is all that enters into the ratio n_1/n_3; thus the mean free path does not actually enter into the crossing statistics, which is the simplification noted at the end of sec. 1a above. Blandford and Ostriker (1978) obtain a power-law spectrum directly from the Boltzmann equation (which, if integrated over momentum space, gives eq. 4). Their derivation is repeated in slightly modified form in sec. 4, below.

2. The Power-Law Index

Given the spectrum of cosmic rays before and after compression, one can calculate the average energy increase per particle. Because this increase is also specified by an adiabatic compression by the same factor, the spectral index is given directly from the equation of state. However, a little care is necessary because the conversion of an input delta-function spectrum of particles all with energy E_0

$$\frac{dN}{dE} = \delta(E - E_0) \tag{9}$$

into a power-law spectrum

$$\frac{dN}{dE} = K \left(\frac{E}{E_0}\right)^{-\mu} \quad (E > E_0),$$
$$- 0\,(E < E_0), \tag{10}$$

involves an irreversible increase in entropy, whereas by definition an adiabatic compression conserves entropy. An adiabatic compression can often be described by $P = \rho^\gamma$, where the equation of state is reduced simply to the value of the *Poisson adiabatic index*, γ. Since P/ρ is directly proportional to the average energy per particle, we have

$$\frac{E'}{E} = \left(\frac{\rho'}{\rho}\right)^{\gamma-1} = R^{1/3} \left(\gamma = \frac{4}{3}\right), \tag{11}$$

where $R \equiv \rho'/\rho$ is the compression ratio and the primes denote values after compression. For relativistic cosmic rays, γ should be 4/3 to an excellent approximation because such particles do not collide with magnetic field irregularities in a manner that could excite internal degrees of freedom, so the only degrees of freedom are the three kinetic ones. By the same token, the above two spectra give an energy increase

$$\frac{E'}{E} = \frac{\int E dN}{E_0 \int dN} = \frac{1 - \mu}{2 - \mu}. \tag{12}$$

If we could simply eliminate E'/E between equations 11 and 12, we would have a relation directly giving the spectral index (μ) from the shock compression ratio (R). However, it is necessary to take into account the fact that some particles cross the shock numerous times while others cross only a few times. For each situation the apparent compression is different. For a particle trapped between two pistons, the number of crossings made during a slow compression is

$$m = \int \dot{m}dt = \int \left(\frac{c}{\lambda}\right)\frac{d\lambda}{u} = \frac{c}{u}\ln r_m, \tag{13}$$

where r_m is the compression ratio corresponding to the m crossings. By hypothesis, the probability of m crossings can be written in the form

$$P_m = \frac{1}{M}e^{-m/M}, \tag{14}$$

where M is some (large) characteristic mean number of crossings ($\approx c/u$); it turns out that it will not be necessary to explicitly evaluate M. This equation is just the normalized form of equation 1. Equation 13 can be rewritten in the form

$$r_m = e^{\alpha m/M}, \tag{15}$$

where a second unknown, α, has been introduced. It is not necessary to evaluate directly either the coefficient c/u in equation 13 or M in equation 14, although the two are of the same order (these quantities are explicitly evaluated in the calculations of Bell). Finally, we calculate from the apparent compressions of each particle population (labeled according to the number of crossings) the actual physical compression of the total system. Thus for particles that cross but once ($m = 1$) a volume proportional to P_1 is set aside. These particles are now compressed by a factor r_1. Continuing in this manner for each particle population, subdivided according to m, one assigns initial volumes $P_1, P_2, P_3, \cdots$ (each having the same density, of course) but each compressed by different amounts $r_1, r_2, r_3, \cdots$. Thus the net compression is by a factor (see figure 9.9)

$$\frac{1}{R} = \frac{\sum P_m/r_m}{\sum P_m} \approx \frac{1}{1+\alpha}. \tag{16}$$

From equation 11, we see that each category of particle is accelerated by a factor

$$E_m = r_m^{1/3}; \tag{17}$$

hence we obtain an energization

$$\frac{E'}{E} = \frac{\sum E_m P_m}{\sum P_m} \approx \frac{1}{1-\frac{\alpha}{3}}. \tag{18}$$

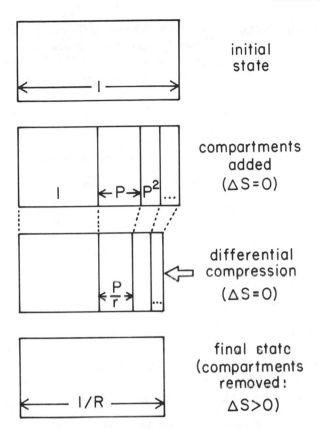

initial
state

compartments
added
($\Delta S = 0$)

differential
compression
($\Delta S = 0$)

final state
(compartments
removed:
$\Delta S > 0$)

FIG. 9.9. Thermodynamic equivalent path of shock acceleration. A volume is compartmentalized into subvolumes in the ratios $1:P^1:P^2:P^3 \cdots$ and the compartments are compressed respectively by the ratios $1:r^1:r^2:r^3: \cdots$ giving a net compression R. The hot and cold gases are then intermixed to yield a power-law spectrum and the increased entropy. From F. C. Michel, 1983a, *Ap. J.*, 247, 664 (figure 2).

The unknown α can now be eliminated between equations 16 and 18 to give, using equation 13,

$$\mu = \frac{R+2}{R-1}. \tag{19}$$

It is customary to rewrite this relationship in terms of the velocity into the shock (u_1) versus the velocity out of the shock (u_2), which from conservation of particle flux is $R = u_1/u_2$, giving

$$\mu = \frac{u_1 + 2u_2}{u_1 - u_2}, \tag{20}$$

which is the result given by Bell and by Blandford and Ostriker. For the strong gas-dynamic shock limit, $R = 4$ and $\mu = 2$. Deriving R from the *apparent compressions* in equation 16 is possible because the shock passage takes the system from one macroscopic thermodynamic state to another, and one is free to choose any convenient equivalent thermodynamic path. The choice is to divide the overall compression into separate differential compressions and use the fact that the latter will be distributed as P_m. One then obtains the spectral index μ by requiring simply that the equivalent system end up at the same volume and energy as the shocked system. Thus the apparent compressions may be regarded as a set of individual real compressions of an *equivalent* compression cycle and must add up to give the same compression as the actual system.

3. Intrinsic Time Dependence

In shock acceleration, a spectrum flatter than (or as flat as) $\mu = 2$ is formally divergent (infinite average energy per particle), which is physically impossible. Most likely, shock acceleration would populate only the lower-energy portion of the spectrum with such a spectral slope, with the spectrum cut off at higher energies because (a) it takes more time to populate the high-energy region, owing to the large number of crossings required, or (b) the mean free path of the higher-energy cosmic rays becomes comparable to the size of the system, and they are no longer accelerated or trapped near the shock. Alternatively, a spectrum with $\mu \leq 2$ cannot be attained, simply because unrestrained energization of the cosmic rays would drain energy from the system, weaken the shock, reduce R, and hence increase μ until it equaled or exceeded 2. In other words, it would then be necessary to include the cosmic ray pressure in the shock relations (Axford et al. 1977; Drury and Völk 1981).

4. Blandford and Ostriker's Formulation

Blandford and Ostriker (1978) derive the power-law behavior directly from the Boltzmann equation, written in the form

$$\frac{\partial f}{\partial t} + u\frac{\partial f}{\partial x} + a\frac{\partial f}{\partial p} = D\frac{\partial^2 f}{\partial x^2}, \tag{21}$$

where $f = f(x, p, t)$ is the particle density in ordinary space (x) and in momentum space (p). Assuming that all the energization is at the shock, they then replace the acceleration (a) with an impulsive term

$$a \rightarrow \frac{1}{3}(u_2 - u_1)p\delta(x), \tag{22}$$

where u_1 is the preshock gas velocity and u_2 is the postshock gas velocity; the factor of 1/3 averages over crossing angles. They then balance equation 21 against the change in momentum by the shock, $(f_2 - g_1)u_1\delta(x)$, where f_1 is the asymptotic preshock distribution (i.e., far upstream) and f_2 is the far downstream distribution. This argument then gives

$$\frac{1}{3}(u_2 - u_1)p\frac{df_2}{dp} = (f_2 - f_1)u_1, \tag{23}$$

which can be solved for f_2 in terms of f_1. Note that the diffusion term (D) does not enter unless we wish to know what distance scales are involved. In theory the value of D makes no difference in an infinite one-dimensional system, whereas in practice one would get negligible acceleration if the mean free path were larger than the radius of a spherical shock wave. If f_1 is simply a delta function in momentum at $p = p_0$, it can be set equal to zero in equation 23 and thereby is the source coupling the two solutions

$$f_2 = \begin{cases} 0 & p < p_0 \\ p^q & p > p_0 \end{cases},$$

(24)

where from equation 23

$$q = \frac{3u_1}{u_2 - u_1}.$$

(25)

Because f is the density in momentum space, $dn/dp \approx fp^2$; thus $\mu - (2+q)$, which gives exactly equation 20. This elegant calculation poses a number of extremely subtle and delicate questions (see Blandford 1980), however, such as why one can take the gas velocity to represent the average velocity of cosmic rays locally (distances short compared to a mean free path). Also, it is not obvious that the "delta" functions in equation 22 and equation 23 are the same. For this reason, a more pedestrian derivation is presented here, in the hope that it would be more accessible to the general reader and student. Moreover, we needed to show for completeness that the exponential spectrum follows from equation 21 to bridge the pedagogical gap between Bell's derivation and that of Blandford and Ostriker.

5. The Bell Formulation

Because the particles should be accelerated in finite energy steps, the compression is not perfectly adiabatic in the sense of being arbitrarily slow. The formulation by Bell avoids this difficulty. The sums in equation 16 and equation 18 can be evaluated exactly and the compression and energization steps replaced by the forms (Bell, eq. 4)

$$r_m = r^m; \ r = \frac{1+\beta}{1-\beta},$$

(26)

$$E_m = E^m; \ E = \frac{3+\beta}{3-\beta},$$

(27)

where $\beta = u/c$ is the velocity change per crossing (in the limit $\beta \to 0$, $E \to r^{1/3}$ as required). The sums can be evaluated easily, writing $P_m = kP^m$, with the normalization

$$\sum P_m = \frac{k}{1-P} \equiv 1, \tag{28}$$

while the compression ratio is given from

$$\frac{1}{R} = \sum \frac{P_m}{r_m} = \frac{1-P}{1-\dfrac{P}{r}}. \tag{29}$$

Because P is given either from r and R or from E and E', it can be eliminated just as α was eliminated between equation 16 and equation 18. After some elementary algebra, one obtains

$$\mu = \frac{2+R+\beta(R-2)}{(R-1)(1+\beta)} \approx 2 - \frac{4\beta}{3} + \cdots \quad (R=4), \tag{30}$$

which indicates that stronger shocks also tend to flatten (*harden*) the spectrum.

10

Pulsating X-ray Sources

10.1 Introduction

1. X-ray Pulsars

Pulsar has come to be the generic term for *radio* pulsar. There are other types of objects that equally deserve to be called pulsars, such as the binary X-ray pulsars, which in contrast do not usually emit detectable radio pulses. X-ray pulsars are typically modeled to be neutron stars onto which matter is falling (Pringle and Rees 1972), the X-rays produced directly from heating as the result of free-fall energy release (plausibly, in the model) and the beaming attributed to an intrinsic magnetic field which controls where the matter falls (e.g., at the magnetic poles). The in-falling matter itself is often attributable to Roche-lobe overflow from a giant star companion which is undergoing an expansive evolutionary stage. Alternatively, a wind from the companion might deposit matter on the neutron star. Although such a broad brush treatment is certainly sketchy, it is a more concrete description than one can yet give for the radio pulsars. The X-ray pulsars are interesting objects in their own right, and may even provide valuable clues to how the radio pulsars function as well as information about neutron stars. Vasyliunas (1979) and Lewin and Joss (1981) give early reviews of X-ray emission from neutron stars, which are constantly being updated in such review papers as Lewin and van den Heuvel (1983), Trümper (1986), Mason et al. (1987), and Lamb (1989). Aficionados distinguish between high-mass X-ray binaries (HMXBs) and low-mass X-ray binaries (LMXBs). Accretion is often assumed to be from a stellar wind in the case of an HMXB and from an accretion disk in the case of a LMXB. We will here concentrate on quantitative discussions that can be made regarding magnetospheres in such systems.

Statistically the X-ray pulsars are quite rare with less than 100 present in the entire galaxy whereas there could easily be 100,000 radio pulsars. The difference in the number of detected objects is not nearly as large because the radio pulsars rarely have radio luminosities in excess of 10^{31} ergs/s while the X-ray objects typically have luminosities of 10^{38} ergs/s, allowing them to be seen in neighboring galaxies whereas radio pulsars so far have not been detected beyond the Magellanic Clouds.

2. The Long Period Problem

If we examine the periods of the binary X-ray pulsars, we notice something rather striking, namely that these stars have very *long* periods, much larger in fact than any known neutron star would spin down to within a Hubble time (typically a limiting period of around 25 s). If magnetic field decay were important, pulsars would further be limited to periods of a few seconds at most. The other known torque of significance is that from accretion, which can apparently be much larger than the electromagnetic spin-down torques, given that characteristic spin-up times of the binary X-ray pulsars can be of the order of 10^3 years or even less. Accretion torques are generally invoked to *spin up* the neutron star, and it takes some footwork to arrange for the very same torques to *spin down* the neutron star when needed. What is observed in many of the binary X-ray pulsars is that they sometimes spin up and sometimes spin down. One simple picture of such behavior is that there are two torques on the neutron star, a electromagnetic coupling to the disk that acts to slow down the neutron star and an accretion torque from the same disk that acts to spin up the neutron star, with the two torques being in some time-averaged balance but not in short-term balance. If such a picture is accurate, the long periods do not have to be explained as an evolutionary sequence; one need not start with a long-period neutron star at the outset of X-ray activity, but rather end up with a long-period neutron star as a *consequence* of that activity. There was a tendency to generalize on the behavior of the first few binary X-ray pulsars (which happened to be slowing down) and assume that the accretion torques always predominated. Indeed, it is the long-period binary X-ray pulsars that have the clearest episodes of spin-up and spin-down. A good review of the situation can be found in Henrichs (1983). The periods of some binary X-ray sources are given in table 10.1.

3. Quasi-Periodic Oscillations (QPOs)

The discovery of quasi-periodic oscillations (QPOs: order 5-50 Hz) in certain X-ray sources (van der Klis et al. 1985) led to the suggestion that new mechanisms are required in addition to simple accretion from a disk (Lamb et al. 1985). In this view, the accretion disk supports "blobs" of plasma which are accreted sequentially to provide a characteristic mean time between events. The idea of blobs of matter in an accretion disk is difficult to reconcile with the prevalent theory of the disks themselves, which requires a very large effective viscosity to drive the disk to spread inward owing to viscous stresses. The very same stresses would obliterate any blobs. These stresses are so large that they have to transfer matter inward across the entire disk in a few days, yet the "blobs" required to give the quasi-periodic oscillations of about 30 hertz or so must correspond to a correspondingly small size on the disk (about 10^7 or less of the

TABLE 10.1—Binary X-ray Pulsar Periods

Source Name	Pulse Period (s)	Orbital Period	Companion Name
1E 1024.0 − 5732	0.061	—	Wack 2134
A0538 − 66	0.069	16.7 d	—
SMC X − 1	0.714	3.89 d	Sk 160
Her X − 1	1.24	1.70 d	HZ Her
1E 2259 + 586	3.49	—	—
4U 0115 + 63	3.61	24.3 d	Johns; V 635 Cas (Be star)
Cen X − 3	4.84	2.09 d	Krzeminski; V 779 Cen
4U 1626 − 67	7.68	41 min	—
2S 1553 − 54	9.26	—	—
LMC X − 4	13.5	1.41 d	Sk/Ph
2S 1417 − 62	17.6	> 15 d	—
OAO 1653 − 40	38.2	—	—
EXO 2030 + 375	41.7	46 d	—
4U 1700 − 377	67.4	3.41 d	HD 153919; V 884 Sco
A0535 + 26	104	> 40 d	HD 245770; V 725 Tau (Be star)
GX 1 + 4	122	—	—
4U 1230 − 61	191	—	—
GX 304 − 1	272	> 13 d	MMV (Be star)
4U 0900 − 40	283	8.98 d	HD 77581; Vela X − 1; GP Vel
4U 1145 − 61	292	187.5? d	HD 102567; Hen 715 (Be star)
1E 1145.1 6141	297	> 25 d	—
A1118 − 61	405	—	Hen 3 − 640 (Be star)
4U 1538 − 52	529	3.73 d	Cowley; QV Nor
GX 301 − 2	696	41.4 d	WRA 977; BP Cru
4U 0352 + 30	835	580? d	X Per (Be star)

disk size). Diffusive mechanisms systematically remove small-scale structures within a time proportional to their size squared. If there are such blobs, they are not incidental features but must be actively formed by a strong dynamical process. Alternatively, the expected impulsive accretion from polar cap regions could provide the same general timing effect (sec. 10.3.6, below). This subject has be reviewed recently (Lamb 1989).

4. Be Stars

The so called Be stars (for stellar type B with emission lines, and not "Be" as in beryllium) are particularly interesting as X-ray objects owing to the supposition that Be stars have neutron stars in eccentric orbit about them, which in turn have, for some reason, disks of matter about them as well. Outbursts of X-ray activity could then be attributed to orbital encounters of the neutron star with the disk. The lack of radio emission from these neutron stars is attributed to smothering by so much plasma (Illarionov and Sunyaev 1975); then a sufficiently eccentric orbit could take the neutron star free enough of the obscuration for radio signals to be detected. Such an object, alternating between being an accretion-powered X-ray source and a radio pulsar, would be very exciting indeed.

10.2 Symmetric Accretion into a Gravitational Well (e.g., Neutron Stars)

The accreted matter must eventually reach the neutron star. The simplest model for such accretion is spherically symmetric accretion, although accretion from a flow would not have this symmetry, of course. Nevertheless, it sets out the relevant physics in an understandable and analytic way. The actual mass transfer should be more complicated if the mass resides in an accretion disk and is controlled by the magnetic field of the neutron star.

The most general spherically symmetric gravitational source would be a Schwarzschild metric, because simply by placing a surface at the appropriate distance it can represent any spherical object. Thus we can develop the equations general relativistically, although even accretion onto black holes is essentially Newtonian in behavior (mainly because the critical points in the fluid inflow occur at large distances). In any case, the solutions are in no way complicated if a full relativistic treatment is included (Michel 1972). Petrich et al. (1988) give *exact* solutions to accretion onto *moving* black holes (for an isothermal equation of state) and in the subsonic limit.

The basic equations of motion are conservation of flux

$$\nabla_k J^k = 0, \tag{1}$$

and energy flux

$$\nabla_k T_t^k = 0, \tag{2}$$

where ∇_k is the covariant derivative as usual (often denoted by semicolons, here $\nabla_k J^k \equiv J^k_{;k}$). These equations have simple forms for steady spherically symmetric accretion, becoming

$$\frac{d}{dr}(J^r g^{1/2}) = 0 \tag{3}$$

and

$$\frac{d}{dr}(T_t^r g^{1/2}) = 0, \tag{4}$$

where $g \equiv -\det g_{ij}$ (minus the determinant of the metric tensor; in most texts g is taken to be negative, leading to expressions repeatedly containing the real quantity $\sqrt{-g}$) and t and r are the time and radial distance coordinates with k above a dummy index (many authors prefer to use numbers as aliases for the spherical coordinates, with time t represented by 0 or 4, radial distance r by

1, etc.). Integration is immediate, and for a perfect fluid we can write

$$\rho U^r g^{1/2} = C_1 \tag{5}$$

and

$$(P + \mu)U_t U^r g^{1/2} = C_2, \tag{6}$$

where the C's are constants, ρ is the proper density of the fluid (i.e., that seen by an observer moving with the flow for which $U^r = 0$), P is the proper pressure, and μ is the total proper energy density. For the Schwarzschild metric we have (just one of a number of equivalent representations)

$$ds^2 = c^2 e^\nu dt^2 - e^\lambda dr^2 - r^2(d\theta^2 + \sin^2 \theta \, d\phi^2), \tag{7}$$

and the remaining terms in equation 6 are the contravariant proper velocity in the radial direction $U^r \equiv u = dr/ds$, the covariant proper time-like velocity $U_t \equiv \sum g_{tk} U^k = c^2 e^\nu dt/ds$, and the (negative) determinant $g = c^2 r^4 \sin^2 \theta \, e^{\nu + \lambda}$. Given an equation of state that relates P and μ to ρ, we have two equations (eqs. 5 and 6) and two unknowns (ρ and u). For the standard exterior solution, $e^\nu = e^{-\lambda} = 1 - 2m/r$, with $m \equiv GM/c^2$. The conservation equations become

$$\rho u r^2 = C_1 \tag{8}$$

and

$$\left(\frac{P + \mu}{\rho}\right)^2 \left(1 - \frac{2m}{r} + \beta^2\right) = (C_2/C_1)^2 = C_3, \tag{9}$$

where $\beta \equiv u/c$. The total energy density is

$$\mu = \rho c^2 + \epsilon, \tag{10}$$

where ϵ is the internal (thermal) energy of the fluid. In the nonrelativistic limit, $C_3 \to 1$ and the (nonrelativistic) solar wind equations are obtained by subtracting off unity from the left-hand side of equation 9 and dropping second-order terms.

To find the critical points, one differentiates equations 8 and 9 and eliminates the $d\rho$ terms to get

$$\frac{d\beta}{\beta}\left[2V^2 - \frac{m}{r\left(1 - \dfrac{2m}{r} + \beta^2\right)}\right] + \frac{dr}{r}\left[V^2 - \frac{\beta^2}{\left(1 - \dfrac{2m}{r} + \beta^2\right)}\right] = 0,$$

$$\tag{11}$$

with

$$V^2 \equiv \frac{d \ln (P + \mu)}{d \ln \rho} - 1. \tag{12}$$

The critical point is determined by the requirement that both bracketed factors in equation 11 vanish together; thus

$$\beta_c^2 = \frac{m}{2r_c} \tag{13}$$

and

$$V_c^2 = \frac{\beta_c^2}{1 - 3\beta_c^2}. \tag{14}$$

No solutions exist if $\beta_c^2 > 1/3$ or $r_c < 6m$.

1. Polytropic Solutions

For the equation of state

$$P = K\rho^\gamma$$

we have

$$P + \epsilon = \frac{\gamma}{\gamma - 1} P, \tag{16}$$

and will define

$$T \equiv \frac{P}{\rho c^2}, \tag{17}$$

which is essentially temperature in units of 10^{13} K for protons ($m_p c^2 / k = 1.09 \times 10^{13}$ K). Then

$$T^n u r^2 = C_4, \tag{18}$$

$$[1 + (1 + n)T]^2 \left(1 - \frac{2m}{r} + u^2\right) = C_5, \tag{19}$$

and

$$V^2 = \frac{(1 + n)T}{n[1 + (1 + n)T]}, \tag{20}$$

where

$$n \equiv \frac{1}{\gamma - 1} \tag{21}$$

is the historical relationship between the polytropic index (n) and the Poisson adiabatic index (γ). We can always use n if there is danger of confusion between the Poisson adiabatic index and the Lorentz factor. Because u enters quadratically, it can have either sign, but typically the boundary conditions of large values close to the object and small values at large distance correspond to infall, and vice versa correspond to outflow.

The natural choices for the Poisson adiabatic index are $\gamma = 5/3$ for a monatomic cold gas and $\gamma = 4/3$ for a hot relativistic gas (monatomic or otherwise).

2. Hot Gas Accretion

We expect at large distances that $u \to 0$, but T corresponds to the temperature of the surrounding medium. Thus for the hot gas case

$$C_5 \to 1 + 8T_\infty \approx 1 + 3\beta_c^2, \tag{22}$$

and

$$4T_c \approx 3\beta_c^2, \tag{23}$$

giving

$$T_c = 2T_\infty. \tag{24}$$

We can now determine the remaining constant

$$C_4 = T_c \beta_c r_c^2 = \frac{3\sqrt{6}}{16} T_\infty^{3/2} m^2, \tag{25}$$

and the temperature close to the object scales as

$$T(r) = \frac{\sqrt{6}}{4} T_\infty^{1/2} \left(\frac{m}{r}\right)^{2/3}. \tag{26}$$

Even for a neutron star immersed in a relatively hot environment ($\approx 10^4$ K), the infall temperature would only be about 13 million degrees by the time the flow reached the surface (the critical point parameters would be $r_c = 2$ AU and $u_c = 14$ km/s). The fluid density does not influence the flow properties but simply scales as

$$\rho = \rho_\infty \left(\frac{T}{T_\infty}\right)^3. \tag{27}$$

3. Cold Gas Accretion

The cold gas case (one appropriate for gas in a 10^4 K cloud) is interesting because C_4 becomes independent of *both* temperature and density at large distances, but instead

$$\frac{5}{3}T_c \approx \frac{3}{2}\beta_c^2,$$

(28)

and therefore

$$C_4 = T_c^{3/2}\beta_c r_c = \frac{3\sqrt{15}}{100} m^2.$$

(29)

Here, the temperature close to the star is

$$T(r) = 0.3 \left(\frac{m}{r}\right)^{4/3}$$

(30)

and the critical points are given from

$$C_5 \approx 1 + 5T_\infty \approx 1 + \frac{27}{4}\beta_c^4,$$

(31)

or

$$\beta_c^4 = \frac{20}{27}T_\infty.$$

(32)

For $T = 10^{-9}$ as before, $u_c \equiv c\beta_c = 1600$ km/s and $r_c = 2.7 \times 10^4$ km.

These equations are valid even for accretion into a Schwarzschild metric (*black hole*). Note that no shock forms in such infall solutions. There is a subsonic to supersonic transition, but that is a rarefaction wave, not a shock wave. It is often assumed that there is a standoff shock front near the neutron star where the flow must halt, but such a shock front is not self-evident. The physical reason for a shock wave is to halt the flow in front of an obstacle, and in general the shock front would propagate outward to do exactly that. Thus the shock wave would change the initial conditions. In other words, an adiabatic flow onto an obstruction would form a shock which would propagate outward and halt the inflow, with the system relaxing toward a hot halo/exosphere configuration, the flow having been halted by the outward propagating shock (all of this presupposes a rather odd initial condition in which the gas was in-falling through an initial vacuum surrounding the object). The resultant *settling flows* have been discussed by Gillman (1979). If the obstruction is not complete, one may have the usual (but usually disregarded) subsonic to subsonic flow solutions. If radiation is introduced, the flow may be more complicated, with radiative cooling causing something like a radiative shock near the surface. It is unlikely that significant radiation from black holes would directly follow from spherical infall, because the plasma becomes significantly heated only near such an object, and the gas would not have time to radiate significantly before entering the hole. For more realistic models, see Shapiro (1973a,b, 1974).

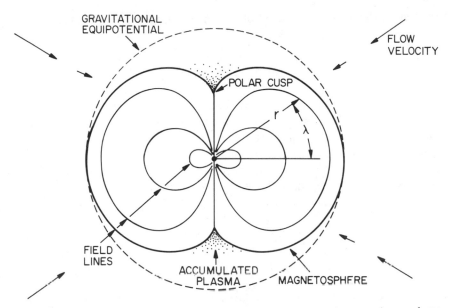

GRAVITATIONAL
EQUIPOTENTIAL

FLOW
VELOCITY

POLAR CUSP

r

λ

FIELD
LINES

ACCUMULATED
PLASMA

MAGNETOSPHERE

FIG. 10.1. Confined magnetosphere. Pressure of in-falling (or resident) plasma produces a characteristically shaped magnetosphere.

10.3 Accretion onto a Magnetized Neutron Star

Even the oversimplified spherical accretion discussed above will eventually become asymmetrically perturbed by the dipole nature of the magnetic field near a neutron star. A number of models of increasing sophistication have been devised to deal with the resulting distortion of the magnetopause and the way that the plasma gets past this boundary to reach the neutron star. We will use *magnetopause* in the sense used in space physics, namely the furthest extent of magnetic field lines from the star; in astrophysics it is sometimes taken more generally to be the region within which the magnetic fields become important, or sometimes simply the region in which something interesting is happening involving magnetic fields.

1. Hydrostatic Models

Consider a stationary gas accumulated about a magnetized gravitational well. One of the earliest calculations was by Midgley and Davis (1962) for a surrounding plasma having constant uniform pressure on the magnetic field (see also Beard 1960 and Slutz 1962). Our treatment is comparable to that of Midgley and Davis owing to the symmetry and the fact that the currents on the magnetopause (see figure 10.1) have constant amplitude. The condition for a solution is that currents flowing on the boundary at the magnetopause must produce a pure external dipole moment to cancel the dipole moment of the star without at the same time adding any other multipole moments. The general expression for the multipole moment of currents flowing on a magnetopause with axial symmetry is

$$I_n = \int_{-1}^{+1} J(\theta) r^{n+1} (r^2 + r_\theta^2)^{1/2} P_n^1(x) \, dx, \tag{1}$$

and the solutions require *all* I_n to vanish except I_1, which must be equal and opposite to a pure dipole source. This latter condition is trivial, given a solution that sets the other multipoles equal to zero, because $J(\theta)$ can simply be scaled up or down accordingly. Here $x = \cos\theta$, $r_\theta = dr/d\theta$, $r = r(\theta)$, $J(\theta)$ is the azimuthal current flowing at colatitude θ, and P_n^1 is the associated Legendre polynomial of order unity and index n (Abramowitz and Stegun 1968). Although there are two free functions, $J(\theta)$ and $r = r(\theta)$, only the latter has to be determined because the first is fixed by the discontinuity in magnetic field across the magnetopause, which in turn is fixed by the pressure balance across the magnetopause. For a constant external pressure, $J = \text{constant}$ independent of θ, leaving only the shape to be determined, while for a hydrostatic equilibrium $P \approx r^{-K}$, we have $J \approx r^{-K/2}$. A straightforward approach, for any given K, is to expand $r(\theta)$ in a series and adjust the first N coefficients to set the first $N - 1$ multipoles beyond the dipole equal to zero (there is no need to set the dipole moment itself to any particular value because that can be accomplished afterward by linear rescaling).

The physical significance of K can be seen by considering a gas with a Poisson adiabatic index γ. If we place a fluid element of such a gas in a magnetosphere, the external magnetic pressure varies with distance as

$$\frac{dP_m}{dr} = \frac{B^2}{r} = r^{-7}, \tag{2}$$

which increases the density and gives a gravitational force density

$$\rho g = r^{-\left(2 + \frac{6}{\gamma}\right)}. \tag{3}$$

The two balance for the critical value $\gamma_c = 6/5$. As a consequence, for any Poisson adiabatic index $> \gamma_c$ (i.e., for most known physical systems), the fluid element will find an equilibrium level at which it would float. Otherwise, the fluid element either will be ejected or will sink. An incompressible sphere, for example, will either be so light as to be ejected or be so heavy as to sink to the origin. Our definition of K is in terms of how the gas pressure changes with distance, not density, and we can even give an effective buoyancy index for the magnetic field:

$$K_m \equiv -\frac{r}{P}\left(\frac{dP}{dr}\right). \tag{4}$$

The relation between K and γ for the fluid is just

$$K \equiv \frac{\gamma}{\gamma - 1},\tag{5}$$

so a monatomic fluid has $\gamma = 5/3$ and $K = 5/2$.

Midgley and Davis numerically approximated the $K = 0$ case and later Arons and Lea (1976a,b) similarly solved for the $K = 2.5$ case, while Michel (1977a) solved for a range of cases from $K = 0$ to $K = 5$ ($K = 6$ can be done analytically, as we will see). Two-dimensional versions of this problem can be solved by conformal mapping (Dungey 1961; Elsner and Lamb 1976). The numerical treatments were of limited precision owing to the cusp over the polar caps, which in principle requires a large number of multipoles if it is to be faithfully reproduced (the shape is illustrated in figure 10.1). The cross-correlated nature of the coefficients in a series can make it increasingly difficult to null out these higher multipoles as more terms are added. For example, if one does not choose an appropriate expansion, adding a succeeding term might change the equatorial radius, which is a first-order change because the radius scales the entire system; thus, adding such a term will immediately require all the remaining terms to be readjusted, hence to be cross-correlated. Such inefficient algorithms can defeat the fastest computer. Some improvement can be obtained by fitting the cusp itself to the exact solution, given by Morozov and Solev'ev (1966). The cusp must begin at a finite height and flairs thereafter with $\delta r \approx \theta^{2/3}$.

It is possible, however, to include *all* multipoles (Michel 1977b) using a simple trick to sum the series in equation 1. First we note that for some arbitrary variable h

$$\sum_{0}^{\infty} I_n h^n = h I_1,\tag{6}$$

because all the multipoles vanish except the dipole. But the generating function for the Legendre polynomials is

$$\sum_{0}^{\infty} P(x)_n h^n = (1 - 2xh + h^2)^{-1/2},\tag{7}$$

and

$$P_n^1 \equiv -\sin\theta \, \frac{dP_n}{dx}.\tag{8}$$

Thus we can create the generating function for P_n^1 by differentiation. After straightforward manipulations, we obtain

$$I(h) = I_1 = \int_{-1}^{+1} J(\theta) r^2 (r^2 + r_\theta^2)^{1/2} \sin\theta R^{-3/2} \, dx,\tag{9}$$

where

$$R \equiv 1 - 2xrh + r^2h^2. \tag{10}$$

Note that the powers of r have been absorbed into R, and n vanishes.

Equation 9 is a nonlinear Fredholm equation of the first kind. In general, such equations need not necessarily even have solutions. Here we have the advantage of knowing from the physics that solutions exist. In addition, we can invert the problem by guessing $r(\theta)$ to see what function $J(\theta)$ is required to give a constant $I(h)$ (as required for a dipole source, although other sources can be treated just as well). The entire problem is reduced to finding a function $f(\theta)$ such that

$$\int_{-1}^{+1} f(\theta)R^{-3/2}\, d\theta = \text{constant}. \tag{11}$$

2. Spherical Confinement

For example, take $r(\theta) = 1$; it is easy to confirm that

$$\int_{-1}^{+1} \frac{(1 - x^2)\, dx}{(1 - 2hx + h^2)^{3/2}} = \frac{4}{3} \quad (h^2 \leq 1), \tag{12}$$

and therefore $f(\theta) = \sin^3\theta$ or $J(\theta) = \sin\theta$. This result is already well known and can be obtained much more simply, but it serves as a useful check. It is far from obvious that the integral on the left could be independent of the value of h, and indeed curious that such integrals even exist.

3. General Multipole Surfaces

The above example is a trivial case of the general one where

$$f(\theta) = \sin^2\theta \left(\frac{1}{r} - \sum a_n r^{n+1} P_n^1 (cos\,\theta) \right) = \text{constant}, \tag{13}$$

where the a_n are arbitrary. The appropriate $J(\theta)$ is given by calculating B from the vector potential, which in this case is given by

$$A_\phi = \frac{f(\theta)}{r \sin\theta}. \tag{14}$$

The magnetic field lines follow lines of constant f. Thus one can construct a solution of the integral equation without even being able to integrate it directly. For example, we can take all the a_n to be zero and the magnetopause becomes one of the dipole magnetic field lines (which is a possible solution if we adjust the external pressure to just match the magnetic pressure along a magnetic field line, hence obtaining a pressure law

$$P = P_0 \left(\frac{4}{r^6} - \frac{3}{ar^5} \right),$$

(15)

where a is the equatorial extent of the magnetopause). In this case, the integral equation is

$$I(h) = \int_0^\pi \frac{(1 + 3\cos^2\theta)\sin\theta\, d\theta}{[(\cos\theta - h\sin^2\theta)^2 + \sin^2\theta]^{3/2}} = 4 \quad \text{(all } h\text{)}.$$

(16)

4. Nontrivial Solutions

Consider a completely different, but simple, case where

$$r = \sin\theta.$$

(17)

In this case, the integral equation can be solved after the transformation $y = \cot\theta$ and the observation that

$$\int_{-\infty}^{+\infty} \frac{dy}{(y^2 - 2hy + 1 + h^2)^{3/2}} = 2 \quad \text{(all } h\text{)}.$$

(18)

The requirement is then that $J(\theta) = (\sin\theta)^{-3}$; therefore the external pressure varies as r^{-6}. Thus we have a simple example which establishes an important physical result, namely that *if $K = 6$, the cusp of the magnetopause can reach down to the surface of the star*. Note that the pressure law for a dipole field line defining the magnetopause ($r = \sin^2\theta$) is *not* $P = r^{-6}$ (see above).

5. Numerical Results

A rather good fit can be obtained with just two parameters, one giving the correct cusp shape and the other controlling the scale of the magnetopause relative to the cusp height:

$$r(\theta) = 1 + A (\sin\theta)^{2/3} + B \sin\theta.$$

(19)

To determine what constitutes a good fit, we differentiate $I(h)$ (which should give zero for a dipole source) and calculate numerically the goodness-of-fit parameter ϵ, where

$$\epsilon \equiv \int_0^1 \left(\frac{dI}{dh} \right)^2 dh \geq 0.$$

(20)

In practice the integral is replaced by a simple sum at equally spaced steps in h (convergence is poor near $h = 1$ and may be avoided without apparently compromising the final results, although assuring convergence at this limit may plausibly make the convergence more robust). The results are shown in table

TABLE 10.2—Magnetopause Parameters

K	A	B	P_0	K_m	r_c
0.0	0.69	-0.05	8.03	2.50	0.610
			7.99	2.48	0.631
2.5	0.736	0.190	7.56	2.71	0.519
			7.54	2.70	0.503
4.0	0.951	0.44	6.98	2.90	0.418
			6.99	2.95	0.357
5.0	0.595	1.44	6.27	3.21	0.330
			6.28	3.28	0.177
6.0	0.	Note a	4.	4.	0.

a) Exact solution $r = \sin\theta$, so formally $B \to \infty$ is required to eliminate the unity in equation 17.

10.2. In table 10.2 are three parameters not yet discussed in detail. First, the second row shows results obtained from a 6 parameter fit (Michel 1977a) for comparison. Second, the quantity r_c is the cusp height normalized to unity for the equatorial extent (i.e., the largest extent of the confined magnetopause). The effect of confinement is to increase the magnetic pressure at the equatorial magnetopause by a factor P_0 over the uncompressed case (in other words, the magnetic field is increased over the free dipole case by a factor $\sqrt{P_0}$). Finally, K_m is the *magnetic* buoyancy index determined by the rate of increase in magnetic field just inside the equatorial magnetopause. The magnetopause is most unstable at the equator and becomes completely stable to interchange at the polar cusps owing to the tight curvature of the magnetic field lines there. Equatorial instability is not the only way the plasma can reach the star, however, because we have seen that when $K = 6$, the cusps simply dip down to the source giving direct flow onto the stellar surface. The local value of the magnetic buoyancy index K_m is entirely determined by the shape of the magnetopause and is given from

$$K_m = \frac{2}{R^2 + R_\theta^2}\,(R^2 + 3R_\theta^2 + RR_\theta \cot\theta - RR_{\theta\theta}), \tag{21}$$

where $R_\theta \equiv dR/d\theta$, etc. Note that we correctly recover the result $K_m = 2$ for a constant circular radius ($R = 1$, $R_\theta = 0$, $R_{\theta\theta} = 0$).

6. Physical Interpretation

If $K < K_m$, the magnetopause is stable to interchange at the equator. An important reason for calculating the magnetospheric shape was, in fact, to calculate this stability or instability. Thus, for the $K = 0$ case, the plasma is incompressible and if a plasma "blob" is pushed into the magnetosphere, it will pop back out like a cork. From table 10.2 we see that equatorial stability extends up to $K \approx 2.7$. Originally, workers assumed that magnetopauses were *unconditionally* interchange unstable, owing to the standard textbook treatments showing that a fluid with nonzero mass density supported by magnetic pressure above a *uniform* magnetic field was unstable just like a heavy fluid floating above a light fluid (i.e., magnetic field is "equivalent to a zero mass fluid"). More careful treatments, however, show that tension along the magnetic field lines plays a stabilizing role. Ironically, the adiabatic index of hydrogen (either gas or plasma) corresponds to $K = 2.5$, which forms a stable magnetopause, at least until cooling can reduce the scale height.

The conventional view is that surrounding matter free-falls toward the magnetopause until brought to a halt by a shock wave. The effect of a shock wave is to decrease K. Thereafter, cooling reduces the scale height to Rayleigh-Taylor unstable values and plasma blobs enter the magnetosphere (see Wang and Robertson 1985). At this point these blobs cannot reach the polar caps (the usual theoretical goal) because they are now trapped in a *minimum B* geometry. It is therefore supposed that some rapid instability allows the blobs to disappear, with the particles diffusing onto the magnetic field lines. Because the particles should radiate away their perpendicular energy, albeit slowly for protons, which have lifetimes of about 1.6×10^{18} s/B_0 (gauss)^2, they can now follow the magnetic field lines to the polar caps and fall onto the surface to re loose the gravitational energy assumed to power the X-ray luminosity. This picture is essentially that pioneered by Lamb, Pethick, and Pines (1973). Whether these processes take place within the required time scales remains uncertain. The alternative of direct flow onto the polar caps was proposed by Davidson and Ostriker (1973). The latter picture requires cooling of the flow as it approaches the star to effectively give $K \geq 6$.

An alternative mode of entry was cooling of the plasma pooled at the polar caps (Michel 1977c). Because the material is densest there, cooling (e.g., via bremsstrahlung) would increase K locally but not globally; consequently the polar cusp should intermittently descend to the surface rather than leave a permanently open channel. Morfill et al. (1984) have revisited this model, describing these in-falling packets (see figure 10.2) as "broomsticks," because they should become highly compressed transversely by the anisotropic magnetic field pressure. Elsner and Lamb (1984) argue, however, that Compton cooling by the X-rays themselves (with temperatures of 10^6 K) of the magnetopause plasma (at temperatures of perhaps 10^9 K) will be so rapid that equatorial entry will take place before polar cap entry is important (see also Elsner and Lamb 1977; Weyman 1965). This leaves the interesting problem of what happens with the plasma, because it cannot very well flow *uphill* out of the cusps to reach the equatorial zones and it cannot sit indefinitely in the cusps per se. Moreover, these considerations are not at all model independent, with cooling that is effective only for the electrons, not the ions.

FIG. 10.2. Polar cap "drip." Detached cusp reforms behind the original cusp as it descends with its load of dense plasma. The latter is filled in behind with lighter plasma that is unable to radiate fast enough to follow it down. From F. C. Michel, 1977c, *Ap. J.*, 216, 838 (figure 1).

a. *QPOs*

Discovery of quasi-periodic oscillations (QPOs) in certain X-ray sources (van der Klis et al. 1985; sec. 10.1.3) has led to the suggestion that new mechanisms are required in addition to simple accretion from a disk. Indeed, the X-ray burster phenomenon had already stimulated the independent suggestion that the plasma was somehow "dripping" onto the polar caps (Lewin 1981). Otherwise, the dripping idea has not been popular and workers have been forced to suppose that some sort of dynamic clumping mechanism is at work to produce large changes in what was supposed to be an approximately steady state accretion (Lamb et al. 1985). Given the huge phenomenological viscosities invoked to feed the neutron star from an accretion disk in the first place, such a clumping mechanism must be quite effective to overcome those dissipative effects.

The time scales of the possible bursting can be crudely estimated by taking the published (Michel 1977c) distance scale ($r^* = 6 \times 10^7$ cm) and dividing by the characteristic sound velocity ($V^* = 2 \times 10^9$ cm/s) to obtain 33 Hz, in plausible accord with the values of 5 to 50 Hz reported (Lamb et al. 1985). Here one imagines that the drops descend at the sound velocity, deposit the plasma, and recover to the equilibrium cusp location. The irregularity in such dripping is, of course, a matter of everyday experience. Transport in the accretion disk will be considered in the next section.

7. General Solutions (e.g., flowing plasma)

It is possible that some sources accrete not from a disk but directly from a wind flowing away from the companion, which calls for nonaxisymmetric solutions. Indeed, other magnetopauses, such as that of the Earth, are not at all axisymmetric. It turns out that the above treatment can be extended to the nonaxisymmetric case because the (vector) multipole moments in three dimensions simply have a second index corresponding to their azimuthal dependences, and one can again write down a function whose power series expansion gives the desired multipole expansion (which will again mainly be zeros except for the first dipole term or two). We now have the possibility of tilting the dipole relative to some flow direction in addition to having a flow direction in the first place. In general, the multipole moments can be summed to form the functions

$$\mathbf{I}(h, \lambda) = \sum_{n=0}^{\infty} \sum_{m=0}^{\infty} \mathbf{I}_{nm} h^n \lambda^m \tag{22}$$

and both summations can be done to give the formal result

$$\mathbf{I}(h, \lambda) = \int \mathbf{J} \left(\frac{-\lambda h r z \sin\theta}{R + \lambda h r z \sin\theta} \right) R^{1/2}, \tag{23}$$

where $z \equiv e^{i\phi}$ introduces the azimuthal dependence and, as before, $R \equiv 1 - 2xrh + r^2h^2$. These equations are derived and discussed in more detail in Michel (1977d).

A simple case where the plasma has unidirectional flow over the magnetized source is illustrated by the so-called Uranus problem where the incident flow is parallel to the magnetic moment (it turned out that while the spin axis of Uranus pointed into the solar wind, the magnetic moment was offset by a large angle!). This model is axisymmetric, and the integral equation becomes

$$I(h) = \int_0^1 \frac{\rho^2}{R^{3/2}} \, d\rho = \frac{1}{3} \quad (h^2 \leq 1), \tag{24}$$

where r, ρ, and z are the usual cylindrical coordinates, and $R \equiv 1 + 2zh + r^2h^2$; see Burgess (1978) for numerical solutions.

10.4 Evolution of the Disk

The standard picture of disk evolution is to assume a phenomenological viscosity which is typically scaled against the maximum physically plausible viscosity (the so-called α disks: Shakura and Sunyaev 1973). Because this view has been developed in numerous places, and because suspicions have arisen as to the physical plausibility of this picture, we will present an alternative view. The alternative view is that the dissipative process necessary for rapid evolution of an accretion disk is driven by spiral hydraulic jump (shock) waves on the surface of the disk (Michel 1984a; Sawada et al. 1986a,b, 1987; Spruit 1987a,b; Spruit et al. 1987; Anthony and Carlberg 1988). See also the recent workshop proceedings *Theory of Accretion Disks* (Meyer et al. 1990). These waves are excited by the asymmetric nature of the central rotator (e.g., neutron star magnetosphere) and spiral out into the disk to form a pattern corotating with the central object. Disk matter in turn is slowed slightly at each encounter with the jump and spirals inward. In this process, the disk is heated by true turbulence produced in the jumps. Analogous phenomena have been proposed to act in the formation of galactic spiral structure.

1. The Model

Historically, the two limiting views on the behavior of a Kepler disk orbiting a central gravitational source have been (1) the central object serves only as the gravitationally confining source; consequently only intrinsic disk processes such as viscosity can drive the disk evolution; and (2) the disk dynamics are driven by perturbations external to the disk, such as direct interaction with the central object or tidal interactions with satellites. The view that the central body is a passive element is implicit in the early use of such models for the presolar nebula, where it seemed natural to view the central Sun as simply an axially symmetric condensation in an initially axially symmetric disk (e.g., Jeffreys 1924). In that case, huge effective viscosities were required for the disk to evolve in times short compared to the age of the Universe, and such viscosities were assumed to arise from some sort of turbulence (Peek 1942; von Weizsacker 1943). Cataclysmic variables (Robinson 1976) and, later, binary X-ray sources came to be modeled in the same way. The alternative view is explored here, namely that it is interaction with the central body that mainly drives the disk evolution. The Roche-lobe overflow and tidal effects from the companion must also play a role, as discussed qualitatively. The essential requirement, of course, is that the central body *not* be axially symmetric. For both cataclysmic variables and binary X-ray sources, there is ample evidence for such asymmetries. One observes oscillations (Robinson 1976) that are almost certainly due to rotation of the central body in the case of the cataclysmic variable DQ Her, and periodic pulsed emissions in the binary X-ray sources. Neither phenomenon is consistent with an axisymmetric central object, and the issue is not whether there is asymmetry but whether the observed asymmetry plays an important role in the accretion process itself.

a. *The Viscosity Problem*

The standard view is not without its own problems. To quote Pringle (1981), "The main failing of accretion disc theory is that it has no predictive power except in certain limiting circumstances. The main reason for the lack of predictive power is the uncertainty as to the nature and magnitude of the viscosity." The idea that internal viscosity is the major mechanism for driving disk evolution is curiously popular in the high-energy astrophysics community, even though it has not completely convinced theorists concerned with cataclysmic variables themselves (e.g., Shu and Lubow 1981), which provided essentially the models taken over for X-ray sources. The high viscosities are the sticking point. The best hope for justifying large viscosities has been to invoke magnetic viscosity (Lynden-Bell 1969; Shakura and Sunyaev 1973; Eardley and Lightman 1975; Ichimaru 1976; Coroniti 1981). A rough consensus might be that it is not out of the question to explain the viscosity on this basis, but it is quite difficult. Moreover, most of the essential information (magnitude and distribution of seed magnetic fields) cannot be observationally confirmed at present. The extension of such assumed high viscosities to other astrophysical settings (e.g., pulsar models with disks: ch. 6) is unwarranted without a clear understanding of where such viscosities come from, if they exist at all. These high putative viscosities would make it difficult for such disks to survive as long as pulsars are thought to live. Even for astrophysical parameters, the gulf between straightforward theory and phenomenology is enormous. The kinematic viscosities of such disks could well be no more than 10^3 cm^2/s, whereas phenomenology requires values of 10^{16} cm^2/s, say. The latter is more like the viscosity of the Earth's mantle! As we saw in sec. 6.3, the straightforward calculations of viscosity give very small values of order unity.

b. *The Magnetic Paddle Wheel*

The inner edge of a disk around a rotating neutron star sees, in effect, an asymmetric rotating object: a *magnetic paddle wheel*. If the two are not in corotation, the motion of the central object creates a disturbance that propagates outward into the disk. This disturbance is closely related to the so-called hydraulic jump. The hydraulic jump is a familiar one-dimensional parallel to a gas-dynamic shock wave. The classic example is water flowing down a channel toward an obstruction. Because the water has a free surface, the flow conservation equations have shock-like solutions even if the water is taken to be incompressible. For a hydraulic jump, the critical flow corresponds to velocities faster than the propagation velocity of a surface wave rather than to velocities faster than a compressional wave in the medium. Returning to the disk problem, the torque on the rotator is essentially given by the drag on the magnetosphere exerted by the local disk medium, which is moving more rapidly. Globally, however, this torque propagates out as a hydraulic jump. Disk fluid passing through the jump slows and therefore falls toward the center. This displacement speeds up the fluid until half an orbit later, when the fluid element is moving faster than Keplerian, but at this point another jump is encountered and the process repeats (figure 10.3). Thus disk material is systematically and rapidly convected inward, regardless of the internal kinematic viscosity of the disk fluid. For fluid or gaseous disks in orbit, the jump conditions are modified by the response of the gas to the heating after the jump as the turbulence is dissipated. As a technical point, the speed of sound is comparable to the speed of surface waves for matter in a disk, so it is largely a matter of taste whether to refer to such waves as hydraulic jumps or as shocks. The former terminology will be used to emphasize the very general dissipative nature of the interaction, which depends on neither the compressibility nor the viscosity of the fluid. The shock wave formed by a fast-moving object such as a supersonic airplane is easily detected even at distances vast compared to the airplane's size, and represents energy dissipated by the object at an earlier moment. An associated effect, although less readily noticeable, is the fact that the air is set into motion in the direction

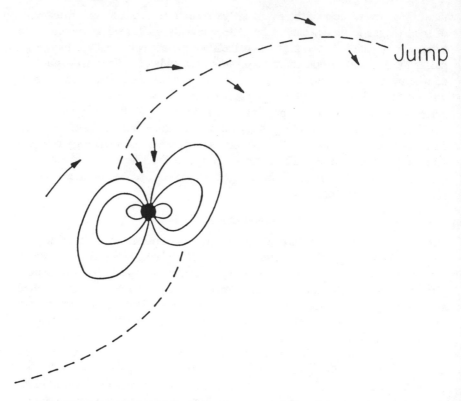

Fɪɢ. 10.3. The model. Rotation of the asymmetric magnetic field configuration of the central object excites a hydraulic jump (essentially a shock wave) that propagates outward into the disk. Repeated passages of the disk fluid through the jump systematically move the fluid inward. From F. C. Michel, 1984a, *Ap. J.*, 279, 807 (figure 1).

of the airplane. This dragging of the air along with the airplane has nothing whatsoever to do with viscous coupling of the air between the skin of the airplane and the distant observer. In the same way, the shock set up by the rotating body gives a small velocity change to the disk fluid, which decreases its orbital velocity and causes it to move inward. Because the fluid encounters two jumps, one from each magnetic pole (figure 10.3) each orbital (synodic) period, the fluid is systematically driven toward the rotator. If for some reason the rotator were unable to accrete, a corotating volume of fluid would accumulate and mask the underlying asymmetry, the jumps would vanish, and a steady state without accretion would be set up. The nomenclature presents some modest problems because a hydraulic jump only occurs when the velocity exceeds that of surface waves, and the corresponding dimensionless flow velocity is termed the Froude number, which we will use although it has an obvious parallel in the Mach number. In addition, such flow is termed *supercritical* in hydrodynamics, which conflicts with the same terminology for accretion flows that exceed the Eddington limit and which we will avoid.

c. *Eddington Limit*

The Eddington limit plays an important role in accretion-powered sources because it is the luminosity at which the radiation pressure becomes comparable to the gravitational attraction of the star. If a star were to exceed this limit, it would blow its outer layers away and undergo mass loss at some equilibrium rate; the calculation assumes the surroundings to be optically thin (a planet would not be blown away, for example). One simply balances the incident momentum flux $L/4\pi r^2 c$ at some distance r times the Thomson cross section of an electron $\sigma_T = 6.65 \times 10^{-25}$ cm^{-2} against the gravitation attraction on an ion, $m_i GM/r^2$, and solves for L, given that the radial dependence drops out.

$$L_{Eddington} = 4\pi(m_i c^2)\frac{GM}{c^2}\frac{c}{\sigma_T} = 1.26 \times 10^{38} \text{ ergs/s}. \tag{1}$$

We have grouped the terms to show that the energy is scaled by the ion mass and the rate is scaled by the gravitational length times c divided by the cross section. Although the radiation pressure is exerted on the electrons, any attempt at charge separation creates an electric field that drags the ions along, so in steady state the force is applied to the ions through the electric field (corresponding to a net negative charge on the star of about 167 coulombs).

2. The Hydraulic Jump

Hydraulic jumps are thoroughly discussed in the literature (see Li and Lam 1964; Massey 1975; Lighthill 1978; etc.), but the discussion is usually restricted to incompressible unidirectional flow in a uniform gravitational field. For orbiting disks, one has compressible gas flow in a tidal gravitational field, but the resulting equations are nevertheless rather similar.

a. *Incompressible Fluid, Uniform Field*

The standard hydraulic jump model incorporates an incompressible fluid and a uniform gravitational field. The flow equations can be written in one-dimensional form conserving the mass flow,

$$\rho h U = \text{constant}, \tag{2}$$

where the flow velocity U is taken to be constant along the vertical height h of the flow sheet, as well as the density ρ. For a gas, one must instead use the height-integrated density. Equation 2 is constant across a hydraulic jump, but h and U change, the velocity slowing as the flow plows into a higher *hydraulic head* (h). The flow is necessarily turbulent because the centers of mass of the two flows do not match, as shown in figure 10.4. This induced turbulence has even been exploited as a mechanism to thoroughly mix fluids (Massey 1975). Momentum must also be conserved, which means that the net force on the transition region must vanish in the rest frame of that discontinuity; thus the static force (here the jump can be considered to be replaced by a dam, and we

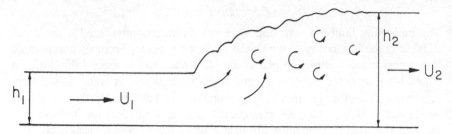

FIG. 10.4. The hydraulic jump. Fluid of one height flows uniformly until an obstruction is encountered. The resulting jump in height is intrinsically turbulent owing to the couple between the two displaced centers of momentum. From F. C. Michel, 1984a, *Ap. J.*, 279, 807 (figure 2).

are calculating the pressure against our dam on each side) is

$$P = \int \rho g h \, dh = \rho g h^2 / 2, \tag{3a}$$

where g is the constant gravitational acceleration, and the dynamic pressure is

$$\int \rho U^2 \, dh = \rho h U^2, \tag{3b}$$

the sum of which is conserved to give

$$\rho h (g h + 2 U^2) = \text{constant}. \tag{4}$$

The equation for the flow of energy is now conserved only if the exit flow has an energy component in addition to its gravitational and kinetic energies. This component is the turbulent energy produced at the jump and later, after dissipation, it is thermal energy. Here ρ is a constant that factors out of these equations. The velocities can then be rewritten in terms of the Froude number

$$F = \frac{U}{(g h)^{1/2}} \tag{5}$$

to give the final results

$$h_2 = \frac{h_1}{2} \left[(1 + 8 F_1^2)^{1/2} - 1 \right], \tag{6}$$

where F_1 is the Froude number in the fluid before it encounters the jump. Just as for shock waves, $F > 1$ is required on the incident flow side because only then is there mechanical energy available to dissipate. As long as the Froude numbers are not too large, an approximate solution is

$$F_1 \approx 1/F_2 \approx 1 + f \quad (f < 1). \tag{7}$$

The energy dissipation in a hydraulic jump is of order $\delta h \cdot \delta U$, hence of the order f^2 (as opposed to f^3 for gas-dynamic shocks, the difference being that some of the energy spent in the shock compression of a gas can be recovered in an adiabatic reexpansion to the original density, whereas conversion of mechanical energy into turbulence in an incompressible fluid is entirely irreversible). In the limit $f \to 0$, the dissipation may be ignored and the limiting waveform is exemplary of a soliton.

b. Compressible Gas, Uniform Field

It would be possible to have flows of, say, a planetary atmosphere into such a jump wave (e.g., following an asteroid impact). In this case, the density is no longer constant; but since ρ is temperature dependent, the energy equation is required to give the effects of heating on the density. These details do not affect equation 7, however, so the extension of the above equations to this case is not presented, although that would be straightforward.

c. Compressible Gas, Tidal Field

In the compressible gas, tidal field model the uniform acceleration g is replaced by the tidal acceleration of particles back to the disk plane, $h\Omega^2$, where Ω is the Keplerian orbital angular velocity and h is the height above the midplane. This modification merely introduces additional numerical coefficients that multiply the height-integrated density and average pressure. The characteristic velocity for jump formation is now of the order Ωh and is also a measure of the sound velocity in the medium as well as the surface wave velocity. Consequently the physical distinction between Froude number and Mach number, as well as between hydraulic jump and shock, becomes one of emphasis. The terms Froude number and hydraulic jump will continue to be used to emphasize the basic hydraulic nature of the jump, because the mechanism discussed would operate just as effectively for an incompressible fluid in orbit (for which case the sound speed would be irrelevant). As above, the modifications to the jump relations are not essential to the rough estimates below. It is interesting that the jump conditions should actually be more accurate for an orbiting disk than for the terrestrial counterpart for which they were derived; viscous drag between the flow and the surface invalidates the assumption that U is a constant independent of height in the latter case (Dryden et al. 1956). For an orbiting disk, any residual viscosity acts instead to ensure the above assumption.

3. Equation of Motion

The traditional disk equations assume viscous dissipation in an axially symmetric disk. Given the spiral pattern shown in figure 10.3, the equations are no longer azimuthally symmetric. More important, the fluid motion is not diffusive but is a systematic spiraling of the matter into the center. The fluid trajectory at successive encounters with the jump is shown in figure 10.5. Note that the fluid flow between R_n (the radius, R, at which jump crossing n takes place) and R_{n+2} must be conserved regardless of n, giving the mass flux conservation

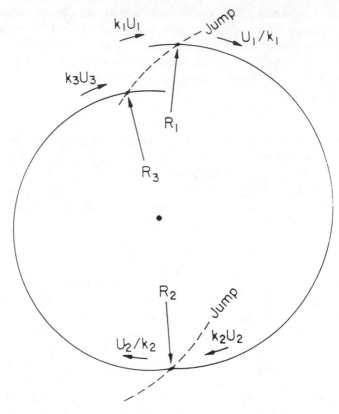

FIG. 10.5. Spiral fluid motion. At each encounter (shown in the frame of the jump, which is assumed to be inertially fixed for simplicity) fluid enters at slightly higher than Keplerian velocity (U_1) and exits at slightly lower velocity, by some factor k_1, which is determined by the jump strength. Flow within spiral flow lines must be conserved. From F. C. Michel, 1984a, *Ap. J.*, 279, 807 (figure 3).

law,

$$\Sigma_{n+1} U_{n+1}(R_n - R_{n+2}) = \dot{M} = \text{constant}, \tag{8}$$

where Σ is the height-integrated density of the disk ($=\rho h$ to a first approximation). It should be recognized that for most cases of interest the spiral angle will be extremely small. Consequently the velocity change per encounter with each hydraulic jump will be small and to an excellent approximation

$$U_n = \left(\frac{GM}{R_n}\right)^{1/2}, \tag{9}$$

namely just the Keplerian orbital velocity. However, the precise velocities change across the shock, with the prejump and postjump velocities given by

$$U(\text{preshock}) = (1 + f)U, \tag{10a}$$

$$U(\text{postshock}) = (1 - f)U, \tag{10b}$$

where f is the small quantity (eq. 7) describing the jump strength. To account for the effect of this velocity change, the fluid elements are assumed to follow a ballistic orbit between encounters with the jumps; thus

$$U_n(\text{postshock})R_n = U_n(\text{preshock})R_{n+1}, \tag{11}$$

which relates the spiral angle to the velocity change per jump, owing to the conservation of angular momentum. We can now solve for Σ_{n+1} in terms of Σ_n and f and rewrite the conservation law (eq. 8) as

$$8fUR\Sigma = \dot{M}. \tag{12}$$

Except for the coefficient of 8, this equation could have been written down by inspection; fU is essentially the inward radial velocity of mass at density Σ/h across the surface area hR. In the standard disk theory, $\dot{M} = 3\pi\nu\Sigma$ (e.g., Pringle 1981); thus the effective viscosity is

$$\nu_{\text{effective}} = \frac{8fUR}{3\pi}. \tag{13}$$

In the *alpha disk* models (Shakura and Sunyaev 1973), the viscosity is estimated to be of order Ωh^2 whereas here the small quantity f multiples the much larger coefficient ΩR^2.

a. *The Jump Strength*

The difficult part of the problem is to calculate the variation of f with distance. It is here that the detailed energy and momentum equations would enter. Instead, we make some rough illustrative estimates. A shock wave in a two-dimensional system propagates at fixed strength if the perturbation is maintained once applied (e.g., a wedge), but for a finite perturbation (a short weir under the water, for example), the perturbation vanishes once the fluid has moved past and the effect will be attenuated with distance R by a factor which can be taken to be S/R, because that is the displacement required to make room for the fluid displaced by the obstacle (the latter having a size of order S). Thus the limits follow:

$$f = f(S/R), \quad 1 \geq f(x) \geq x, \tag{14}$$

which unfortunately embraces the range in which Σ either increases or decreases with distance. Actually the intermediate choice $f \approx x^{1/2}$ is probably not too far off and corresponds to constant values for both Σ and $\nu_{\text{effective}}$. The torque on the object is the rate at which angular momentum crosses at the effective inner boundary distance (S); hence

$$T = SU(S)\dot{M} \qquad (15)$$

for a nonrotating central paddle wheel.

4. Conditions near the Paddle Wheel

As discussed in sec. 3, the magnetospheric boundary between a dipole magnetic field is quite asymmetric, and even in the simplest and most conservative case (uniform confining pressure) there is approximately a 30% difference in equatorial versus polar radius. For more realistic cases, this difference can essentially approach 100%. A rotating inclined dipole therefore presents an asymmetric obstacle to a surrounding plasma disk. Two additional considerations may further increase this asymmetry. For one, the shape of the magnetopause can be unstable to deformation. Arons and Lea (1976a,b) examined this instability but concentrated on the short-wavelength components (with the view of finding a rapid diffusion mechanism to transport particles across the magnetosphere; see also Barnard et al. 1983). However, the same instability can act at long wavelengths as well, which simply correspond to the magnetic field bulging out on one side and concentrating the plasma flow on the other, an aneurysm of sorts, which would accentuate any interaction with the disk. Elsner and Lamb (1976, 1977) also discuss this instability, but again only in the short-wavelength limit. The second effect would be orthogonal (mis)alignment of the rotator. In the case of pulsars the slowing-down torques on a magnetized neutron star also have a component that acts to align the rotation and magnetic dipole axes (Michel and Goldwire 1970; Davis and Goldstein 1970; Soper 1972). In the same way, if the magnetic torques are reversed to spin the neutron star up, as is evidently the case for pulsating X-ray sources, there should also be a torque component that acts to increase the inclination axis (Wang 1981). Unfortunately, it is not obvious that the time-averaged torque on the neutron star is enough to accelerate it, because many neutron stars are still observed to have quite long periods despite rapid spin-up on short time scales (Elsner et al. 1980). Observationally, however, appreciable asymmetries must be present.

5. Conditions in the Outer Disk

a. *The Accretion Disk*

The way in which matter is added to the disk can itself be an important perturbation. In the cataclysmic variables Robinson (1976) even attributes a flickering component of the optical emission to a hot spot formed on the disk. It seems quite likely that a standing hydraulic jump/shock wave is produced on the outer disk by the Roche-lobe overflow (Lubow and Shu 1976), the effect of which is again to induce a mechanical transport of matter. The agitation implied by such time variability as revealed in the flickering seems also a possible extrinsic source of effective viscosity.

b. *The External Tides*

Alternatively, we have the tidal effects of the companion star that contribute the matter in the first place. These tides must surely limit the extent of the disk itself. Tidal effects could therefore provide effective matter transport in the outer regions of the disk (Lubow 1981; Kriz 1982). Such effects would be expected to become less important close to the neutron star because the tidal effects from the companion become small compared to the gravitational forces of the accreting star itself. However, it is in this regime that the paddle wheel would play a role.

6. Numerical Estimates

The normal assumption, which we adopt, is that $\dot{M}$ is imposed on the system by Roche-lobe overflow from a companion star. Thus the central luminosity is of the order of

$$L = \phi_s \dot{M}, \tag{16}$$

where ϕ_s is the surface gravitational potential of the central object, typically 10^{-1} for neutron stars and 10^{-3} for white dwarfs in units of c^2. The possibility (indeed, probability) of mass loss from the system introduces important uncertainties as to exactly how much of the mass loss from the companion actually reaches the neutron star, so that estimated from the luminosity as above refers only to the accreted fraction. The torque on the object can be measured from its spin-up rate and period, assuming a moment of inertia of order 10^{45} g cm^2 for neutron stars and 10^{49} g cm^2 for white dwarfs. The torque is given roughly from equation 15, which together with equation 16 and the measured rates gives the interaction distance S. Finally, the radius S in turn can be used to infer the magnetic field of the central source (i.e., the asymmetric part of the rotator) by balancing magnetic pressure there against the dynamic pressure,

$$\Sigma U(S)^2 = hB^2/2\mu_0, \tag{17}$$

which is equation 6 of Elsner and Lamb (1977) if h is scaled as R. These estimates constitute a quite general dimensional analysis of the accretion mechanics, although they have been previously presented in the context of highly specific accretion models. Thus the deductions are largely model-independent (Vasyliunas 1979), and do not necessarily serve to single out the dominant physical processes. Ghosh and Lamb (1979), for example, give a more sophisticated analysis to include a second source of torque, namely that of the magnetic coupling between the disk and the neutron star. At large distances (beyond the corotation distance), the disk orbits less rapidly than the star and such coupling (assuming the field lines to penetrate the disk as opposed to being excluded from it) acts to eject the disk and slow down the neutron star. Consequently one can change the sign of the total (accretion plus magnetic) torque if the magnetic torque is large and the accretion takes place near the corotation distance. Such cancellation nicely explains how some objects such as Her X-1 might have high luminosities but comparatively small spin-up rates. Alternatively, if the magnetic torque is negligible, the accretion must take place near the star for the spin-up rate to be small (there, the specific angular momentum of the disk particles is small). Wang (1987) points out that the Ghosh and Lamb solutions assume $\mathbf{J} = \sigma_{\text{eff}}(\mathbf{E} + \mathbf{V} \times \mathbf{B})$ in one place and effectively use $\mathbf{E} + \mathbf{V} \times \mathbf{B} = 0$ elsewhere with the result that the conductivities are essentially arbitrary and the resultant current system is not necessarily self-consistent (see also Kaburaki 1986, 1987).

In general, then, there are two possible values of S. The solution for the larger value of S is termed the *fast* solution and corresponds to near cancellation by the two torques (see also Kundt 1983a). The other solution is termed the *slow* solution and corresponds to the neglect of magnetic torques. Actually, Ghosh and Lamb (1979) do include the magnetic torque, but the net magnetic torque at small distances acts to spin up the star as well and one is simply extracting the angular momentum of the particles to be accreted at a slightly earlier stage. Moreover, this torque is an exceedingly sensitive function of distance to the inner edge of the disk (roughly as the inverse 5th power, depending on the details of the scaling), and since the extracting angular momentum causes the particles to move inward, they are quickly swept up. Once the magnetic coupling to the disk is important, that portion of the disk has already been accreted for all practical purposes. Indeed, one should in principle be careful not to count contributions to the spin-up torque twice, although factors of 2 are not at present too critical. In sum, one is basically quibbling over the exact definition of S, which is only roughly determined anyway. Because we neglect the magnetic torques entirely, the above solutions then correspond to the slow solutions of Ghosh and Lamb (1979). If the fast solutions are appropriate, one needs an additional mechanism to produce disk viscosity in order to keep the disk linked to the magnetic field, as discussed by these authors. The turbulence produced in the jumps might serve such a role. However, we are only considering the direct mechanical effects of these jumps, and these are directly relevant to the slow solutions.

a. *The Magnetic Field*

The slow rotator solutions of Ghosh and Lamb (1979) give a distribution of magnetic surface fields very similar to that given by Michel (1983b) for pulsars, namely a geometric mean field of about 4×10^{10} gauss. It has become traditional to assume that fields of order 10^{12} gauss or higher are typical. Whether this value is in fact typical is of considerable interest. The basis for adopting such large values has been strengthened by the observation of a feature in the X-ray spectrum of several sources corresponding to the energy expected for cyclotron emission in fields of several times 10^{12} gauss. Even given such an identification (ideally, spectroscopic identifications are not based on a single line), comparison between magnetic field and magnetic moment involves the third power of the neutron star radius (which could well be significantly different from the usually adopted value of 10 km). Also, selectional effects may pick out atypically strong field candidates; only half of all radio pulsars have fields in excess of 10^{12} (Manchester and Taylor 1981), yet the correspondingly lower energy lines have not yet been seen in X-ray spectra, possibly because the continuum radiation typically rises rapidly toward lower energy and may swamp any lines. The cyclotron line interpretation has been strengthened by the observation of what appear to be a line at about 11.5 keV and a harmonic at 23 keV in the source 4U 0115+63 (White et al. 1983; see also data published by Rose et al. 1979). However, the $2\hbar\omega_c$ transition is usually forbidden in harmonic-oscillator types of spectra although relativistic effects may boost the latter to about 10% of the fundamental (see Bekefi 1966). Nevertheless, the resultant magnetic field of about 10^{12} gauss is in good agreement with Ghosh and Lamb (1979), who give 1.4×10^{12} (fast solution). Her X-1, on the other hand, would have a magnetic field of 5×10^{12} gauss (Trümper et al. 1978), which is a full order of magnitude larger than even the fast solution, so the situation remains in a state of some flux. These large magnetic fields, if real, are a puzzle. Binary pulsars, in contrast, seem to have weak fields, not strong ones.

b. *Outbursts in Cataclysmic Variables*

A promising development in viscous disk theory is the possibility that the integrated viscosity is not a monotonic function of Σ and consequently outbursts can be attributed to a hysteresis effect wherein on increasing Σ the low viscosity suddenly jumps to high viscosity (i.e., high accretion rates; the outburst), which then depletes the disk material and drops Σ to the point where the viscosity again drops to a small value (Meyer and Meyer-Hofmeister 1981; Smak 1982; Abramowicz et al. 1988). Bath and Pringle (1982) calculate quantitative light curves and find that, while the simple model outbursts do not necessarily give faithful simulations of actual outbursts, one indeed obtains an outburst-like behavior. It is not yet known if the hydraulic jump model displays an analogous instability, partially because the two-dimensional shock structure is more difficult to calculate than one-dimensional viscous disk accretion.

7. Discussion

The hydraulic jump model would produce disks that are truly turbulent (in the sense of having large vorticity fluctuations, as opposed to the casual use of "turbulent" as being synonymous with nonuniform or inhomogeneous). Here, however, the turbulence is a consequence of the dissipative process, not the cause of it. If accretion does take place by such a mechanism, one can understand why disks around pulsating X-ray binaries appear to be so viscous whereas those (postulated) around pulsars would not be. An isolated disk can be driven beyond the corotation distance by electromagnetic torques but cannot be ejected from the system. In an accreting system, however, matter is relentlessly added to the disk and once the inner edge of the disk intrudes inside the corotation distance, the neutron star can accrete the ionized component of the disk matter. Excitation of the disk at the outer edge may be the agent that induces this intrusion in the first place. A related subject is the spiral structure seen in galactic disks. Kormendy and Norman (1979) point out that this spiral structure may well result from perturbations by either internal bars or external companions, essentially the very same two components expected to coexist in the case of binary X-ray sources and cataclysmic variables, where the magnetosphere forms something akin to the bar in a barred galaxy and the companion is required anyhow as a source of plasma to accrete. If so, one has a rather natural theoretical interconnection between the two types of disk. In the galactic case, the analogous theoretical discussion involves the so-called two-arm spiral shock (TASS) model recently reviewed by Rohlfs (1977). If disks do in fact require extrinsic perturbations to evolve rapidly, a problem may arise in models featuring the feeding of matter into black holes, because black holes cannot (theoretically) exhibit a central axially asymmetric perturbation.

11

Gamma-Ray Burst Sources

11.1 Introduction

Gamma-ray bursts were discovered by Klebesadel et al. (1973) to be abrupt (order s) pulses of gamma-rays (considerable emission in excess of 1 MeV). Since then several hundred of these events have been observed (e.g., Mazets et al. 1981; Klebesadel et al. 1982; Baity et al. 1984; Attcia et al. 1987). Although generally of the order of 1 s, gamma-ray bursts can have durations of milliseconds to minutes. Furthermore, they apparently consist of several spikes of emission within the event, and the event can be very impulsive or slowly varying. The bulk of the energy, however, is in gamma-rays, although even here there are exceptions (Mazets et al. 1982). A famous exceptional burst (Cline et al. 1980) was GRB 790305b, which was the brightest ever seen, had an extraordinary rise time (ms), had a complicated pulse shape (a sharp spike followed by broad emission followed by exponentially decaying 8 s oscillations), and was located superimposed (or at!) the position of the N49 supernova remnant in the Large Magellanic Cloud (55 kpc distant!). Here the notation 790305b is for the second [b] gamma-ray burst [GRB] observed on 5 [05] March [03] 1979 [79], although the designation GBS 0526-66 (paralleling that for pulsars) is also used. It also had an extremely soft spectrum, almost an X-ray burst.

A second unusual burst (if any of them can be considered to be usual), GRB 841215 (Laros et al. 1985), had a more typical spectrum but was very bright and of short duration (0.3 s), which allowed its structure to be examined with relatively high statistical accuracy. A least seven components could be resolved (such spikes have been seen before, but with questionable statistical significance). The total energy received (*fluence*) was not atypical of this event, about 4×10^{-4} ergs/cm^2. If the burst were at a distance of 100 pc (somewhat close perhaps), the energy output would have been 4×10^{38} ergs. The luminosity was rather high, however, 2.5×10^{-3} ergs/cm^2 s, corresponding in the same way to a source luminosity of 2.5×10^{39} ergs/s, which is large compared to the Eddington limit of about 10^{38} ergs/s for a 1 $M_\odot$ object.

Imamura and Epstein (1987) have emphasized the paucity of X-rays in gamma-ray bursters. However, Harding and Preece (1987) point out that a synchrotron spectrum cuts off below the cyclotron frequency, which is of order 25 keV for near pulsar surfaces. Actual spectra are much softer, of course, so it can't just be synchrotron radiation in a fixed magnetic field strength.

Evidence for cyclotron absorption features (Murakami et al. 1988) at 40 and 20 keV in GRB 880205 has further strengthened the case for strong magnetic fields like those seen in the pulsating X-ray binaries. Evidence for a 2.2 s periodicity in GRB 1988 August 5 (Kouveliotou et al. 1988) is suggestive of a rotating neutron star. The clear 8 s periodicity seen in the March 5 1979 event is highly suggestive of old neutron stars. Fenimore et al. (1988) report a clearly resolved fundamental and second harmonic corresponding to magnetic fields of 2×10^{12} gauss in GRB 880205 and 870303 (see also Murakami et al. 1988).

Historical optical emissions (recorded on photographic plates) near modern-day gamma-ray burst cites were first reported by Schaefer (1981) and then by others (see review in Hartmann et al. 1988, who also examine the reprocessing of the gamma-rays to produce such optical emission, presumably in the form of flashes). However, no intentional ground-based observation has seen such flashes in association with a gamma-ray burst, so the situation remains fluid.

Theories for gamma-ray bursts have recently been reviewed by Liang (1987), and seem to be centering on neutron stars (Usov 1984) for rather simple reasons. First, the bursts are essentially isotropic in the sky, which means they must be either very close (like the nearby stars which fill the sky) or very far (like the distant galaxies which fill the sky). The very far alternative is unpalatable because the energetics to be explained grow as the square of the distance. Furthermore there seems to be no correlation with the nearby galaxies, which means that one must move the likely origin out to where such correlation would not be evident (probably beyond 10 Mpc). Thus we would have to explain events 10^{10} more energetic than if they were of local origin! On the other hand we would be sampling something like 10^{15} stars or more, so the objects in question could be quite exotic and improbable. If one takes the more conservative view of nearby sources, neutron stars are the only likely nearby objects whose energetics (a proton falling onto the surface would liberate about 100 MeV) even vaguely resemble that of the gamma-ray bursters. And the only confident source of neutron stars is pulsars. But existing pulsars do not seem to be sources (Thompson et al. 1983). However, pulsars are observable for only a few million years, so there must be many times more unseen ones (extinct or too dim to be seen), which is just what one needs to explain why at the present rate of 10 to 100 bursts per year there are not any obvious repeats (the unusually strong 5 March 1979 event is reported to repeat weakly: Golenetskii et al. 1984). Thus old neutron stars have the right statistics if we only knew why they might make gamma-ray bursts.

A handful of what might be distinct objects, the soft gamma-ray repeaters (SGR), are known at the moment: SGR 1806-20, with numerous outbursts (Atteia et al. 1987; Laros et al. 1987; previously GB790107); SGR 0526-66, identified with the 5(b) March 1979 burst and site of about 16 bursts; and SGR 1900 + 44, with a few repetitions (Mazets et al. 1980; previously GB790324). These objects are characterized by rise times of about 1 ms, durations longer than 0.1 s, energies of about 30 keV, and apparently stochastic time distributions of the events.

1. Models: Old Pulsars

A useful summary of models is given by Hurley (1986) and adapted in table 11.1. Most are special event models (typically something catastrophic happening in the neutron star core, or something hitting the neutron star). Otherwise the models appeal to a natural evolution of pulsars into old age when they are no longer seen as radio sources, but now either accrete interstellar medium (which might be stored above the polar caps; Ruderman and Cheng 1988) or have a disk source of matter already in the system (Kafka and Meyer 1984; Colgate et al. 1984; Michel 1985c; Epstein 1985; Melia 1988a,b). A third evolutionary scenario involves putative explosions upon the occurrence of accretion-induced collapse of neutron stars (Michel 1988c). S. E. Woosley has observed that a given theorist is entitled to a maximum of one theory per lifetime for gamma-ray burst sources, to which the present author has little to rejoin beyond noting that supernovae are extraordinary phenomena and there are nevertheless at least two distinct types of these too (also, neither model was originally developed to account for gamma ray bursters). There are also other repeat authors to be found in table 11.1.

As a class, gamma-ray bursters and pulsars have some interesting parallels. They are both represented by hundreds of examples, they are both like fingerprints in that each is distinctive yet identifiable, they have both been known for about two decades, and they have both inspired a large amount of important theoretical effort that has yet to produce a confident picture of what they are.

11.2 Gamma-Ray Bursts from Extinct Radio Pulsars

Neutron stars with accretion disks are a standard model for many X-ray sources (sec. 10.1). Disks may also be required for pulsar action (sec. 6.2). What happens to these disks long after the neutron star becomes unobservable (and perhaps inactive) as a pulsar? Deposition of disk matter onto the surface seems an unavoidable eventuality, and could account for gamma-ray bursts. At the moment, the nature of this phenomenon is quite uncertain and any theory correspondingly speculative. We include the following discussion because it represents a possible link between the physics of radio pulsars and these sources.

Kafka and Meyer (1984) suggested on the basis of the pulsar disk idea a scenario whereby a disk with an assumed mass of 10^{18} g undergoes *sudden magnetic coupling* and is deposited onto the neutron star surface. Michel (1985c) addressed the issue of where such a disk might come from, why it has such a mass, and what the physical nature of such a sudden coupling might be; he also gave some estimates for the occurrence rate of such magnetic coupling. A

TABLE 11.1—Some Gamma-ray Burst Models

Object	Mechanism	Author(s)
Normal star	Flare	Brecher and Morrison 1974, Mullan 1976
Normal star	Antimatter collision	Sofia and Van Horn 1974
Flare star	Flare	Stecker and Frost 1973
White dwarf	Sudden accretion	Chanmugam 1974
White dwarf	Thermonuclear runaway	Hoyle and Clayton 1974
Neutron star	Flare	Liang and Antiochos 1984
Neutron star	Sudden accretion-companion	Lamb, Lamb, and Pines 1973
Neutron star	Sudden accretion-generic disk	Kafka and Meyer 1984, Colgate, Petscheck, and Sarracino 1984, Melia 1988a, b
Neutron star	Sudden accretion-thin disk/old PSR	Michel 1985c, Liang 1989
Neutron star	Sudden accretion-thick disk	Epstein 1985
Neutron star	Sudden accretion-pooled ISM plasma	Ruderman and Cheng 1988
Neutron star	Thermonuclear runaway-companion	Woosley (unpublished 1985)
Neutron star	Thermonuclear ranaway-ISM	Hameury et al. 1982
Neutron star	Pulsar-like glitch	Pacini and Ruderman 1974, Shklovsky and Mitrofanov 1985
Neutron star	Comet collision	Harwit and Salpeter 1973, Shklovsky 1974, Guseinov and Vanysek 1974
Neutron star	Comet tidal disruption	Tremaine and Zytkow 1986
Neutron star	Asteroid from belt	Van Buren 1981
Neutron star	Asteroid formed in disk	Joss and Rappaport 1984b
Neutron star	Crustquake	Fabian, Icke, and Pringle 1976
Neutron star	Corequake	Ellison and Kanzanas 1983, Brecher 1982, Blaes et al. 1989
Neutron star	Exploding superdense matter	Zwicky 1974,
Neutron star	Collapsing and then exploding	Michel 1988a, b, c
Neutron star	Coalescing binary pairs	Goodman 1986, Eichler et al. 1989
Supernova	Intrinsic	Colgate 1974
Stellar collapse	Intrinsic	Bisnovatyi-Kogan et al. 1975, Baan 1982
Black hole	Exploding primordial	Page and Hawking 1976
Black hole	Sudden accretion	Piran and Shaham 1975
White hole	Intrinsic	Narlikar, Apparao, and Dadnich 1974

closely related model was suggested by Tremaine and Zytkow (1986) wherein the source of the matter is a distant *Oort cloud* of comets that experience an occasional close approach. Given that the comets would be tidally disrupted far from the neutron star, the final event would depend on the evolution of the disk of matter left over from tidal disruption, which is discussed below. Livio and Taam (1987) have proposed a comet model for the rapid repetitive transient GB790107.

1. A Model

Let us model an old, cool extinct pulsar as a magnetized neutron star circled by a ring of cold, dense matter with an orbital period roughly equal to the rotational period of the neutron star (presumably periods of the order of a few s, the periods at which pulsars fade from view). The dense ring is assumed to have the low viscosity expected of degenerate matter and is therefore not importantly heated by either the central neutron star or internal dissipation. The low-density surface of the ring, however, is expected to have a much larger viscosity than that in the core and therefore spreads away from the ring to form a thin disk, half of which approaches the neutron star and half of which recedes from it and the dense ring (see figure 11.1). As long as the interaction with the neutron star is weak, the disk diffusing toward the star will remain cool. On the other hand, a strong interaction between the two would heat both, which in turn would cause the inner disk to be ionized, and the resulting plasma, now inside the corotation distance and trapped into corotation by the stellar magnetic field, should essentially free-fall to the surface and thereby give a gamma-ray burst. Such a model requires that (1) the ring exists in the first place, (2) the pulsar is in a permanent null state (or at least is largely devoid of plasma), and (3) precipitation of disk ions on the surface produces a gamma-ray burst. The third condition is nontrivial because the energy transport must be such as not seriously to degrade the gamma-rays, which is difficult in strong magnetic fields and at high photon flux densities.

2. Preburst Conditions

Even if still active as a pulsar ($L_{radio} \leq 10^{25}$ ergs/s), the neutron star surface has probably cooled to well below 10^4 K, and is itself no longer an important source of ionizing radiation. Recent realistic calculations (Glen and Sutherland 1980; Van Riper and Lamb 1981) show cooling curves with similar characteristics; namely the surface temperature declines slowly at first, but after about 10^5 to 10^7 years it begins to plummet and drops steeply through 10^5 K. Albeit still somewhat uncertain, these cooling times are quite short compared with the age of the galaxy. Consequently blackbody radiation from the neutron star need not be an important source of heat to a disk. (The scaling is such that the neutron star as viewed from the disk would appear about as large and bright as the Sun as viewed from Earth.)

In disk theories, the corotation distance plays an important role because the

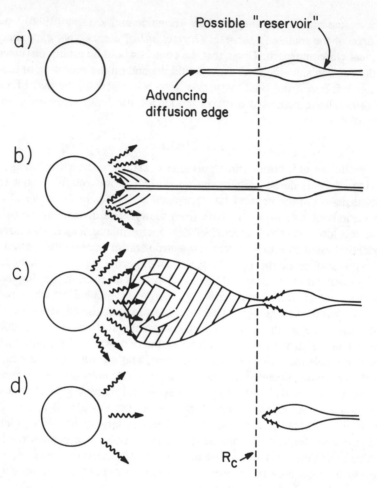

FIG. 11.1. Gamma-ray burst model. Disk (a) approaches star owing to internal viscous drag. Currents start to flow 9b) between the neutron star and the disk, causing the magnetic field to rotate with the star. Runaway accretion (c), enhanced by the electromagnetic acceleration, deposits disk material with the corotation distance onto surface. In (d), disk and star return to nearly original state. From F. C. Michel, 1985c, *Ap. J.*, 290, 721 (figure 1).

plasma is readily accreted from inside this distance; charged particles corotating with the stellar magnetic field are insufficiently supported centrifugally against gravity and therefore fall along field lines to the surface (e.g., Michel and Sturrock 1974). Thus the inner edge of the disk was probably near a corotation distance corresponding to a period of a few s. Once the neutron star cools, any orbiting material becomes predominantly neutral and there is no longer any important electromagnetic coupling to the disk (essentially the situation with Saturn's disk). Consequently, there should be a profound change in the electric and magnetic field structure around the neutron star. Without an abundance of plasma circulating about the pulsar, the plasma should collapse to a small *electrosphere* shielding an inactive neutron star (sec. 4.3.2). Elsewhere one should have an extraordinarily hard vacuum. Accordingly, the magnetic field lines of the neutron star are no longer equipotentials, since there is no plasma to provide the necessary conductivity, and the electric field of the neutron star now declines as a vacuum (rotationally induced) quadrupole. Because the corotation distance is now of the order of 10^2 neutron star radii, the electric field at the orbiting ring is now 10^8 times smaller, or only a few V/m. Similarly, a dipolar magnetic field is 10^6, smaller hence 10^6 gauss for a nominal 10^{12} gauss surface field. Because the field lines are now open-circuited, the disk (which must reasonably have some small level of ionization) orbits with a rate independent of the rotation rate of the neutron star.

The part of the disk of interest here is that diffusing inward from the ring: the inner disk. As time passes, this thin disk slowly approaches the neutron star. These should be cool (≈ 200 K) disks of ordinary neutral matter, probably in the form of solid particles, having surface densities $\Sigma \approx 10^2$ g/cm^2 and kinematic viscosities of $\nu \approx 10^3$ cm^2/s. The disks of Saturn are approximately of such a character. Like Saturn's disk, there should be only a trace of a gas phase at most, with this gas puffed up about the thin particulate disk. As can be seen from stellar evolution calculations (e.g., Woosley and Weaver 1982) the mass elements outside the neutronized core of the presupernova massive star are mainly silicon, sulfur, and oxygen. Consequently the most plausible composition of the fossil disk is just such matter, with perhaps a bit of the core itself mixed in (which would add iron), because the more distant mass shells are ejected in the supernova event. None of these elements are particularly volatile except oxygen, which is probably tied up in silicate grains. A few percent by mass of argon seems the most likely gas phase. The effect of this possible gas phase is discussed after equation 5 below. A simple criterion for the thin disk settling onto a given viscosity and surface density is that the collision rate of the disk particles should be roughly Ω, which corresponds to a disk about one optical depth thick. The reasoning here is that the random motion of the particles normal to the disk (which gives the disk finite thickness, h) corresponds to Kepler orbits inclined at a tiny angle to the mean disk plane, whereas the random motion in the disk plane corresponds to slightly elliptical orbits compared to the mean circulation. Particles in a disk of unit optical depth will then make an average of one collision per orbit, because particles in Kepler orbit must pass from one side of the disk to the other twice per orbit. The collisions rearrange the orbital parameters so as to interchange motions normal to the disk of order h into radial motions of order h, and vice versa. Radial transport via random walk (diffusion) is most efficient if the particle moves from periapsis to apoapsis between collisions (i.e., a collision rate $\approx \Omega$). At very high densities, one would have a more or less normal fluid again characterized by fairly low kinematic viscosity as well (e.g., Landau and Lifshitz 1958; it is the dynamic viscosity that tends to be independent of density). At very much lower

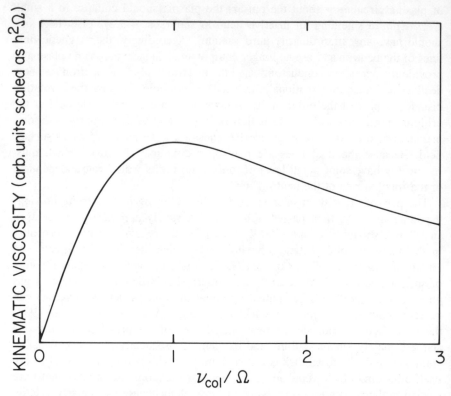

FIG. 11.2. Viscosity as a function of surface density. At high densities the disk acts as a classical fluid, with kinematic viscosity declining with increasing density (the dynamic viscosity $\eta = \rho\nu$ is more nearly constant). At low densities the viscosity in a Keplerian disk declines because increasing the mean free path no longer increases the radial transport given by collisions, which themselves become less frequent. A maximum therefore exists, roughly where the collision rate equals the orbital rate.

collision rates the radial excursion is limited to $\approx h$ and even tends to average out over many orbits. Consequently there is a natural maximum in the viscosity versus surface density (figure 11.2, after figure 1b of Bath and Pringle 1982). It is here that the so-called α (Shakura and Sunyaev 1973) of a disk would approach unity. A physical analog is the Pederson (cross-field) conductivity of a magnetized plasma. At low collision rates, the particles generally experience $\mathbf{E} \times \mathbf{B}$ drift and this conductivity is small. At high collision rates, the collisions inhibit particle motion regardless of the magnetic field and again the conductivity is low. As a result, the conductivity is a maximum in between, essentially when the particles collide at about the cyclotron frequency (Johnson 1965), for the reasons just discussed.

For a particulate disk of thickness h, we have from dimensional considerations a kinematic viscosity of the order

$$\nu \approx h^2\Omega; \tag{1}$$

the random velocity of the particles is then

$$V \approx h\Omega; \tag{2}$$

most of the kinetic energy is dissipated by inelastic collisions and reradiated as thermal radiation, giving (a disk of unit optical depth is essentially a blackbody for our purposes)

$$\sigma T^4 \approx \Sigma h^2 \Omega^3, \tag{3}$$

and a net disk luminosity

$$L \approx \sigma T^4 R^2 \tag{4}$$

during which the particles take a time

$$\tau \approx \frac{R^2}{\nu} \tag{5}$$

to diffuse from radius R to the near vicinity of the neutron star (most of the time is taken to go from R to $R/2$, so the exact proximity is unimportant). If we take $h \approx 30$ cm, $\Omega \approx 1$ radian/s, and $\Sigma \approx 10^2$ g/cm^2, we get $\nu \approx 10^3$ cm^2/s, $V \approx 30$ cm/s, $T \approx 200$, $L \approx 10^{21}$ ergs/s, and $\tau \approx 10^5$ years. The same analysis applied to Saturn's disk system would suggest a stability over the age of the solar system (mainly owing to their much larger size). In principle, a gas phase could increase the overall system viscosity above the value of 10^3 estimated here for the particulate disk itself. The gas itself would only have a small molecular viscosity, but some differential (non-Kepler) motion owing to the finite gas pressure could boost the effective disk viscosity. Thus the above viscosity estimate may be a lower limit.

The intense gamma-rays from a burst will heat and perturb the reservoir (i.e., the main disk) at the corotation distance. However, the reservoir is beyond the corotation distance and therefore any magnetic coupling acts to repel the ionization formed, not accrete it, since the centrifugal force there exceeds gravity by definition.

a. *The Magnetic Field*

It should be clear that in such a model, indeed in such a class of models, the magnetic field should play an important role in controlling and confining any radiating plasma. Otherwise it would be very difficult to see how enough matter could accumulate in the vicinity of the neutron star without becoming opaque (some of these difficulties are not entirely alleviated *with* a magnetic field, owing in part to magnetic opacity to gamma-rays). If the magnetic field were actually to decay within a few million years, the old pulsars would be uninterestingly magnetized (counter to some observational evidence presented in the previous section).

3. The Burst

As the disk evolves toward the neutron star, it will eventually reach the electrosphere. The relative velocities between particles in the corotating electrosphere and the Kepler disk will be of the order of 0.1 c, and direct collisions will ionize the disk particles and also produce energetic photons which can cause photoionization. Currents can now flow because there is finally a source of particles to supply conduction along the magnetic field lines. Consequently the magnetic field starts to rotate with the neutron star and the disk suddenly finds itself bathed in corotating plasma. Each energetic ion ($\approx$10 MeV) can produce many thousands of ion-electron pairs, each of which then becomes another energetic ion when it is picked up and accelerated into corotation by the electromagnetic fields. Ionization should quickly exponentiate and totally ionize the disk. At the same time, scattering among the ions rapidly increases as their concentration increases and they can now enter the loss cone to be precipitated to the surface. To add to the already runaway ionization of the disk, the particles reaching the neutron star deposit their energy at the surface, heating it and creating energetic ionizing photons. The disk would be flash-ionized to fill the magnetic flux tubes within the corotation distance with energetic plasma.

It is very unlikely that, in contrast, a stable state can be attained whereby the electrosphere quiescently accretes and deposits the disk particles as they diffuse inward. When the thin disk is far from the neutron star, very little ionization is expected, whereas once the disk approaches (say) the electrosphere, it will be ionized as fast as it can diffuse forward. The disk will be unstable if at some point enough ionization is present for the field lines to corotate. The critical number is certainly reached when the minimum space charge is present, about 10^{29} particles, which is supplied in the standard pulsar model scalings with space-charge limited flow out of the polar caps in about one rotation of the star. It is less clear how rapidly the particles of the opposite sign are lost, presumably by conduction along the disk and then by corona discharge from the outer edge of the disk (figure 6.2), but the polar cap emission in any case forms an upper limit. Thus the system will fill with ionization if supplied at a rate of about 10^{29} particles/s. The equilibrium ionization rate of the disk is simply given from the standard accretion formula,

$$\dot{M} = 3\pi\nu\Sigma, \tag{6}$$

which is here 10^{30} particles/s. The near coincidence of these two numbers is entirely fortuitous, because they are quite unrelated. Both ν and Σ may well be underestimated, while the discharging rate is probably an upper limit. Thus a thin disk can provide enough ionization for the field lines to enforce corotation, at which point the entire disk should be quickly and fully ionized.

a. *The Rise Time of the Burst*

The rise time would be characterized by the time it takes a relativistic particle to travel from the inner edge of the disk to the neutron star. For the sake

TOTAL POWER OUTPUT (erg)

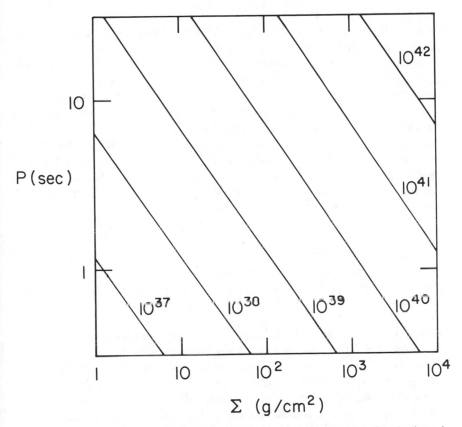

FIG. 11.3. Energy release from sudden disk accretion. Plotted as function of surface density and neutron star rotation period. From F. C. Michel, 1985c, *Ap. J.*, 290, 721 (figure 2).

of argument, consider an inner disk edge at 10 neutron star radii (i.e., the geometrical mean distance to the outer ring) and a particle velocity 0.1 c, in which case the transit time is around 3 ms. The very first particles would move even faster, being accelerated in the very large electric fields of the electrosphere near the neutron star, which would shorten this time to 0.3 ms. The duration of the burst need not be equal to the rise time. One controlling factor is the magnetic field strength of the neutron star, which can force accretion only at a limited rate.

b. Energy Release and Repetition Rate

The mass deposited on the surface is just that of the inner disk, or about

$$\Delta M = \Sigma R_c^2. \tag{7}$$

Using $\Sigma \approx 10^2$ g/cm^2 gives a nominal deposition of 10^{18} g of material and therefore an energy release

$$\Delta E = \Delta M \phi_{NS}, \qquad (8)$$

or about 10^{38} ergs, as shown in figure 11.3 (assuming ϕ_{NS}, the gravitational potential at the neutron star surface, is about $c^2/10 \approx 10^{20}$ cm^2/s^2). The time for a disk to diffuse to the near vicinity of the neutron star (eq. 5) is at most 10^5 years. In the burst, the disk is ionized and consumed, and therefore the system is roughly back where it started. Another diffusion time (eq. 5) is required for a new inner disk from the reservoir to diffuse back into the critical distance. A source as old as the galaxy (10^{10} years) could have produced 10^5 bursts. How many neutron stars could be providing such bursts? The isotropic distribution of observed gamma-ray sources suggests that these neutron stars would have to be the nearby ones closer than the galactic disk scale height. There are about 20 active pulsars within such a volume. If the average active pulsar lifetime is 10^6 years and we see about 1 in 5 owing to beaming effects, then there should be $N \approx 10^6$ inactive neutron stars within this volume. Alternatively, we could use the estimate from nucleosynthesis arguments that there are 10^8 neutron stars in the galaxy, which if uniformly distributed through a disk of 10 kpc would give 10^5 neutron stars within 300 pc, essentially consistent with the pulsar estimates above if one discards the problematical correction for beaming. The resultant burst rate would then be only 1 per year. However, this number comes directly from our estimate for the viscosity which, as noted above, could reasonably be considered a lower limit. It would be nice if the repetition rate were indeed faster than 1 per 10^5 years, insofar as any hope of seeing repeats is concerned. There seems little constraint insofar as the available mass is concerned: a 10^{-5} $M_\odot$ reservoir could supply 1 burst per year over a Hubble time, far more than required by observation. One could increase the distance scale by a factor of 10 without too much difficulty. Owing to their tendency to have high spatial velocities, the pulsars when extinct should have had time to form a thicker galactic disk than that for the field stars. The volume density could then be much less and still provide the same isotropic event rate.

c. Accretion Limited by Torque Transfer

If particle precipitation is caused by the magnetic field forcing the charged particles into non-Keplerian velocities, one must consider whether the magnetic field is strong enough to overpower the particle inertia. The magnetic energy beyond 10 neutron star radii is about 10^{38} ergs for a 10^{12} gauss surface field, which is sufficient to control a typical burst if we adopt typical neutron star fields. If the magnetic moment is much smaller, the accretion rate is reduced and the burst duration is proportionately lengthened. Confirmation of rapid neutron star magnetic field decay would therefore appear to rule out such a class of models as well as several other models.

The radiation in the form of gamma-rays per se is gratuitously assumed here. The likely processes have been discussed in Hartmann et al. (1988), in particular the likely optical reprocessing.

12

Other Phenomena Driven by Neutron Stars

12.1 Introduction

Given the large gravitational energy release possible from matter accreted onto neutron stars, together with their large rotational and magnetic field energies, neutron stars naturally draw the attention of workers looking for an explanation for peculiar objects, particularly ones that emit very energetic particles or radiation. As one can see from the radio pulsars, even having a large data set has not necessarily served to narrow the possible models. For objects which form only a small set, the difficulties in constraining the models are even greater. We know by and large what is typical about pulsars and what is unique to only a few. For these objects such a distinction is much more difficult. Finally, almost any model that can deliver the required energy can also apparently be contorted to fit the phenomenology. In the case of the gamma-ray burst sources, for example, models included neutron stars in a binary system exploding off surface layers of accreted material, solitary neutron stars experiencing starquakes, asteroidal or comet impact, magnetic flares, and pooled accretion of interstellar matter. Clearly most of these models must be square pegs forcibly inserted into round holes, and to some extent these are "untheories" in the sense that, at the point where the model would clearly fail, some special pleading is introduced. Of course there is always the legitimate concern that something unusual is happening, given that these are indeed unusual objects, and these theories are in effect guesses at where some magical process must enter. Discovering the weak points is usually left to the reader.

At this point there is not too much quantitative theory for most of these objects, with the exceptions already discussed of the pulsating X-ray sources (ch. 10) and a severe subset of the gamma-ray burst models (ch. 11).

12.2 Objects Possibly Containing Neutron Stars

A number of astrophysical objects might harbor neutron stars. The pulsating X-ray sources are one large class (ch. 10) which convincingly do, while the gamma-ray burst sources (ch. 11) are strong possibilities. One is uncertain how to classify the soft gamma-ray repeater SGR 1806-20, which is unlike the typical gamma-ray burst source in that it repeats frequently, but not periodically, and emits at the lower energy ranges (Atteia et al. 1987; Laros et al. 1986, 1987). This object is also argued to be a neutron star (Liang 1989). Here we simply touch on a number of other odd objects that are under suspicion. A recent compilation of such objects made by Seward (1985) included CTB 80, which was later discovered to contain a pulsar (Kulkarni et al. 1988a; Hester and Kulkarni 1988, 1989; Fesen et al. 1988).

1. COS-B Sources

A source list of these possible gamma-ray point sources is given by Swanenburg et al. 1981; see also Buccheri et al. 1983). The source identification numbers below indicate galactic longitude and latitude rather than right ascension and declination.

a. *Geminga (2CG* 195 + 04*)*

It has been suggested that an X-ray source (1E 0630 + 178) detected inside the Geminga error box (Bignami et al. 1983) is a pulsar (Strong 1983; Chadwick et al. 1985b report a 12.6 ms pulsar based on gamma-ray detections) or a neutron star in a close binary (Nulsen and Fabian 1984). A blue candidate identified by Halpern and Tytler (1988) has been interpreted as an off-beam object similar to the Vela pulsar. Bignami et al. (1984) think that there might be a 59 ms periodicity, although this has been questioned by Buccheri et al. (1985).

b. *2CG* 013 + 00

The suggestion that the steep-spectrum source 2CG 013 + 00 is the result of a beam of particles coming from the low-mass X-ray binary GX 13 + 1 (Ozel and Ormes 1989) is similar to suggestions for Cyg X-3.

2. Galactic Center Positron Source

The proposal by Mastichiadis et al. (1987) that a young ($\leq$200 years) pulsar produces enough positrons to explain the observed 511 keV annihilation line originally observed by the Haymes group (Johnson, Harnden, and Haymes 1972; Johnson and Haymes 1973) and confirmed by Leventhal et al. (1980). Lingenfelter and Ramaty (1989) reviewed the data and also discussed quite provocatively why the original three observations of the line were thought to be at 480 and at 530 keV, not at 511. The plausible arguments are that a high-energy tail shifts the centroid to higher energies when the object is active and that Comptonization shifts the centroid to lower energies when the object is not injecting fresh positrons. Benford (1988) and Kluźniak et al. (1988b) propose a neutron star plus disk model for this object. McClintock and Leventhal (1989) argue on the basis of similar activity patterns and location that this object is GX 1 + 4, a 2 minute binary X-ray pulsar associated with an M6 giant V2116 Oph. (In the heteroclite notation of astronomy, this source near the galactic center is located by its rough position in galactic longitude and latitude; a pulsar at the galactic center would be designated PSR 1742-28.)

3. TeV Muon/Gamma-Ray Sources

Ultra-high-energy particles entering the atmosphere produce showers and resultant Cherenkov radiation.

a. *Her X-1*

Dowthwaithe et al. (1984) found evidence for such showers associated with the periodicity of Her X-1. The evidence is not entirely compelling, given that the showers are spread over very long times relative to the pulsation period inferred by folding the data, the total number of counts is often rather small, and identification is provided by finding a statistically enhanced candidate period at the known period (Chardin 1987).

b. *Cyg X-3*

Cyg X-3 has similarly been observed at energies above 10^{15} eV (Samorski and Stamm 1983; Lloyd-Evans et al. 1983), possibly the extension of a Crab-like power-law spectrum (Samorski and Stamm 1983, figure 5). This periodic 4.8 hour source (presumably the orbital period) had previously been observed in the 30-100 McV range (Lamb et al. 1977) and in the TeV range (Danaher et al. 1981; Lamb et al. 1982). The source of such gamma-rays is argued to be a neutron star producing an energetic proton beam that is bombarding the atmosphere (if very extended) of its companion (Vestrand 1983 and Vestrand and Eichler 1984; see also Kundt 1983b, Eichler and Vestrand 1984, Hillas 1984b, Chanmugam and Brecher 1985, Fichtel and Linsey 1986, Kanzanas and Ellison 1986, Király and Mészáros, 1988). Molnar (1988) draws a parallel with PSR 1957 + 20 in the possible evolutionary history of this system, both of which may involve light companions in orbits of order 1 $R_\odot$, a view elaborated by Tavani et al. (1989), who propose a self-sustained mass transfer wherein radiation from a neutron star drives mass loss from its companion, which is itself the fuel driving the neutron star radiation. There is something of a problem with gamma-rays being the incident particles at Earth given that muons have been seen in coincidence (Marshak et al. 1985), consistent with an incident cosmic ray but not an incident gamma-ray (which is expected to produce muon-poor showers both observationally and theoretically). The usual suggestions have been put forward: an exotic new particle (Baym et al. 1985), energetic neutrons, anomalous behavior of gamma-rays at high energy, etc. (summarized by Schwarzschild 1988). As for the case of SS 433, there is possible evidence for relativistic jets from this object (Geldzahler et al. 1983). In addition, this object is characterized by radio outbursts (see the entire 23 October 1972 issue of *Nature* and recently Molnar et al. 1984) and is seen at most wavelengths (e.g., Becklin et al. 1974).

c. *Vela X-1*

Very energetic gamma-rays have been reported in Vela X-1 with energies in excess of 3×10^{15} eV (Protheroe et al. 1984). Again, energetic beams are invoked (Protheroe 1984). Results of searches of other sources are given in

Protheroe and Clay (1985), which reports a detection of LMC X-4 (source 0532-664). A study of the period variations (Deeter et al. 1989) suggests a random walk in neutron star spin frequency about the nominal period of about 283 s.

4. SS 433

SS 433, which displays relativistic jets, has been reviewed in Margon (1984). Shukre et al. (1983) have suggested a neutron star/disk model for this object.

5. Black Hole Candidates

All of these systems are put forward as candidates on the basis of the assumption that the systems are binary and that the mass of the companion can be estimated with reasonable accuracy. A list of candidates is given by White et al. 1984.

a. *Cyg X-1*

The prime black hole candidate is Cyg X-1 and is reviewed in Liang and Nolan (1984; see also Blumenthal and Tucker 1974). The weak points in such determinations involve having to assume (1) that the star nearest the X-ray source is indeed a companion star (in this case, HDE 226868 with a 5.6 day period), (2) that the spectral properties of the companion are not greatly changed by being in such an energetic system (which leads to the mass estimate), and (3) that there are only two stars in the system. The first weak point is illustrated by the recent difficulties in pinpointing the progenitor of SN 1987A (about three stars are packed closely near the likely point of origin) and in obtaining the light curve of the companion of PSR 1957 + 20, which has another star almost exactly along its line of sight.

b. *LMC X-3*

With a mass function of 2.3, LMC X-3 also seems a good black hole candidate (van den Heuvel and Habets 1984).

c. *GX 339-4*

The source GX 339-4 (Dolan et al. 1987) has similar behavior to Cyg X-1 and A0620-00, also a leading black hole candidate (Johnston et al. 1989) However, Cowley et al. (1987) determine a mass <2.5 $M_\odot$ for the compact star in GX 339-4. Imamura et al. (1987) report a possible 1.13 ms periodicity in the optical, suggesting of course a neutron star in the system.

6. IC 443

A particularly interesting supernova remnant is IC 443, which apparently involves a supernova in the close vicinity of (perhaps in) a dense molecular cloud. If such an event can create cosmic rays, one would expect enhanced ionization in the cloud. Searches to date, however, have failed to confirm such effects (Ziurys, private communication). Petre et al. (1988) give a comprehensive study of the structure and spectrum of this object.

7. Active Galactic Nuclei/Quasars

If anything, quasars are more poorly understood than pulsars. However, they are currently considered to be galaxies with extremely active nuclei. Two suggestions based on parallels with pulsars have been made: (1) that a quasar is, in effect, one huge pulsar (Morrison 1969; Cavaliere et al. 1969; Sturrock 1971c; Stecker 1971; Fowler 1971; Ozernoi and Usov 1973a,b; but see Bisnovatyi-Kogan and Blinnikov 1973), and (2) that a cloud of pulsars excites the quasar activity (Rees 1971a; Arons et al. 1975; Kulsrud 1975; Kulsrud and Arons 1975), which has been proposed as an alternative to the multiple supernovae model (see Petschek et al. 1976). In some models the electromagnetic activity is postulated to be dominated by an accretion disk (Lovelace 1976; Blandford 1976), in which case the nature of the central object can remain open. Present day conventional wisdom insists that the central object is a massive balck hole.

Supernovae in a dense cluster, even if frequent, are expected to be inefficient power sources because most of their energy is expected to be produced in the form of neutrinos and observationally the expansion velocity of the remnants is fast ($\approx 10^4$ km/s) but very nonrelativistic, so the energy liberation per unit mass is small compared to the high efficiency ($\approx 10\%$ of the total rest mass) often inferred. Michel (1988d) has suggested instead that the collapse of the neutron stars themselves could produce energetic neutrino-poor explosions to power the activity.

References

Ables, J. G., Jacka, C. E., McConnell, D., Hamilton, P. A., and McCulloch, P. M. 1988, *IAU Circ.*, No. 4602.

Ables, J. G., McConnell, D., Jacka, C. E., McCulloch, P. M., Hall, P. J., and Hamilton, P. A. 1989, *Nature*, 342, 158.

Abramowicz, M. A., Czerny, B., Losata, J. P., and Szuszkiewicz, E. 1988, *Ap. J.*, 332, 646.

Abramowitz, M., and Stegun, I. A. 1968, *Handbook of Mathematical Functions* (Washington: Government Printing Office), p. 332.

Acuna, M. M., Ness, N. F., and Connerney, J. E. P. 1980, *J. Geophys. Res.*, 85, 5675.

Adler, S. 1971, *Ann. Phys.*, 67, 599.

Adler, S. L, Bahcall, J. N., Callan, C. G., and Rosenbluth, M. N. 1970, *Phys. Rev. Letters*, 25, 1061.

Ahluwalia, D. V., and Wu, T.-Y. 1978, *Letters Nuovo Cimento*, 23, 406.

Akasofu, S.-I. 1978, *Space Sci. Rev.*, 21, 489.

Al'ber, Ya. I., Krotova, Z. N., and Eidman, V. Ya. 1975, *Astrophysics* (USA), 11, 189.

Alcock, C., Farhi, E., and Olinto, A. 1986, *Ap. J.*, 310, 261.

Aldridge, F. T. 1968, *Proc. Nat. Acad. Sci. USA*, 60, 743.

Alfvén, H. 1983, *Ap. Space Sci.*, 97, 79.

Alfvén, H., Ip, W.-I., and Burkenroad, M. D. 1974, *Nature*, 250, 634.

Allen, C. W. 1963, *Astrophysical Quantities* (London: Athlone Press).

Alpar, M. A., Anderson, P. W., Pines, D., and Shaham, J. 1981, *Ap. J. (Letters)*, 249, L29.

Alpar, M. A., Cheng, A. F., Ruderman, M. A., and Shaham, J. 1982, *Nature*, 300, 728.

Alpar, M. A., and Ho, C. 1983, *M.N.R.A.S.*, 204, 655.

Alpar, M. A., and Shaham, J. 1985, *Nature*, 316, 239.

Alsop, D., and Arons, J. 1988, *Phys. Fluids*, 31, 839.

Aly, J. J. 1980, *Astr. Ap.*, 86, 192.

Ambartsumyan, V. A., and Saakyan, G. S. 1960, *Sov. Astr. AJ*, 4, 187.

Anderson, B., and Lyne, A. G. 1983, *Nature*, 303, 597.

455

Anderson, J. L., and Cohen, J. M. 1970, *Ap. Space Sci.*, 9, 146.

Anderson, S., Gorham, P., Kulkarni, S., and Prince, T. 1989a, *IAU Circ.*, No. 4762.

Anderson, S., Gorham, P., Kulkarni, S., and Prince, T. 1989b, *IAU Circ.*, No. 4772.

Anderson, S., Kulkarni, S., and Prince, T. 1989c, *IAU Circ.*, No. 4819.

Angelie, C., Deutsch, C., and Signore, M. 1980, *J. Phys. Colloq.* (France), 41, 133.

Anthony, D. M., and Carlberg, R. G. 1988, *Ap. J.*, 332, 637.

Apparao, K. M. V. 1969, *Nature*, 223, 385.

Apparao, K. M. V. 1974, *Ap. Space Sci.*, 31, L9.

Apparao. K. M. V., and Chitre, S. M. 1970, *Proc. Indian Acad. Sci.*, A 72, 285.

Apparao, K. M. V., and Hoffman, J. 1970, *Ap. Letters*, 5, 25.

Ardavan, H. 1976a, *Ap. J.*, 203, 226.

Ardavan, H. 1976b, *Ap. J.*, 204, 889.

Ardavan, H. 1976c, *M.N.R.A.S.*, 175, 645.

Ardavan, H. 1976d, *Ap. J.*, 206, 822.

Ardavan, H. 1976e, *M.N.R.A.S.*, 177, 661.

Ardavan, H. 1982, *Ap. J.*, 251, 674.

Arnett, W. D. 1969, *Nature*, 222, 359.

Arnett, W. D. 1987, *Ap. J.*, 319, 136.

Arnett, W. D., and Bowers, R. L. 1977, *Ap. J. Suppl.*, 33, 415.

Arnett, W. D., Branch, D., and Wheeler, J. C. 1985, *Nature*, 314, 337.

Arnett, W. D., and Schramm, D. N. 1973, *Ap. J. (Letters)*, 184, L47 (Erratum 187, L47).

Arons, J. 1972, *Ap. J.*, 177, 395.

Arons, J. 1979, *Space Sci. Rev.*, 24, 437.

Arons, J. 1981a, in *Pulsars*, ed. W. Sieber and R. Wielebinski (IAU Conf. No. 95) (Dordrecht: Reidel), p. 69.

Arons, J. 1981b, *Ap. J.*, 248, 1099.

Arons, J. 1981c, in *Origins of Cosmic Rays*, ed. G. Setti, G. Spada, and A. W. Wolfendale (IAU Conf. No. 94) (Dordrecht: Reidel), p. 175.

Arons, J. 1983a, *Ap. J.*, 266, 215.

Arons, J. 1983b, *Nature*, 302, 301.

Arons, J. 1983c, in *Positron-Electron Pairs in Astrophysics*, ed. M. Burns, A. K. Harding, and R. Ramaty (New York: AIP), p. 163.

Arons, J. 1984, *Adv. Space Res.*, 3, 287.

Arons, J., and Barnard, J. J. 1983, *J. Ap. Astr.*, 4, 191.

Arons, J., and Barnard, J. J. 1986, *Ap. J.*, 302, 120.

Arons, J., Kulsrud, R. M., and Ostriker, J. P. 1975, *Ap. J.*, 198, 687.

Arons, J., and Lea, S. M. 1976a, *Ap. J.*, 207, 914.

Arons, J., and Lea, S. M. 1976b, *Ap. J.*, 210, 792.

Arons, J., and Scharlemann, E. T. 1979, *Ap. J.*, 231, 854.

Arons, J., and Smith, D. F. 1979, *Ap. J.*, 229, 728.

Aschenbach, B., and Brinkmann, W. 1975, *Astr. Ap.*, 41, 147.

Ashour-Abdalla, M., Leboeuf, J. N., Tajima, T., Dawson, J. M., and Kennel, C. F. 1981, *Phys. Rev.*, A23, 1906.

Ashworth, M., Lyne, A. G., and Smith, F. G. 1983, *Nature*, 301, 313.

Asséo, E., Beaufils, D., and Pellat, R. 1984, *M.N.R.A.S.*, 209, 285.

Asséo, E., Kennel, C. F., and Pellat, R. 1978, *Astr. Ap.*, 65, 401.

Asséo, E., Kennel, C. F., and Pellat, R. R. 1975, *Astr. Ap.*, 44, 31.

Asséo, E., Llobet, X., and Schmidt, G. 1980a, *Phys. Rev.*, A22, 1293.

Asséo, E., Pellat, R., and Rosado, M. 1980b, *Ap. J.*, 239, 661.

Asséo, E., Pellat, R., and Sol, H. 1983, *Ap. J.*, 266, 201.

Atteia, J.-L., and 15 other authors. 1985, *Proc. 19th Int. Cosmic Ray Conf.*, 1, 33.

Atteia, J.-L., and 15 other authors. 1987, *Ap. J. (Letters)*, 320, L105.

Avakyan, R. M., Arutyunyan, G. G., and Saakyan, G. S. 1972, *Astrophysics (USA)*, 8, 282.

Avetisyan, A. K. 1979, *Astrophysics (USA)*, 15, 80.

Axford, W. I., Johnson, H. E., Mendis, D. A., and Yeh, T. 1970, *Comments Ap. Space Phys.*, 2, 53.

Axford, W. I., Leer, E., and Skadron, G. 1977, *Proc. 15th Int. Cosmic Ray Conf.*, 11, 132.

Ayasli, S., and Ogelman, H. 1980, *Ap. J.*, 237, 227.

Baade, W. 1942, *Ap. J.*, 96, 188.

Baade, W., and Zwicky, F. 1934a, *Proc. Nat. Acad. Sci.*, 20, 259.

Baade, W., and Zwicky, F. 1934b, *Phys. Rev.*, 45, 138.

Baan, W. 1982, *Ap. J. (Letters)*, 261, L71.

Backer, D. C. 1970, *Nature*, 228, 42.

Backer, D. C. 1976, *Ap. J.*, 209, 895.

Backer, D. C., Kulkarni, S. R., Heiles, C., Davis, M. M., and Goss, W. M. 1982, *Nature*, 300, 615.

Backer, D. C., Kulkarni, S. R., and Taylor, J. H. 1983, *Nature*, 301, 314.

Backer, D. C., and Rankin, J. M. 1980, *Ap. J. Suppl.*, 42, 143.

Backer, D. C., Rankin, J. M., and Campbell, D. B. 1976, *Nature*, 263, 202.

Baglin, A., and Heyvaerts, J. 1969, *Nature*, 222, 1258.

Bahcall, J. N., Rees, M. J., and Salpeter, E. E. 1970, *Ap. J.*, 162, 737.

Baity, W. A., Hueter, G. J., and Lingenfelter, R. E. 1984, *AIP Conf. No. 115*, p. 434.

Baker, D. N., Borovsky, J. E., Benford, G., and Eilek, J. A. 1988, *Ap. J.*, 326, 110.

Banerjee, B., Constantinescu, D. H., and Rehak, P. 1974, *Phys. Rev.*, D10, 2384.

Barbieri, R. 1971, *Nucl. Phys.* (Netherlands), A161, 1.

Barker, B. M., and O'Connell, R. F. 1975, *Ap. J. (Letters)*, 199, L25.

Barker, B. M., and O'Connell, R. F. 1976, *Phys. Rev.*, D14, 861.

Barnard, D. J., Lea, S. M., and Arons, J. 1983, *Ap. J.*, 266, 175.

Barnard, J. J., and Arons, J. 1982, *Ap. J.*, 254, 713.

Barnard, J. J., and Arons, J. 1986, *Ap. J.*, 302, 138.

Baron, E., Cooperstein, J., and Kahana, S. 1985, *Phys. Rev. Letters*, 44, 126.

Baroni, L., Callegari, G., Gualdi, C., and Fortini, P. 1980, *Letters Nuovo Cimento*, 27, 509.

Bath, G. T., and Pringle, J. E. 1982, *M.N.R.A.S.*, 199, 267.

Baym, G., and Chin, S. A. 1976, *Phys. Letters*, 62B, 2451.

Baym, G., Kolb, E. W., McLerran, L., Walker, T. P., and Jaffe, R. L. 1985, *Phys. Letters*, 160B, 181.

Baym, G., and Pethick, C. J. 1975, *Ann. Rev. Nuclear Sci.*, 25, 27.

Baym, G., and Pethick, C. 1979, *Ann. Rev. Astr. Ap.*, 17, 415.

Baym, G., Pethick, C. J., and Pines, D. 1969a, *Nature*, 224, 674.

Baym, G., Pethick, C., Pines, D., and Ruderman, M. 1969b, *Nature*, 224, 872.

Beard, D. B. 1960, *J. Geophys. Res.*, 65, 3559.

Becker, J. H., and Helfand, D. J. 1983, *Nature*, 302, 688.

Becklin, E. E., and 9 other authors. 1974, *Ap. J. (Letters)*, 192, L119.

Bekefi, G. 1966, *Radiation Processes in Plasmas* (New York: John Wiley).

Belcher, J. W., and MacGregor, K. B. 1976, *Ap. J.*, 210, 498.

Bell, A. R. 1978, *M.N.R.A.S.*, 182, 147.

Benford, G. 1975, *Ap. J.*, 201, 419.

Benford, G. 1977, *M.N.R.A.S.*, 179, 311.

Benford, G. 1984, *Ap. J.*, 282, 154.

Benford, G. 1988, *Ap. J.*, 333, 735.

Benford, G., and Buschauer, R. 1977, *M.N.R.A.S.*, 179, 189.

Benford, G., and Buschauer, R. 1980, *M.N.R.A.S.*, 190, 945.

Bennett, K., and 13 other authors. 1977, *Astr. Ap.*, 61, 279.

Bertotti, B., and Anile, A. M. 1973, *Astr. Ap.*, 28, 429.

Bertotti, B., Cavaliere, A., and Pacini, F. 1969a, *Nature*, 221, 624.

Bertotti, B., Cavaliere, A., and Pacini, F. 1969b, *Nature*, 223, 1351.

Beskin, V. S., Gurevich, A. V., and Istomin, Ya. N. 1983, *Sov. Phys. JETP*, 58, 235.

Beskin, V. S., Gurevich, A. V., and Istomin, Ya. N. 1984, *Ap. Space Sci.*, 102, 301.

Beskin, V. S., Gurevich, A. V., and Istomin, Ya. N. 1986, *Sov. Phys. Uspekhi*, 29, 946.

Beskin, V. S., Gurevich, A. V., and Istomin, Ya. N. 1987, *Sov. Phys. JETP*, 65, 715.

Beskin, V. S., Gurevich, A. V., and Istomin, Ya. N. 1988, *Phys. Rev. Letters*, 61, 649.

Bethe, H. A., and Johnson, M. B. 1974, *Nucl. Phys.* (Netherlands), A230, 1.

Bigg, E. K. 1964, *Nature*, 203, 1008.

Biggs, J. D., McCulloch, P. M., Hamilton, P. A., Manchester, R. N., and Lyne, A. G. 1985, *M.N.R.A.S.*, 215, 281.

Bignami, G. F., Caraveo, P. A., and Lamb, R. C. 1983, *Ap. J. (Letters)*, 272. L9.

Bignami, G. F., Caraveo, P. A., and Paul, J. A. 1984, *Nature*, 310, 464.

Bionta, R. M, and 36 other authors. 1987, *Phys. Rev. Letters*, 58, 1494.

Birkeland, K. 1908, *On the Cause of Magnetic Storms and the Origin of Terrestrial Magnetism* (London: Longmans, Green and Co.).

Bisnovatyi-Kogan, G. S. 1970, *Radiophys. Quantum Electron.* (USA), 13, 1441.

Bisnovatyi-Kogan, G. S., and Blinnikov, S. I. 1973, *Sov. Astr. AJ* (USA), 17, 304.

Bisnovatyi-Kogan, G. S., Imshennik, V., Nadyozhin, D., and Chechetkin, V. 1975, *Ap. Space Sci.*, 35, 23.

Bisnovatyi-Kogan, G. S., and Komberg, B. V. 1974, *Sov. Astr. AJ* (USA), 18, 217.

Black, D. C. 1969, *Nature*, 221, 157.

Blackett, P. M. S. 1949, *Nature*, 159, 658.

Blaes, O., Blandford, R., Goldreich, P., and Madau, P. 1989, *Ap. J.*, 343, 839.

Blaizot, J. P., Gogny, D., and Grammaticos, B. 1976, *Nucl Phys.* (Nether lands), A265, 315.

Blanco, V. M., and McCuskey, S. W. 1961, *Basic Principles of the Solar System* (Reading. Addison-Wesley), p. 179ff.

Bland, B. H. 1968, *Nature*, 219, 23.

Blandford, R., and Eichler, D. 1987, *Phys. Reports*, 154, 1.

Blandford, R., and Teukolsky, S. A. 1975, *Ap. J. (Letters)*, 198, L27.

Blandford, R., and Teukolsky, S. A. 1976, *Ap. J.*, 205, 580.

Blandford, R. D. 1975, *M.N.R.A.S.*, 170, 551.

Blandford, R. D. 1976, *M.N.R.A.S.*, 176, 465.

Blandford, R. D. 1980, *Ap. J.*, 238, 410.

Blandford, R. D., Applegate, J. H., and Hernquist, L. 1983, *M.N.R.A.S.*, 204, 1025.

Blandford, R. D., and DeCampli, W. M. 1981, in *Pulsars*, ed. W. Sieber and R. Wielebinski (IAU Conf. No. 95) (Dordrecht: Reidel), p. 371.

Blandford, R. D., and Ostriker, J. P. 1978, *Ap. J. (Letters)*, 221, L29.

Blandford, R. D., and Romani, R. W. 1988, *M.N.R.A.S.*, 234, 57p.

Blandford, R. D., and Scharlemann, E. T. 1976, *M.N.R.A.S.*, 174, 59.

Blumenthal, G. R., and Tucker, W. H. 1974, *Ann. Rev. Astr. Ap.*, 12, 23.

Bodenheimer, P., and Ostriker, J. P. 1974, *Ap. J.*, 191, 465.

Bodmer, A. R. 1971, *Phys. Rev.*, D4, 1601.

Böhm-Vitense, E. 1969, *Ap. J. (Letters)*, 156, L131.

Bonometto, S., and Scrascia, L. 1974, *Astr. Ap.*, 32, 115.

Boriakoff, V. 1983, *Ap. J.*, 272, 687.

Boriakoff, V., Buccheri, R., and Fauci, F. 1983, *Nature*, 304, 417.

Boriakoff, V., Buccheri, R., and Fauci, F. 1984, in *Millisecond Pulsars*, ed. S. P. Reynolds and D. R. Stinebring (Green Bank: National Radio Astronomy Observatory), p. 24.

Born, M., and Wolf, E. 1986, *Principles of Optics* (New York: Pergamon).

Bouchet, P., Danziger, I. J., and Lucy, L. B. 1989, *IAU Circ.*, No. 4933.

Braginskii, S. I. 1964, *Sov. Phys. JETP*, 20, 1462.

Brandi, H. A. 1975, *Phys. Rev.*, A11, 1835.

Braude, S. Ya., and Bruk, Yu. M. 1980, *Pis'ma V. Astr. Z.* (USSR), 6, 301.

Braun, R., Goss, W. M., and Lyne, A. G. 1989, *Ap. J.*, 340, 355.

Braun, R., Gull, S. F., and Perley, R. A. 1987, *Nature*, 327, 395.

Brecher, K. 1975, *Ap. J. (Letters)*, 195, L113.

Brecher, K. 1982, in *Gamma-ray Transients and Related Astrophysical Phenomena*, ed. R. Lingenfelter, H. Hudson, and D. Worrall (New York: AIP), p. 293.

Brecher, K., and Chanmugam, G. 1983, *Nature*, 301, 124.

Brecher, K., and Morrison, P. 1974, *Ap. J. (Letters)*, 186, L97.

Brinkmann, W., Aschenbach, B., and Langmeier, A. 1985, *Nature*, 313, 662.

Broad, W. J. 1981, *Science*, 212, 1116.

Brown, G. E. 1988, *Nature*, 336, 520.

Bruenn, S. W. 1987, *Phys. Rev. Letters*, 59, 938.

Buccheri, R., D'Amico, N., Hermsen, W., and Sacco, B. 1985, *Nature*, 316, 131.

Buccheri, R., and 9 other authors. 1978, *Astr. Ap.*, 69, 141.

Buccheri, R., and 15 other authors. 1983, *Astr. Ap.*, 128, 245.

Buckee, J. W., Grounds, S., Miranda, L. C. M., and Ter Haar, D. 1974, *AGARD Conf. Proc.*, 138 (Neuilly sur Seine: AGARD), p. 18.

Buckley, R. 1976, *M.N.R.A.S.*, 177, 415.

Buckley, R. 1977a, *M.N.R.A.S.*, 180, 125.

Buckley, R. 1977b, *Nature*, 266, 37.

Buckley, R. 1978, *M.N.R.A.S.*, 183, 771.

Bullard, E. 1978, in *Topics in Nonlinear Dynamics, a Tribute to Sir Edward Bullard*, ed. S. Jorna (New York: AIP), p. 373.

Burbidge, G. R. 1956, *Ap. J.*, 124, 416.

Burbidge, G. R., and Strittmatter, P. A. 1968, *Nature*, 218, 433.

Burdyuzha, V. V. 1977, *Pis'ma V. Astr. Z.* (USSR), 3, 121.

Burgess, G. O. 1978, Rice University M.S. thesis.

Burman, R. 1977a, *Ap. Space Sci.*, 51, 239.

Burman, R. 1977b, *Phys. Letters*, A60, 309.

Burman, R. 1980a, *Ap. Space Sci.*, 72, 251.

Burman, R. 1980b, *Austr. J. Phys.*, 33, 771.

Burman, R. R. 1985, *Austr. J. Phys.*, 38, 749.

Burman, R. R., and Mestel, L. 1978, *Austr. J. Phys.*, 31, 455.

Burman, R. R., and Mestel, L. 1979, *Austr. J. Phys.*, 32, 681.

Burns, J. A. 1970, *Nature*, 228, 986.

Burrows, A. 1987, *Phys. Today*, 40 (9), 28.

Burrows, A. 1988, *Ap. J.*, 334, 891.

Buschauer, R. 1977, *M.N.R.A.S.*, 179, 99.

Buschauer, R., and Benford, G. 1976, *M.N.R.A.S.*, 177, 109.

Buschauer, R., and Benford, G. 1977a, *M.N.R.A.S.*, 179, 189.

Buschauer, R., and Benford, G. 1977b, *M.N.R.A.S.*, 179, 99.

Buschauer, R., and Benford, G. 1978, *M.N.R.A.S.*, 185, 493.

Buschauer, R., and Benford, G. 1980, *M.N.R.A.S.*, 190, 945.

Callaway, J. 1972, *Phys. Letters*, A40, 331.

Cameron, A. G. W. 1959a, *Ap. J.*, 130, 452.

Cameron, A. G. W. 1959b, *Ap. J.*, 130, 884.

Cameron, A. G. W., and Iben, I., Jr. 1986, *Ap. J.*, 305, 228.

Candy, B. N., and Blair, D. G. 1983, *M.N.R.A.S.*, 205, 281.

Canuto, V. 1971, in *The Crab Nebula* (IAU Symp. No. 46) (Dordrecht: Reidel), p. 455.

Canuto, V., and Chiu, H.-Y. 1968, *Phys. Rev.*, 176, 1438.

Canuto, V., Chiu, H.-Y., Chiuderi, C., and Lee, H. J. 1970, *Nature*, 225, 47.

Canuto, V., Chiu, H.-Y., and Fassio-Canuto, L. 1969, *Ap. Space Sci.*, 3, 158.

Canuto, V, and Kelley, D. C. 1972, *Ap. Space Sci.*, 17, 277.

Carr, T. D., and Gulkis, S. 1969, *Ann. Rev. Astr. Ap.*, 7, 577.

Carr, T. D., Desch, M. D., and Alexander, J. K. 1983, in *Physics of the Jovian Magnetosphere*, ed. A. J. Dessler (Cambridge: Cambridge University Press), p. 226.

Casperson, L. W. 1977, *Ap. Space Sci.*, 48, 389.

Cassé, M., and Paul, J. A. 1980, *Ap. J.*, 237, 236.

Cavaliere, A., Pacini, F., and Setti, G. 1969, *Ap. Letters*, 4, 103.

Cesarsky, C. 1980, *Ann. Rev. Astr. Ap.*, 18, 289.

Chadwick, P. M., and 11 other authors. 1985b *Nature*, 318, 642.

Chadwick, P. M., Dowthwaite, J. C., Harrison, A. B., Kirkman, I. W., McComb, T. J. L., Orford, K. J., and Turver, K. E. 1985a, *Nature*, 317, 236.

Chanan, G. A., Helfand, D. J., and Reynolds, S. P. 1984, *Ap. J. (Letters)*, 287, L23.

Chandrasekhar, S. 1957, *An Introduction to the Study of Stellar Structure* (New York: Dover), p. 412.

Chandrasekhar, S. 1984, *Rev. Mod. Phys.*, 56, 137.

Chanmugam, G. 1973, *Ap. J. (Letters)*, 182, L39.

Chanmugam, G. 1974, *Ap. J. (Letters)*, 193, L75.

Chanmugam, G. 1978, *Ap. J.*, 221, 965.

Chanmugam, G., and Brecher, K. 1985, *Nature*, 313, 767.

Chanmugam, G., and Brecher, K. 1987, *Nature*, 329, 696.

Chanmugam, G., and Gabriel, M. 1971, *Astr. Ap.*, 11, 268.

Chapline G., and Nauenberg, M. 1977, *Phys. Rev.*, D16, 450.

Chardin, G. 1987, in *Proc. VII Moriond Workshop on New and Exotic Phenomena* (Gif-sur-Yvette, France: Editions Frontières).

Charugin, V. M. 1975, *Ap. Space Sci.*, 37, 449.

Chau, W. Y. 1970, *Nature*, 228, 655.

Chau, W. Y., and Henriksen, R. N. 1970, *Ap. J. (Letters)*, 161, L137.

Chau, W. Y., and Henriksen, R. N. 1971, *Ap. Letters*, 8, 49.

Chau, W. Y., Henriksen, R. N., and Rayburn, D. R. 1971, *Ap. J. (Letters)*, 168, L79.

Chau, W. Y., and Srulovicz, P. 1971, *Phys. Rev.*, D3, 1999.

Chen, H.-H., Ruderman, M. A., and Sutherland, P. G. 1974, *Ap. J.*, 191, 473.

Chen, K., and Shaham, J. 1989, *Ap. J.*, 339, 279.

Chenette, D. L., Conlon, T. F., and Simpson, J. A. 1974, *J. Geophys. Res.*, 79, 3551.

Cheng, A. 1974, *Ap. Space Sci.*, 31, 49.

Cheng, A. F. 1981, in *Pulsars*, ed. W. Sieber and R. Wielebinski (IAU Conf. No. 95) (Dordrecht: Reidel), p. 99.

Cheng, A. F. 1983, *Ap. J.*, 275, 790.

Cheng, A. F. 1985, *Ap. J.*, 299, 917.

Cheng, A. F. 1989, *Ap. J.*, 337, 803.

Cheng, A. F., and Helfand, D. J. 1983, *Ap. J.*, 271, 271.

Cheng, A. F., and Ruderman, M. A. 1977a, *Ap. J.*, 212, 800.

Cheng, A. F., and Ruderman, M. A. 1977b, *Ap. J.*, 214, 598.

Cheng, A. F., and Ruderman, M. A. 1977c, *Ap. J.*, 216, 865.

Cheng, A. F., and Ruderman, M. A. 1979, *Ap. J.*, 229, 348.

Cheng, A. F., and Ruderman, M. A. 1980, *Ap. J.*, 235, 576.

Cheng, A. F., Ruderman, M., and Sutherland, P. 1976, *Ap. J.*, 203, 209.

Cheng, K. S., Alpar, M. A., Pines, D., and Shaham, J. 1988, *Ap. J.*, 330, 835.

Cheng, K. S., Ho, C., and Ruderman, M. 1986a, *Ap. J.*, 300, 500.

Cheng, K. S., Ho, C., and Ruderman, M. 1986b, *Ap. J.*, 300, 522.

Chernoff, D. F., and Djorgovski, S. 1989, *Ap. J.*, 339, 904.

Chevalier, R. A., and Emmerling, R. T. 1986, *Ap. J.*, 304, 140.

Chian, A. C.-L., and Kennel, C. F. 1983, *Ap. Space Sci.*, 97, 8.

Chin, S., and Kerman, A. 1979, *Phys. Rev. Letters*, 43, 1292.

Chitre, D. M., and Hartle, J. B. 1976, *Ap. J.*, 207, 592.

Chiu, H.-Y. 1971, in *The Crab Nebula* (IAU Symp. No. 46) (Dordrecht: Reidel), p. 414.

Chiu, H.-Y., and Canuto, V. 1969a, *Nature*, 221, 529.

Chiu, H.-Y., and Canuto, V. 1969b, *Phys. Rev. Letters*, 22, 415.

Chiu, H.-Y., and Canuto, V. 1970, *Nature*, 225, 1230.

Chiu, H.-Y., and Canuto, V. 1971, *Ap. J.*, 163, 577.

Chiu, H.-Y., Canuto, V., and Fassio-Canuto, L. 1969, *Nature*, 221, 529.

Chiu, H.-Y., and Occhionero, F. 1969, *Nature*, 223, 1113.

Chiuderi, C., and Occhionero, F. 1970, *Nature*, 226, 337.

Clark, D. H., Murdin, P., Wood, R., Gilmozzi, R., Danziger, J., and Furr, A. W. 1983, *M.N.R.A.S.*, 204, 415.

Cline, T. L., and 14 other authors. 1980, *Ap. J. (Letters)*, 237, L1.

Cocke, W. J. 1973, *Ap. J.*, 184, 291.

Cocke, W. J., and Cohen, J. M. 1968, *Nature*, 219, 1009.

Cocke, W. J., Disney, M. J., and Taylor, J. H. 1969, *Nature*, 221, 525.

Cocke, W. J., and Ferguson, D. C. 1974, *Ap. J.*, 194, 725.

Cocke, W. J., Ferguson, D. C., and Muncaster, G. W. 1973, *Ap. J.*, 183, 987.

Cocke, W. J., and Pacholczyk, A. G. 1976, *Ap. J. (Letters)*, 204, L13.

Cohen, J. M., Kegeles, L. S., and Rosenblum, A. 1975, *Ap. J.*, 201, 783.

Cohen, J. M., and Mustafa, E. 1987, *Ap. J.*, 319, 930.

Cohen, J. M., and Rosenblum, A. 1972, *Ap. Space Sci.*, 16, 130.

Cohen, J. M., and Rosenblum, A. 1973, *Ap. J.*, 186, 267.

Cohen, J. M., and Toton, E. T. 1971, *Ap. Letters*, 7, 213.

Cohen, R., Lodenquai, J., and Ruderman, M. 1970, *Phys. Rev. Letters*, 25, 467.

Cohen, R. M., Coppi, B., and Treves, A. 1973, *Ap. J.*, 179, 269.

Cole, T. W. 1970, *Nature*, 227, 788.

Cole, T. W. 1976, *M.N.R.A.S.*, 175, 93p.

Colgate, S. A. 1974, *Ap. J.*, 187, 333.

Colgate, S. A., and Johnson, M. H. 1960, *Phys. Rev. Letters*, 5, 235.

Colgate, S. A., Petschek, A. G., and Sarracino, R. 1984, in *High Energy Transients in Astrophysics* (New York: AIP), p. 548.

Constantinescu, D. M., and Moruzzi, G. 1978, *Phys. Rev.*, D18, 1820.

Cooperstein, F. I, and Lim, P. H. 1985, *Phys. Rev. Letters*, 55, 265.

Coppi, B. 1972, *Verh. Deut. Phys. Ges.* (Germany), 7, 777.

Coppi, B., and Ferrari, A. 1970a, *Letters Nuovo Cimento*, 3, 93.

Coppi, B., and Ferrari, A. 1970b, *Ap. J. (Letters)*, 161, L65.

Coppi, B., and Ferrari, A. 1971, in *The Crab Nebula* (IAU Symp. No. 46) (Dordrecht: Reidel), p. 460.

Cordes, J. M. 1976, *Ap. J.*, 210, 780.

Cordes, J. M. 1978, *Ap. J.*, 222, 1006.

Cordes, J. M. 1979a, *Austr. J. Phys.*, 32, 9.

Cordes, J. M. 1979b, *Space Sci. Rev.*, 24, 567.

Cordes, J. M. 1986, *Ap. J.*, 311, 183.

Cordes, J. M., Downs, G. S., and Krause-Polstorff, J. 1986, *Ap. J.*, 330, 847.

Cordes, J. M., and Greenstein, G. 1981, *Ap. J.*, 245, 1060.

Cordes, J. M., and Stinebring, D. R. 1984, *Ap. J. (Letters)*, 277, L53.

Cordes, J. M., Weisberg, J. M., and Boriakoff, V. 1983, *Ap. J.*, 268, 370.

Coroniti, F. V. 1981, *Ap. J.*, 244, 587.

Coroniti, F. V. 1990, *Ap. J.*, 349, 538.

Cowley, A. P., Crampton, D., and Hutchings, J. B. 1987, *Astr. J.*, 93, 195.

Cowling, S. A. 1983, *M.N.R.A.S.*, 204, 1237.

Cowling, T. G. 1934, *M.N.R.A.S.*, 94, 39.

Cowling, T. G. 1957, *Magnetohydrodynamics* (New York: Interscience), p. 77.

Cowsik, R., Ghosh, P., and Melvin, M. A. 1983, *Nature*, 303, 308.

Cox, J. L., Jr. 1979, *Ap. J.*, 229, 734.

Daishido, T. 1975, *Publ. Astr. Soc. Japan*, 27, 181.

Damashek, M., Backus, P. R., Taylor, J. H., and Burkhardt, R. K. 1982, *Ap. J. (Letters)*, 253, L57.

Damour, T., and Deruelle, N. 1986, *Ann. Inst. H. Poincaré* (Physique Théorique), 44, 263.

Damour, T., Gibbins, G. W., and Taylor, J. H. 1988, *Phys. Rev. Letters*, 61, 1151.

Danaher, S. , Fegan, D. J., Porter, N. A., and Weeks, T. C. 1981, *Nature*, 289, 568.

Darwin, C. 1949, *Nature*, 164, 1112.

Datta, B., and Ray, A. 1983, *M.N.R.A.S.*, 204, 75p.

Daugherty, J. K., and Harding, A. K. 1982, *Ap. J.*, 252, 337.

Daugherty, J. K., and Harding, A. K. 1983, *Ap. J.*, 273, 761.

Daugherty, J. K., and Harding, A. K. 1989, *Ap. J.*, 336, 861.

Daugherty, J. K., and Lerche, I. 1975, *Ap. Space Sci.*, 38, 437.

Daugherty, J. K., and Lerche, I. 1976, *Phys. Rev.*, D14, 340.

Davidson, K., and Ostriker, J. P. 1973, *Ap. J.*, 179, 585.

Davidson, R. C. 1974, *Theory of Nonneutral Plasmas* (Reading: W. A. Benjamin).

Davila, J., Wright, C., and Benford, G. 1980, *Ap. Space Sci.*, 71, 51.

Davis, L., and Goldstein, M. 1970, *Ap. J. (Letters)*, 159, L81.

Davis, M. M., Taylor, J. H., Weisberg, J. M., and Backer, D. C. 1985, *Nature*, 315, 547.

Dean, A. J., and Turner, M. J. L. 1971, *Ap. Letters*, 8, 145.

de Costa, A. A., and Kahn, F. D. 1982, *M.N.R.A.S.*, 119, 211.

Deeter, J. E., Boynton, P. E., Lamb, F. K., and Zylstra, G. 1989, *Ap. J.*, 336, 376.

DeGrand, T., Jaffe, R. L., Johnson, K., and Kiskis, J. 1975, *Phys. Rev.*, D12, 2060.

deGrassie, J. S., and Malmberg, J. H. 1977, *Phys. Rev. Letters*, 39, 1077.

Deich, W. T. S., Cordes, J. M., Hankins, T. H., and Rankin, J. M. 1986, *Ap. J.*, 300, 540.

Demiański, M., and Prószyński, M. 1983, *M.N.R.A.S.*, 202, 437.

de Sabbata, V. 1970, *Mem. Soc. Astr. Ital.*, 41, 65.

Dessler, A. J. 1980a, *Planetary Space Sci.*, 28, 781.

Dessler, A. J. 1980b, *Icarus*, 44, 291.

Dessler, A. J., and Juday, R. D. 1965, *Planetary Space Sci.*, 13, 63.

Dessler, A. J., Sandel, B. R., and Atreya, S. K. 1981, *Planetary Space Sci.*, 29, 215.

Detweiler, S. 1979, *Ap. J.*, 234, 1100.

Deutsch, A. J. 1955, *Ann. d'Astrophysique*, 18, 1.

Dewey, R. J., Taylor, J. H., Weisberg, J. M, and Stokes, G. H. 1985, *Ap. J. (Letters)*, 294, L25.

Dewhirst, D. W. 1983, *Observatory*, 103, 114.

Dicke, R. H. 1964, *Nature*, 202, 432.

Djorgovski, S. 1982, *Nature*, 300, 618.

Djorgovski, S., and Evans, C. R. 1988, *Ap. J. (Letters)*, 335, L61.

Djorgovski, S., and Spinrad, H. 1983, *Nature*, 306, 569.

Dokuchaev, V. P., Tamoikin, V. V., and Chugunov, Yo. V. 1976, *Sov. Astr. AJ* (USA), 20, 299.

Dolan, J. F., Crannell, C. J., Dennis, B. R., and Orwig, L. E. 1987, *Ap. J.*, 322, 324.

Dorman, L. I., Kats, M. Ye., and Yukhimuk, A. K. 1973, *Geomagn. Aeron.* (USA), 13, 171.

Douglas-Hamilton, D. H. 1968, *Nature*, 218, 1035.

Dowden, R. L. 1968, *Proc. Astr. Soc. Austral.*, 1, 159.

Downs, G. S. 1981, *Ap. J.*, 249, 687.

Downs, G. S. 1982, *Ap. J. (Letters)*, 257, L67.

Downs, G. S., and Reichley, P. E. 1983, *Ap. J. Suppl.*, 53, 169.

Dowthwaithe, J. C., Harrison, A. B., Kirkman, I. W., Macrae, H. J., Orford, K. J., Turver, K. E., and Walmsley, M. S. 1984, *Nature*, 309, 691.

Drake, F. D., and Craft, H. D., Jr. 1968, *Nature*, 220, 231.

Drake, J. F., Lee, Y. C., and Tsintsadze, N. L. 1976, *Phys. Rev. Letters*, 36, 31.

Driscoll, C. F., and Malmberg, J. H. 1983, *Phys. Rev. Letters*, 50, 167.

Drury, L. O'C., and Völk, H. J. 1981, *Ap. J.*, 248, 344.

Dryden, H. L., Murnagh, F. D., and Bateman, H. 1956, *Hydrodynamics* (New York: Dover), p. 465.

Dungey, J. W. 1961, *J. Geophys. Res.*, 66, 1043.

Durand, E. 1964, *Electrostatique I* (Paris: Masson et Cie.), p. 336.

Durdin, J. M., Large, M. I., Little, A. G., Manchester, R. N., Lyne, A. G., and Taylor, J. H. 1979, *M.N.R.A.S.*, 186, 39p.

Durisen, R. H. 1973, *Ap. J.*, 244, 587.

Durisen, R. H., and Tohline, J. E. 1985, in *Protostars and Planets II*, ed. D. C. Black (Tucson: University of Arizona Press), p. 543.

Durney, B. R., Faulkner, J., Gribbin, J. R., and Roxburgh, I. W. 1968, *Nature*, 219, 20.

Dyson, F. J. 1969a, *Ap. J.*, 156, 529.

Dyson, F. J. 1969b, *Nature*, 223, 486.

Eardley, D. M. 1975, *Ap. J. (Letters)*, 196, L59.

Eardley, D. M., and Lightman, A. P. 1975, *Ap. J.*, 200, 187.

Eastlund, B. J. 1968, *Nature*, 220, 1293.

Eastlund, B. J. 1970, *Nature*, 225, 430.

Eastlund, B. J. 1971, in *The Crab Nebula* (IAU Symp. No. 46) (Dordrecht: Reidel), p. 443.

Eddington, A. S. 1926, *The Internal Constitution of the Stars* (Cambridge: Cambridge University Press), p. 232.

Eddington, A. S. 1935a, *M.N.R.A.S.*, 95, 194.

Eddington, A. S. 1935b, *M.N.R.A.S.*, 96, 20.

Eddington, A. S. 1940, *M.N.R.A.S.*, 100, 582.

Eichler, D. 1978a, *AIP Conf. Proc.*, 52, 38.

Eichler, D. 1978b, *Ap. J.*, 222, 1109.

Eichler, D. 1978c, *Nature*, 275, 725.

Eichler, D., and Cheng, A. F. 1989, *Ap. J.*, 336, 360?

Eichler, D., and Levinson, A. 1988, *Ap. J. (Letters)*, 335, L67.

Eichler, D., Livio, M., Piran, T., and Schramm, D. N. 1989, *Nature*, 340, 126.

Eichler, D., and Schramm, D. N. 1978, *Nature*, 275, 704.

Eichler, D., and Vestrand, W. T. 1984, *Nature*, 307, 613.

Eidman, V. Ya. 1971, *Astrophysics* (USA), 7, 78.

El-Gowhari, A., and Arponen, J. 1972, *Nuovo Cimento*, 11B, 201.

Elitzur, M. 1974, *Ap. J.*, 190, 673.

Elitzur, M. 1979, *Ap. J.*, 229, 742.

Ellison, D. 1975, Rice University M.S. thesis.

Ellison, D., and Kanzanas, D. 1983, *Astr. Ap.*, 128, 102.

Elsässer, K. 1976, *Astr. Ap.*, 52, 177.

Elsässer, K., and Kirk, J. 1976, *Astr. Ap.*, 52, 449.

Elsner, R. F., Ghosh, P., and Lamb, F. K. 1980, *Ap. J. (Letters)*, 241, L155.

Elsner, R. F., and Lamb, F. K. 1976, *Nature*, 262, 356.

Elsner, R. F., and Lamb, F. K. 1977, *Ap. J.*, 215, 879.

Elsner, R. F., and Lamb, F. K. 1984, *Ap. J.*, 278, 326.

Endean, V. G. 1972a, *M.N.R.A.S.*, 158, 13.

Endean, V. G. 1972b, *Nature* (Phys. Sci.), 237, 72.

Endean, V. G. 1973, *Nature*, 241, 184.

Endean, V. G. 1974, *Ap. J.*, 187, 359.

Endean, V. G. 1976, *M.N.R.A.S.*, 174, 125.

Endean, V. G. 1980, *M.N.R.A.S.*, 193, 213.

Endean, V. G. 1981, *M.N.R.A.S.*, 195, 55p.

Endean, V. G., and Allen, J E. 1970, *Nature*, 228, 348.

Epstein, R. 1977, *Ap. J.*, 216, 92 (Erratum 231, 644).

Epstein, R. I. 1973, *Ap. J.*, 183, 593.

Epstein, R. I. 1985, *Ap. J.*, 291, 822.

Epstein, R. I. 1988, *Ap. J.*, 333, 880.

Epstein, R. I., Colgate, S. A., and Haxton, W. C. 1988, *Phys. Rev. Letters*, 61, 2038.

Epstein, R. I., and Petrosian, V. 1973, *Ap. J.*, 183, 611.

Erber, T. 1966, *Rev. Mod. Phys.*, 38, 626.

Erber, T., and Spector, H. N. 1973, *Ap. J.*, 184, 301.

Erickson, W. C. 1983, *Ap. J. (Letters)*, 264, L13.

Erickson, W. C., Mahoney, M. J., Becker, R. H., and Helfand, D. J. 1987, *Ap. J. (Letters)*, 314, L45.

Esposito, L. W., and Harrison, E. R. 1975, *Ap. J. (Letters)*, 196, L1.

Evangelidis, E. 1977, *Ap. Space Sci.*, 51, 319.

Evangelidis, E. A. 1979, *Ap. Space Sci.*, 60, 213.

Eviatar, A., and Siscoe, G. L. 1980, *Geophys. Res. Letters*, 7, 1085.

Ewart, G. M., Guyer, R. A., and Greenstein, G. 1975, *Ap. J.*, 202, 238.

Fabian, A. C., Icke, V., Pringle, J. E. 1976, *Ap. Space Sci.*, 42, 77.

Fan, C. Y., and Jiping, W. 1982, *Ap. J.*, 260, 353.

Fang, L.-Z., and Liu, Y.-Z. 1976, *Acta Phys. Sin.* (China), 25, 521.

Faulkner, J., and Gribbin, J. R. 1968, *Nature*, 218, 734.

Fawley, W. M., Arons, J., and Scharlemann, E. T. 1977, *Ap. J.*, 217, 227.

Feinberg, G. 1969, *Science*, 166, 879.

Felten, J. E., Dwek, E., and Viegas-Aldrovandi, S. M. 1989, *Ap. J.*, 340, 943.

Fenimore, E. E., and 13 other authors. 1988, *Ap. J. (Letters)*, 335, L71.

Ferguson, D. C. 1973, *Ap. J.*, 183, 977.

Ferguson, D. C. 1976a, *Ap. J.*, 205, 247.

Ferguson, D. C. 1976b, *Ap. J.*, 209, 606.

Ferguson, D. C. 1979, *Nature*, 278, 331.

Ferguson, D. C. 1981a, *Comments Ap.*, 9, 127.

Ferguson, D. C. 1981b, in *Pulsars*, ed. W. Sieber and R. Wielebinski (IAU Conf. No. 95) (Dordrecht: Reidel), p. 141.

Ferguson, D. C., Cocke, W. J., and Gehrels, T. 1974, *Ap. J.*, 190, 375.

Ferrari, A. 1972, *Mem. Soc. Astr. Ital.*, 43, 715.

Ferrari, A., and Trussoni, E. 1971, *Letters Nuovo Cimento*, 1, 137.

Ferrari, A., and Trussoni, E. 1973, *Ap. Space Sci.*, 24, 3.

Ferrari, A., and Trussoni, E. 1974, *Astr. Ap.*, 36, 267.

Ferrari, A., and Trussoni, E. 1975a, *Phys. Letters*, 51A, 304.

Ferrari, A., and Trussoni, E. 1975b, *Ap. Space Sci.*, 33, 111.

Ferraro, V. C. A., and Plumpton, C. 1961, *An Introduction to Magneto-fluid Mechanics* (New York: Oxford University Press), p. 28.

Fesen, R., and Gull, T. 1983, in *Supernova Remnants and Their X-ray Emission*, ed. J. Danziger and P. Gorenstein (IAU Symp. No. 101) (Dordrecht: Reidel), p. 141.

Fesen, R. A., Shull, J. M., and Saken, J. M. 1988, *Nature*, 334, 229.

Fichtel, C., and Linsey, J. 1986, *Ap. J.*, 300, 474.

Filippenko, A. V., Radhakrishnan, V. 1982, *Ap. J.*, 263, 828.

Filippenko, A. V., Readhead, A. C. S., and Ewing, M. S. 1983, in *Positron-Electron Pairs in Astrophysics*, ed. M. Burns, A. K. Harding, and R. Ramaty (New York: AIP), p. 113.

Finzi, A., and Wolf, R. A. 1968, *Ap. J.*, 153, 835.

Finzi, A., and Wolf, R. A. 1969, *Ap. J. (Letters)*, 155, L107.

Fischer, W., and Straumann, N. 1972, *Helv. Phys. Acta*, 45, 1089.

Fishman, G. J., Harnden, F. R., and Haymes, R. C. 1969a, *Ap. J. (Letters)*, 156, L107.

Fishman, G. J., Harnden, F. R., Johnson, W. N., and Haymes, R. C. 1969b, *Ap. J. (Letters)*, 158, L61.

Fisk, L. 1971, *J. Geophys. Res.*, 76, 1662.

Fitzpatrick, R., and Mestel, L. 1988a, *M.N.R.A.S.*, 232, 277.

Fitzpatrick, R., and Mestel, L. 1988b, *M.N.R.A.S.*, 232, 303.

Flowers, E., and Itoh, N. 1976, *Ap. J.*, 206, 218.

Flowers, E., and Ruderman, M. A. 1977, *Ap. J.*, 215, 302.

Flowers, E. G., Lee, J.-F., Ruderman, M. A., Sutherland, P. G., Hillebrandt, W., and Müller, E. 1977, *Ap. J.*, 215, 291.

Fowler, L. A., Cordes, J. M., and Taylor, J. H. 1979, *Austr. J. Phys.*, 32, 35.

Fowler, L. A., and Wright, G. A. E. 1982, *Astr. Ap.*, 109, 279.

Fowler, L. A., Wright, G. A. E., and Morris, D. 1981, *Astr. Ap.*, 93, 54.

Fowler, W. A. 1971, in *The Crab Nebula* (IAU Symp. No. 46) (Dordrecht: Reidel), p. 364.

Freeman, B. A. and McLerran, L. D. 1977, *Phys. Rev.*, D17, 1109.

Friedman, J. L. 1983, *Phys. Rev. Letters*, 51, 11 and 718.

Friedman, J. L., and Ipser, J. R. 1987, *Ap. J.*, 314, 594.

Friedman, J. L., Ipser, J. R., and Parker, L. 1984, *Nature*, 312, 255.

Friedman, J. L., Ipser, J. R., and Parker, L. 1989, *Phys. Rev. Letters*, 62, 3015.

Friedman, J. L., Iwamura, J. N., Durisen, R. H., and Parker, L. 1988, *Nature*, 336, 560.

Fruchter, A. S., and 12 other authors. 1989, *Ap. J.*, submitted.

Fruchter, A. S., Gunn, J. E., Lauer, T. R., and Dressler, A. 1988a, *Nature*, 334, 686.

Fruchter, A. S., Stinebring, D. R., and Taylor, J. H. 1988b, *Nature*, 333, 237.

Fryxell, B. A. 1979, *Ap. J.*, 234, 641.

Fryxell, B. A., and Arnett, W. D. 1981, *Ap. J.*, 243, 994.

Fujimoto, M., and Murai, T. 1972, *Publ. Astr. Soc. Japan*, 24, 269.

Fujimoto, M., and Murai, T. 1973, *Publ. Astr. Soc. Japan*, 25, 75.

Fujimoto, M. Y., Hanawa, T., Iben, I., Jr., and Richardson, M. B. 1987, *Ap. J.*, 315, 198.

Fujimura, F. S., and Kennel, C. F. 1979, *Astr. Ap.*, 79, 299.

Fujimura, F. S., and Kennel, C. F. 1980, *Ap. J.*, 236, 245.

Galtsov, D. V., and Petukhov, V. I. 1978, *Phys. Letters*, 66A, 346.

Gehrels, T. (ed.). 1976, *Jupiter* (Tucson: University of Arizona Press).

Geldzahler, B. J., and 9 other authors. 1983, *Ap. J. (Letters)*, 273, L65.

Ghosh, P. 1984, *J. Ap. Astr.*, 5, 307.

Ghosh, P., and Lamb, F. K. 1979, *Ap. J.*, 234, 296.

Gil, J. 1985, *Ap. J.*, 299, 154.

Gil, J. 1986, *Astr. Ap.*, 123, 7.

Gil, J. 1987, *Ap. J.*, 314, 629.

Gilinsky, V., Hubbard, M., Modesitt, G., and Collas, P. 1970, *Topics in High Energy Physics* (P-4494) (Santa Monica: The Rand Corp.), p. 188.

Gillet, S. L. 1988, *Astr. J.*, 96, 1988.

Gillman, A. 1979, *Ap. J.*, 234, 632.

Ginzburg, V. L., and Zheleznyakov, V. V. 1970a, *Comments Ap. Space Phys.*, 2, 167.

Ginzburg, V. L., and Zheleznyakov, V. V. 1970b, *Comments Ap. Space Phys.*, 2, 197.

Ginzburg, V. L., and Zheleznyakov, V. V. 1975, *Ann. Rev. Astr. Ap.*, 13, 511.

Ginzburg, V. L., Zheleznyakov, V. V., and Zaitsev, V. V. 1968, *Nature*, 220, 355.

Ginzburg, V. L., Zheleznyakov, V. V., and Zaitsev, V. V. 1969a, *Ap. Space Sci.*, 4, 464.

Ginzburg, V. L., Zheleznyakov, V. V., and Zaitsev, V. V. 1969b, *Sov. Phys. Uspekhi*, 12, 378.

Glasser, M. L. 1975, *Ap. J.*, 199, 206.

Glatzel, W., Fricke, K. J., and El Eid, M. 1981, *Astr. Ap.*, 93, 395.

Glen, G., and Sutherland, P. G. 1980, *Ap. J.*, 239, 671.

Glencross, W. M. 1972, *Nature*, 237, 157.

Glendenning, N. K. 1986, *Phys. Rev. Letters*, 57, 1120.

Gold, T. 1968, *Nature*, 218, 731.

Gold, T. 1969a, *Nature*, 221, 25.

Gold, T. 1969b, *Nature*, 223, 162.

Gold, T. 1974, *Philos. Trans.*, A277, 453.

Gold, T. 1979, *Science*, 206, 1071.

Goldhaber, A. S., and Nieto, M. M. 1971, *Rev. Mod. Phys.*, 43, 277.

Goldman, I. 1989, Tel-Aviv University preprint.

Goldman, T., Maltman, K., Stephenson, G. J., Jr., Schmidt, K. E., and Wang, F. 1987, *Phys. Rev. Letters*, 59, 627.

Goldreich, P. 1969, *Proc. Astr. Soc. Austral.*, 1, 227.

Goldreich, P. 1970, *Ap. J. (Letters)*, 160, L17.

Goldreich, P., and Julian, W. H. 1969, *Ap. J.*, 157, 869.

Goldreich, P., and Julian, W. H. 1970, *Ap. J.*, 160, 971.

Goldreich, P., and Keeley, D. A. 1971, *Ap. J.*, 170, 463.

Goldreich, P., and Lynden-Bell, D. 1969, *Ap. J.*, 156, 59.

Goldreich, P., Pacini, F., and Rees, M. J. 1971, *Comments Ap. Space Sci.*, 3, 185.

Golenetskii, S. V., Ilyinskii, V. N., and Mazets, E. P. 1984, *Nature*, 307, 41.

Good, M. L. 1969, *Nature*, 221, 250.

Good, M. L., and Ng, K. K. 1985, *Ap. J.*, 299, 706.

Goodman, J. 1986, *Ap. J. (Letters)*, 308, L47.

Gott, J. R., III, Gunn, J. E., and Ostriker, J. P. 1970, *Ap. J. (Letters)*, 160, L91.

Greenstein, G. 1972, *Ap. J.*, 177, 251.

Greenstein, G. 1979, *Ap. J.*, 231, 880.

Greenstein, G., and Hartke, G. J. 1983, *Ap. J.*, 271, 283.

Greenstein, G. S., and Cameron, A. G. W. 1969, *Nature*, 222, 862.

Grewing, M., and Heintzmann, H. 1971, *Ap. Letters*, 8, 167.

Grindlay, J. E., and Bailyn, C. D. 1988, *Nature*, 336, 48.

Groth, E. J. 1975a, *Ap. J.*, 200, 278.

Groth, E. J. 1975b, *Ap. J. Suppl.*, 29, 453.

Gruber, D. E., and Primini, F. A. 1982, in *Accreting Neutron Stars*, ed. W. Brinkman and J. Trümper (MPE Report 177) (Garching), p. 41.

Gudmundsson, E. H., Pethick, C. J., and Epstein, R. I. 1983, *Ap. J.*, 272, 286.

Gull, T. R., and Fesen, R. A. 1982, *Ap. J. (Letters)*, 260, L75.

Gullahorn, G. E., and Rankin, J. M. 1982, *Ap. J.*, 260, 520.

Gunn, J. E., and Ostriker, J. P. 1969, *Nature*, 221, 454.

Gunn, J. E., and Ostriker, J. P. 1970, *Ap. J.*, 160, 979.

Guseinov, O. Kh., and Novruzova, Kh. I. 1974, *Astrophysics* (USA), 10, 163.

Guseinov, O. Kh., and Vanysek, V. 1974, *Ap. Space Sci.*, 28, L11.

Halpern, J. P., and Tytler, D. 1988, *Ap. J.*, 330, 201.

Hamada, T., and Salpeter, E. E. 1961, *Ap. J.*, 134, 683.

Hameury, J.-M., Bonazzola, S., Heyvaerts, J., and Ventura, J. 1982, *Astr. Ap.*, 111, 242.

Hamilton, P. A., King, E. A., McConnell, D., and McCulloch, P. M. 1989, *IAU Circ.*, No. 4708.

Hamilton, T. T., Helfand, D. J., and Becker, R. H. 1985, *Astr. J.*, 90, 606.

Hankins, T. H., and Cordes, J. M. 1980, in *Pulsars*, ed. W. Sieber and R. Wielebinski (IAU Conf. No. 95) (Dordrecht: Reidel), p. 207.

Hankins, T. H., and Fowler, L. A. 1986, *Ap. J.*, 304, 256.

Hankins, T. H., and Rickett, B. J. 1986, *Ap. J.*, 311, 684.

Hankins, T. H., Stinebring, D. R., and Rawley, L. A. 1987, *Ap. J.*, 315, 149.

Hara, T., and Sato, H. 1979, *Prog. Theo. Phys.*, 62, 969.

Hardee, P. E. 1977, *Ap. J.*, 216, 873.

Hardee, P. E. 1979, *Ap. J.*, 227, 958.

Hardee, P. E., and Morrison, P. J. 1979, *Ap. J.*, 227, 252.

Hardee, P. E., and Rose, W. K. 1974, *Ap. J. (Letters)*, 194, L35.

Hardee, P. E., and Rose, W. K. 1976, *Ap. J.*, 210, 533.

Hardee, P. E., and Rose, W. K. 1978, *Ap. J.*, 219, 274.

Harding, A. K. 1981, *Ap. J.*, 245, 267.

Harding, A. K. 1983, *Nature*, 303, 683.

Harding, A. K., and Preece, R. 1987, *Ap. J.*, 319, 939.

Harding, A. K., Sturrock, P. A., and Daugherty, J. K. 1989, Stanford University preprint CSSA-ASTRO-89-01.

Harding, A. K., and Tademaru, E. 1979, *Ap. J.*, 233, 317.

Harding, A. K., and Tademaru, E. 1980, *Ap. J.*, 238, 1054.

Harding, A. K., and Tademaru, E. 1981, *Ap. J.*, 243, 597.

Harding, A. K., Tademaru, E., and Esposito, L. W. 1978, *Ap. J.*, 225, 226.

Hari Dass, N. D., and Radhakrishnan, V. 1975, *Ap. Letters*, 16, 135.

Harnden, F. R., and Seward, F. D. 1984, *Ap. J.*, 283, 279.

Harnden, F. R., Jr., Giacconi, R., Grindlay, J., Hertz, P., Schreier, E., Steward, F., Tannenbaum, H., and Van Speybroeck, L. 1980, *Bull. Am. Astr. Soc.*, 12, 799.

Harrison, E. R. 1970, *Nature*, 225, 44.

Harrison, E. R., and Tademaru, E. 1975, *Ap. J.*, 201, 447.

Harrison, E. R., Wakano, M., and Wheeler, J. A. 1958, *La Structure et l'evolution de l'universe* (Brussels: R. Stoops).

Hartmann, D., Woosley, S. E., and Arons, J. 1988, *Ap. J.*, 332, 777.

Harwit, M., and Salpeter, E. E. 1973, *Ap. J. (Letters)*, 186, L37.

Haugan, M. P. 1988, preprint.

Hawking, S. W. 1974, *Nature*, 248, 30.

Haxton, W. C., and Johnson, C. W. 1988, *Nature*, 333, 325.

Hegyi, D., Novick, R., and Thaddeus, P. 1969, *Ap. J. (Letters)*, 158, L77.

Heiles, C., Kulkarni, S. R., Stevens, M. A., Backer, D. C., Davis, M. M., and Goss, W. N. 1983, *Ap. J. (Letters)*, 273, L75.

Heintzmann, H., and Grewing, M. 1972, *Z. Phys.*, 250, 254.

Heintzmann, H., Kundt, W., and Lasota, J. P. 1975a, *Phys. Letters*, 51A, 105.

Heintzmann, H., Kundt, W., and Lasota, J. P. 1975b, *Phys. Rev.*, A12, 204.

Heintzmann, H., and Nitsch, J. 1972, *Astr. Ap.*, 21, 291.

Heintzmann, H., and Schrüfer, E. 1977, *Phys. Letters*, A60, 79.

Helfand, D. J. 1979, *Nature*, 278, 720.

Helfand, D. J. 1983, in *Supernova Remnants and Their X-ray Emission*, ed. J. Danziger and P. Gorenstein (IAU Symp. No. 101) (Dordrecht: Reidel), p. 471.

Helfand, D. J., Ruderman, M. A., and Shaham, J. 1983, *Nature*, 304, 423.

Helfand, D. J., and Tademaru, E. 1977, *Ap. J.*, 216, 842.

Helliwell, R. A., Carpenter, D. L., and Miller, T. R. 1980, *J. Geophys. Res.*, 85, 3360.

Henrichs, H. F. 1983, in *Accretion-driven Stellar X-ray Sources*, ed. W. H. G. Lewin and E. P. J. van den Heuvel (Cambridge: Cambridge University Press), p. 393.

Henricks, H. F., and van den Heuvel, E. P. J. 1983, *Nature*, 303, 213.

Henriksen, R. N. 1970, *Ap. Letters*, 7, 89.

Henriksen, R. N., and Norton, J. A. 1975a, *Ap. J.*, 201, 431.

Henriksen, R. N., and Norton, J. A. 1975b, *Ap. J.*, 201, 719.

Henriksen, R. N., and Rayburn, D. R. 1971, *M.N.R.A.S.*, 152, 323.

Henriksen, R. N., and Rayburn, D. R. 1972, *Ap. Letters*, 11, 107.

Henriksen, R. N., and Rayburn, D. R. 1974, *M.N.R.A.S.*, 166, 409.

Henry, G. R. 1968, *Phys. Rev. Letters*, 21, 468.

Herold, H., Ruder, H., and Wunner, G. 1984, *Ap. J.*, 285, 870.

Herold, H., Ruder, H., and Wunner, G. 1985, *Phys. Rev. Letters*, 54, 1452.

Hester, J. J., and Kulkarni, S. R. 1988, *Ap. J. (Letters)*, 331, L121.

Hester, J. J., and Kulkarni, S. R. 1989, *Ap. J.*, 340, 362.

Hewish, A. 1981, in *Pulsars*, ed. W. Sieber and R. Wielebinski (IAU Conf. No. 95) (Dordrecht: Reidel), p. 1.

Hewish, A., Bell, S. J., Pilkington, J. D. M., Scott, P. F., and Collins, R. A. 1968, *Nature*, 217, 709.

Hill, T. W. 1980, *Ap. Letters*, 21, 11.

Hill, T. W., Dessler, A. J., and Maher, L. J. 1981, *J. Geophys. Res.*, 86, 9020.

Hill, T. W., and Michel, F. C. 1975, *Rev. Geophys. Space Phys.*, 13, 967.

Hillas, A. M. 1984a, *Ann. Rev. Astr. Ap.*, 22, 425.

Hillas, A. M. 1984b, *Nature*, 312, 50.

Hillebrandt, W., and Müller, E. 1976, *Ap. J.*, 207, 589.

Hills, J. G. 1970, *Nature*, 226, 730.

Hinata, S. 1973, *Ap. J.*, 186, 1027.

Hinata, S. 1976a, *Ap. J.*, 203, 223.

Hinata, S. 1976b, *Ap. Space Sci.*, 44, 389.

Hinata, S. 1976c, *Ap. J.*, 206, 282.

Hinata, S. 1977a, *Ap. J.*, 216, 101.

Hinata, S. 1977b, *Ap. Space Sci.*, 51, 303.

Hinata, S. 1978, *Ap. J.*, 221, 1003.

Hinata, S. 1979, *Ap. J.*, 227, 275.

Hinata, S., and Jackson, E. A. 1974, *Ap. J.*, 192, 703.

Hirakawa, M., Tsubono, K., and Fujimoto, M.-K. 1978, *Phys. Rev.*, D17, 1919.

Hirata, K, and 22 other authors. 1987, *Phys. Rev. Letters*, 58, 1490.

Hjorth, P. G., and O'Neil, T. M. 1987, *Phys. Fluids*, 30, 2612.

Hoffman, J. A., Lewin, W. H. G., and Doty, J. 1977, *Ap. J. (Letters)*, 217, L23.

Hog, E., and Lohsen, E. 1970, *Nature*, 227, 1229.

Holloway, N. J. 1973, *Nature* (Phys. Sci.), 246, 6.

Holloway, N. J. 1975, *M.N.R.A.S.*, 171, 619.

Holloway, N. J. 1977, *M.N.R.A.S.*, 181, 9p.

Holloway, N. J., and Pryce, M. H. L. 1981, *M.N.R.A.S.*, 194, 95.

Holt, S. S., and Ramaty, R. 1970, *Nature*, 228, 351.

Horowitz, P., Papaliolios, C., and Carleton, N. P. 1972, *Ap. J. (Letters)*, 172, L51.

Hoyle, F. 1960, *M.N.R.A.S.*, 120, 338.

Hoyle, F., and Clayton, D. D. 1974, *Ap. J.*, 191, 705.

Hoyle, F., and Narlikar, J. 1968, *Nature*, 218, 123.

Huguenin, G. R., Manchester, R. N., and Taylor, J. H. 1971, *Ap. J.*, 169, 97.

Huguenin, G. R., Taylor, J. H., and Helfand, D. F. 1973, *Ap. J. (Letters)*, 181, L139.

Hulse, R. A., and Taylor, J. H. 1975, *Ap. J. (Letters)*, 195, L51.

Hurley, K. 1986 in *Gamma-Ray Bursts*, ed. E. P. Liang and V. Petrosian (New York: AIP), p. 56.

Hylton, D. J., and Rau, A. R. P. 1980, *Phys. Rev.*, A22, 321.

Ichimaru, S. 1970, *Nature*, 226, 731.

Ichimaru, S. 1976, *Ap. J.*, 208, 701.

Ichimaru, S. 1982, *Rev. Mod. Phys.*, 54, 1017.

Ignat'ev, Yu. G. 1975, *Acta Phys.* (Poland), B6, 203.

Illarionov, A. F., and Sunyaev, R. A. 1975, *Astr. Ap.*, 39, 185.

Imamura, J. N., and Epstein, R. 1987, *Ap. J.*, 313, 711.

Imamura, J. N., Steiman-Cameron, T. Y., and Middleditch, J. 1987, *Ap. J. (Letters)*, 314, L11.

Imoto, M., and Kanai, M. 1971, *Publ. Astr. Soc. Japan*, 23, 363.

Imoto, M., and Kanai, M. 1972, *Accad. Nazion. dei Lincei*, 162, 271.

Inglis, D. R. 1955, *Rev. Mod. Phys.*, 27, 248.

Ingraham, R. L. 1973, *Ap. J.*, 186, 625.

Ipser, J. R. 1971, *Ap. J.*, 166, 175.

Ipser, J. R., and Lindblom, L. 1989, *Phys. Rev. Letters*, 62, 2777.

Isern, J., Canal, R., Hernanz, M., and Labay, J. 1987, *Ap. Space Sci.*, 131, 665.

Israel, W. 1968, *Nature*, 218, 1235.

Itoh, N., Kohyama, Y., and Takeuchi, H. 1987, *Ap. J.*, 317, 733.

Izvekova, V. A., Malov, I. F., and Malofeev, V. M. 1977, *Sov. Astr. Letters* (USA), 3, 442.

Jackson, E. A. 1976a, *Nature*, 259, 25.

Jackson, E. A. 1976b, *Ap. J.*, 206, 831.

Jackson, E. A. 1978a, *Ap. J.*, 222, 675.

Jackson, E. A. 1978b, *M.N.R.A.S.*, 183, 445.

Jackson, E. A. 1979, *Ap. J.*, 227, 266.

Jackson, E. A. 1980a, *Ap. J.*, 237, 198.

Jackson, E. A. 1980b, *Ap. J.*, 238, 1081.

Jackson, E. A. 1981a, *Ap. J.*, 247, 650.

Jackson, E. A. 1981b, *Ap. J.*, 251, 665.

Jackson, J. D. 1975, *Classical Electrodynamics* (New York: John Wiley).

Jeffreys, H. 1924, *The Earth* (New York: Cambridge University Press), p. 55ff.

Jelley, J. V., and Willstrop, R. V. 1969, *Nature*, 224, 568.

Johnson, F. S., Jr. 1965, *Satellite Environment Handbook* (Stanford: Stanford University Press).

Johnson, W. N., Harnden, F. R., and Haymes, R. C. 1972, *Ap. J. (Letters)*, 172, L1.

Johnson, W. N., and Haymes, R. C. 1973, *Ap. J.*, 184, 103.

Johnston, H. M., Kulkarni, S. R., and Oke, J. B. 1989, *Ap. J.*, 345, 492.

Jones, D. H., Smith, F. G., and Nelson, J. E. 1980, *Nature*, 283, 50.

Jones, P. B. 1975, *Ap. Space Sci.*, 33, 215.

Jones, P. B. 1976a, *Nature*, 262, 120.

Jones, P. B. 1976b, *Ap. J.*, 209, 602.

Jones, P. B. 1976c, *Ap. Space Sci.*, 45, 369.

Jones, P. B. 1977a, *M.N.R.A.S.*, 178, 879.

Jones, P. B. 1977b, *Nature*, 270, 37.

Jones, P. B. 1978, *M.N.R.A.S.*, 184, 807.

Jones, P. B. 1979, *Ap. J.*, 228, 536.

Jones, P. B. 1980a, *Ap. J.*, 236, 661.

Jones, P. B. 1980b, *M.N.R.A.S.*, 192, 847.

Jones, P. B. 1980c, *Ap. J.*, 237, 590.

Jones, P. B. 1981, *M.N.R.A.S.*, 197, 1103.

Jones, P. B. 1982, *M.N.R.A.S.*, 200, 1081.

Jones, P. B. 1983, *M.N.R.A.S.*, 153, 337.

Jones, P. B. 1985, *Phys. Rev. Letters*, 55, 1338.

Jones, P. B. 1988, *M.N.R.A.S.*, 233, 875.

Joshi, C., Mori, W. B., Katsouleas, T., Dawson, J. M., Kindel, J. M., and Forslund, D. W. 1984, *Nature*, 311, 525.

Joss, P. C. 1977, *Nature*, 270, 310.

Joss, P. C. 1978, *Ap. J. (Letters)*, 225, L123.

Joss, P. C., and Rappaport, S. A. 1983, *Nature*, 304, 419.

Joss, P. C., and Rappaport, S. A. 1984a, *Ann. Rev. Astr. Ap.*, 22, 537.

Joss, P. C., and Rappaport, S. A. 1984b, in *High Energy Transients in Astrophysics* (New York: AIP), p. 555.

Julian, W. H. 1973, *Ap. J.*, 183, 967.

Kaburaki, O. 1978, *Ap. Space Sci.*, 58, 427.

Kaburaki, O. 1980, *Ap. Space Sci.*, 67, 3.

Kaburaki, O. 1981, *Ap. Space Sci.*, 74, 333.

Kaburaki, O. 1982, *Ap. Space Sci.*, 82, 441.

Kaburaki, O. 1983, *Ap. Space Sci.*, 92, 113.

Kaburaki, O. 1986, *M.N.R.A.S.*, 220, 321.

Kaburaki, O. 1987, *M.N.R.A.S.*, 229, 165.

Kaburaki, U., and Itoh, M. 1987, *Astr. Ap.*, 172, 191.

Kadomtsev, B. B., and Kudryavtsev, V. S. 1971, *JETP Letters* (USA), 13, 42.

Kafka, P., and Meyer, F. 1984, in *High Energy Transients in Astrophysics* (New York: AIP), p. 578.

Kaiser, M. L., Desch, M. D., Warwick, J. W., and Pearce, J. B. 1980, *Science*, 209, 1238.

Kanbach, G., and 16 other authors. 1977, in *Proceedings of the 12th ESLAB Symposium Frascati* (ESA SP-124), p. 21.

Kanzanas, D., and Ellison, D. 1986, *Nature*, 319, 380.

Kaplan, S. A., and Eidman, V. Ya. 1969, *JETP Letters* (USA), 10, 203.

Kaplan, S. A., and Tsytovich, V. N. 1973a, *Nature* (Phys. Sci.), 241, 122.

Kaplan, S. A., and Tsytovich, V. N. 1973b, *Phys. Letters*, 7C, 1.

Kaplan, S A., Tsytovich, V. N., and Chikhachev, A. S. 1970, *Astrophysics* (USA), 6, 253.

Kaplan, S. A., Tsytovich, V. N., and Eidman, V. Ya. 1974, *Sov. Astr. AJ* (USA), 18, 211.

Kardashev, N. S. 1970, *Sov. Astr. AJ* (USA), 14, 375.

Karpman, V. I., Norman, C. A., ter Haar, D., and Tsytovich, V. N. 1975, *Phys. Scripta* (Sweden), 11, 271.

Katsouleas, T., and Dawson, J. M. 1983, *Phys. Rev. Letters*, 51, 392.

Kawamura, K., and Suzuki, I. 1977, *Ap. J.*, 217, 832.

Keinigs, R. 1984, *Phys. Fluids*, 27, 1427.

Kellogg, O. D. 1967, *Foundations of Potential Theory* (New York: Springer-Verlag), p. 262.

Kennel, C. F., and Coroniti, F. V. 1975, *Space Sci. Rev.*, 17, 857.

Kennel, C. F., and Coroniti, F. V. 1984a, *Ap. J.*, 283, 694.

Kennel, C. F., and Coroniti, F. V. 1984b, *Ap. J.*, 283, 710.

Kennel, C. F., Fujimura, F. S., and Okamoto, I. 1983, *Geophys. Ap. Fluid Dyn.*, 26, 147.

Kennel, C. F., and Pellat, R. 1976, *J. Plasma Phys.*, 15, 335.

Kennel, C. F., Schmidt, G., and Wilcox, T. 1973, *Phys. Rev. Letters*, 31, 1364.

Khakimova, M., Khakimov, F. K., and Tsytovich, V. N. 1976, *Astrophysics* (USA), 12, 348.

Király, P., and Mészáros, P. 1988, *Ap. J.*, 333, 719.

Kirk, J. G., and ter Haar, D. 1978, *Astr. Ap.*, 66, 359.

Kirk, J. G., and Trümper, J. 1982, in *Accretion-driven Stellar X-ray Sources*, ed. W. H. G. Lewin and E. P. J. van den Heuvel (Cambridge: Cambridge University Press).

Kirshner, R. P., and Uomoto, A. 1986, *Ap. J.*, 308, 685.

Klebesadel, R. W., and 13 other authors. 1982, *Ap. J. (Letters)*, 259, L51.

Klebesadel, R. W., Strong, I., and Olson, R. A. 1973, *Ap. J. (Letters)*, 182, L85.

Kluźniak, W., Lindblom, L., Michelson, P., and Waggoner, R. V. 1989, Columbia University preprint.

Kluźniak, W., Ruderman, M., Shaham, J., and Tavani, M. 1988a, *Nature*, 334, 225.

Kluźniak, W., Ruderman, M., Shaham, J., and Tavani, M. 1988b, *Nature*, 336, 558.

Klyakotko, M. A. 1977, *Sov. Astr. Letters* (USA), 3, 129.

Knight, F. K. 1982, *Ap. J.*, 260, 538.

Ko, H. C. 1979, *Ap. J.*, 231, 589.

Ko, H. C., and Chuang, C. W. 1978, *Ap. J.*, 222, 1012.

Kochanek, C. S., Shapiro, S. L., and Tuekolsky, S. A. 1987, *Ap. J.*, 320, 73.

Kochhar, R. K. 1977, *Nature*, 270, 38.

Kolbenstvedt, H. 1977, *Phys. Rev.*, D15, 975.

Komesaroff, M. M. 1970, *Nature*, 225, 612.

Komesaroff, M. M., Ables, J. G., and Hamilton, P. A. 1971, *Ap. Letters*, 9, 101.

Kormendy, J., and Norman, C. 1979, *Ap. J.*, 233, 539.

Kouveliotou, C., Desai, U. D., Cline, T. L., Dennis, B. R., Fenimore, E. E., Klebesadel, R. W., and Laros, J. G. 1988, *Ap. J. (Letters)*, 330, L101.

Kovalev, Yu. A. 1979, *Ap. Space Sci.*, 63, 3.

Kovalev, Yu. A. 1980, *Ap. Space Sci.*, 67, 387.

Krause-Polstorff, J., and Michel, F. C. 1985a, *Astr. Ap.*, 144, 72.

Krause-Polstorff, J., and Michel, F. C. 1985b, *M.N.R.A.S.*, 213, 43p.

Krimsky, V. 1977, *Dokl. Akad. Nauk SSSR*, 243, 1306.

Krishnamohan, S., and Downs, G. S. 1983, *Ap. J.*, 265, 372.

Kristian, J., and 14 other authors. 1989, *Nature*, 338, 234.

Kristian, J., Visvanathan, N., Westphal, J., and Snellen, G. H. 1970, *Ap. J.*, 162, 475.

Krivdik, V. G., and Jukhimuk, A. K. 1977, *Geofiz.*, SB (USSR), 79, 78.

Kriz, S. 1982, *M.N.R.A.S.*, 199, 725.

Kroll, N. M., and McMullin, W. A. 1979, *Ap. J.*, 231, 425.

Kronberg, P. P, Biermann, P., and Schwab, F. R. 1981, *Ap. J.*, 246, 751.

Kronberg, P. P, Biermann, P., and Schwab, F. R. 1985, *Ap. J.*, 291, 693.

Kulkarni, S. R. 1986, *Ap. J. (Letters)*, 306, L85.

Kulkarni, S. R., Clifton, T. R., Backer, D. C., Foster, R. S., Fruchter, A. S., and Taylor, J. H. 1988a, *Nature*, 331, 50.

Kulkarni, S. R., Djorgovski, S., and Fruchter, A. S. 1988b, *Nature*, 334, 504.

Kulkarni, S. R., and Hester, J. J. 1988b, *Nature*, 335, 801.

Kulkarni, S. R., and Narayan, R. 1988, *Ap. J.*, 335, 755.

Kulsrud, R. M. 1972, *Ap. J. (Letters)*, 174, L25.

Kulsrud, R. M. 1975, *Ap. J.*, 198, 709.

Kulsrud, R. M., and Arons, J. 1975, *Ap. J.*, 198, 709.

Kumar, N. 1969, *Phys. Letters*, 30A, 199.

Kundt, W. 1981a, *Astr. Ap.*, 98, 207.

Kundt, W. 1981b, in *Pulsars*, ed. W. Sieber and R. Wielebinski (IAU Conf. No. 95) (Dordrecht: Reidel), p. 57.

Kundt, W. 1982, in *Extragalactic Radio Sources*, ed. D. S. Heeschen and C. M. Wade (IAU Symp. No. 97) (Dordrecht: Reidel), p. 265.

Kundt, W. 1983a, *Astr. Ap.*, 121, L15.

Kundt, W. 1983b, *Proc. 18th Int. Cosmic Ray Conf.*, 12, 135.

Kundt, W. 1986, in *The Evolution of Galactic X-ray Binaries*, ed. J. Trümper et al. (Dordrecht: Reidel), p. 263.

Kundt, W., and Krotscheck, E. 1980, *Astr. Ap.*, 83, 1.

Kundt, W., and Robnik, M. 1980, *Astr. Ap.*, 91, 305.

Kundu, M. R., and Chitre, S. M. 1968, *Nature*, 218, 1037.

Kuo-Petravic, L. G., and Petravic, M. 1976, *Phys. Rev. Letters*, 36, 686.

Kuo-Petravic, L. G., Petravic, M., and Roberts, K. V. 1974, *Phys. Rev. Letters*, 32, 1019.

Kuo-Petravic, L. G., Petravic, M., and Roberts, K. V. 1975, *Ap. J.*, 202, 762.

Lamb, D. Q., and Lamb, F. K. 1976, *Ap. J.*, 204, 168.

Lamb, D. Q., and Lamb, F. K. 1978, *Ap. J.*, 220, 291.

Lamb, D. Q., Lamb, F. K., and Pines, D. 1973, *Nature* (Phys. Sci.), 246, 52.

Lamb, F. K. 1981, in *Pulsars*, ed. W. Sieber and R. Wielebinski (IAU Conf. No. 95) (Dordrecht: Reidel), p. 303.

Lamb, F. K. 1988, *Adv. Space Res.*, 8, (2) 421.

Lamb, F. K. 1989, in *Timing Neutron Stars*, ed. H. Ögelman and E. P. J. van den Heuvel (Norwell, MA: Kluwer Academic Publishers), p. 649.

Lamb, F. K., Pethick, C. J., and Pines, D. 1973, *Ap. J.*, 184, 271.

Lamb, F. K., Shibazaki, N., Alpar, M. A., and Shaham, J. 1985, *Nature*, 317, 681.

Lamb, R. C., Fichtel, C. E., Hartman, R. C., Kniffen, D. A., and Thompson, D. J. 1977, *Ap. J. (Letters)*, 212, L63.

Lamb, R. C., Godfrey, C. P., Wheaton, W. A., and Tümer, T. 1982, *Nature*, 296, 543.

Landau, L. 1932, *Phys. Z. Soviet*, 1, 88.

Landau, L. D., and Lifshitz, E. M. 1958, *Physics of Fluids* (Reading: Addison-Welsey).

Landstreet, J. D., and Angel, J. R. P. 1971, *Nature*, 230, 103.

Langdon, A. B., Arons, J., and Max, C. 1988, *Phys. Rev. Letters*, 61, 779.

Laros, J. G., and 12 other authors. 1986, *Nature*, 322, 152.

Laros, J. G., and 16 other authors. 1987, *Ap. J. (Letters)*, 320, L111.

Laros, J. G., Fenimore, E. E., Fikani, M. M., Klebesadel, R. W., van der Klis, M., and Gottwald, M. 1985, *Nature*, 318, 448.

Larroche, O., and Pellat, R. 1988a, *Phys. Rev. Letters*, 59, 1104.

Larroche, O., and Pellat, R. 1988b, *Phys. Rev. Letters*, 61, 650.

Lasker, B. M. 1976, *Ap. J.*, 203, 193.

Lattimer, J. M., and Schramm, D. N. 1976, *Ap. J.*, 210, 549.

Layzer, D. 1968, *Nature*, 220, 247.

Leboeuf, J. N., Ashour-Abdalla, M, Tajima, T., Kennel, C. F., Coroniti, F. V., and Dawson, J. M. 1982, *Phys. Rev.*, A25, 1023.

Lee, H. J., Canuto, V., Chiu H.-Y., and Chiuderi, C. 1969, *Phys. Rev. Letters*, 23, 390.

Lee, M. A. 1974, *Ap. J.*, 194, 165.

Lee, M. A., and Lerche, I. 1974, *Ap. J.*, 194, 409.

Lee, M. A., and Lerche, I. 1975, *Ap. J.*, 198, 477.

Le Guillou, J. C., and Zinn-Justin, J. 1983, *Ann. Phys.*, 147, 57.

Lerche, I. 1970a, *Ap. J.*, 159, 229.

Lerche, I. 1970b, *Ap. J.*, 160, 1003.

Lerche, I. 1970c, *Ap. J.*, 162, 153.

Lerche, I. 1971, in *The Crab Nebula* (IAU Symp. No. 46) (Dordrecht: Reidel), p. 449.

Lerche, I. 1974a, *Ap. J.*, 187, 589.

Lerche. I. 1974b, *Ap. J.*, 187, 597.

Lerche, I. 1974c, *Ap. J.*, 188, 627.

Lerche, I. 1974d, *Ap. J.*, 191, 191.

Lerche, I. 1974e, *Ap. J.*, 191, 753.

Lerche, I. 1974f, *Ap. J.*, 191, 759.

Lerche, I. 1974g, *Ap. J.*, 191, 763.

Lerche, I. 1974h, *Ap. J.*, 194, 177.

Lerche, I. 1974i, *Ap. J.*, 194, 403.

Lerche, I. 1975a, *Phys. Rev.*, D11, 740.

Lerche, I. 1975b, *Ap. J. Suppl.*, 29, 113.

Lerche, I. 1975c, *Ap. Space Sci.*, 35, 363.

Lerche, I. 1975d, *Ap. J.*, 199, 734.

Lerche, I. 1976, *Ap. Space Sci.*, 41, 387.

Lerche, I., and Schramm, D. N. 1977, *Ap. J.*, 216, 881.

Leventhal, M., MacCallum, C. J., Hutters, A. F., and Stang, P. D. 1980, *Ap. J.*, 240, 338.

Levine, J., and Stebbins, R. 1972, *Phys. Rev.*, D6, 1465.

Levy, E. H. 1976, *Ann. Rev. Earth Plant. Sci.*, 4, 159.

Lewin, W. H. G. 1981, *Sci. Am.*, 244 (5), 72.

Lewin, W. H. G., and Joss, P. C. 1981, *Space Sci. Rev.*, 28, 3.

Lewin, W. H. G., and van den Heuvel, E. P. J. (eds.). 1983, *Accretion-driven Stellar X-ray Sources* (Cambridge: Cambridge University Press).

Lewin, W. H. G., and van Paradijs, J. 1986, *Comments Ap.*, 11, 127.

L'Heureux, J., and Mayer, P. 1976, *Ap. J.*, 209, 955.

Li, W. H., and Lam, S. H. 1964, *Principles of Fluid Mechanics* (Reading: Addison-Wesley), p. 160.

Liang, E. 1987, *Comments Ap.*, 12, 35.

Liang, E. 1989, *GRO Science Workshop Proc.* (April 10-12, Greenbelt MD), in press.

Liang, E., and Antiochos, S. K. 1984, *Nature*, 310, 121.

Liang, E., and Nolan, P. 1984, *Space Sci. Rev.*, 38, 353.

Lighthill, J. 1978, *Waves in Fluids* (New York: Cambridge University Press).

Lindblom, L. 1984, *Ap. J.*, 278, 368.

Lingenfelter, R. E., and Ramaty, R. 1989, in *The Galactic Center*, ed. M. R. Morris (IAU Symp. No. 136) (Dordrecht: Reidel).

Linscott, J. R., and Backer, P. B. 1982, *Bull. Am. Astr. Soc.*, 14, 661.

Lipunov, V. M. 1978, *Sov. Astr. AJ* (USA), 22. 702.

Livio, M., and Taam, R. E. 1987, *Nature*, 398, 398.

Lloyd-Evans, J., Coy, R. N., Lambert, A., Lapikens, J., Patel, M., Reid, R. J. O., and Watson, A. A. 1983, *Nature*, 305, 784.

Lodenquai, J. 1984, *Ap. Space Sci.*, 24, 91.

Lohsen, E. H. G. 1981, *Astr. Ap. Suppl.*, 44, 1.

Lominadze, J. G., Machabeli, G. Z., and Mikhailovskii, A. B. 1979a, *Sov. J. Plasma Phys.* (USA), 5, 1337.

Lominadze, J. G., Machabeli, G. Z., and Mikhailovskii, A. B. 1979b, *J. Phys. Colloq.* (France), 40, C7/713.

Lominadze, J. G., and Mikhailovskii, A. B. 1979, *Sov. Phys. JETP*, 49, 483.

Long, K. S., Helfand, D. J., and Grabelsky, D. A. 1981, *Ap. J.*, 248, 925.

Loskutov, Yu. M., and Skobelev, V. V. 1976, *Theor. Math. Phys.*, 29, 932.

Loskutov, Yu. M., and Skobelev, V. V. 1980, *Sov. J. Nucl. Phys.* (Netherlands), 31, 1279.

Lovelace, R. V. 1973, *Bull. Am. Astr. Soc.*, 5, 426.

Lovelace, R. V. E. 1976, *Nature*, 262, 649.

Lu, D.-J., and Gao, J.-G. 1976, *Acta Phys. Sin.* (China), 25, 181.

Lu, K. U. 1976, *Int. J. Theor. Phys.*, 15, 411.

Lubow, S. H. 1981, *Ap. J.*, 245, 274.

Lubow, S. H., and Shu, F. H. 1976, *Ap. J. (Letters)*, 207, L53.

Luheshi, M., and Stewart, P. 1979, *Astr. Ap.*, 75, 185.

Lynden-Bell, D. 1969, *Nature*, 223, 690.

Lyne, A. G. 1984, *Nature*, 310, 300.

Lyne, A. G. 1987, *Nature*, 326, 569.

Lyne, A. G., Anderson, B., and Salter, M. J. 1982, *M.N.R.A.S.*, 201, 503.

Lyne, A. G., and Ashworth, M. 1983, *M.N.R.A.S.*, 204, 519.

Lyne, A. G., Biggs, J. D., Brinklow, A., Ashworth, M., and McKenna, J. 1988a, *Nature*, 332, 45.

Lyne, A. G., Brinklow, A., Middleditch, J., Kulkarni, S. R., Backer, D. C., and Clifton, T. R. 1987, *Nature*, 328, 399.

Lyne, A. G., and Manchester, R. N. 1988, *M.N.R.A.S.*, 234, 477.

Lyne, A. G., Manchester, R. N., and Taylor, J. H. 1985, *M.N.R.A.S.*, 213, 613.

Lyne, A. G., Pritchard, R. S., and Smith, F. G. 1988b, *M.N.R.A.S.*, 233, 667.

Lyne, A. G., Ritchings, R. T., and Smith, F. G. 1975, *M.N.R.A.S.*, 171, 579.

Lyne, A. G., and Smith, F. G. 1979, *M.N.R.A.S.*, 188, 675.

Lyne, A. G., and Smith, F. G. 1982, *Nature*, 298, 825.

Lyne, A. G., Smith, F. G., and Graham, D. A. 1971, *M.N.R.A.S.*, 153, 337.

Machabeli, G. Z., and Usov, V. V. 1979, *Sov. Astr. Letters* (USA), 5, 445.

MacMillan, W. D. 1958, *The Theory of the Potential* (New York: Dover), pp. 17 and 45.

Macy, W. W., Jr. 1974, *Ap. J.*, 190, 153.

Madore, B. F. 1980, in *Globular Clusters*, ed. D. Hanes and B. Madore (Cambridge: Cambridge University Press), p. 28.

Madsen, J. 1988, *Phys. Rev. Letters*, 61, 2909.

Mahoney, M. J., and Erickson, W. C. 1985, *Nature*, 317, 154.

Mahoney, W. A., Ling, J. C., and Jacobson, A. S. 1984, *Ap. J.*, 278, 784.

Malofeev, V. M., and Shitov, Yu. P. 1981, *Astr. Ap.*, 78, 45.

Malov, I. F. 1979, *Sov. Astr. Letters* (USA), 5, 177.

Malov, I. F., and Malofeev, V. M. 1977, *Sov. Astr. AJ* (USA), 21, 55.

Manchester, R N. 1978, *Proc. Astr. Soc. Austral.*, 3, 200.

Manchester, R. N. 1972, *Ap. J.*, 172, 43.

Manchester, R. N. 1974, *Science*, 186, 66.

Manchester, R. N., and Durdin, J. M. 1983, in *Supernova Remnants and Their X-ray Emission*, ed. J. Danziger and P. Gorenstein (IAU Symp. No. 101) (Dordrecht: Reidel), p. 421.

Manchester, R. N., Durdin, J. M., and Newton, L. M. 1985, *Nature*, 313, 374.

Manchester, R. N., Newton, L. M., Cooke, D. J., and Lyne, A. G. 1980, *Ap. J. (Letters)*, 236, L25.

Manchester, R. N., Newton, L. M., Cooke, D. J., Backus, P. R., Damashek, M., Taylor, J. H., and Condon, J. J. 1983, *Ap. J.*, 268, 832.

Manchester, R. N., Newton, L. M., Goss, W. M., and Hamilton, P. A. 1978, *M.N.R.A.S.*, 184, 35p.

Manchester, R. N., and Peterson, B. A. 1989, *Ap. J. (Letters)*, 342, L23.

Manchester, R. N., and Tademaru, E. 1971, *Nature* (Phys. Sci.), 232, 164.

Manchester, R. N., and Taylor, J. H. 1977, *Pulsars* (San Francisco: Freeman).

Manchester, R. N., and Taylor, J. H. 1981, *Astr. J.*, 86, 1953.

Manchester, R. N., Tademaru, E., Taylor, J. H., and Huguenin, G. R. 1973, *Ap. J.*, 185, 951.

Manchester, R. N., Taylor, J. H., and Van, Y. Y. 1974, *Ap. J. (Letters)*, 189, L119.

Manchester, R. N., Tuohy, I. R., and D'Amico, N. 1982, *Ap. J. (Letters)*, 262, L31.

Maraschi, L., and Cavaliere, A. 1977, *Highlights in Astronomy*, 4, 127.

Maraschi, L., and Treves, A. 1974, *Proc. Int. Conf. Supernovae and Supernova Remnants* (Dordrecht: Reidel), p. 307.

Margolis, S. H., Schramm, D. N., and Silberberg, R. 1978, *Ap. J.*, 221, 990.

Margon, B. 1984, *Ann. Rev. Astr. Ap.*, 22, 507.

Marshak, M. L., and 13 other authors. 1985, *Phys. Rev. Letters*, 54, 2079.

Mason, K. O., Watson, M. G., and White, N. E. (eds.). 1987, in *Proc. ESA Workshop on the Physics of Accetion onto Compact Objects* (Hamburg: Springer).

Massaro, E., and Salvati, M. 1979, *Astr. Ap.*, 71, 51.

Massaro, E., Salvati, M., and Buccheri, R. 1979, *M.N.R.A.S.*, 189, 823.

Massey, B. S. 1975, *Mechanics of Fluids* (3rd ed., New York: Van Nostrand and Reinhold), p. 345.

Massnou, J. 1980, private communication.

Mast, T. S., Nelson, J. E., and Saarloos, J. A. 1974, *Ap. J. (Letters)*, 187, L49.

Mast, T. S., Nelson, J. E., Saarloos, J., Muller, R. A., and Bolt, B. A. 1972, *Nature*, 240, 140.

Mastichiadis, A., Brecher, K., and Marscher, A. P. 1987, *Ap. J.*, 314, 88.

Matese, J. J., and Whitmire, D. P. 1980, *Ap. J.*, 235, 587.

Max, C. 1973, *Phys. Fluids*, 16, 1277.

Max, C., and Perkins, F. 1971, *Phys. Rev. Letters*, 27, 1342.

Max, C., and Perkins, F. 1972, *Phys. Rev. Letters*, 29, 1731.

Maxwell, O. V. 1979, *Ap. J.*, 231, 201.

Mazets, E. P., and 11 other authors. 1981, *Ap. Space Sci.*, 80, 3.

Mazets, E. P., Goletnetskii, S. V., and Gur'yan, Yu. A. 1980, *Sov. Astr. Letters*, 5, 343.

Mazets, E. P., Golenetskii, S. V., Gur'yan, Yu. A., and Ilyinski, V. N. 1982, *Ap. Space Sci.*, 84, 173.

McClintock, J. E., and Leventhal, M. 1989, *Ap. J.*, 346, 143.

McCrea, W. H. 1972, *M.N.R.A.S.*, 157, 359.

McCulloch, P. M., Hamilton, P. A., Ables, J. G., and Hunt, A. J. 1983, *Nature*, 303, 307.

McIlraith, A. H. 1968, *Nature*, 220, 461.

McKee, C. F., and Ostriker, J. P. 1977, *Ap. J.*, 218, 148.

McKenna, J., and Lyne, A. G. 1988, *Nature*, 336, 326.

Melia, F. 1988a, *Ap. J. (Letters)*, 334, L9.

Melia, F. 1988b, *Ap. J.*, 335, 965.

Melia, F. 1988c, *Nature*, 336, 658.

Melosh, H. L. 1969, *Nature*, 224, 781.

Melrose, D. B. 1978, *Ap. J.*, 225, 557.

Melrose, D. B. 1979, *Austral. J. Phys.*, 32, 61.

Melrose, D. B., and Parle, A. J. 1983a, *Austral. J. Phys.*, 36, 755.

Melrose, D. B., and Parle, A. J. 1983b, *Austral. J. Phys.*, 36, 775.

Melrose, D. B., and Parle, A. J. 1983c, *Austral. J. Phys.*, 36, 799.

Melrose, D. B., and Stoneham, R. J. 1977, *Proc. Astr. Soc. Austral.*, 3, 120.

Melzer, D. W., and Thorne, K. S. 1966, *Ap. J.*, 145, 514.

Mertz, L. 1974, *Ap. Space Sci.*, 30, 43.

Mestel, L. 1966, *Congré Colloq. Univ. Liège*, 41, 351.

Mestel, L. 1968, *M.N.R.A.S.*, 140, 177.

Mestel, L. 1971, *Nature* (Phys. Sci.), 233, 149.

Mestel, L. 1973, *Ap. Space Sci.*, 24, 289.

Mestel, L., Phillips, P., and Wang, Y.-M. 1979, *M.N.R.A.S.*, 188, 385.

Mestel, L., Robertson, J. A., Wang, Y.-M., and Westfold, K. C. 1985, *M.N.R.A.S.*, 217, 443.

Mestel, L., and Wang, Y.-M. 1979, *M.N.R.A.S.*, 188, 799.

Mestel, L., and Wang, Y.-M. 1982, *M.N.R.A.S.*, 198, 405.

Mestel, L., Wright, G. A. E, and Westfold, K. C. 1976, *M.N.R.A.S.*, 175, 257.

Meyer, B. S. 1989, *Ap. J.*, 343, 254.

Meyer, F., Duschl, W., Frank, J., and Meyer-Hofmeister, E. (eds.). 1990, *Theory of Accretion Disks* (Munich: Kluwer).

Meyer, F., and Meyer-Hofmeister, E. 1981, *Astr. Ap.*, 104, L10.

Michel, F. C. 1969a, *Phys. Rev. Letters*, 23, 247.

Michel, F. C. 1969b, *Ap. J.*, 158, 727.

Michel, F. C. 1970a, *Nature*, 228, 1072.

Michel, F. C. 1970b, *Comments Ap. Space Phys.*, 2, 227.

Michel, F. C. 1970c, *Ap. J. (Letters)*, 159, L25.

Michel, F. C. 1971, *Comments Ap. Space Phys.*, 3, 80.

Michel, F. C. 1972, *Ap. Space Sci.*, 15, 153.

Michel, F. C. 1973a, *Ap. J. (Letters)*, 180, L133.

Michel, F. C. 1973b, *Ap. J.*, 180, 207.

Michel, F. C. 1974a, *Phys. Rev. Letters*, 33, 1521.

Michel, F. C. 1974b, *Ap. J.*, 187, 585.

Michel, F. C. 1974c, *Ap. J.*, 192, 713.

Michel, F. C. 1975a, *Ap. J. (Letters)*, 195, L69.

Michel, F. C. 1975b, *Ap. J.*, 196, 579.

Michel, F. C. 1975c, *Ap. J.*, 197, 193.

Michel, F. C. 1975d, *Ap. J.*, 198, 683.

Michel, F. C. 1977a, *Ap. J.*, 214, 261.

Michel, F. C. 1977b, *Ap. J.*, 213, 836.

Michel, F. C. 1977c, *Ap. J.*, 216, 838.

Michel, F. C. 1977d, *J. Geophys. Res.*, 82, 5181.

Michel, F. C. 1978a, *Ap. J.*, 220, 1101.

Michel, F. C. 1978b, *Ap. J.*, 224, 988.

Michel, F. C. 1979a, *Space Sci. Rev.*, 24, 381.

Michel, F. C. 1979b, *Ap. J.*, 227, 579.

Michel, F. C. 1980, *Ap. Space Sci.*, 72, 175.

Michel, F. C. 1981, *Ap. J.*, 251, 654.

Michel, F. C. 1982, *Rev. Mod. Phys.*, 54, 1.

Michel, F. C. 1983a, *Ap. J.*, 247, 664.

Michel, F. C. 1983b, *Ap. J.*, 266, 188.

Michel, F. C. 1984a, *Ap. J.*, 279, 807.

Michel, F. C. 1984b, *Ap. J.*, 284, 384.

Michel, F. C. 1985a, *Ap. J.*, 288, 138.

Michel, F. C. 1985b, *Bull. Am. Astr. Soc.*, 17, 908.

Michel, F. C. 1985c, *Ap. J.*, 290, 721.

Michel, F. C. 1985d, *Proc. Astr. Soc. Austral.*, 6, 127.

Michel, F. C. 1986, *Phys. Today*, 39 (10), 9.

Michel, F. C. 1987a, *Ap. J.*, 312, 271.

Michel, F. C. 1987b, *Nature*, 329, 310.

Michel, F. C. 1987c, *Ap. J.*, 321, 714.

Michel, F. C. 1987d, *Ap. J.*, 322, 822.

Michel, F. C. 1988a, *Phys. Rev. Letters*, 60, 677.

Michel, F. C. 1988b, *Ap. J. (Letters)*, 327, L81.

Michel, F. C. 1988c, in *Nuclear Spectroscopy of Astophysical Sources* (New York: AIP), p. 307.

Michel, F. C. 1988d, *Comments Ap.*, 12, 191.

Michel, F. C. 1988e, *Nature*, 333, 644.

Michel, F. C. 1989, *Nature*, 337, 236.

Michel, F. C., and Dessler, A. J. 1981a, *Ap. J.*, 251, 654.

Michel, F. C., and Dessler, A. J. 1981b, *Proc. 17th Int. Cosmic Ray Conf.* (Paris), 2, 340.

Michel, F. C., and Dessler, A. J. 1983, *Nature*, 303, 48.

Michel, F. C., and Goldwire, H. C., Jr. 1970, *Ap. Letters*, 5, 21.

Michel, F. C., Kennel, C. F., and Fowler, W. A. 1987, *Science*, 238, 938.

Michel, F. C., and Pellat, R. 1981, in *Pulsars*, ed. W. Sieber and R. Wielcbinski (IAU Conf. No. 95) (Dordrecht: Reidel), p. 37.

Michel, F. C., and Sturrock, P. A. 1974, *Planetary Space Sci.*, 22, 1501.

Michel, F. C., and Tucker, W. H. 1969, *Nature*, 223, 277.

Michel, F. C., and Yahil, A. 1973, *Ap. J.*, 179, 771.

Middleditch, J., and 13 other authors. 1989, *IAU Circ.*, No. 4735.

Middleditch, J., and 9 other authors. 1983a, *Nature*, 306, 163.

Middleditch, J., and Kristian, J. 1984, *Ap. J.*, 279, 157.

Middleditch, J., and Pennypacker, C. 1985, *Nature*, 313, 659.

Middleditch, J., Pennypacker, C., and Burns, M. S. 1983b, *Ap. J.*, 273, 261.

Middleditch, J., Pennypacker, C., and Burns, M. S. 1987, *Ap. J.*, 315, 142.

Midgley, J. E., and Davis, L., Jr. 1962, *J. Geophys. Res.*, 67, 499.

Mikhailovskii, A. B. 1980, *Sov. J. Plasma Phys.* (USA), 22 133.

Miller, R. H., Lasker, B M., Hesser, J. E., and Bracker, S. B. 1975, *Ap. J.*, 196, 121.

Minkowski, R. 1942, *Ap. J.*, 96, 199.

Misner, C. W., Thorne, K. S., and Wheeler, J. A. 1973, *Gravitation* (San Francisco: Freeman).

Modisette, J. L. 1967, *J. Geophys. Res.*, 72, 1521.

Møller, C. 1952, *The Theory of Relativity* (London: Oxford Press), p. 328.

Møller, C., and Chandrasekhar, S. 1935, *M.N.R.A.S.*, 95, 673.

Molnar, L. A. 1988, *Adv. Space Res.*, 8, 605.

Molnar, L. A., Reid, M. J., and Grindlay, J. E. 1984, *Nature*, 310, 662.

Morfill, G. E., Trümper, J., Bodenheimer, P., and Tenorio-Tagle, G. 1984, *Astr. Ap.*, 139, 7.

Morozov, A. I., and Solev'ev, L. S. 1966, *Reviews of Plasma Physics* (New York: Plenum), p. 95.

Morris, C. D., Jr. 1975, Rice University Ph.D. thesis.

Morris, D., Graham, D. A., and Bartel, N. 1981, *M.N.R.A.S.*, 194, 7p.

Morris, D., Radhakrishnan, V., and Shukre, C. 1976, *Nature*, 260, 124.

Morris, D., Radhakrishnan, V., and Shukre, C. 1978, *Astr. Ap.*, 68, 289.

Morrison, P. 1969, *Ap. J. (Letters)*, 157, L73.

Morton, D. C. 1964, *Ap. J.*, 140, 460.

Mueller, R. O., Rau, A. R. P., and Spruch, L. 1971, *Phys. Rev. Letters*, 26, 1136.

Mueller, R. O., Rau, A. R. P., and Spruch, L. 1975, *Phys. Rev.*, A11, 789.

Mulholland, J. D. 1971, *Ap. J.*, 165, 105.

Mullan, D. 1976, *Ap. J.*, 208, 199.

Murakami, T., and 12 other authors. 1988, *Nature*, 335, 234.

Nagel, W. 1981, *Ap. J.*, 251. 278.

Nakamura, Y. 1980, *Sci. Rep. Tohoku Univ.* I Ser. (Japan), 62, 121.

Nakazato, and 14 other authors. 1989, *Phys. Rev. Letters*, 63, 1245.

Nandkumar, R., and Pethick, C. J. 1984, *M.N.R.A.S.*, 209, 511.

Narayan, R. 1984, in *Millisecond Pulsars*, ed. S. P. Reynolds and D. R. Stinebring (Green Bank: National Radio Astronomy Observatory), p. 279.

Narayan, R., and Ostriker, J. P. 1989, *Ap. J.*, submitted.

Narayan, R., and Vivekanand, M. 1981, *Nature*, 290, 571.

Narayan, R., and Vivekanand, M. 1982, *Astr. Ap.*, 113, L3.

Narayan, R., and Vivekanand, M. 1983a, *Ap. J.*, 274, 771.

Narayan, R., and Vivekanand, M. 1983b, *Astr. Ap.*, 122, 45.

Narlikar, J. V., Apparao, K. M. V., and Dadhich, N. 1974, *Nature*, 251, 591.

Nerney, S. 1980, *Ap. J.*, 242, 723.

Neuhauser, D., Koonin, S. E., and Langanke, K. 1987, *Phys. Rev.*, 36A, 4163.

Newton, R. G. 1971, *Phys. Rev.*, D3, 626.

Nomoto, K. 1981, in *Fundamental Problems in the Theory of Stellar Evolution*, ed. D. Sugimoto, D. Q. Lamb, and D. N. Schramm (IAU Symp. No. 93) (Dordrecht: Reidel), p. 295.

Nomoto, K., and Tsuruta, S. 1981, *Ap. J. Letters*, 250, L19.

Nordtvedt, K., Jr. 1975, *Ap. J.*, 202, 248.

Novick, R., Weisskopf, M. C., Angel, J. R. P., and Sutherland, P. G. 1977, *Ap. J. (Letters)*, 215, L117.

Nulsen, P. E. J., and Fabian, A. C. 1984, *Nature*, 312, 48.

Ochelkov, Yu. P., Rozental', I. L. and Shukalov, I. B. 1972, *Sov. Astr. AJ* (USA), 16, 244.

Ochelkov, Yu. P., and Usov, V. V. 1979, *Sov. Astr. Letters* (USA), 5, 180.

Ochelkov, Yu. P., and Usov, V. V. 1980a, *Ap. Space Sci.*, 69, 439.

Ochelkov, Yu. P., and Usov, V. V. 1980b, *Sov. Astr. Letters* (USA), 6, 414.

Ochelkov, Yu. P., and Usov, V. V. 1984, *Nature*, 309, 332.

O'Connell, R. F. 1975, *Ap. J.*, 195, 751.

O'Connell, R. F., and Roussel, K. M. 1972, *Astr. Ap.*, 18, 198.

Ögelman, H., Koch-Miramond, L., and Auriére, M. 1989, *Ap. J. (Letters)*, submitted (MPE preprint 151).

Ögelman, M., Fitchel, C. E., Kniffen, D. A., and Thompson, D. J. 1976, *Ap. J.*, 209, 548.

Oide, K., Hirakawa, H., and Fujimoto, M.-K. 1979, *Phys. Rev.*, D20, 2480.

Okamoto, I. 1974, *M.N.R.A.S.*, 167, 457.

Okamoto, I. 1975, *M.N.R.A.S.*, 170, 81.

Okamoto, I. 1978, *M.N.R.A.S.*, 185, 69.

Olson, W. P. (ed.). 1979, *Quantitative Modeling of Magnetospheric Processes* (Washington: American Geophysical Union).

O'Neil, T. M. 1980, *Phys. Fluids*, 23, 2216.

O'Neil, T. M., and Hjorth, P. G. 1985, *Phys. Fluids*, 28, 3241.

Onishchenko, G. 1975, *Sov. Astr. AJ* (USA), 19, 171.

Onishchenko, O. G. 1979, *Astrophysics* (USA), 15, 169.

Oppenheimer, J. R., and Snyder, H. 1939, *Phys. Rev.*, 56, 455.

Oppenheimer, J. R., and Volkoff, G. M. 1939, *Phys. Rev.*, 55, 374.

Osborne, J. L., and Wolfendale, A. W. (eds.). 1975, *Origin of Cosmic Rays* (Dordrecht: Reidel).

Oster, L. 1975, *Ap. J.*, 196, 571.

Oster, L., Hilton, D. A., and Sieber, W. 1976a, *Astr. Ap.*, 57, 1.

Oster, L., Hilton, D. A., and Sieber, W. 1976b, *Astr. Ap.*, 57, 323.

Oster, L., and Sieber, W. 1976a, *Ap. J.*, 203, 233.

Oster, L., and Sieber, W. 1976b, *Ap. J.*, 210, 220.

Oster, L., and Sieber, W. 1978, *Astr. Ap.*, 65, 179.

Ostriker, J. 1968, *Nature*, 217, 1227.

Ostriker, J. P. 1987, *Nature*, 327, 287.

Ostriker, J. P., and Gunn, J. E. 1969a, *Ap. J.*, 157, 1395.

Ostriker, J. P., and Gunn, J. E. 1969b, *Ap. J.*, 160, 1395.

Ostriker, J. P., and Gunn, J. E. 1969c, *Ap. J. (Letters)*, 164, L95.

Ostriker, J. P., and Gunn, J. E. 1969d, *Ap. J.*, 165, 523.

Ostriker, J. P., and Gunn, J. E. 1969e, *Nature*, 223, 813.

Ostriker, J. P., and Gunn, J. E. 1971, *Ap. J. (Letters)*, 164, L95.

Ostriker, J. P., Rees, M. J., and Silk, J. 1970, *Ap. Letters*, 6, 179.

Ostriker, J. P., and Tassoul, J.-L. 1968, *Nature*, 219, 577.

Özel, M. E., and Ormes, J. F. 1989, *Astr. Ap.*, in press.

Ozernoi, L. M., and Usov, V. V. 1972, *Sov. Astr. AJ* (USA), 17, 270.

Ozernoi, L. M., and Usov, V. V. 1973a, *Ap. Space Sci.*, 25, 149.

Ozernoi, L. M., and Usov, V. V. 1973b, *Ap. Letters*, 13, 151.

Ozernoi, L. M., and Usov, V. V. 1977, *Sov. Astr. AJ* (USA), 21, 425.

Pacini, F. 1967, *Nature*, 216, 567.

Pacini, F. 1968, *Nature*, 219, 145.

Pacini, F. 1969, *Nature*, 224, 160.

Pacini, F., and Rees, M. J. 1970, *Nature*, 226, 622.

Pacini, F., and Ruderman, M. 1974, *Nature*, 251, 399.

Pacini, F., and Salpeter, E. E. 1968, *Nature*, 218, 733.

Pacini, F., and Salvati, M. 1973, *Ap. J.*, 186, 249.

Pacini, F., and Salvati, M. 1983, *Ap. J.*, 274, 369.

Paczyński, B. 1976, *Comments Ap.*, 6, 495.

Paczyński, B. 1983, *Nature*, 304, 421.

Paczyński, B. 1986, *Ap. J. (Letters)*, 308, L43.

Page, D., and Hawking, S. 1976, *Ap. J.*, 206, 1.

Panofsky, W. K. H., and Phillips, M. 1955, *Classical Electricity and Magnetism* (Reading: Addison-Wesley).

Papini, G., and Valluri, S.-R. 1975, *Can. J. Phys.*, 53, 2312.

Papoyan, V. V., Sedrakyan, D. M., and Chubaryan, E. V. 1973, *Sov. Astr. AJ* (USA), 16, 615.

Parish, J. L. 1974, *Ap. J.*, 193, 225.

Parker, E. N. 1955, *Ap. J.*, 122, 293.

Parker, E. N. 1958, *Phys. Rev.*, 109, 1325.

Parker, L., and Tiomno, J. 1972a, *Ap. J.*, 178, 809.

Parker, L., and Tiomno, J. 1972b, *Nature* (Phys. Sci.), 238, 57.

Pavlov, G. G., and Shibanov, Yu. A. 1978, *Sov. Astr. AJ* (USA), 22, 214.

Pavlov, G. G., and Shibanov, Yu. A. 1979, *Sov. Phys. JETP* (USA), 49, 741.

Pechenick, K. R., Ftaclas, C., and Cohen, J. M. 1983, *Ap. J.*, 274, 846.

Peek, B. M. 1942, *J. Brit. Astr. Assoc.*, 53, 23.

Peierls, R. 1936, *M.N.R.A.S.*, 96, 780.

Pelizzari, M. A. 1975, Rice University M.S. thesis.

Pelizzari, M. A. 1976, Rice University Ph.D. thesis.

Peratt, A. L., and Dessler, A. J. 1988, *Ap. Space Sci.*, 144, 451.

Peterson, B. A., and 7 other authors. 1978, *Nature*, 276, 475.

Petravic, M. 1976, *Comput. Phys. Commun.* (Netherlands), 12, 9.

Petre, R., Szymkowiak, A. E., Seward, F. D., and Willingale, R. 1988, *Ap. J.*, 335, 215.

Petrich, L. I., Shapiro, S. L., and Teukolsky, S. A. 1988, *Phys. Rev. Letters*, 60, 1781.

Petschek, A. G., Colgate, S. A., and Colvin, J. D. 1976, *Ap. J.*, 209, 356.

Pfarr, J. 1972, *Z. Phys.*, 251, 152.

Pfarr, J. 1976, *Gen. Rel. Grav.*, 7, 459.

Phinney, E. S., and Blandford, R. D. 1981, *M.N.R.A.S.*, 194, 137.

Phinney, E. S., Evans, C. R., Blandford, R. D., and Kulkarni, S. R. 1988, *Nature*, 333, 832.

Piddington, J. H. 1957, *Austral. J. Phys.*, 10, 530.

Piddington, J. H. 1969, *Nature*, 222. 965.

Pilipp, W. G. 1974, *Ap. J.*, 190, 391.

Pineault, S. 1986, *Ap. J.*, 301, 145.

Pines, D., and Shaham, J. 1974, *Comments Ap. Space Sci.*, 6, 37.

Piran, T., and Shaham, J. 1975, *Nature*, 256, 112.

Pneumann, G. W., and Kopp, R. A. 1971, *Solar Phys.*, 18, 258.

Pollack, J. B., Guthrie, P. D., and Shen, B. S. P. 1971, *Ap. J. (Letters)*, 169, L113.

Prakash, M., Ainsworth, T. L., and Lattimer, J. M. 1988, *Phys. Rev. Letters*, 61, 2518.

Pringle, J. E. 1981, *Ann. Rev. Astr. Ap.*, 19, 137.

Pringle, J. E., and Rees, M. J. 1972, *Astr. Ap.*, 21, 1.

Prószyński, M. 1979, *Astr. Ap.*, 79, 8.

Prószyński, M., and Przybycień, D. F. 1984, in *Millisecond Pulsars*, ed. S. P. Reynolds and D. R. Stinebring (Green Bank: National Radio Astronomy Observatory), p. 151.

Protheroe, R. J. 1984, *Nature*, 310, 296.

Protheroe, R. J., and Clay, R. W. 1985, *Nature*, 315, 205.

Protheroe, R. J., Clay, R. W., and Gerhardy, P. R. 1984, *Ap. J. (Letters)*, 280, L47.

Purvis, A. 1983, *M.N.R.A.S.*, 202, 605.

Pustil'nik, L. A. 1977, *Sov. Astr. AJ* (USA), 21, 432.

Quinlan, G. D., and Shapiro, S. L. 1989, *Ap. J.*, 343, 725.

Radhakrishnan, V. 1971, in *The Crab Nebula* (IAU Symp. No. 46) (Dordrecht: Reidel), p. 441.

Radhakrishnan, V., and Cooke, D. J. 1969, *Ap. Letters*, 3, 225.

Rankin, J. M. 1983a, *Ap. J.*, 274, 344.

Rankin, J. M. 1983b, *Ap. J.*, 274, 359.

Rankin, J. M., Campbell, D. B., Isaacman, R. B., and Payne, R. R. 1988a, *Astr. Ap.*, 202, 166.

Rankin, J. M., and Counselman, C. C. 1973, *Ap. J.*, 181, 161.

Rankin, J. M., Payne, R. R., and Campbell, D. B. 1974, *Ap. J. (Letters)*, 193, L71.

Rankin, J. M., and Roberts, J. A. 1971, in *The Crab Nebula* (IAU Symp. No. 46) (Dordrecht: Reidel), p. 114.

Rankin, J. M., Wolszczan, A., and Stinebring, D. R. 1988b, *Ap. J.*, 324, 1048.

Ratnatunga, K. U., and van den Bergh, S. 1989, *Ap. J.*, 343, 713.

Rau, A. R. P., Mueller, R. O., and Spruch, L. 1975, *Phys. Rev.*, A11, 1865.

Rau, A. R. P., and Spruch, L. 1976, *Ap. J.*, 207, 671.

Raubenheimer, B. C., VandenBerg, D. A., Smith, G. H., Fahlman, G. G., Richter, H. B., Hesser, J. E., Harris, W. E., Stetson, P. B., and Bell, R. A. 1986, *Ap. J. (Letters)*, 307, L49.

Rawley, L. A., Taylor, J. H., and Davis, M. M. 1987, *Science*, 238, 761.

Rawley, L. A., Taylor, J. H., and Davis, M. M. 1988, *Ap. J.*, 326, 947.

Rawls, J. M. 1972, *Phys. Rev.*, D5, 487.

Ray, A. 1980, *Phys. Fluids*, 23, 898.

Ray, A., and Chitre, S. M. 1983, *Nature*, 303, 409.

Rees, M. 1971a, *Nature*, 229, 312.

Rees, M. J. 1971b, *Nature* (Phys. Sci.), 230, 55.

Rees, M. J., and Gunn, J. E. 1974, *M.N.R.A.S.*, 167, 1.

Rees, M. J., Trimble, V. L., and Cohen, J. M. 1971, *Nature*, 229, 395.

Rhoades, C. E., and Ruffini, R. 1974, *Phys. Rev. Letters*, 32, 324.

Richardson, M. B. 1980, State University of New York, Albany, Ph.D. thesis.

Rickard, J. J., Erickson, W. C., Perley, R. A., and Cronyn, W. M. 1983, *M.N.R.A.S.*, 204, 647.

Rickett, B. J. 1975, *Ap. J.*, 197, 185.

Rickett, B. J., and Cordes, J. M. 1981, in *Pulsars*, ed. W. Sieber and R. Wielebinski (IAU Conf. No. 95) (Dordrecht: Reidel), p. 107.

Riffert, H., and Mészáros, P. 1978, *Ap. J.*, 325, 207.

Ritchings, R. T. 1976, *M.N.R.A.S.*, 176, 249.

Rivlin, L. A. 1980, *Sov. J. Quantum Electron.*, 10, 612.

Roberts, D. H. 1976, *Ap. J.*, 207, 949.

Roberts, D. H., and Sturrock, P. A. 1972a, *Ap. J.*, 172, 435.

Roberts, D. H., and Sturrock, P. A. 1972b, *Ap. J. (Letters)*, 173, L33.

Roberts, D. H., and Sturrock, P. A. 1973, *Ap. J.*, 181, 161.

Roberts, J. A., and Fahlman, G. G. 1969, *Nature*, 222, 862.

Roberts, P. H., and Stix, M. 1971, Technical Note 60, NCAR, Boulder, CO.

Robinson, E. L. 1976, *Ann. Rev. Astr. Ap.*, 14, 119.

Rohlfs, K. 1977, *Lectures on Density Wave Theory* (New York: Springer-Verlag), p. 124.

Romani, R. W., Kulkarni, S. R., and Blandford, R. D. 1987, *Nature*, 329, 309.

Rose, L. A., Pravdo, S. H., Kaluzienski, L. J., Marshall, F. E., Holt, S. S., Boldt, E. A., Rothschild, R. E., and Serlemitsos, P. J. 1979, *Ap. J.*, 231, 919.

Rosen, L. C., and Cameron, A. G. W. 1972, *Ap. Space Sci.*, 15, 137.

Rosen, N. 1978, *Ap. J.*, 221, 284.

Rosi, L. A., and Zimmerman, R. L. 1976, *Ap. Space Sci.*, 45, 447.

Rossi, B., and Olbert, S. 1970, *Introduction to the Physics of Space* (New York: McGraw-Hill).

Rozental, I. L., and Usov, V. V. 1984, *Ap. Space Sci.*, 109, 365.

Ruderman, M. A. 1969, *Comments Nuclear Particle Phys.*, 3, 37.

Ruderman, M. A. 1971, *Phys. Rev. Letters*, 27, 1306.

Ruderman, M. A. 1976, *Ap. J.*, 203, 206.

Ruderman, M. A. 1979, *J. Magn. and Magn. Mater.* (Netherlands), 11, 269.

Ruderman, M. A. 1981, in *Pulsars*, ed. W. Sieber and R. Wielebinski (IAU Conf. No. 95) (Dordrecht: Reidel), p. 87.

Ruderman, M. A. 1985, in *Cosmogonical Processes*, ed. W. D. Arnett et al. (Utrecht: VNU Science Press), p. 199.

Ruderman, M. A., and Cheng, K. S. 1988, *Ap. J.*, 335, 306.

Ruderman, M. A., and Shaham, J. 1983, *Nature*, 304, 425.

Ruderman, M. A., and Shaham, J. 1985, *Ap. J.*, 289, 244.

Ruderman, M. A., and Sutherland, P. G. 1975, *Ap. J.*, 196, 51.

Ruffini, R. 1971, in *The Crab Nebula* (IAU Symp. No. 46) (Dordrecht: Reidel), p. 382.

Rylov, Yu. A. 1976, *Sov. Astr. AJ* (USA), 20, 23.

Rylov, Yu. A. 1977, *Ap. Space Sci.*, 51, 59.

Rylov, Yu. A. 1978, *Ap. Space Sci.*, 53, 377.

Rylov, Yu. A. 1982, *Ap. Space Sci.*, 88, 173.

Rylov, Yu. A. 1984, *Ap. Space Sci.*, 107, 381.

Rylov, Yu. A. 1988, *Ap. Space Sci.*, 143, 269.

Saakyan, G. S. 1963, *Sov. Astr. AJ*, 1, 60.

Sadeh, D. 1972, *Nature*, 240, 139.

Sadeh, D., Hollinger, J. P., Knowles, S. H., and Youmans, A. B. 1968, *Science*, 162, 897.

Saggion, A. 1975, *Astr. Ap.*, 44, 285.

Salpeter, E. E. 1960, *Ann. Phys.*, 11, 393.

Salvati, M. 1973, *Astr. Ap.*, 27, 413.

Salvati, M., and Massaro, E. 1978, *Astr. Ap.*, 67, 55.

Samorski, M., and Stamm, W. 1983, *Ap. J. (Letters)*, 268, L17.

Sang, Y., and Chanmugam, G. 1987, *Ap. J. (Letters)*, 323, L61.

Saslaw, W. C., Faulkner, J., and Strittmatter, P. A. 1968, *Nature*, 217, 1222.

Sato, H. 1977, *Prog. Theo. Phys.*, 58, 549.

Savonije, G. J. 1983, *Nature*, 304, 422.

Sawada, K., Matsuda, T., and Hachisu, I. 1986a, *M.N.R.A.S.*, 219, 75.

Sawada, K., Matsuda, T., and Hachisu, I. 1986b, *M.N.R.A.S.*, 221, 679.

Sawada, K., Matsuda, T., Inoue, M., and Hachisu, I. 1987 *M.N.R.A.S.*, 224, 307.

Sazhin, M. V. 1978, *Vestn. Mosk. Univ. Fiz. Astr.*, 19, 118.

Sazonov, V. N. 1973, *Sov. Astr. AJ* (USA), 16, 971.

Scargle, J. D. 1969, *Ap. J.*, 156, 401.

Scargle, J. D., and Harlan, E. A. 1970, *Ap. J. (Letters)*, 159, L143.

Scargle, J. D., and Pacini, F. 1971, *Nature* (Phys. Sci.), 232, 144.

Schaefer, B. E. 1981, *Nature*, 294, 722.

Scharlemann, E. T. 1974, *Ap. J.*, 193, 217.

Scharlemann, E. T., Arons, J., and Fawley, W. M. 1978, *Ap. J.*, 222, 297.

Scharlemann, E. T., and Wagoner, R. V. 1973, *Ap. J.*, 182, 951.

Schiff, L. I. 1939, *Proc. Nat. Acad. Sci.*, 25, 391.

Schlickeiser, R. 1980, *Ap. J.*, 236, 945.

Schmalz, R., Ruder, H., and Herold, H. 1979, *M.N.R.A.S.*, 189, 709.

Schmalz, R., Ruder, H., Herold, H., and Rossmanith, C. 1980, *M.N.R.A.S.*, 192, 409.

Schmid-Burgk, J. 1973, *Astr. Ap.*, 26, 335.

Schwarzschild, B. 1988, *Phys. Today*, 41 (11), 17.

Schweizer, M., and Straumann, N. 1979, *Phys. Letters*, 71A, 493.

Sedrakyan, D. M. 1970a, *Nature*, 228, 1074.

Sedrakyan, D. M. 1970b, *Astrophysics* (USA), 6, 339.

Sedrakyan, D. M., and Shakhabasyan, K. M. 1972, *Astrophysics* (USA), 8, 326.

Sedrakyan, D. M., Shakhabasyan, K. M., and Mücket, Y. A. 1975, *Astrophysics* (USA), 10, 154.

Sedrakyan, D. M., Shakhabasyan, K. M., and Rudolph, R. 1977, *Astrophysics* (USA), 13, 78.

Segelstein, D. J., Rawley, L. A., Stinebring, D. R., Fruchter, A. S., and Taylor, J. H. 1986, *Nature*, 322, 714.

Setti, G., and Woltjer, L. 1970, *Ap. J. (Letters)*, 159, L87.

Seward, F. D. 1985, *Comments Ap.*, 11, 15.

Seward, F. D., and Wang, Z.-R. 1988, *Ap. J.*, 332, 199.

Shabad, A. E., and Usov, V. V. 1982, *Nature*, 295, 215.

Shabad, A. E., and Usov, V. V. 1985, *Ap. Space Sci.*, 117, 309.

Shabad, A. E., and Usov, V. V. 1986, *Proc. Int. School of Plasma Physics at Sukhumi, USSR* (ESA SP-251), p. 399.

Shakura, N. I., and Sunyaev, R. A. 1973, *Astr. Ap.*, 24, 337.

Shang-Hui, G., Yun-Zao, G., Leung, Y. C., Zong-Wei, L., and Shao-Rang, L. 1981, *Ap. J.*, 245, 1110.

Shapiro, M. M., and Silberberg, R. 1979, *Report Nat. Res. Lab. Prog.* (July), 1.

Shapiro, S. L. 1973a, *Ap. J.*, 180, 531.

Shapiro, S. L. 1973b, *Ap. J.*, 185, 69.

Shapiro, S. L. 1974, *Ap. J.*, 189, 343.

Shapiro, S. L., and Lightman, A. P. 1976, *Ap. J.*, 207, 263.

Shapiro, S. L., and Teukolsky, S. A. 1983, *Black Holes, White Dwarfs, and Neutron Stars* (New York: John Wiley).

Shapiro, S. L., Teukolsky, S. A., and Wasserman, I. 1983, *Ap. J.*, 272, 702.

Shapirovskaya, N. Y., and Sieber, W. 1984, *Astr. Ap.*, 136, 171.

Shaposhnikov, V. E. 1976, *Astrophysics* (USA), 12, 43.

Shatzman, E. 1963, *Ann. d'Astrophysique*, 26, 234.

Shibata, S. 1985, *Ap. Space Sci.*, 108, 337.

Shibata, S., and Kaburaki, O. 1985, *Ap. Space Sci.*, 108, 203.

Shier, L. M. 1990, in *The Magnetospheric Structure and Emission Mechanisms of Radio Pulsars* (IAU Coll. No. 128), in press.

Shklovsky, I. S. 1968, *Supernovae* (New York: Wiley-Interscience), p. 318.

Shklovsky, I. S. 1970a, *Ap. J. (Letters)*, 159, L77.

Shklovsky, I. S. 1970b, *Nature*, 225, 251.

Shklovsky, I. S. 1974, *Astr. Z.*, 51, 665.

Shklovsky, I. S. 1977, *Sov. Astr. AJ* (USA), 21, 371.

Shklovsky, I. S., and Mitrofanov, I. 1985, *M.N.R.A.S.*, 212, 545.

Shu, F. H., and Lubow, S. H.,1981, *Ann. Rev. Astr. Ap.*, 19, 277.

Shukre, C., Manchester, R. N., and Allen, D. A. 1983 *Nature*, 303, 501.

Shvartsman, V. F. 1970, *Radiophys. Quantum Electron.* (USA), 13, 1428.

Shvartsman, V. F. 1971, *Sov. Astr. AJ* (USA), 15, 342.

Sieber, W. 1982, *Astr. Ap.*, 113, 311.

Sieber, W., and Oster, L. 1977, *Astr. Ap.*, 61, 445.

Silk, J. 1971, *Ap. J. (Letters)*, 166, L39.

Silverstein, S. D. 1969, *Phys. Rev. Letters*, 23, 139.

Simard-Normandin, M., and Kronberg, P. P. 1980, *Ap. J.*, 242, 74.

Simola, J., and Virtamo, J. 1978, *J. Phys. B: Atom. Molec. Phys.*, 11, 3309.

Simon, M., and Strange, D. L. P. 1969, *Nature*, 224, 49.

Simon, N. R., and Sastri, V. K. 1971, *Bull. Am. Astr. Soc.*, 3, 479.

Skilling, J. 1968, *Nature*, 218, 923.

Skobelev, V. V. 1976, *Sov. Phys. JETP*, 44, 660.

Slutz, R. J. 1962, *J. Geophys. Res.*, 67, 505.

Smak, J. 1982, *Acta Astr.*, 32, 199.

Smith, D. F., Muth, L., and Arons, J. 1985, *Ap. J.*, 289, 165.

Smith, E. J., Davis, L., Jr., Jones, D. E., Coleman, P. J., Jr., Colburn, D. S., Dyal, P., and Sonett, C. P. 1980, *J. Geophys. Res.*, 85, 5655.

Smith, E. R., Henry, R. J. W., Surmelian, G. L, O'Connell, R. F., and Hajagopal, A. K. 1972, *Phys. Rev.*, D6, 3700.

Smith, F. G. 1969, *Nature*, 223, 934.

Smith, F. G. 1970, *M.N.R.A.S.*, 149, 1.

Smith, F. G. 1971a, in *The Crab Nebula* (IAU Symp. No. 46) (Dordrecht: Reidel), p. 431.

Smith, F. G. 1971b, *M.N.R.A.S.*, 154, 5p.

Smith, F. G. 1971c, *Nature* (Phys. Sci.), 231, 191.

Smith, F. G. 1971d, *Nature* (Phys. Sci.), 232, 164.

Smith, F. G. 1973a, *Nature*, 243, 207.

Smith, F. G. 1973b, *M.N.R.A.S.*, 161, 9p.

Smith, F. G. 1974, *M.N.R.A.S.*, 167, 43p.

Smith, F. G. 1976, *Q. J. R. Astr. Soc.*, 17, 383.

Smith, F. G. 1977, *Pulsars* (Cambridge: Cambridge University Press).

Smith, F. G., Jones, D. H. P, Dick, J. S. B., and Pike, C. D. 1988, *M.N.R.A.S.*, 233, 305.

Smoluchowski, R. 1972, *Nature* (Phys. Sci.), 240, 54.

Sofia, S., and Van Horn, H. 1974, *Ap. J.*, 194, 593.

Sokolov, A. A., and Ternov, I. M. 1968, *Synchrotron Radiation* (New York: Pergamon).

Soper, S. R. K. 1972, *Ap. Space Sci.*, 19, 249.

Spruit, H. C. 1987a, *Astr. Ap.*, 184, 173.

Spruit, H. C. 1987b, *M.N.R.A.S.*, 229, 517.

Spruit, H. C., Matsuda, T., Inoue, M., and Sawada, K. 1987, *M.N.R.A.S.*, 229, 517.

Staelin, D. H., and Reifenstein, E. C. 1968, *Science*, 162, 1481.

Stecker, F. W. 1971, *Nature*, 229, 105.

Stecker, F. W., and Frost, K. 1973, *Nature* (Phys. Sci.), 245, 70.

Stenflo, L. 1980, *Phys. Scripta* (Sweden), 21, 831.

Stevenson, D. J. 1983, *Reports Prog. Phys.*, 46, 555.

Stewart, P. 1974, *Astr. Ap.*, 32, 13.

Stewart, P. 1975, *Astr. Ap.*, 41, 169.

Stewart, P. 1977, *Astr. Ap.*, 55, 387.

Stinebring, D., and Cordes, J. M. 1981, *Ap. J.*, 249, 704.

Stinebring, D. R., Cordes, J. M., Rankin, J. M., Weisberg, J. M., and Boriakoff, V. 1984a, *Ap. J. Suppl.*, 55, 247.

Stinebring, D. R., Cordes, J. M., Rankin, J. M., Weisberg, J. M., and Boriakoff, V. 1984b, *Ap. J. Suppl.*, 55, 279.

Stokes, G., Segelstein, D. J., Taylor, J. H., and Dewey, R. J. 1986, *Ap. J.*, 311, 694.

Stokes, G., Taylor, J. H., and Dewey, R. J. 1985, *Ap. J. (Letters)*, 294, L21.

Stothers, R. 1969, *Nature*, 223, 279.

Strong, A. W. 1983, *Nature*, 303, 476.

Sturrock, P. A. 1970, *Nature*, 227, 465.

Sturrock, P. A. 1971a, *Ap. J.*, 164, 529.

Sturrock, P. A. 1971b, *Ap. J. (Letters)*, 169, L7.

Sturrock, P. A. 1971c, *Ap. J.*, 170, 85.

Sturrock, P. A., and Baker, K. B. 1979, *Ap. J.*, 234, 612.

Sturrock, P. A., Baker, K., and Turk, J. S. 1976, *Ap. J.*, 206, 273.

Sturrock, P. A., Bracewell, R. N., and Switzer, P. 1971, *Nature*, 229, 186.

Sturrock, P. A., Petrosian, V., and Turk, J. S. 1975, *Ap. J.*, 196, 73.

Sunyach, C. 1980, private communication.

Sutherland, P. G. 1979, *Fund. Cosmic Phys.*, 4, 95.

Sutherland, P. G., and Wheeler, J. C. 1984, *Ap. J.*, 280, 282.

Sutton, J. M., Staelin, D. H., Price, R. M., and Weimer, R. 1970, *Ap. J. (Letters)*, 159, L89.

Suvorov, E. V., and Chugunov, Yu. V. 1973, *Ap. Space Sci.*, 23, 189.

Suzuki, H., and Sato, K. 1989, *Prog. Theo. Phys.*, 79, 725.

Swanenburg, B. N., and 13 other authors. 1981, *Ap. J. (Letters)*, 243, L69.

Swank, J. H., Becker, R. M., Boldt, E. A., Holt, S. S., Pravdo, S. H., and Serlemitsos, P. J. 1977, *Ap. J. (Letters)*, 212, L73.

Sweeney, G. S. S., and Stewart, P. 1974, *Astr. Ap.*, 37, 201.

Synge, J. L. 1969, *Nature*, 223, 161.

Taam, R. E., and Picklum, R. E. 1978, *Ap. J.*, 224, 210.

Taam, R. E., and van den Heuvel, E. P. J. 1986, *Ap. J.*, 305, 235.

Tademaru, E. 1971, *Ap. Space Sci.*, 12, 193.

Tademaru, E. 1973, *Ap. J.*, 183, 625.

Tademaru, E. 1974, *Ap. Space Sci.*, 30, 179.

Tademaru, E. 1976, *Ap. J.*, 209, 245.

Tademaru, E. 1977, *Ap. J.*, 214, 885.

Tademaru, E., and Harrison, E. R. 1975, *Nature*, 254, 676.

Tajima, T., and Dawson, J. M. 1979, *Phys. Rev. Letters*, 43, 267.

Takakura, T. 1969, *Nature*, 224, 252.

Tavani, M., Ruderman, M., and Shaham, J. 1989, *Ap. J. (Letters)*, 342, L31.

Taylor, J. H., Fowler, L. A., and McCulloch, P. M. 1979, *Nature*, 277, 437.

Taylor, J. H., and Huguenin, G. R. 1971, *Ap. J.*, 167, 273.

Taylor, J. H., Hulse, R. A., Fowler, L. A., Gullahorn, G. E., and Rankin, J. M. 1976, *Ap. J. (Letters)*, 206, L53.

Taylor, J. H., Manchester, R. N., and Huguenin, G. R. 1975, *Ap. J.*, 195, 513.

Taylor, J. H., and Stinebring, D. R. 1986, *Ann. Rev. Astr. Ap.*, 24, 285.

Taylor, J. H., and Weisberg, J. M. 1982, *Ap. J.*, 253, 908.

Taylor, J. H., and Weisberg, J. M. 1989, *Ap. J.*, 345, 434.

Tennakone, K. 1972, *Letters Nuovo Cimento*, 3, 583.

Terazawa, H. 1981, *Genshikaku Kenkyu*, 25, 51.

Terazawa, H. 1989, *J. Phys. Soc. Japan*, 58, 3555.

Thompson, D. J., Bertsch, D. L., Hartman, R. C., and Hunter, S. D. 1983, *Astr. Ap.*, 127, 220.

Thorne, K. S., and Ipser, J. R. 1968, *Ap. J. (Letters)*, 152, L71.

Todeschunk, J. P., Crossley, D. J., and Rochester, M. G. 1981, *Geophys. Res. Letters*, 8, 505.

Tohline, J. E. 1984, *Ap. J.*, 285, 721.

Tohline, J. E., and Williams, H. A. 1988, *Ap. J.*, 334, 449.

Tremaine, S., and Zytkow, A. N. 1986, *Ap. J.*, 301, 155.

Treves, A. 1971a, *Astr. Ap.*, 15, 471.

Treves, A. 1971b, *Nuovo Cimento*, B4, 88.

Trimble, V. 1968, *Astr. J.*, 73, 535.

Trimble, V. 1982, *Rev. Mod. Phys.*, 54, 1183.

Trimble, V. 1983, *Rev. Mod. Phys.*, 55, 511.

Trimble, V., and Rees, M. 1970, *Ap. Letters*, 5, 93.

Trimble, V., and Rees, M. J. 1971a, in *The Crab Nebula* (IAU Symp. No. 46) (Dordrecht: Reidel), p. 273.

Trimble, V., and Rees, M. J. 1971b, *Ap. J. (Letters)*, 166, L85.

Trümper, J. (ed.). 1986, *Origin and Evolution of X-ray Binaries* (Dordrecht: Reidel).

Trümper, J., Pietsch, W., Reppin, C., Voges, W., Staubert, R., and Kendziorra, E. 1978, *Ap. J. (Letters)*, 219, L105.

Tsai, W.-Y., and Erber, T. 1974, *Phys. Rev.*, D10, 492.

Tsuruta, S. 1974, in *Physics of Dense Matter* (IAU Symp. No. 53) (Dordrecht: Reidel), p. 209.

Tsuruta, S. 1975, *Ap. Space Sci.*, 34, 199.

Tsuruta, S. 1980, in *X-ray Astronomy*, ed. R. Giacconi and G. Setti (Dordrecht: Reidel), p. 72.

Tsuruta, S. 1986, *Comments Ap.*, 9, 151.

Tsuruta, S. 1989, *Nature*, 339, 669.

Tsuruta, S., Canuto, V., Lodenquai, J., and Ruderman, M. 1972, *Ap. J.*, 176, 739.

Tsygan, A. I. 1977, *Sov. Astr. Letters* (USA); *Pis'ma V. Astr. Z.*, 3, 531.

Tsytovich, V. N., Buckee, J. W., and Ter Haar, D. 1970, *Phys. Letters*, 32A, 471.

Tsytovich, V. N., and Chikhachev, A. S. 1969, *Sov. Astr. AJ* (USA), 13, 385.

Tsytovich, V. N., and Kaplan, S. A. 1972, *Astrophysics* (USA), 8, 260.

Tucker, W. H. 1969, *Nature*, 223, 1250.

Tümer, O. T., Dayton, B., Long. J., O'Neill, T., Zych, A., and White, R. S. 1984, *Nature*, 310, 214.

Unwin, S. C., Readhead, A. C. S., Wilkinson, P. N., and Ewing, M. S. 1978, *M.N.R.A.S.*, 182, 711.

Urpin, V. A., Levshakov, S. A., and Yakovlev, D. G. 1986, *M.N.R.A.S.*, 219, 703.

Urpin, V. A., and Yakovlev, D. G. 1980, *Sov. Astr. AJ* (USA), 24, 126 and 425.

Usov, V. V. 1975, *Ap. Space Sci.*, 32, 375.

Usov, V. V. 1983, *Nature*, 305, 409.

Usov, V. V. 1984, *Ap. Space Sci.*, 107, 191.

Usov, V. V. 1988, *Proc. Varenna-Abastrumani Int. School and Workshop Plasma Ap.* (Italy) (ESA SP-285), p. 257.

Van Buren, D. 1981, *Ap. J.*, 249, 297.

Vandakurov, Yu. V. 1972, *Sov. Astr. AJ* (USA), 16, 265.

van den Bergh, S., and Kamper, K. W. 1984, *Ap. J. (Letters)*, 280, L51.

van den Heuvel, E. P. J. 1981, in *Pulsars*, ed. W. Sieber and R. Wielebinski (IAU Conf. No. 95) (Dordrecht: Reidel), p. 379.

van den Heuvel, E. P. J., and Bonsema, P. F. J. 1984, *Astr. Ap.*, 139, L16.

van den Heuvel, E. P. J., and Habets, G. M. H. 1984, *Nature*, 309, 598.

van den Heuvel, E. P. J., and van Paradijs, J. 1988, *Nature*, 334, 227.

van den Heuvel, E. P. J., van Paradijs, J., and Taam, R. E. 1986, *Nature*, 322, 153.

van der Klis, M., Jansen, F., van Paradijs, J., Lewin, W. H. G., van den Heuvel, E. P. J., Trümper, J. E., and Sztajno, M. 1985, *Nature*, 316, 225.

Van Horn, H. M. 1968, *Nature*, 220, 762.

Van Horn, H. M. 1980, *Ap. J.*, 236, 899.

van Paradijs, J. 1978, *Nature*, 274, 650.

van Paradijs, J., Allington-Smith, J., Callanan, P., Charles, P. A., Hassall, B. J. M., Machin, G., Mason, K. O., Naylor, T., and Smale, A. P. 1988, *Nature*, 334, 684.

Van Riper, K. A. 1988, *Ap. J.*, 329, 339.

Van Riper, K. A., and Lamb, D. Q. 1981, *Ap. J. (Letters)*, 244, L13.

Vasyliunas, V. M. 1972. *J. Geophys. Res.*, 77, 6271.

Vasyliunas, V. M. 1979, *Space Sci. Rev.*, 24, 609.

Vasyliunas, V. M. 1987, *Geophys. Res. Letters*, 14, 171.

Vasyliunas, V. M., and Dessler, A. J. 1981, *J. Geophys. Res.*, 86, 8435.

Ventura, J. 1979, *Phys. Rev.*, D19, 1684.

Vestrand, W. T. 1983, *Ap. J.*, 271, 304.

Vestrand, W. T., and Eichler, D. 1984, *Ap. J.*, 261, 251.

Vila, S. C. 1969, *Nature*, 224, 157.

Vila, S. C. 1978, *Ap. J.*, 223, 979.

Virtamo, J., and Jauho, P. 1973, *Ap. J.*, 182, 935.

Virtamo, J., and Jauho, P. 1975, *Nuovo Cimento*, 26B, 537.

Vivekanand, M., and Narayan, R. 1981, *J. Ap. Astr.*, 2. 315.

Vivekanand, M., Narayan, R., and Radhakrishnan, V. 1982, *J. Ap. Astr.*, 3, 237.

Vladimirskii, V. V. 1969, *Z. Eksper, Teor. Fiz. Pis'ma* (USSR), 9, 116.

von Weizsacker, C. F. 1943, *Z. Ap.*, 22, 19.

Wadehra, J. M. 1983, *Ap. J.*, 271, 879.

Wagoner, R. V. 1969, *Ap. J.*, 158, 739.

Wagoner, R. V. 1975, *Ap. J. (Letters)*, 196, L63.

Wagoner, R. V., and Will, C. M. 1976, *Ap. J.*, 210, 764.

Wallace, P. T., and 11 other authors. 1977, *Nature*, 266, 692.

Wang, Q., Chen, K., Hamilton, T. T., Ruderman, M., and Shaham, J. 1989, *Nature*, 338, 319.

Wang, Y.-M. 1978, *M.N.R.A.S.*, 182, 157.

Wang, Y.-M. 1981, *Space Sci. Rev.*, 30, 341.

Wang, Y.-M. 1987, *Astr. Ap.*, 183, 257.

Wang, Y.-M., and Robertson, J. A. 1985, *Ap. J.*, 299, 85.

Warner, B., and Nather, R. E. 1969, *Nature*, 222, 157.

Warwick, J. W. 1969, *Science*, 163, 959.

Warwick, J., and 12 other authors. 1981, *Science*, 212, 239.

Wasserman, I., and Cordes, J. M. 1988, *Ap. J. (Letters)*, 333, L91.

Watson, M. G., Stanger, V., and Griffiths, R. E. 1984, *Ap. J.*, 286, 144.

Weber, E. J., and Davis, L., Jr. 1967, *Ap. J.*, 148, 217.

Wegmann, R. 1987, *Math. Meth. in the Appl. Sci.*, 9, 367.

Weinberg, S. 1972, *Gravitation and Cosmology* (New York: Wiley).

Weisberg, J. M., Boriakoff, V., Ferguson, D. C., Backus, P. R., and Cordes, J. M. 1981, *Astr. J.*, 86, 1098.

Weisberg, J. M., Romani, R. W., and Taylor, J. H. 1989, *Ap. J.*, 347, 1030.

Weisberg, J. M., and Taylor, J. H. 1984, *Phys. Rev. Letters*, 52, 1348.

Weisskopf, M. C., Elsner, R. F., Darbro, W., Leahy, D., Naranan, S., Sutherland, P. G., Grindlay, J. E., Harnden, F. R., and Seward, F. D. 1983, *Ap. J.*, 267, 711.

Wendell, C. W. 1988, *Ap. J. (Letters)*, 333, L95.

Weyman, R. 1965, *Phys. Fluids*, 8, 2112.

Wheaton, W. A., and 17 other authors. 1979, *Nature*, 282, 240.

Wheeler, J. A. 1966, *Ann. Rev. Astr. Ap.*, 4, 393.

Wheeler, J. C. 1975, *Ap. J. (Letters)*, 196, L67.

White, N. E., Kaluzienski, J. L., and Swank, J. H. 1984, in *High Energy Transients in Astrophysics* (New York: AIP), p. 578.

White, N. E., and Stella, L. 1988, *Nature*, 332, 416.

White, N. E., Swank, J. H., and Holt, S. S. 1983, *Ap. J.*, 270, 711.

Wiggins, R. A., and Press, F. 1969, *J. Geophys. Res.*, 74, 5351.

Wigner, E. P. 1934, *Phys. Rev.*, 46, 1002.

Will, C. M. 1975, *Ap. J. (Letters)*, 196, L3.

Will, C. M. 1976, *Ap. J.*, 205, 861.

Will, C. M. 1977, *Ap. J.*, 214, 826.

Will, C. M., and Eardley, D. M. 1977, *Ap. J. (Letters)*, 212, L91.

Williams, A. C., Weisskopf, M. C., Elsner, R. F., Darbo, W., and Sutherland, P. G. 1986, *Ap. J.*, 305, 759.

Wilson, L. W. 1974, *Ap. J.*, 188, 349.

Wilson, R. B., and Fishman, G. J. 1983, *Ap. J.*, 269, 273.

Witten, E. 1984, *Phys. Rev.*, D30, 272.

Witten, T. A., Jr. 1974, *Ap. J.*, 188, 615.

Wolszczan, A. 1983, *M.N.R.A.S.*, 204, 591.

Wolszczan, A., and Cordes, J. M. 1987, *Ap. J. (Letters)*, 320, L35.

Wolszczan, A., Kulkarni, S. R., Middleditch, J., Backer, D. C., Fruchter, A. S., and Dewey, R. J. 1989, *Nature*, 337, 531.

Woodward, J. F. 1978, *Ap. J.*, 225, 574.

Woosley, S. E., Pinto, P. A., Martin, P. G., and Weaver, T. A. 1987, *Ap. J.*, 318, 664.

Woosley, S. E., and Taam, R. E. 1976, *Nature*, 263, 101.

Woosley, S. E., and Weaver, R. K. 1982, *Ap. J.*, 2358, 716.

Woosley. S. E and Weaver, R. K. 1986, in *Nucleosynthesis and Its Implications on Nuclear and Particle Physics*, ed. J. Audouze and N. Mathieu (Dordrecht: Reidel), p. 145.

Wright, C. C. H., and Forester, J. R. 1980, *Ap. J.*, 239, 873.

Wright, G. A. E. 1979, *Nature*, 277, 363.

Wright, G. A. E., and Fowler, L. A. 1981, *Astr. Ap.*, 101, 356.

Wuerker, R. F., Shelton, H., and Langmuir, R. V. 1959, *J. Appl. Phys.*, 30, 342.

Wunner, G. 1980, *Ap. J.*, 240, 971.

Wunner, G., and Ruder, H. 1980, *Ap. J.*, 242, 828.

Wunner, G., and Ruder, H. 1981, *Astr. Ap.*, 95, 204.

Wunner, G., Ruder, H., and Herold, J. 1981, *Ap. J.*, 247, 374.

Yu, M. Y., Shukla, P. K., and Rao, N. N. 1984, *Ap. Space Sci.*, 107, 327.

Yukhimuk, A. K. 1971, *Astrophysics* (USA), 7, 366.

Zaumen, W. T. 1976, *Ap. J.*, 210, 776.

Zel'dovich, Ya. B., and Novikov, I. D. 1971, *Relativistic Astrophysics*, vol. 1, *Stars and Relativity* (Chicago: University of Chicago Press), p. 47.

Zheleznyakov, V. V. 1971, *Ap. Space Sci.*, 13, 87.

Zheleznyakov, V. V., and Shaposhnikov, V. E. 1972, *Ap. Space Sci.*, 18, 166.

Zheleznyakov, V. V., and Shaposhnikov, V. E. 1975, *Ap. Space Sci.*, 33, 141.

Zheleznyakov, V. V., and Shaposhnikov, V. E. 1979, *Austr. J. Phys.*, 32, 49.

Zheleznyakov, V. V., and Suvorov, E. V. 1972, *Ap. Space Sci.*, 15, 24.

Zimmerman, M. 1978, *Nature*, 271, 524.

Zlobin, V. N., and Udal'tsov, V. A. 1975, *Sov. Astr. AJ* (USA), 19, 683.

Zwicky, F. 1938, *Ap. J.*, 88, 522.

Zwicky, F. 1974, *Ap. Space Sci.*, 28, 111.

Author Index

Subject Index